Laser Science and Applications

Laser Science and Applications

Edited by **Juan Landers**

Willford Press
New York

Published by Willford Press,
118-35 Queens Blvd., Suite 400,
Forest Hills, NY 11375, USA
www.willfordpress.com

Laser Science and Applications
Edited by Juan Landers

International Standard Book Number: 978-1-68285-194-4 (Hardback)

Printed in China.

Contents

Preface

I am honored to present to you this unique book which encompasses the most up-to-date data in the field. I was extremely pleased to get this opportunity of editing the work of experts from across the globe. I have also written papers in this field and researched the various aspects revolving around the progress of the discipline. I have tried to unify my knowledge along with that of stalwarts from every corner of the world, to produce a text which not only benefits the readers but also facilitates the growth of the field.

This book provides a global overview of lasers and their applications, with contributions from scientists and technologists from across the world. It covers topics related to laser development and applications in the field of fiber optics, remote sensing, ultra-fast phenomena, laser spectroscopy, laser photobiology, laser photochemistry, etc. The concepts, techniques and case studies in this book will help the readers in keeping pace with the rapid changes in this field.

Finally, I would like to thank all the contributing authors for their valuable time and contributions. This book would not have been possible without their efforts. I would also like to thank my friends and family for their constant support.

Editor

1

Non-scanning systems for far-infrared radiation detection from laser-induced plasmas

Chun Li[1], Guo-Qian Liao[1], and Yu-Tong Li[1,2]

[1] *Beijing National Laboratory of Condensed Matter Physics, Institute of Physics, CAS, Beijing 100190, PR China*

[2] *IFSA Collaborative Innovation Center, Shanghai Jiao Tong University, Shanghai 200240, PR China*

Abstract

With the development of ultraintense terawatt (TW) and petawatt (PW) laser systems, powerful terahertz (THz) radiation from laser–plasma interactions has been reported. Plasma-based THz systems, which are usually operated at extremely low repetition rates, call for single-shot diagnostics. In this paper, various state-of-the-art single-shot detection methods are introduced or designed for measurements and applications involved in high-power plasma-based THz sciences.

Keywords: laser–plasma interaction; single-shot detection; terahertz radiation

1. General review on THz detectors

Nowadays, detection of radiation in the terahertz (THz) frequency domain is of great interest in active research[1, 2]. Generally speaking, the methods involved can be mainly categorized into two types: incoherent (direct) detection and coherent detection. In order to meet the requirements of different applications, designs for specific THz radiation detection systems need to take the following three factors into account: (1) whether the system has sufficiently high sensitivity to allow signals to be extracted from thermal background radiation; (2) whether the data sampling rate allows real-time measurements of ultrafast processes; and (3) whether full characterization of the THz radiation, including amplitude and phase information, can be provided with no or little distortion. The last requirement cannot be achieved other than by using coherent detectors/methods.

Spectral information is always considered essential in real applications. For applications that require very high spectral resolution within a known frequency region, heterodyne detection systems based on frequency mixing are preferred. At room temperature, a semiconductor-based detector, such as a planar Schottky-diode mixer, is combined with a local oscillator for frequency downconversion[3, 4]. The created downshifted signal – the intermediate-frequency (IF) signal – is then filtered and amplified to obtain high sensitivity of detection, the noise equivalent power (NEP) of which is typically on the order of 10^{-19} W/$\sqrt{}$Hz. Cryogenic-cooling techniques are used in heterodyne systems to provide even smaller NEPs of 10^{-21}–10^{-20} W/$\sqrt{}$Hz. The most widely used superconducting heterodyne detector includes the superconductor–insulator–superconductor (SIS) tunnel junction mixer[5]. In principle, the detectable frequency range of a heterodyne system is determined by the operating frequency of the local oscillator, which is normally in the sub-THz or low-THz frequency region. The corresponding response time is on the level of picoseconds.

For broadband detection, direct detectors based on thermal absorption are widely used. Room-temperature thermal-type detectors include pyroelectric detectors and Golay cells. Bolometers can also be operated at room temperatures, though helium-cooling technologies are always provided to reduce background thermal noise. Compared with heterodyne devices, direct detectors have typical response times of milliseconds and suffer from a much higher NEP value (typ. 10^{-10} W/$\sqrt{}$Hz for room temperature; 10^{-12}–10^{-15} W/$\sqrt{}$Hz for helium-cooled systems). The responsivity of a thermal-type direct detector usually shows little dependence on radiation wavelength; hence bandpass filters or interferometric measurements are generally involved in providing spectral information of the measured signals. A Fourier transform infrared (FTIR) spectrometer, which is commonly used for the identification and analysis of materials with frequency structures in the THz domain, is operated in conjunction with a far-infrared (FIR) optical interferometer and a direct detector. Moreover, since FIR interferometry is based on the autocorrelation effects of THz

Correspondence to: Y.-T. Li, Laboratory of Optical Physics, Institute of Physics, Chinese Academy of Sciences, Beijing 100190, PR China. Email: ytli@iphy.ac.cn

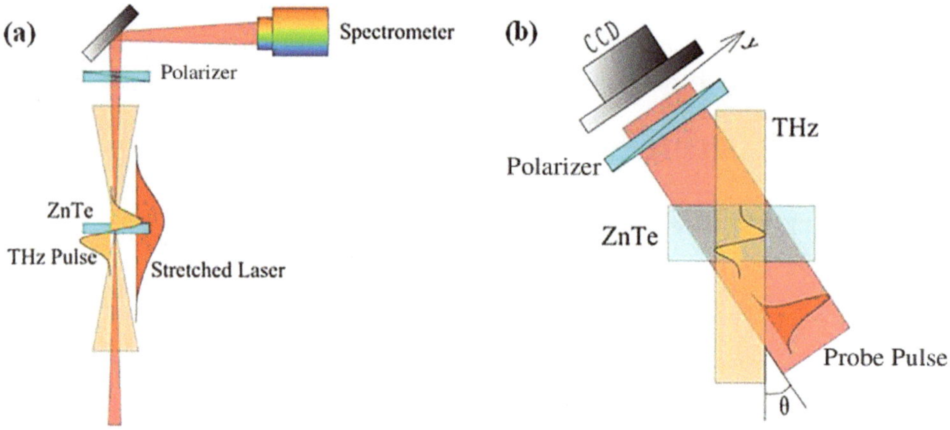

Figure 1. Schematic of (a) spectral encoding and (b) spatial encoding scheme.

pulses, it can also be used to measure the pulse duration of FIR signals[6].

With the rapid development of ultrafast laser systems, pulsed detectors based on photoconductive or free-space electro-optic (EO) sampling are generally used for coherent detection of broadband THz radiation. The detection scheme excludes the majority of background noise from the time measurement window, and the signal–noise ratio (SNR) of detection surpasses 10^4 using lock-in technologies[7]. Comparing photoconductive- and EO-based detection systems, the spectral response of the former is affected by the carrier lifetime as well as by antenna structures[8], while the latter, as a pure optical technique, is considered more reliable in providing a flat frequency response over a wider frequency domain by using short probe pulses. Detectable frequencies in excess of 20 and 100 THz have been demonstrated using photoconductive-[9] and EO-sampling[10] detectors, respectively. THz time–domain spectroscopy (THz–TDS), which has been developed on the basis of pulsed THz techniques, is now widely used to study material properties and transient processes in the FIR region.

The effective spectral response range of EO detection systems is ultimately determined by[11]: (1) the group-velocity mismatch (GVM) between THz and probe pulses; and (2) the phonon absorption of EO crystals. ZnTe and GaP are two of the most commonly used sensors for EO sampling. The former shows excellent detection sensitivity around 2 THz due to the small GVM of optical/THz pulses within the tuning range of Ti:sapphire lasers[12], while the latter has a high fundamental transverse-optical (TO) phonon resonance frequency near 11 THz[13]. Generally speaking, thinner EO crystals tend to provide a wider spectral response at the cost of detection sensitivity. The effective use of a ⟨110⟩-cut ZnTe crystal is narrowed by about 40% (from 4 to 2.5 THz) as the crystal thickness is changed from 0.2 to 1 mm[14, 15]. In considering the trade-off between high detection sensitivity and broadband spectral response, the type and thickness of EO crystals should be chosen carefully for specific research.

2. Non-scanning schemes based on EO sampling

One of the main drawbacks of conventional EO-sampling techniques is the low acquisition rate. Normally, the readout time for a single THz waveform would be at least seconds using serial acquisition schemes, which limits the application of this technique for real-time measurements of fast-moving objects or for real-time imaging. Moreover, how to maintain a high SNR of detection while operating the lock-in amplifier with a low time constant has always been an issue in fast scanning[7].

To increase the data acquisition rate, conventional single-channel delay-scanning detection should be replaced by parallel data collection and multi-channel detector arrays. Much effort has been made in developing 'single-shot' (non-scanning) EO detection systems since 1998, and in general the techniques are progressing in two directions: spectral encoding[16] and spatial encoding[17]; see Figures 1(a) and 1(b), respectively. The main idea is to record the entire temporal profile of a THz pulse onto the spectral or spatial profile of a probe pulse for parallel data acquisition by a spectrometer or a charge-coupled device (CCD) camera. However, both schemes have obvious shortcomings. The spectral encoding method, which relies on measurement of the spectral modulation of a chirped probe pulse, has a fundamental temporal resolution limit related to the chirped pulse duration[18]. The spatial encoding method, which is based on the non-collinear geometry of a collimated short probe pulse along with a collimated THz pulse, retains an excellent time resolution but suffers from a much weaker sensitivity of detection. In addition, the spatial encoding method relies only on the energy/pointing stability of the laser system, while the spectral encoding scheme is also sensitive to spectral instability of the laser.

To achieve both high sensitivity and temporal resolution, a third 'single-shot' EO-cased technique has been introduced as a cross-correlation technique by second-harmonic generation (SHG)[19], as depicted in Figure 2, which can

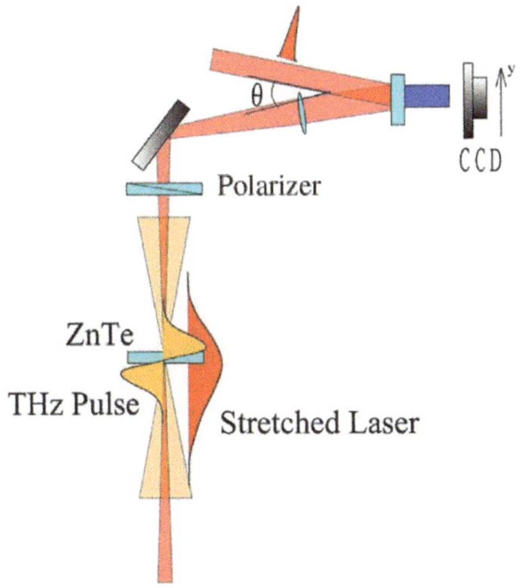

Figure 2. Schematic of 'single-shot' cross-correlation scheme.

be considered as a combination of the spectral and spatial encoding techniques. However, since a second probe pulse is necessary, the complexity of the setup limits the applicability of this technique in real applications. In order to improve the temporal resolution while retaining the simplicity of measurements included in the spectral encoding technique, a different retrieval algorithm has been applied to the experimental setup depicted in Figure 1(a). The corresponding technique is referred to as 'single-shot' in-line spectral interferometry[20, 21], by which the THz field will be derived from the measured optical spectral modulation through a matrix inversion. However, the algorithm cannot be applied in experiments which involve substantial noise.

Based on the schemes (including algorithms) shown in Figures 1 and 2, proof-of-principle experiments have been conducted on quasi-continuous-wave lasers[16, 17, 19, 20] (operated at kHz or MHz). Since at least two laser shots are needed to identify the modulation caused by the THz field, the accuracy of signal recovery depends strongly on the stability of the laser system. Sequential acquisition of the signal and reference pulses will prevent the use of such techniques in high-energy laser systems, which usually operate at low repetition rates and are accordingly less stable.

3. True single-shot THz diagnostics in intense laser–plasma interactions

The development of THz sciences is in great need of high-power THz sources. With the rapid development of ultra-short laser systems, laser–plasma interactions as a new field of research have received considerable attention for strong THz radiation generation[6, 22–26]. However, with increasing laser energies, the repetition rates and stability of laser systems are both decreased. The sensitivity of laser–plasma interactions to small laser fluctuations has motivated the development of true single-shot THz-diagnostic techniques.

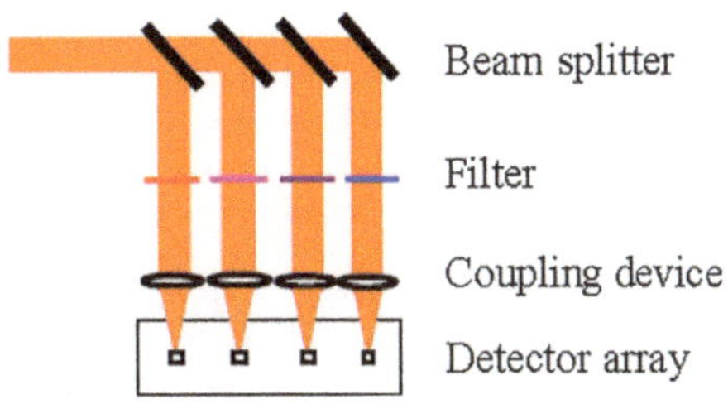

Figure 3. Schematic of a multi-channel THz-diagnostic system based on a direct detector array.

3.1. Incoherent detection

Direct measurements based on thermal absorption are still by far the most convenient and widely used methods for the detection of broadband plasma-based THz sources. In order to obtain spectral information in one laser shot, multi-channel detection with bandpass or low-pass filters is preferred, as depicted in Figure 3. The detector array usually consists of a number of identical pyroelectric elements; or in the case of sufficient radiation flux, a pyroelectric camera can be considered as a receiver. Selectable beam splitters include silicon wafers and pellicles; coupling devices can be from a wide range of THz lenses and waveguides. By deliberate calibration of the split ratios, coupling coefficients, filter transmission curves and detector response curves, the essential spectral information of THz pulses can be obtained from one laser shot. This multi-channel THz-diagnostic system provides a convenient way to avoid shot-to-shot laser fluctuations, and offers the flexibility of selecting the frequency range of interest simply by changing the THz filters.

3.2. Coherent detection

THz pulses generated from ultrafast laser–plasma interactions are often characterized by pump and probe measurements. Traditional pump–probe schemes such as EO sampling, which relies on lock-in techniques, are very sensitive to laser noise (which can be quite substantial at low repetition rates) and low-frequency disturbance. For intense laser–plasma interactions, the low repetition rate and relatively large laser fluctuations have motivated the development of true single-shot EO-based measurements. In principle, the non-scanning schemes introduced in Section 2 can be used in a single-shot measurement if a reference beam be provided along with the probe beam in one laser shot for direct comparison[25]. In this section, two further single-shot techniques via temporal–spectral or temporal–spatial conversion will be introduced.

3.2.1. Single-shot interferometric scheme using frequency-domain holography

Spectral interferometry, which is also called Fourier-domain interferometry (FDI), is a well-known linear technique for

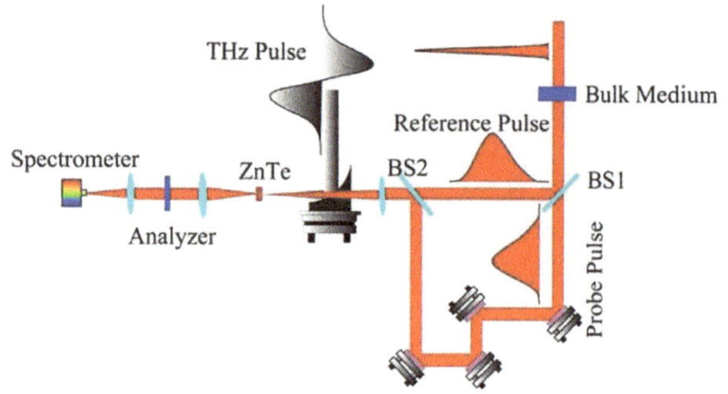

Figure 4. Spectral interferometry scheme using twin-chirped-pulse FDH (quoted from Ref. [28]).

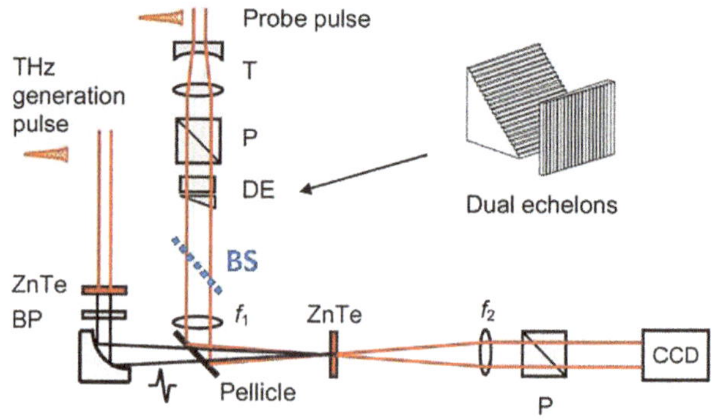

Figure 5. Schematic of single-shot THz measurement using dual echelon optics (quoted from Ref. [29]).

phase and amplitude retrieval from an unknown field in the femtosecond domain[27]. Normally, multiple shots are required to obtain the complete temporal field, but the acquisition rate can be greatly increased by using linearly chirped probe pulses. The in-line interferometric scheme is considered unsuitable for true single-shot THz field diagnostics, since a reference shot has to be recorded as the background before performing the signal retrieval[20]. We proposed a modified scheme using single-shot frequency-domain holography (FDH)[28], as depicted in Figure 4. The original femtosecond optical pulse is frequency chirped and time stretched to several picoseconds before splitting by a 50/50 beam splitter (BS1). A fixed time delay τ is introduced to overlap the probe pulse within the duration of the THz pulse in the ZnTe crystal. The delayed and modulated probe pulse then interferes with the reference pulse in an imaging spectrometer, and the interferometric fringes are recorded on a CCD camera. The interference pattern, with a period proportional to $2\pi/\tau$, which contains information on the phase and amplitude modulation caused by the THz field, will then be analyzed to retrieve the THz waveform in a single shot. The temporal resolution of this technique relies on the retrieval algorithm. If the signal recovery is based on the linear chirped time–frequency relation, a temporal limit identical to that of the spectral encoding method will be given by the uncertainty principle[18]. The implementation of such a technique requires use of a high-resolution spectrometer.

3.2.2. Single-shot THz-pulse characterization by dual echelon optics

In the 2000s, dual echelon optics was first employed to single-shot THz diagnostics[29]. The main idea is to divide the transverse profile of a probe beam by stair-step (echelon) structures into multiple delayed beamlets, which are then focused onto the same spot on an EO crystal to monitor the phase modulation caused by a copropagating THz pulse. The portions of probe beamlets with THz temporal information encoded on them are sequentially imaged onto different positions of a CCD camera for single-shot image processing. To maximize the number of sampling points within one laser shot, two orthogonally oriented echelons are used (rather than one) to fully subdivide the transverse probe beam profile, which has been pre-magnified in diameter. The so-called echelon optics technique, as depicted in Figure 5, which uses a collinear geometry with focused copropagating THz and probe beams, maintains a high detection sensitivity while providing high time resolution and a large time window. For a true single-shot measurement, a beam splitter (blue dot line in Figure 5) should be applied in the optical path to provide an additional reference pulse which bypasses the ZnTe crystal. The reference image is used for direct comparison with the signal image within one laser shot.

3.3. Infrared streak camera

Most electronic devices cannot be used for the direct measurement of fast transient processes in the picosecond region, with the exception of streak cameras. The spectral response of a streak camera is determined by the photocathode material, most of which are sensitive only to electromagnetic radiation of wavelengths shorter than 1.5 μm. By using gas-phase Rydberg atoms as the cathode, the measurable wavelength can be extended to the FIR region[30]. Alternatively, by placing an EO crystal as a converter before an optical streak camera, THz pulses can be retrieved indirectly by measuring copropagating long probe pulses[31]. Comparing an atomic streak camera and an EO-based infrared streak camera, the former can provide information only on the time envelope of THz waves, while the latter is in principle a coherent measurement (but requires an additional reference shot for THz field diagnosis). The temporal resolution of infrared streak camera systems is not comparable to that which can be achieved by single-shot EO-based optical systems.

4. Summary and outlook

In this paper, we present an overview on the state-of-the-art techniques for single-shot THz diagnostics involved in intense laser–plasma interactions. For incoherent detection, direct detectors can be used with THz filters to provide general spectral information. Atomic infrared streak cameras, which operate in the infrared regime (typ. from 1 to 100 μm), are able to measure the intensity profile of THz pulses directly, with a time resolution of about one picosecond. For coherent detection, a THz field can be retrieved via single-shot EO-sampling techniques, either by time-to-space or time-to-spectrum conversion. Associated with specific applications, THz-diagnostic systems should be carefully designed according to the experimental and laser conditions. Practically speaking, high detection sensitivity is always the first concern, followed by high time resolution. Efforts toward higher-power THz sources and an improved understanding of THz radiation under extreme conditions, would promise continuing advances in single-shot THz technologies for intense laser–plasma interactions.

Acknowledgements

This work is supported by the National Basic Research Program of China (grant no. 2013CBA01501), the National Nature Science Foundation of China (grant nos. 11135012 and 11375262), and the Key Research Program of the Chinese Academy of Science (grant no. KGZD-EW-T05).

References

1. B. Ferguson and X.-C. Zhang, Nat. Mater. **1**, 26 (2002).
2. P. H. Siegel, IEEE Trans. Microwave Theory Technol. **50**, 910 (2002).
3. S. Tzortzakis, G. Méchain, G. Patalano, Y.-B. André, B. Prade, M. Franco, A. Mysyrowicz, J.-M. Munier, M. Gheudin, G. Beaudin, and P. Encrenaz, Opt. Lett. **27**, 1944 (2002).
4. M. C. Gaidis, H. M. Pickett, C. D. Smith, R. P. Smith, S. C. Martin, and P. H. Siegel, IEEE Trans. Microwave Theory Technol. **48**, 733 (2000).
5. G. J. Dolan, T. G. Phillips, and D. P. Woody, Appl. Phys. Lett. **34**, 347 (1979).
6. H. Hamster, A. Sullivan, S. Gordon, W. White, and R. W. Falcone, Phys. Rev. Lett. **71**, 2725 (1993).
7. Y. Cai, I. Brener, J. Lopata, J. Wynn, L. Pfeiffer, J. B. Stark, Q. Wu, X. C. Zhang, and J. F. Federici, Appl. Phys. Lett. **73**, 444 (1998).
8. S.-G. Park, M. R. Melloch, and A. M. Weiner, Appl. Phys. Lett. **73**, 3184 (1998).
9. S. Kono, M. Tani, P. Gu, and K. Sakai, Appl. Phys. Lett. **77**, 4104 (2001).
10. C. Kübler, R. Huber, and A. Leitenstorfer, Semicond. Sci. Technol. **20**, S128 (2005).
11. G. Gallot and D. Grischkowsky, J. Opt. Soc. Am. B **16**, 1204 (1999).
12. Q. Wu and X.-C. Zhang, Appl. Phys. Lett. **68**, 1604 (1996).
13. Q. Wu and X.-C. Zhang, Appl. Phys. Lett. **70**, 1784 (1997).
14. J. van Tilborg, C. B. Schroeder, C. V. Filip, Cs. Tóth, C. G. R. Geddes, G. Fubiani, R. Huber, R. A. Kaindl, E. Esarey, and W. P. Leemans, Phys. Rev. Lett. **96**, 014801 (2006).
15. A. Nahata, A. S. Weling, and T. F. Heinz, Appl. Phys. Lett. **69**, 2321 (1996).
16. Z. Jiang and X.-C. Zhang, Appl. Phys. Lett. **72**, 1945 (1998).
17. J. Shan, A. S. Weling, E. Knoesel, L. Bartels, M. Bonn, A. Nahata, G. A. Reider, and T. F. Heinz, Opt. Lett. **25**, 426 (2000).
18. F. G. Sun, Z. Jiang, and X.-C. Zhang, Appl. Phys. Lett. **73**, 2233 (1998).
19. S. P. Jamison, J. Shen, A. M. MacLeod, W. A. Gillespie, and D. A. Jarpszynski, Opt. Lett. **28**, 1710 (2003).
20. K. Y. Kim, B. Yellampalle, G. Rodriguez, R. D. Averitt, A. J. Taylor, and J. H. Glownia, Appl. Phys. Lett. **88**, 041123 (2006).
21. B. Yellampalle, K. Y. Kim, G. Rodriguez, J. H. Glownia, and A. J. Taylor, Appl. Phys. Lett. **87**, 211109 (2005).
22. W. P. Leemans, C. G. R. Geddes, J. Faure, Cs. Tóth, J. van Tilborg, C. B. Schroeder, E. Esarey, G. Fubiani, D. Auerbach, B. Marcelis, M. A. Carnahan, R. A. Kaindl, J. Byrd, and M. C. Martin, Phys. Rev. Lett. **91**, 074802 (2003).
23. Z.-M. Sheng, K. Mima, J. Zhang, and H. Sanuki, Phys. Rev. Lett. **94**, 095003 (2005).
24. K. Y. Kim, A. J. Taylor, J. H. Glownia, and G. Rodriguez, Nat. Photonics **2**, 605 (2008).
25. Y. T. Li, C. Li, M. L. Zhou, W. M. Wang, F. Du, W. J. Ding, X. X. Lin, F. Liu, Z. M. Sheng, X. Y. Peng, L. M. Chen, J. L. Ma, X. Lu, Dong Z. H. Wang, Z. Y. Wei, and J. Zhang, Appl. Phys. Lett. **100**, 254101 (2012).
26. A. Gopal, S. Herzer, A. Schmidt, P. Singh, A. Reinhard, W. Ziegler, D. Brömmel, A. Karmakar, P. Gibbon, U. Dillner, T. May, H.-G. Meyer, and G. G. Paulus, Phys. Rev. Lett. **111**, 074802 (2013).
27. J.-P. Geindre, P. Audebert, S. Rebibo, and J.-C. Gauthier, Opt. Lett. **26**, 1612 (2001).
28. M. L. Zhou, F. Liu, C. Li, F. Du, Y.-T. Li, W.-M. Wang, Z.-M. Sheng, L.-M. Chen, J.-L. Ma, X. Lu, Q.-L. Dong, and J. Zhang, Chinese Phys. Lett. **29**, 015202 (2012).
29. K. Y. Kim, B. Yellampalle, A. J. Taylor, G. Rodriguez, and J. H. Glownia, Opt. Lett. **32**, 1968 (2007).
30. M. Drabbels and L. D. Noordam, Appl. Phys. Lett. **74**, 1797 (1999).
31. Z. Jiang, F. G. Sun, and X.-C. Zhang, Opt. Lett. **24**, 1245 (1999).

2

Beam guiding system geometric arrangement in the target area of high-power laser drivers

Lei Ren, Dongfeng Zhao, and Jianqiang Zhu

National Laboratory on High Power Laser and Physics, Shanghai Institute of Optics and Fine Mechanics, Chinese Academy of Science, Shanghai 201800, China

Abstract

With the increasing number of laser beams, the main difficulty in arranging beam guiding systems (BGSs) involves determining the corresponding relationships between the output and input ports to realize the identified light path length of all beams. Given the basic constraints of geometric arrangement, a BGS model is established, and a base-line algorithm is proposed to address the difficulty mentioned above. Boundary conditions of target area and target chamber are discussed to increase the number of laser beams, and a maximum value exists for a specific target area. Finally, the compatibility of a cylindrical hohlraum target chamber with a spherical hohlraum is analyzed, and a moveable final optics assembly is proposed to execute the switch between the two different targets.

Keywords: beam guiding system; laser drivers; spherical hohlraum; target areas

1. Introduction

A high-power laser driver is a massive facility aimed at inertial confinement fusion (ICF)[1, 2]. It is a promising way to realize ICF if the output power is about 1–2 MJ at 351 nm with a pulse width of 1–5 ns[3]. Based on experiments[4] carried out at the National Ignition Facility (NIF)[5] with laser powers reaching 520 TW and laser energies of up to 1.9 MJ, although the ignition failed, it was suggested that the achievement of higher laser powers and energies to increase the x-ray drive and the margin for ICF has become possible. Besides trying to improve the output power per laser beam, the other approach to obtain higher energy is to increase the number of beam lines. Increasing the number of laser beams is considerably safer considering that laser damage to optical components requires less attention. In addition, the risk for regular operation is lower, given that the single-beam energy will decrease because of the fixed total operation energy. Furthermore, better confinement symmetry can be ensured for the indirect and direct methods if many more laser beams are used to shoot the target.

The earliest study of a multi-beam laser driver beam guiding system (BGS) can be traced back to Shiva[6]. A 20-laser-beam facility with a total output energy fluence of 10 kJ was invented at the Lawrence Livermore National Laboratory (LLNL) in the 1970s. Shiva's BGS is relatively simple and involves only two mirrors. After amplification, the laser beams are guided by these two mirrors, which redirect their angles to meet the physical requirements and shoot the target. In the 1980s, almost simultaneously, when LLNL was upgrading the Nova facility[3] and the French Commissariat à l'Energie Atomique (CEA) was upgrading the Phébus facility[7], both research bodies considered increasing their laser beam numbers to 288 to obtain a laser energy that exceeded 1 MJ laser energy. Two BGS concept design configurations were proposed, namely, crosswise orientation and in-line orientation[8]. Given the disadvantages of in-line orientation, the BGS configurations of high-power laser drivers that have subsequently been built have all been based on crosswise orientation, but with varying shapes. For example, NIF possesses 192 laser beams, and every quad consists of four beams, with 48 quads in total. In the switchyard of NIF, every quad is guided by four or five mirrors into the target chamber. The BGS and target chamber make the shape of the character 'U'[9, 10]. The Laser Mégajoule (LMJ) Project[11] built by CEA possesses 240 laser beams, and one quad also includes four beams. Its BGS is composed of six mirrors and directly inherits the configuration of crosswise orientation, which looks like the character 'I'[12]. The in-built SG-III[13] in China consists of 60 laser beams, and each line is guided by four mirrors, making the BGS L-shaped[14]. The evolution of the BGS configuration discussed above is pictured in Figure 1.

Correspondence to: J. Zhu, No. 390, Qinghe Rd., Jiading, Shanghai 201800, China. Email: jqzhu@mail.shcnc.ac.cn

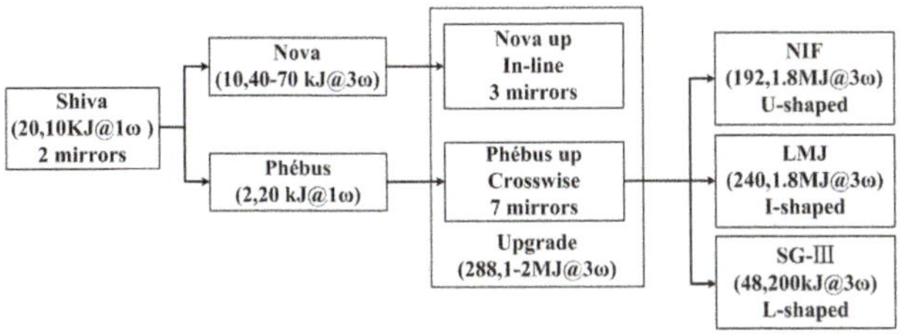

Figure 1. The evolution of the entire BGS configuration.

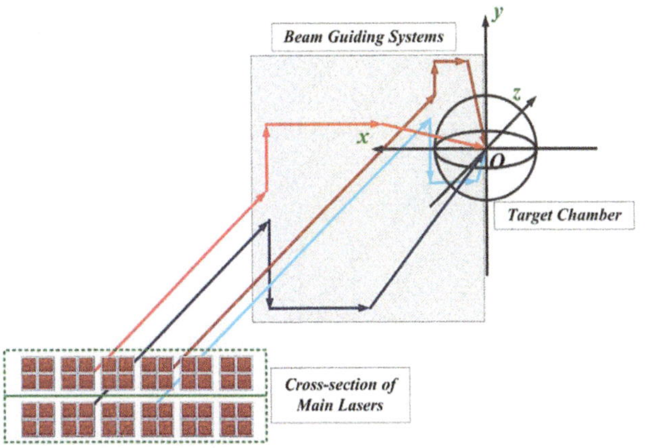

Figure 2. Simplified scheme of a light path from the spatial filter to the target chamber center.

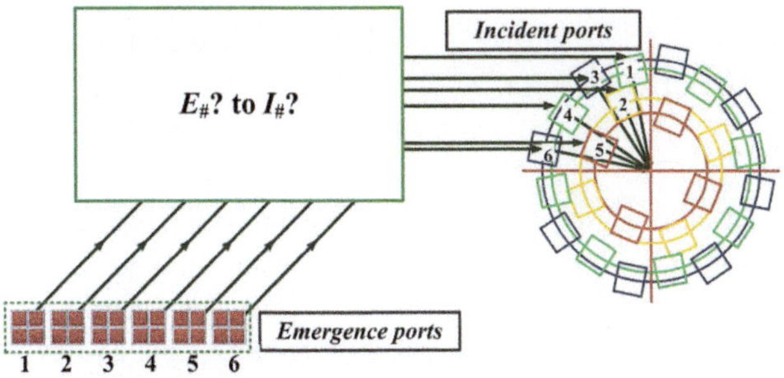

Figure 3. The fundamental problem of BGS arrangement.

2. Modeling of BGSs

2.1. Basic function of BGSs

In the target area of high-power laser drivers, the process of main laser propagation can be simplified as follows: after exiting from the last spatial filter, the main lasers are guided by several mirrors, redirected to meet the angle requirement of the physics experiments, and then shot through the final optics assembly (FOA) to the center of the target chamber, as shown in Figure 2. Thus, the BGS maps the rectangular arrangement at the laser output to a spherical geometry configuration at the target chamber. In addition, all the laser beams share the same light path length (LPL) when they arrive at the target. Furthermore, all reflections at the mirrors in the BGS are in-plane (either S or P) and the incident light must be P polarized. Finally, no obscuring or intersection occurs among laser beams when propagating.

2.2. Fundamental problems of BGS arrangement

Taking six laser beams in Figure 2 as an example, after being redirected by the BGS, they will propagate to the target through six incident ports in the chamber. As shown in Figure 3, the distribution of incident ports is determined by the physics requirement, and the emergence ports are determined by the laser amplifiers. Therefore, the fundamental problem of BGS arrangement involves determining the correspondence between the emergence and incident ports, namely, the $E_{\#}$ and $I_{\#}$ shown in Figure 3, so that all the laser beams can meet the requirements of the physics experiment.

2.3. Typical constraints in BGS modeling

From the viewpoint of laser driver construction, the first constraint is the space of the target area and target chamber. Then, considering the mirror coatings, all incidence angles of each mirror must be less than 45°. Moreover, the number

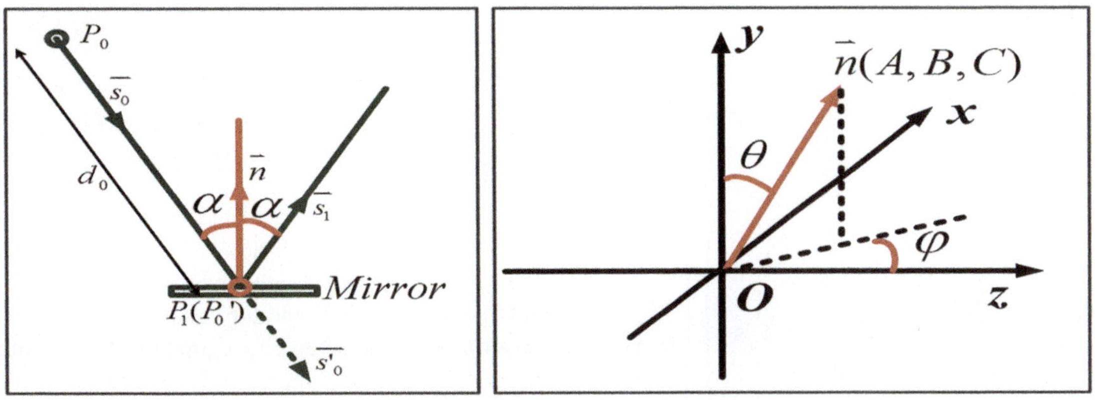

Figure 4. The process of a light propagating a certain distance and being reflected by one mirror.

and types of mirrors must be as small as possible so that the BGS will be simple and easy to realize. The final constraint is the total BGS configuration, and the following modeling is based on a U-shaped configuration.

2.4. Multi-beam BGS modeling

Given that the BGS is constructed from several mirrors, analysis of the process of light reflected by a mirror is crucial. As pictured in Figure 4, a light can be expressed as an emergence point and a direction vector. One light travels along the direction vector $s_0(a_0, b_0, c_0)$ from an emergence point $P_0(x_0, y_0, z_0)$ to a point P_0' in a mirror, and the length of the light path is d_0. The propagating matrix can be calculated through Equation (1*a*), where the vector s_0 is equal to $s_{0'}$. After being reflected by a mirror, whose unit normal vector is $n(A, B, C)$, the emergence point and direction vector of the light become P_1 and s_1, respectively. The propagating matrix can be calculated through Equation (1*b*), where the emergence point P_1 is equal to P_0'. By combining the two processes, we obtain the transforming matrix from incident light to emergent light after propagating a certain distance and being reflected by one mirror, as shown in Equation (1*c*), where $A = \sin\theta\sin\varphi$, $B = \cos\theta$, $C = \sin\theta\cos\varphi$,

$$\begin{pmatrix} x_0' \\ y_0' \\ z_0' \\ a_0' \\ b_0' \\ c_0' \end{pmatrix} = M_1 \begin{pmatrix} x_0 \\ y_0 \\ z_0 \\ a_0 \\ b_0 \\ c_0 \end{pmatrix} = \begin{pmatrix} 1 & 0 & 0 & d_0 & 0 & 0 \\ 0 & 1 & 0 & 0 & d_0 & 0 \\ 0 & 0 & 1 & 0 & 0 & d_0 \\ 0 & 0 & 0 & 1 & 0 & 0 \\ 0 & 0 & 0 & 0 & 1 & 0 \\ 0 & 0 & 0 & 0 & 0 & 1 \end{pmatrix} \begin{pmatrix} x_0 \\ y_0 \\ z_0 \\ a_0 \\ b_0 \\ c_0 \end{pmatrix}, \tag{1a}$$

$$\begin{pmatrix} x_1 \\ y_1 \\ z_1 \\ a_1 \\ b_1 \\ c_1 \end{pmatrix} = M_2 \begin{pmatrix} x_0' \\ y_0' \\ z_0' \\ a_0' \\ b_0' \\ c_0' \end{pmatrix} = \begin{pmatrix} 1 & 0 & 0 & 0 & 0 & 0 \\ 0 & 1 & 0 & 0 & 0 & 0 \\ 0 & 0 & 1 & 0 & 0 & 0 \\ 0 & 0 & 0 & 1-2A^2 & -2BA & -2CA \\ 0 & 0 & 0 & -2AB & 1-2B^2 & -2CB \\ 0 & 0 & 0 & -2AC & -2BC & 1-2C^2 \end{pmatrix} \begin{pmatrix} x_0' \\ y_0' \\ z_0' \\ a_0' \\ b_0' \\ c_0' \end{pmatrix}, \tag{1b}$$

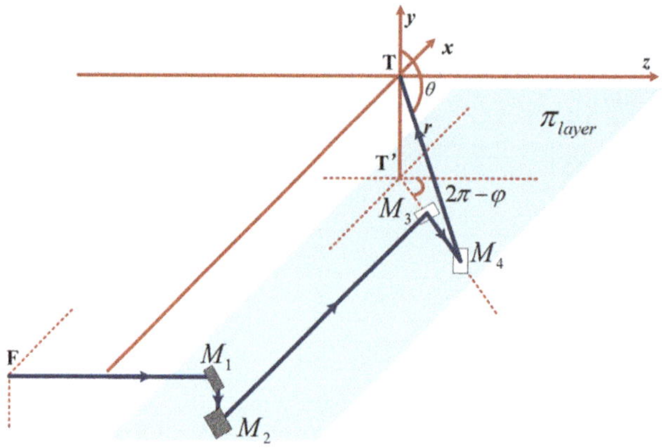

Figure 5. Model of a single light guided by a BGS.

$$\begin{pmatrix} x_1 \\ y_1 \\ z_1 \\ a_1 \\ b_1 \\ c_1 \end{pmatrix} = T \begin{pmatrix} x_0 \\ y_0 \\ z_0 \\ a_0 \\ b_0 \\ c_0 \end{pmatrix} = M_2 M_1 \begin{pmatrix} x_0 \\ y_0 \\ z_0 \\ a_0 \\ b_0 \\ c_0 \end{pmatrix} = \begin{pmatrix} 1 & 0 & 0 & d_0 & 0 & 0 \\ 0 & 1 & 0 & 0 & d_0 & 0 \\ 0 & 0 & 1 & 0 & 0 & d_0 \\ 0 & 0 & 0 & 1-2A^2 & -2BA & -2CA \\ 0 & 0 & 0 & -2AB & 1-2B^2 & -2CB \\ 0 & 0 & 0 & -2AC & -2BC & 1-2C^2 \end{pmatrix} \begin{pmatrix} x_0 \\ y_0 \\ z_0 \\ a_0 \\ b_0 \\ c_0 \end{pmatrix}. \tag{1c}$$

Based on Equation (1*c*), with only two mirrors, one laser beam can be transported from one point to any other point in any direction without any constraints. However, according to the incident angle and polarization requirements and all the constraints discussed above, the configuration of single-light propagation can be designed in Figure 5. The main lasers emerge from the center point F of the spatial filter and reach the target chamber center T after being reflected by four mirrors M_j $(j = 1, 2, 3, 4)$. Mirror 1 only changes the light path in the y direction and mirror 2 does so in the x direction. The light path from mirror 3 to mirror 4 is at the same altitude, and the projection of the line M_4T in the XZ plane is the line $T'M_3M_4$ to ensure that the incident light is P polarized.

The angle between the incident light and the y-axis positive direction is defined as θ $(0 \leqslant \theta \leqslant \pi)$, the angle between the projection of incident light and the z-axis positive direction is φ $(0 \leqslant \varphi \leqslant 2\pi)$ and the counterclockwise direction is positive. Mirrors 2, 3 and 4 are supposed to be in the same plane π_{layer}. The clear radius is defined to be the distance between the target chamber center T and the fourth mirror M_4 and its length is r. In order to avoid obscuring and

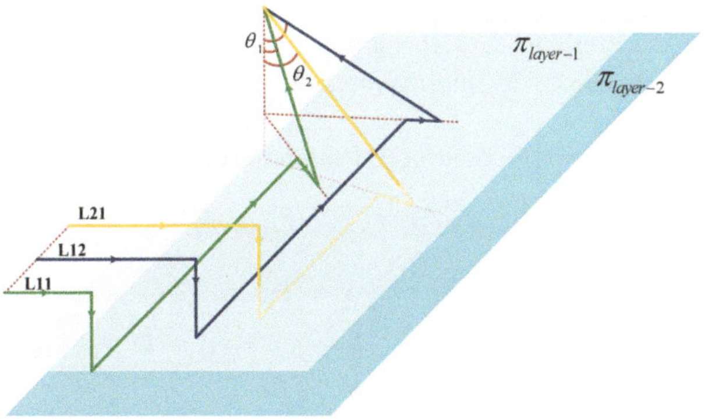

Figure 6. Scheme of different lights guided by a BGS. L11 and L12 possess the same incident angle so they are in plane layer 1. Meanwhile, L21 with a different incident angle is in plane layer 2.

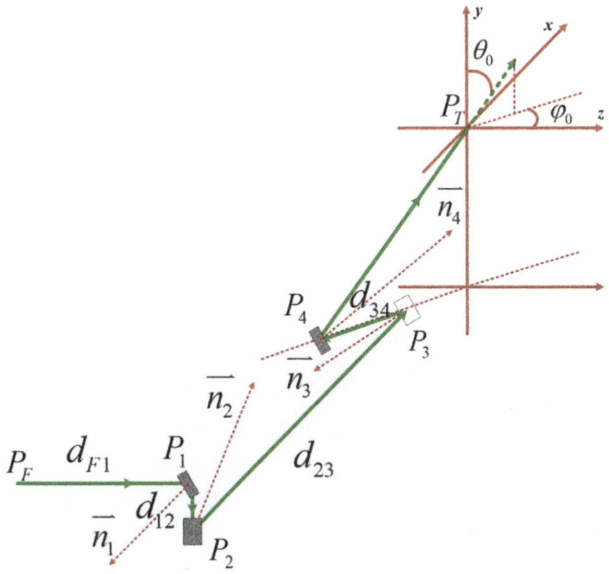

Figure 7. The process of a light propagating a certain distance and being reflected by four mirrors.

intersection among the laser beams when propagating, the clear radii of incident light rays with the same θ should be equal. In this manner, laser beams with different incident angles will travel in different planes or layers. The model of multiple beams guided by a BGS is plotted in Figure 6, which shows that all lights are arranged in different layers according to their incident angles, and no intersection should occur among the beams in the different layers.

According to Equation (1*c*) and Figure 5, the propagation process can be pictured in Figure 7. Given the original emergence direction vectors and unit normal vectors of the four mirrors in Equation (2*a*), the expression of each emergence point and direction vector can be calculated with Equation (2*b*). Because P_F, θ_0 and φ_0 are constant variables that are determined by the laser amplifiers and target chamber, supposing that the clear radius r is fixed, then the position of mirror 4, P_4, and the direction vector s_4 can be calculated. According to Equation (2*c*), the distance between two adjacent mirrors and the total LPL can be expressed in Equation (2*d*),

$$\begin{cases} a_f = 0, \quad b_f = 0, \quad c_f = 1 \\ A_1 = 0, \quad B_1 = \dfrac{-\sqrt{2}}{2}, \quad C_1 = \dfrac{-\sqrt{2}}{2} \\ A_2 = \dfrac{\sqrt{2}}{2}, \quad B_2 = \dfrac{\sqrt{2}}{2}, \quad C_2 = 0 \\ A_3 = -\sin\left[\dfrac{\pi}{4} + \dfrac{\phi_0}{2}\right], \quad B_3 = 0, \\ \quad C_3 = -\cos\left[\dfrac{\pi}{4} + \dfrac{\phi_0}{2}\right] \\ A_4 = \sin\left[\dfrac{\pi}{4} + \dfrac{\theta_0}{2}\right]\sin[\phi_0], \quad B_4 = \cos\left[\dfrac{\pi}{4} + \dfrac{\theta_0}{2}\right], \\ \quad C_4 = \sin\left[\dfrac{\pi}{4} + \dfrac{\theta_0}{2}\right]\cos[\phi_0] \end{cases} \tag{2a}$$

$$\begin{bmatrix} x_1 \\ y_1 \\ z_1 \\ a_1 \\ b_1 \\ c_1 \end{bmatrix} = \begin{bmatrix} x_f \\ y_f \\ d_{F1} + z_f \\ 0 \\ -1 \\ 0 \end{bmatrix}, \quad \begin{bmatrix} x_2 \\ y_2 \\ z_2 \\ a_2 \\ b_2 \\ c_2 \end{bmatrix} = \begin{bmatrix} x_f \\ -d_{12} + y_f \\ d_{F1} + z_f \\ 1 \\ 0 \\ 0 \end{bmatrix},$$

$$\begin{bmatrix} x_3 \\ y_3 \\ z_3 \\ a_3 \\ b_3 \\ c_3 \end{bmatrix} = \begin{bmatrix} d_{23} + x_f \\ -d_{12} + y_f \\ d_{F1} + z_f \\ -\sin[\phi_0] \\ 0 \\ -\cos[\phi_0] \end{bmatrix},$$

$$\begin{bmatrix} x_4 \\ y_4 \\ z_4 \\ a_4 \\ b_4 \\ c_4 \end{bmatrix} = \begin{bmatrix} d_{23} - \sin[\phi_0]d_{34} + x_f \\ -d_{12} + y_f \\ -\cos[\phi_0]d_{34} + d_{F1} + z_f \\ \sin[\theta_0]\sin[\phi_0] \\ \cos[\theta_0] \\ \cos[\phi_0]\sin[\theta_0] \end{bmatrix}, \tag{2b}$$

$$\begin{bmatrix} x_4 \\ y_4 \\ z_4 \\ a_4 \\ b_4 \\ c_4 \end{bmatrix} = \begin{bmatrix} d_{23} - \sin[\phi_0]d_{34} + x_f \\ -d_{12} + y_f \\ -\cos[\phi_0]d_{34} + d_{F1} + z_f \\ \sin[\theta_0]\sin[\phi_0] \\ \cos[\theta_0] \\ \cos[\phi_0]\sin[\theta_0] \end{bmatrix}$$

$$= \begin{bmatrix} -r_{\theta_0}\sin[\theta_0]\sin[\phi_0] \\ -r_{\theta_0}\cos[\theta_0] \\ -r_{\theta_0}\sin[\theta_0]\cos[\phi_0] \\ \sin[\theta_0]\sin[\phi_0] \\ \cos[\theta_0] \\ \cos[\phi_0]\sin[\theta_0] \end{bmatrix}, \tag{2c}$$

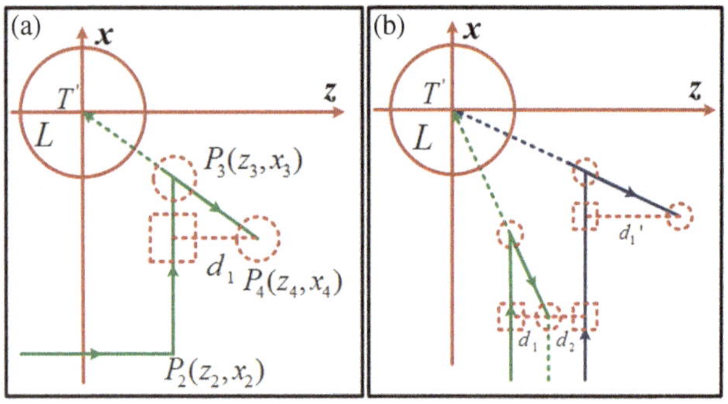

Figure 8. Feasible region of P_3: (a) single light (b) two lights.

$$\left\{\begin{aligned}
&d_{23} = x_4 - x_f + \sin[\phi_0](z_f + d_{F1} - z_4)/\cos[\phi_0]\\
&d_{12} = y_f - y_4\\
&d_{34} = (z_f + d_{F1} - z_4)/\cos[\phi_0]\\
&LPL_{total} = d_{F1}(1 + \sec[\phi_0] + \tan[\phi_0])\\
&\quad + x_4 - x_f - y_4 + y_f + (-z_4 + z_f)\\
&\quad \times (\sec[\phi_0] + \tan[\phi_0]) + r_{\theta_0}.
\end{aligned}\right. \tag{2d}$$

Equations (2*c*) and (2*d*) show that all distances and positions in Figure 7 can be expressed as a function of d_{F1}, which can be related to the z position of mirror 3. Besides d_{F1}, the total LPL is decided by two parts, namely, the position of the emergence ports and the distribution angles of the incident ports. Again, the same conclusion can be obtained from Section 2.2. Therefore, the correspondence relationship between the emergence and incident ports is the core problem that the algorithm needs to solve. Considering that the LPL varies with P_3, fixing the position of mirror 3 is also necessary.

3. Base-line algorithm

3.1. Feasible region of P_3 in the XZ plane

The multi-beam BGS model has ensured that there will be no intersection among the layers; in order to avoid intersection among beams in the same layer, the position of P_3 is constrained in a feasible region. Considering the situation in Figure 8(a), in which only one light exists in the layer, the feasible region of P_3 is confined by the following conditions. First, as discussed above, P_3 should be in the projection line segment of P_4T'. Second, considering the sizes of the laser beams and mirrors as the red dotted circle and square marked in the figure and supposing that they equal D, then the distance between P_4 and P_2P_3, d_1, must be larger than D to ensure that no obstruction exists in the light paths of the laser beams in the same layer. Third, according to the actual physical conditions, the room must be constructed for a diagnostic instrument in the polar areas of the target chamber. Assuming that this room is cylindrical with the radius L, all lights and mirrors must be situated outside of this cylinder, that is, the distance between P_3 and T' should be larger than $L + D$. All three conditions are listed in Equation (3*a*). In this manner, the feasible region of P_3, in which two lights are in the same layer as shown in Figure 8(b), can be found from Equation (3*b*),

$$\left\{\begin{aligned}
&|x_3| < |x_4|, \quad x_3 < 0, \quad z_3 = x_3 \cot\varphi\\
&\sqrt{x_3^2 + z_3^2} = |x_3/\sin\varphi| > L + D\\
&|z_3| < |z_4|, \quad z_3 > 0, \quad d_1 = |z_3 - z_4| > D
\end{aligned}\right. \tag{3a}$$

$$\left\{\begin{aligned}
&|x_3'| < |x_4'|, \quad x_3' < 0\\
&|z_3'| < |z_4'|, \quad z_3' > 0, \quad d_1' = |z_3' - z_4'| > D\\
&|z_3'| > |z_4|, \quad d_2 = |z_3' - z_4| > D\\
&z_4' = x_3' \cot\varphi, \quad |x_3'/\sin\varphi'| > L + D.
\end{aligned}\right. \tag{3b}$$

3.2. Base-line algorithm

We take NIF as an example to interpret the algorithm. The incident ports are distributed in the target chamber at four altitudes, namely, 23.5°, 30°, 44.5° and 50°[15]. The clear radius r is properly selected to render the lights incident to the ports at 23.5° and 30° in the same layer. The target chamber is divided into four parts, namely, top, bottom, left and right. Then, in each section of the target chamber, the incident ports are distributed in three layers with four ports. In addition, the emergence ports are divided into three groups according to the incident layers, as shown in Figure 9. Consideration of the corresponding relations of the four lights in the same layer is unnecessary. $P11$ is connected to $B11$, $P12$ to $B12$, $P13$ to $B13$ and $P14$ to $B14$. In the base-line algorithm, the corresponding relations are fixed, whereas the positions of the emergence ports are variables. The core of the base-line algorithm is to adjust the relative positions of emergence ports to ensure equal LPLs among all beam quads. In this way, the correspondence between the emergence and incident ports is transformed to the calculation of the positions of all the emergence ports, which will make construction and calculation of the model much easier.

The detailed process is illustrated in Figure 10. Taking the first layer as an example, a base-line $BL01$ is defined for this layer, and then the distances between the base-line and the center of each beam quad are designated as $Dq1i$ ($i = 1, 2, 3, 4$). The LPLs of the four lights are calculated, and the different $Dq1i$ are adjusted to ensure the same path length $d1$ for all of them. By this method, the shared path lengths in the second and third layers will be $d2$ and $d3$. What should be noticed is that the shared LPL of the layer varies with the base-line of the layer. By comparing the three shared LPLs, $d1$, $d2$ and $d3$, and adjusting the base-lines of the respective layers, then the 12 beam quads in the same group will all share the same LPL. Given that $BL0j$ and

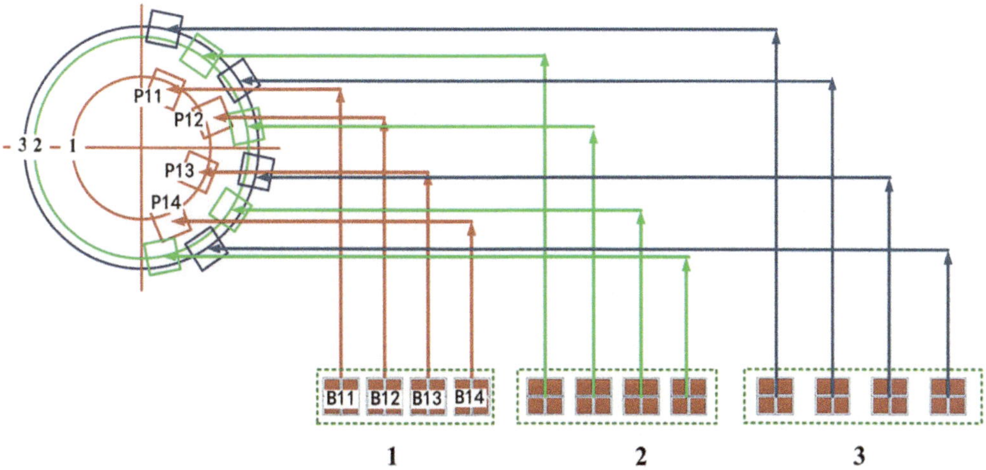

Figure 9. The emergence ports are divided into three groups according to the incident layers.

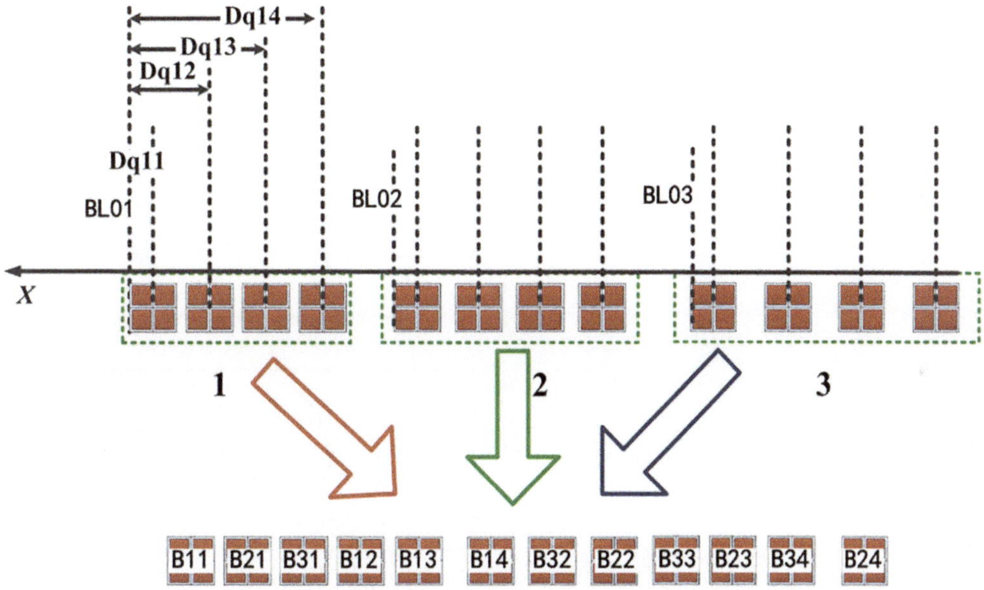

Figure 10. The calculation process of the base-line algorithm.

$Dqji$ ($j = 1, 2, 3; i = 1, 2, 3, 4$) are all fixed, the actual position of the center of the 12 quads can be calculated. Then, the quads can be combined and the distribution of the emergence ports in this part can be obtained.

According to the base-line algorithm, the BGS of a 192-beam-line laser driver is arranged. Figure 11 shows the right and bottom parts of the target area, whereas the whole BGS is plotted in Figure 12.

Comparing the calculated BGS arrangement with the NIF BGS, the calculated LPL varies from 57.2 to 63.6 m, which is considerably shorter than that of NIF, which varies from 69.7 to 81.7 m[10]. Considering the radiation shielding that is built in the target area, as shown in Figure 13, the first two mirrors in each beam line must be outside, causing a longer LPL at NIF. Besides the LPL, the BGS arranged through the base-line algorithm is very similar to the real NIF BGS. The base-line algorithm provides the possibility of avoiding heavy calculations of the corresponding relations. Thus, the arrangement of the BGS of a laser driver is simple and efficient.

4. Boundary conditions and maximum laser beams

Due to the base-line algorithm, the arrangement of a multi-beam laser driver BGS is disregarded. However, discounting this type of BGS does not mean that the number of beam lines can be increased to any expected value to obtain higher laser energy and power. The practical boundary conditions of target area and chamber need to be considered. The

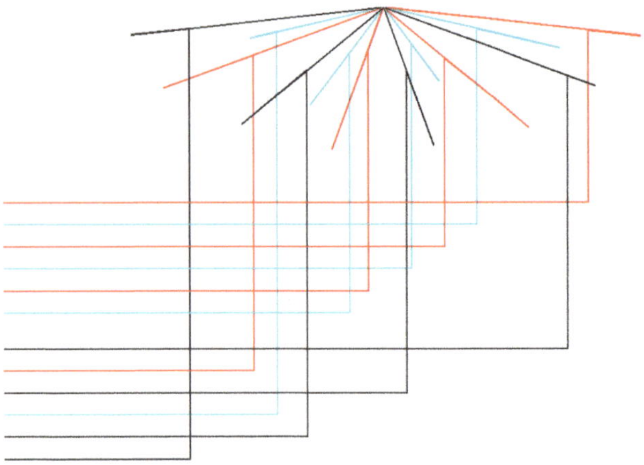

Figure 11. The calculated arrangement of the BGS in the right and bottom parts of the target area.

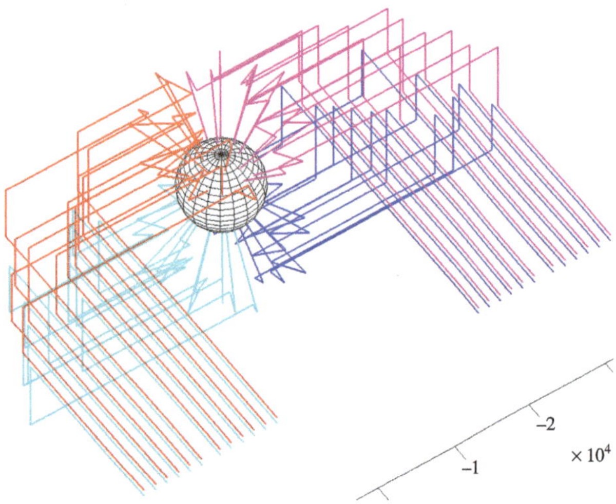

Figure 12. The whole BGS of a 192-beam-line laser driver.

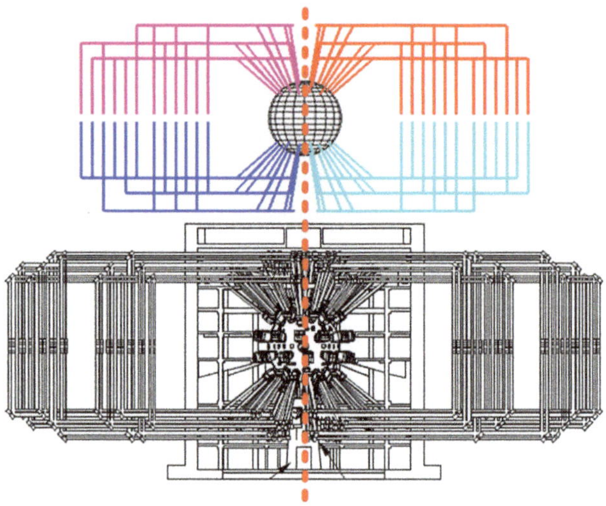

Figure 13. Comparison of the calculated BGS and the NIF BGS.

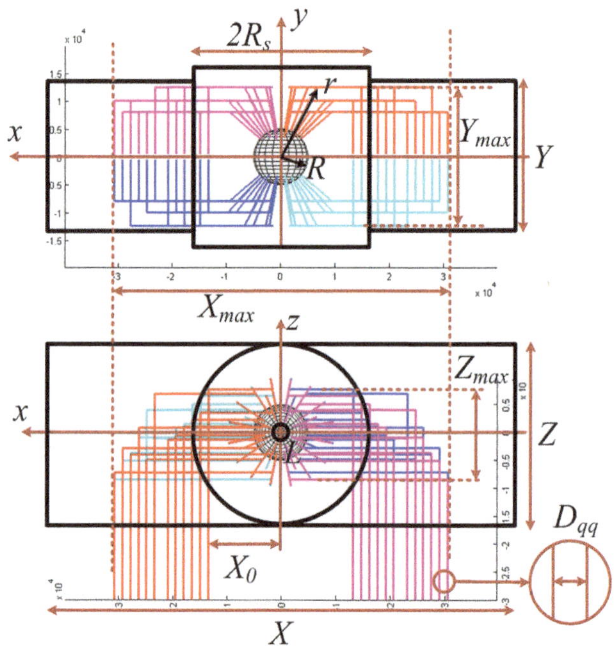

Figure 14. Parameters involved in BGS arrangement.

following relative parameters of the real target area involved in BGS arrangement can be generalized and classified, as shown in Figure 14: the radius of the target chamber R, the radius of the cylindrical room L, the radius of the radiation shielding building R_s, the distribution of the incident ports (θ, φ), the clear radius r, the volume of the target area $X \times Y \times Z$, the starting position of the emergence ports X_0, the total number of laser beams N, the size of a beam quad D_q and the distance between beam quads D_{qq}.

According to Figure 14, as the number of laser beams increases, all the beams and mirrors must be inside the target area. Considering radiation shielding, the starting position of the emergence ports must be outside the building, and the last two or three mirrors must be inside. In addition, the distance between beam quads must be larger than a beam quad. From the side of the target area, the boundary conditions can be expressed as shown in Equation (4),

$$\begin{cases} X_{\max} = 2|X_{f\,\max}| < X, \\ Z_{\max} = 2r_\theta \sin\theta \cos\varphi < Z, \\ Y_{\max} = 2r_\theta \cos\theta < Y, \\ r_{xz\,\max} = r_\theta \sin\theta < R_s, \\ X_0 = |X_{f\,\min}| > R_s, \\ D_{qq} = 1.2\text{–}1.5D_{\text{quad}}. \end{cases} \tag{4}$$

In Equation (4), $X_{f\max}$ and $X_{f\min}$ are the maximum and minimum values of the emergence port position in the x direction. Still taking the parameters of the NIF target

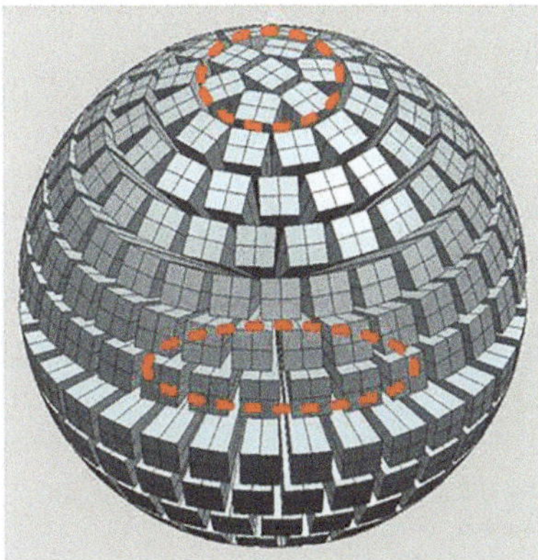

Degree	Ports #
0	1
11.5	6
23	12
34.5	17
46	22
57.5	26
69	29
80.5	30

Figure 15. Extreme situation: the whole target chamber surface is used for incident light ports.

area[16], the maximum number of laser beams will vary from 266 to 304 based on different values of D_{qq}.

From the side of the target chamber, we consider the extreme situation plotted in Figure 15. Considering a 10-m-diameter target chamber, if the incident light ports are distributed over the whole chamber surface, we obtain 16 cones of ports with different incident angles and different ports in separate cones, and the total number of laser beams will be 1144. Given the constraints of indirect drivers, only incident angles that are located between 20° and 60° will be suitable[16]. Thus, the four cones in the middle latitude will be selected in the upper half-sphere, as shown in the table in Figure 15. In this manner, 154 ports will be suitable. However, considering the mechanic stability of the target chamber, distribution of the ports in the whole cone is impossible. Furthermore, certain ports will be spared for the physical diagnostic equipment. Thus, half of the 154 ports can be used for incident quads, and the maximum number of laser beams that a 10-m-diameter chamber can accommodate is 308.

According to the boundary conditions of both the target chamber and the target area, and considering that the beam count must be divisible by 48 for a direct-drive cylindrical hohlraum[16], the maximum number of laser beams is 288 for the NIF target chamber and target area.

Based on the discussion above, we can draw several conclusions. A maximum count of laser beams exists for a specific target area and chamber, so increase of the total laser energy by increasing the beam lines is feasible to a certain extent. More laser beams can be obtained by enlarging the target area and chamber. In addition, for 336 laser beams, the chamber will be about 12 m in diameter. In larger chambers, the focus of the wedged focus lens in the FOA needs to be redesigned to retain the focus. However, with a larger focus, the focus spots in the target will also be enlarged, which may introduce several complicated problems in the target hohlraum[17, 18]. Apart from enlarging the target area and chamber, which is costly, the development of other drive

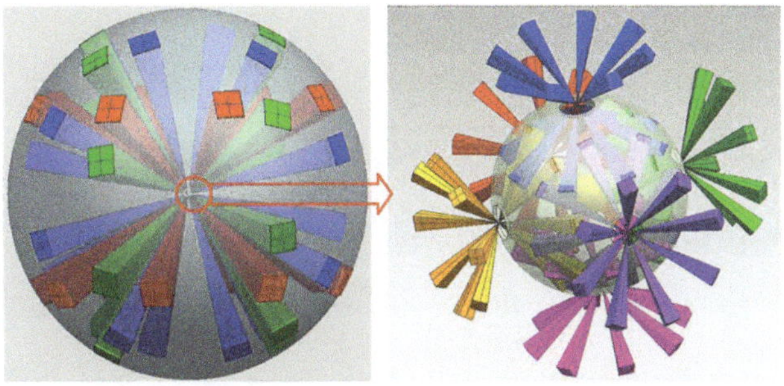

Figure 16. Port distribution in the target chamber and the 6LEH spherical hohlraum.

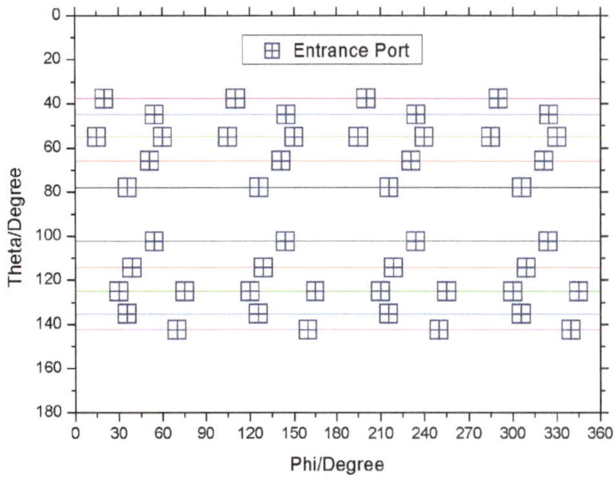

Figure 17. Scheme of the incident port distribution in the target chamber in the theta/phi plane.

targets, such as direct-drive targets[19] and hohlraums with multi-laser entrance holes[20, 21], is beneficial.

5. Six-laser-entrance-hole spherical hohlraum compatibility

In 2014, Lan *et al.*[22] proposed a spherical hohlraum with six laser entrance holes that possess high symmetry and low backscatter. According to the laser arrangement and constraints of the octahedral hohlraum, the incident port distribution in the target chamber for 192 laser beams and the lasers entering the six-laser-entrance-hole (6LEH) hohlraum are plotted in Figure 16. In addition, the incident port distribution can be calculated and plotted in the theta/phi plane of the target chamber, as shown in Figure 17. From Figure 17, it can be seen that the distribution is very similar to that of an indirect-drive cylindrical hohlraum and the ports are situated at different altitudes with different longitudes. by placing both port distributions in the same reference system, the expected results can be obtained. As seen in Figure 18, most of the 6LEH ports are found near the existing NIF

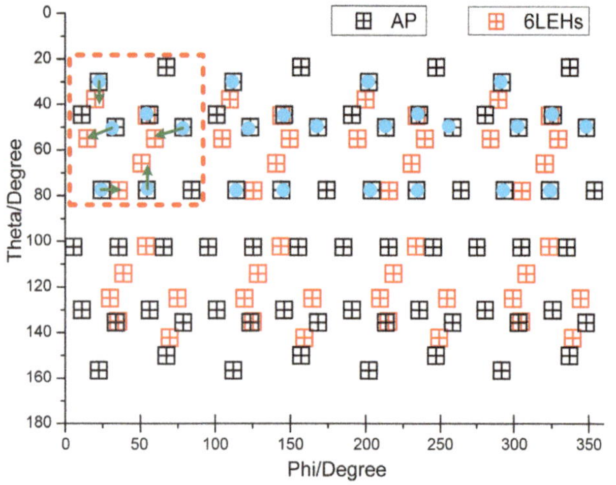

Figure 18. NIF indirect/direct drive and 6LEH incident ports in the theta/phi plane. AP means all ports, which includes direct and indirect ports. Ports with dots indicate that the FOA needs to move to transform.

ports[23], whereas several are coincident, which means that these ports in the NIF chamber can be directly used for a 6LEH hohlraum. For most of the 6LEH ports, if the FOAs of neighboring ports are rotated, as represented by the dots in Figure 18, by small angles while maintaining the aim toward the center of the target chamber, the incident angles of the 6LEH hohlraum can be met easily. Figure 19 is one eighth of the port distribution of Figure 18, indicated by the red dotted rectangle marked in Figure 18. As the figure shows that the longest rotation curve in the target chamber is smaller than the size of two beam quads, a larger hole, probably twice the size of the origin one, can be opened in the dotted ports to accommodate the rotation of the FOA and realize the switch between these two different types of targets. The rotation direction should be along the latitude or longitude so that the movement of the mirrors in the BGS will be constrained in the x and z directions and the change of the BGS will be as small and easy as possible.

6. Conclusions

In summary, a BGS model is constructed, and a baseline algorithm is proposed to solve the BGS arrangement in multi-beam laser drivers. According to the boundary conditions, the maximum number of laser beams to obtain higher laser powers and energies for NIF is 288. In addition, more laser beams require a larger chamber, which will introduce an undefined factor to ignition. Through rotation of the FOAs, the target chamber becomes highly compatible with the 6LEH spherical hohlraum. BGS transformation is also relatively easy to achieve. However, considerable work still needs to be accomplished to design a moveable FOA and the detailed steps of switching the BGS between two different types of targets.

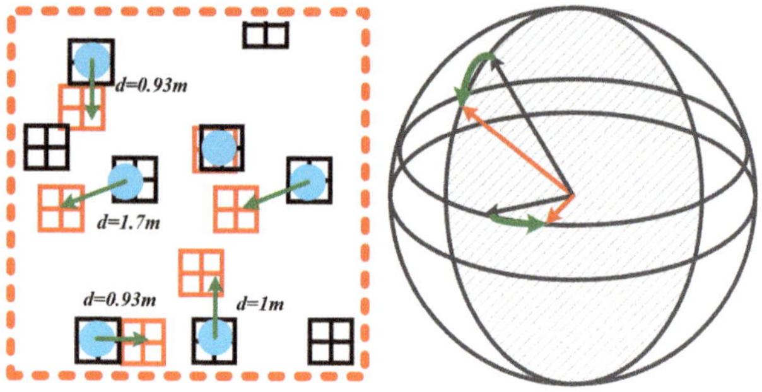

Figure 19. The FOAs are rotated by small angles while maintaining the aim towards the center of the target chamber.

Acknowledgements

The authors would like to acknowledge Dr. B. Wang, Dr. K. You, and Dr. X. Jiao for their helpful discussion and assistance in BGS code writing and model drawing.

References

1. E. I. Moses, Nucl. Fusion **49**, 104022 (2009).
2. S. E. Bodner, A. J. Schmitt, and J. D. Sethian, High Power Laser Sci. Eng. **1**, 2 (2013).
3. E. M. Campbell, in *Proceedings of International Conference on Plasma Physics and Controlled Nuclear Fusion Research* CONF-9210315-2 (1992).
4. J. L. Kline, D. A. Callahan, S. H. Glenzer, N. B. Meezan, J. D. Moody, D. E. Hinkel, O. S. Jones, A. J. MacKinnon, R. Bennedetti, R. L. Berger, D. Bradley, E. L. Dewald, I. Bass, C. Bennett, M. Bowers, G. Brunton, J. Bude, S. Burkhart, A. Condor, J. M. Di Nicola, P. Di Nicola, S. N. Dixit, T. Doeppner, E. G. Dzenitis, G. Erbert, J. Folta, G. Grim, S. Glenn, A. Hamza, S. W. Haan, J. Heebner, M. Henesian, M. Hermann, D. G. Hicks, W. W. Hsing, N. Izumi, K. Jancaitis, O. S. Jones, D. Kalantar, S. F. Khan, R. Kirkwood, G. A. Kyrala, K. LaFortune, O. L. Landen, L. Lagin, D. Larson, S. L. Pape, T. Ma, A. G. MacPhee, P. A. Michel, P. Miller, M. Montincelli, A. S. Moore, A. Nikroo, M. Nostrand, R. E. Olson, A. Pak, H. S. Park, J. P. Patel, L. Pelz, J. Ralph, S. P. Regan, H. F. Robey, M. D. Rosen, J. S. Ross, M. B. Schneider, M. Shaw, V. A. Smalyuk, D. J. Strozzi, T. Suratwala, L. J. Suter, R. Tommasini, R. P. J. Town, B. Van Wonterghem, P. Wegner, K. Widmann, C. Widmayer, H. Wilkens, E. A. Williams, M. J. Edwards, B. A. Remington, B. J. MacGowan, J. D. Kilkenny, J. D. Lindl, L. J. Atherton, S. H. Batha, and E. Moses, Phys. Plasmas **20**, 056314 (2013).
5. C. A. Haynam, R. A. Sacks, P. J. Wegner, M. W. Bowers, S. N. Dixit, G. V. Erbert, G. M. Heestand, M. A. Henesian, M. R. Hermann, K. S. Jancaitis, K. R. Manes, C. D. Marshall, N. C. Mehta, J. Menapace, M. C. Nostrand, C. D. Orth, M. J. Shaw, S. B. Sutton, W. H. Williams, C. C. Widmayer, R. K. White, S. T. Yang, and B. M. V. Wonterghem, J. Phys.: Conf. Ser. **112**, 032004 (2008).
6. J. A. Glaze and R. O. Godwin, in *Proceedings of Topical Meeting on Inertial Confinement Fusion* WE1/1 (1977).
7. M. Rabeau, J. H. Pitts, J. F. Mengue, and G. Maurin, in *Proceedings of 14th IEEE/Npss Symposium: Fusion Engineering* 1175 (1992).

8. M. Rabeau, J. H. Pitts, J. F. Mengue, and G. Maurin, Fusion Technol. **23**, 337 (1993).
9. R. E. English, C. W. Laumann, J. L. Miller, and L. G. Seppala, in *Proceedings of International Optical Design Conference 1998* 726 (1998).
10. J. L. Miller, R. E. English, R. J. Korniski, and J. M. Rodgers, in *Proceedings of Third International Conference on Solid State Lasers for Application to Inertial Confinement Fusion* 294 (1999).
11. N. Fleurot, C. Cavailler, and J. L. Bourgade, Fusion Eng. Des. **74**, 147 (2005).
12. M. L. Andre, Fusion Eng. Des. **44**, 43 (1999).
13. W. Zheng, X. Zhang, X. Wei, F. Jing, Z. Sui, K. Zheng, X. Yuan, X. Jiang, J. Su, H. Zhou, M. Li, J. Wang, D. Hu, S. He, Y. Xiang, Z. Peng, B. Feng, L. Guo, X. Li, Q. Zhu, H. Yu, Y. You, D. Fan, and W. Zhang, J. Phys.: Conf. Ser. **112**, 032009 (2008).
14. M.-Z. Zhu, M.-C. Wang, X.-J. Chen, W.-K. Wu, and G. Chen, Opt. Precis. Eng. **21**, 701 (2013).
15. P. Bell, P. Di Nicola, G. P. Grim, D. Kalantar, T. McCarville, J. Klingmann, S. Alvarez, R. Lowe-Webb, J. Lawson, P. Datte, P. Danforth, M. Schneider, J. M. Di Nicola, J. Jackson, C. Orth, S. Azevedo, R. Tommasini, A. Manuel, and R. Wallace, Proc. SPIE **8505**, 85050B (2012).
16. U.S. Lawrence Livermore Natl Lab, Inertial Confinement Fusion Annual Report (1997).
17. J. D. Lindl, P. Amendt, R. L. Berger, S. G. Glendinning, S. H. Glenzer, S. W. Haan, R. L. Kauffman, O. L. Landen, and L. J. Suter, Phys. Plasmas **11**, 339 (2004).
18. Z. Jiao, Y. Zhang, J. Zhang, and J. Zhu, High Power Laser Sci. Eng. **1**, 88 (2013).
19. D. Eimerl, J. Rothenberg, M. Key, S. Weber, C. Verdon, S. Skupsky, J. Soures, and S. Craxton, in *Proceedings of 1st Annual International Conference on Solid State Lasers for Application to Inertial Confinement Fusion* 170 (1995).
20. D. W. Phillion and S. M. Pollaine, Phys. Plasmas **1**, 2963 (1994).
21. J. M. Wallace, T. J. Murphy, N. D. Delamater, K. A. Klare, J. A. Oertel, G. R. Magelssen, E. L. Lindman, A. A. Hauer, P. Gobby, J. D. Schnittman, R. S. Craxton, W. Seka, R. Kremens, D. Bradley, S. M. Pollaine, R. E. Turner, O. L. Landen, D. Drake, and J. J. MacFarlane, Phys. Rev. Lett. **82**, 3807 (1999).
22. K. Lan, J. Liu, D. Lai, W. Zheng, and X.-T. He, Phys. Plasmas **21**, 010704 (2014).
23. U.S. Lawrence Livermore Natl Lab, National Ignition Facility User Guide (2012).

3

Temperature dependence of parametric instabilities in the context of the shock-ignition approach to inertial confinement fusion

S. Weber[1] and C. Riconda[2]

[1]Institute of Physics of the ASCR, ELI-Beamlines, 18221 Prague, Czech Republic

[2]LULI, Sorbonne Universités - Université Pierre et Marie Curie - Ecole Polytechnique - CNRS - CEA, 75005 Paris, France

Abstract

The role of the coronal electron plasma temperature for shock-ignition conditions is analysed with respect to the dominant parametric processes: stimulated Brillouin scattering, stimulated Raman scattering, two-plasmon decay (TPD), Langmuir decay instability (LDI) and cavitation. TPD instability and cavitation are sensitive to the electron temperature. At the same time the reflectivity and high-energy electron production are strongly affected. For low plasma temperatures the LDI plays a dominant role in the TPD saturation. An understanding of laser–plasma interaction in the context of shock ignition is an important issue due to the localization of energy deposition by collective effects and hot electron production. This in turn can have consequences for the compression phase and the resulting gain factor of the implosion phase.

Keywords: inertial confinement fusion; shock ignition; laser–plasma interaction; parametric instabilities

1. Introduction

The principal constituents of shock ignition (SI), converging shocks and their returns in spherical or cylindrical geometry, are based on ideas suggested some time ago[1–3], before it was applied for concrete applications in inertial confinement fusion (ICF)[4–8]. SI is an alternative scheme to ignite pre-compressed fuel in an efficient and robust way, relaxing conditions on the degree of compression and illumination symmetry with respect to standard direct or indirect drive[9–11]. Differently from standard approaches it relies on a short, high-intensity pulse to drive an additional shock which collides with the rebound of the compression shock and allows the creation of a hotspot and thereby ignition of the fuel. Basically SI is a redistribution of the available driver energy by shaping the temporal pulse structure (see Figure 1). This approach decouples target compression from ignition, like fast ignition. It provides a higher target gain/yield for a given laser drive energy. The lower implosion velocity is advantageous for hydrodynamic instabilities. Moreover, SI is less sensitive to hot electrons and fuel preheat. Several experiments regarding the hydrodynamics of SI have been performed already[12–16]. The intensity of the additional shock-inducing spike is of the order of $I\lambda_o^2 \approx 10^{15}$–$10^{16}$ W μm^2 cm^{-2}. This is at least a factor of 10 higher then the laser intensities employed in direct or indirect drive ICF. Strong parametric instability[17] activity such as stimulated Raman backscattering (SRS), stimulated Brillouin backscattering (SBS) and two-plasmon decay (TPD) is likely to be triggered (see Figure 2). The occurrence of inflationary SRS (iSRS)[18] is also likely. Laser–plasma interaction (LPI) under SI conditions has been little studied up to now. At these high laser intensities hydrodynamic modelling becomes questionable as the processes involved are of kinetic nature. It is therefore necessary to use, e.g., particle-in-cell (PIC) codes to better understand the laser absorption process in the plasma corona.

One of the important results of kinetic simulations of LPI in the framework of SI is the fact that the laser energy is absorbed not at the critical density via inverse Bremsstrahlung but by collective effects in the low-density plasma corona[19–24]. This affects the gain considerably as far as ignition is concerned[11]. More detailed investigations are required concerning the coupling of kinetic simulation results from LPI and radiation hydrodynamic simulations for the compression and ignition phase. This paper considers in some detail the role of the coronal electron plasma temperature as far as the SI scenario is concerned. It is complementary to previous publications[19, 20], which have mostly concentrated on the fundamental scenario of

Correspondence to: S. Weber. Email: stefan.weber@eli-beams.eu

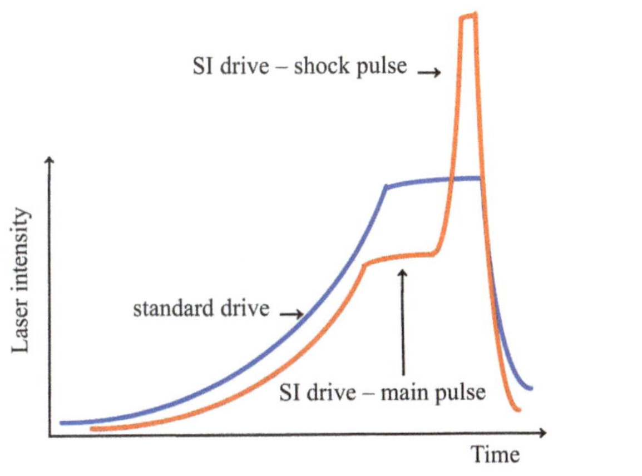

Figure 1. The temporal evolution of the intensity in the case of conventional drive (blue curve) and SI drive (red curve). In the standard approach to ICF the driver is responsible for fuel assembly and a high velocity, $\sim 4 \times 10^7$ cm s^{-1}, for igniting the fuel due to the creation of a hotspot. In the SI scenario the main drive is responsible for fuel assembly but at a lower velocity, $\sim 2 \times 10^7$ cm s^{-1}, preventing ignition. The short high-intensity shock-inducing pulse launched at a later time will reach the fuel at stagnation and ignite it. (Note: the curves in this cartoon drawing are not to scale.)

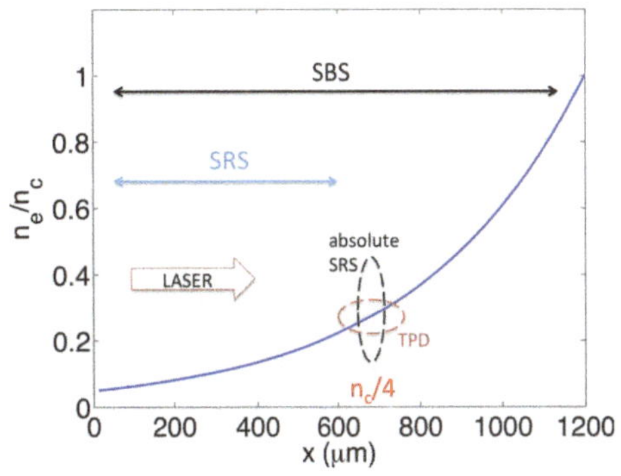

Figure 2. Localization of the various parametric instabilities in the plasma profile. The figure represents a realistic profile. The one used in the simulations is smaller (see Section 2).

parameteric instabilities in the framework of SI. Originally, cavitation at the quarter critical density was identified as a fundamental aspect of LPI for the kind of parameter regime used in the simulations. The present work shows that this phenomenon is strongly temperature dependent. Similarly, the relative production of hot electrons by SRS and TPD is affected by the electron temperature. The coronal plasma temperature is therefore an important issue as present-day preliminary experiments for SI-LPI operate in a regime of lower electron temperature. The coronal plasma electron temperature is expected to be of the order of $\sim$5 keV, whereas the installations used for LPI in the context of SI achieve much less at present (LULI: $\sim$1 keV; PALS: $\leqslant$1 keV; LIL: $\sim$2 keV; OMEGA: $\geqslant$2 keV). Care has to be taken when extrapolating physics behaviour from low temperature to the higher operating temperature expected for SI.

Most of the experimental activity initially concentrated on the hydrodynamic aspects of SI. However, in recent years, several experiments have been performed of relevance or at least related to LPI aspects of SI[10]. Experiments have been performed that concern beam propagation[25–27], SBS[25, 28–35], SRS[36–39] and TPD (Refs. [40, 41] and references therein). It seems that in general energy losses due to backscattering are dominated by Brillouin rather than Raman. The creation of hot electrons by TPD and SRS at the quarter critical density has been observed. The main problem is that one has to fulfil simultaneously the conditions imposed by plasma scale length, electron temperature and laser intensity, as given by possible future experiments on ignition-scale facilities such as NIF and LMJ. Therefore, it is difficult to be conclusive as far as the present experimental effort is concerned. The present experiments should be compared carefully with multi-dimensional kinetic simulations in order to benchmark the codes and to make predictive simulations for the SI parameters.

Table 1. Summary of the simulations. Here, I_o refers to the laser intensity, T_e is the electron plasma temperature. All simulations are at $3\omega_o$, i.e., a laser wavelength of $\sim$0.3 μm. The fully relativistic PIC code emi2D[42] was used for all simulations; $ZT_i/T_e = 0.2$ for all simulations. The reduced intensity case i8 will not be discussed in the text as the results show the same scenario as the corresponding high-intensity case h8.

Case	$I_o\lambda_o^2$ (W μm^2 cm^{-2})	T_e (keV)
c8	1.2×10^{15}	0.5
h8	1.2×10^{15}	2
h7	1.2×10^{15}	5
h9	1.2×10^{15}	10
i8	4.7×10^{14}	2

The remainder of this paper is organized as follows. Section 2 describes the setup for the kinetic simulations using a PIC approach. The results are presented in Section 3 with respect to reflectivity, parametric instabilities, cavitation and the generation of hot electrons. A conclusion is presented in Section 4.

2. Simulation setup

The simulations used in this paper are summarized in Table 1. All simulations are two-dimensional using a full speckle with a Gaussian transverse size of $\sim$24 μm FWHM. A mass ratio $m_i/m_e = 3672$ is used and the ion charge is $Z = 1$. The total simulation time is of the order of $1.5 \times 10^4 \omega_o^{-1} \approx 5$ ps. Radiation hydrodyamic simulations show that the temperature varies little in the plasma corona. Therefore the same temperature is assigned for the whole plasma profile.

The simulation box is $3600\,k_o^{-1}$ in the parallel direction (laser propagation direction) and $2160\,k_o^{-1}$ in the transverse direction. For a wavelength of $\sim$0.3 μm this corresponds to $\sim$172 μm by $\sim$103 μm. The simulated plasma fills the box completely in the transverse direction but has a length of only $\sim$148 μm in the parallel direction, being surrounded by a vacuum region on both side. The plasma profile is exponential in the propagation direction of the laser with a scale length of $L_n = (\mathrm{d}\ln n/\mathrm{d}x)^{-1} \approx 60$ μm and extends from $0.04\,n_c$ up to $0.4\,n_c$ with 60 particles per cell at the highest density. It should be noted that in reality the gradient scale length is of the order of a few hundred microns. However, previous simulations in 1D have shown that L_n does not affect the underlying scenario and physics strongly but only affects the time scales. The plasma temperature is a much more stringent parameter for LPI-SI simulations then the gradient scale length. The boundary conditions are periodic in the transverse direction and open in the parallel direction. The incident laser light is p-polarized in normal incidence. The pulse length is infinite with a ramp-up time of a few laser cycles. The spatial discretization is $\Delta x = 0.18\,k_o^{-1}$ (in the laser propagation direction), $\Delta y = 0.18\,k_o^{-1}$ (in the transverse direction) and $\Delta t = 0.18\,\omega_o^{-1}$, respecting the Courant–Friedrichs–Lewy condition for explicit integration of the PIC equations. Relativistic normalization is used throughout, i.e., the time t and spatial coordinate x, y are normalized by the laser frequency ω_o and vacuum k-vector k_o, respectively. Collisions are not accounted for. The simulation times are too short and the temperatures too high for collisions to have much of an effect on the parametric instabilities analysed in the following simulations. The typical electron–ion collision frequency is given as $\gamma_{\mathrm{coll}}/\omega_o = 2.74 \times 10^{-5}(n_e/n_c)\ln \Lambda_{ei} Z/(\lambda_o T_e^{3/2})$, with λ_o in μm and T_e in keV. The use of, as worst-case scenario, the highest plasma density, $0.4\,n_c$, and the lowest temperature, 500 eV, employed in the simulations results in a collision time of $\tau_{\mathrm{coll}} = 1/\gamma_{\mathrm{coll}} \approx 3 \times 10^4\,\omega_o^{-1}$. This is twice the simuation time. Moreover, for the considered intensities the growth rates of the instabilities are also much larger than the typical collision frequencies.

Figure 3 shows the plasma profile and the transverse cut of the full speckle used in the simulations. The Gaussian profile is cut at 2.67 × FWHM in order to avoid numerical boundary effects.

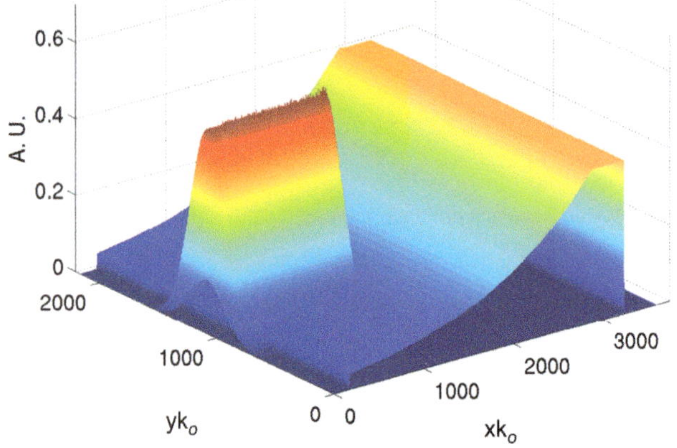

Figure 3. The profiles of the plasma and the incident laser beam. The parameters are given in Section 2.

3. Analysis of simulation results

3.1. Characterization of the parametric instabilities involved

The two main instabilities of interest here are SRS and TPD instability. The resonant SRS process consists of the decomposition of a laser photon (ω_o) into a backscattered frequency-downshifted photon (ω_1) and a forward travelling electron plasma wave (ω_p) (EPW). It fulfils the following conditions for the frequency and wavevector:

$$\begin{aligned}\omega_o &= \omega_1 + \omega_p \\ \mathbf{k}_o &= \mathbf{k}_1 + \mathbf{k}_p.\end{aligned} \tag{1}$$

The frequency of the EPW follows from the dispersion relation as

$$\omega_p \approx \omega_{\mathrm{pe}}(1 + 3k_{epw}^2\lambda_D^2)^{1/2}, \tag{2}$$

where $\omega_{\mathrm{pe}} = (4\pi ne^2/m_e)^{1/2}$ is the local electron plasma frequency and $\lambda_D = v_{\mathrm{the}}/\omega_{\mathrm{pe}} = (kT_e/4\pi ne^2)^{1/2}$ is the Debye length, which in practical units is given as $\lambda_D = 7.43 \times 10^2 T_e^{1/2} n^{-1/2}$ cm, with T_e in eV and n in cm^3; $v_{\mathrm{the}} = (T_e/m_e)^{1/2}$ is the electron thermal velocity. The electromagnetic k-vector of the backscattered light is given as $k_1 = (\omega_1/c)\sqrt{1 - \omega_{\mathrm{pe}}^2/\omega_1^2}$.

The TPD is the decomposition of the laser photon (ω_o) into two plasmons (ω_{p1}, ω_{p2}). The plasmon frequencies are determined by the dispersion relation for the electron plasma waves, Equation (2), and the matching conditions:

$$\begin{aligned}\omega_o &= \omega_{p1} + \omega_{p2} \\ \mathbf{k}_o &= \mathbf{k}_{p1} + \mathbf{k}_{p2}.\end{aligned} \tag{3}$$

Figure 4 shows the geometry of the wavevectors for TPD. The wavevectors of the two plasmons ($\mathbf{k}_1, \mathbf{k}_2$) obey the relations

$$\begin{aligned}\mathbf{k}_{1y} &= -\mathbf{k}_{2y} \equiv \mathbf{k}_y \\ \mathbf{k}_{1,2x} &= \frac{k_o}{2} \pm \Delta\mathbf{k} \\ \Delta\mathbf{k}^2 &= \frac{\mathbf{k}_o^2}{4} + \mathbf{k}_y^2.\end{aligned} \tag{4}$$

Figure 4. Geometry of the k-vectors involved in the TPD instability. The decay of a photon into two plasmons can be realized in two possible ways while preserving energy and momentum. This particular geometry applies in 2D and helps with the interpretation of the phase space diagrams. In reality, 3D, the number of possible k-vectors is infinite lying on an asymmetric cone around the laser k-vector.

The threshold for the TPD instability in an inhomogeneous plasma profile[43, 44] is temperature dependent and is given as

$$I_{16}^{\mathrm{TPD}} \approx 0.5 T_e L_n^{-1} \lambda_o^{-1}. \tag{5}$$

Here, T_e, is in units of kilo electron volts and I_{16} is the laser intensity in units of 10^{16} W cm^{-2}; L_n and λ_o are given in units of microns. The TPD instability is excited in the vicinity of the quarter critical density and develops as an absolute instability as the slow plasma waves do not escape the resonance. In contrast, the threshold for SRS excitation near $n_c/4$ is not a function of T_e and is given by[17, 45]

$$I_{16}^{\mathrm{SRS}} \approx 12 L_n^{-4/3} \lambda_o^{-2/3}. \tag{6}$$

Depending on the electron temperature, the thresholds for the two instabilities can be quite similar and develop in competition. A high-frequency hybrid instability (HFHI) can develop[46, 47], leading to a co-existence of electrostatic and electromagnetic modes. The growth rate for Raman is given as[48] $\gamma_{\mathrm{SRS}} = \gamma_o \min(1, \gamma_o/\gamma_{\mathrm{tot}})$, with $\gamma_{\mathrm{tot}} = \gamma_L + \gamma_{\mathrm{coll}}$. Here, γ_L and γ_{tot} are the damping rates due to Landau damping and collisions, respectively. In the case when Raman is above threshold, i.e., $\gamma_{\mathrm{SRS}}^2 > \gamma_L \gamma_{\mathrm{coll}}$, the growth rate reduces to $\gamma_{\mathrm{SRS}} = \gamma_o$. Even for the lowest temperature used in the simulations, 500 eV, the collisional damping rate is two orders of magnitude smaller than the Landau damping rate for EPWs, which, for backward SRS, is given as

$$\gamma_L/\omega_o = 0.14 \frac{(n_e/n_c)^{1/2}}{(k_{epw}\lambda_D)^3} \exp(-1/(2k_{epw}^2 \lambda_D^2)). \tag{7}$$

Here, $k_{epw} = k_o(\sqrt{1 - n_e/n_c} + \sqrt{1 - 2\sqrt{n_e/n_c}})$ is the electron plasma wavevector and one has the following expression for $k_o^2 \lambda_D^2$ in practical units:

$$k_o^2 \lambda_D^2 = 1.9 \times 10^{-3} T_e/(n_e/n_c), \tag{8}$$

with T_e in keV. It should be noted that k_o is the value of the electromagnetic wavevector in vacuum.

Evaluation of these expressions at a density of $0.04 n_c$ and for $T_e = 500$ eV results in a Landau damping rate of $\sim 1 \times 10^{-2} \omega_o^{-1}$, i.e., much larger than the collision rate quoted in Section 2. The largest damping rate will correspond to the largest temperature, 10 keV, and is of the order of $\sim 1 \times 10^{-3} \omega_o^{-1}$ at the quarter critical density $n_e/n_c = 0.25$. However, even in the most damped case, since the collision frequency is so low as discussed in Section 2, the growth rate of Raman remains significantly above threshold and is therefore given by[48]

$$\gamma_{\mathrm{SRS}}/\omega_o = \gamma_o/\omega_o = \frac{(k_{epw}/k_o)(v_{osc}/c)}{2(\sqrt{n_c/n_e} - 1)^{1/2}}, \tag{9}$$

where $v_{osc} = eE_o/\omega_o m_e$ is the quiver velocity of the electrons in the laser electric field. For an intensity of $I\lambda_o^2 = 1.2 \times 10^{15}$ W µm^2 cm^{-2} and a density of $n_e/n_c = 0.25$ one obtains a growth rate for SRS of $\gamma_{\mathrm{SRS}}/\omega_o \approx 1.5 \times 10^{-2}$.

At very low electron plasma temperature, a few hundred eV, the Langmuir decay instability (LDI) can play an important role as it saturates TPD and SRS activity. Although of limited importance in the case of SI which operates in multi-keV conditions, it could be relevant for present-day LPI experiments for SI, as these take place at a much lower temperature. Therefore it is presented in some more detail.

The LDI was predicted in the 1960s[49, 50] and in LPI experiments it was verified more recently by direct observation[36, 51–53], although its experimental existence had already been conjectured before[54], and it has been observed indirectly due to the ion-acoustic wave (IAW) damping on SRS[55–59].

The LDI induces a decay of the pump plasma wave (k_{epw}, ω_p) into an IAW characterized by $(2k_{epw} - \delta k, \omega_{cs})$ (travelling in the direction of the original EPW) and an anti-Stokes daughter EPW (travelling in the opposite direction) having a frequency close to the original frequency (down-shifted by ω_{cs}) and a wavevector given by $k_{epw1} = k_{epw} + \delta k$, where the correction δk has the form

$$\delta k = \tfrac{2}{3}(Zm_e/m_i)^{1/2}/\lambda_D. \tag{10}$$

Here, $\lambda_D = (kT_e/4\pi n e^2)^{1/2}$ is the Debye length, which in practical units is given as $\lambda_D = 7.43 \times 10^2 T_e^{1/2} n^{-1/2}$ cm, with T_e in eV and n in cm^3. A necessary condition for LDI to take place is therefore that[49, 60]

$$k_{epw}\lambda_D > \frac{2}{3}\left(\frac{Zm_e}{m_i}\right)^{1/2} \approx 0.03. \tag{11}$$

For all the cases considered in Table 1 this is clearly the case. However, the higher the temperature, the more the EPWs are damped. The LDI is in general most active at low temperature, provided Equation (11) is fulfilled.

The maximum growth rate is given by

$$\gamma_{\mathrm{LDI}} = \frac{1}{2}\frac{eE_p}{m_e\omega_p}\frac{1}{v_{\mathrm{the}}}\sqrt{\omega_{p1}\omega_{cs}}. \tag{12}$$

Here, E_p is the electrostatic field associated with the EPWs, ω_p, ω_{p1} represent the two plasmons, and

$$\omega_{cs}^2 = k_{cs}^2 c_s^2 (1/(1 + k_{cs}^2 \lambda_D^2) + 3T_i/ZT_e) \quad (13)$$

is the IAW frequency. In order for LDI to develop, the growth rate has to be above a threshold given by the linear damping of electron plasma and ion-plasma waves: $\gamma_{\text{LDI}} > \sqrt{\gamma_{epw}\gamma_{iaw}}$. In the regime where electron kinetic effects are neglegible, LDI, LDI cascade and Langmuir wave collapse tend to dominate the nonlinear evolution of the EPW, participating in the saturation of the SRS instability. If the EPW density fluctuations are below the wavebreaking limit, neglecting the kinetic effect is equivalent to considering $k_{epw}^2 \lambda_D^2 \ll 1$. In such regimes it is thus expected to observe LDI, and its influence on SRS[61–66].

For the intensity considered in this study SBS takes place in the weak-coupling regime even for the lowest temperature used, i.e., Equation (13) is the correct dispersion relation for the IAWs with $k_{cs} \approx 2k_{po}$, with $k_{po} = k_o\sqrt{1 - n_e/n_c}$. The growth rate for SBS in this regime is given as $\gamma_{SBS} = k_{po} v_{osc} \omega_{pi}/\sqrt{2\omega_{cs}\omega_o}$, which in practical units is

$$\gamma_{SBS}/\omega_o = 3.1 \times 10^{-2} \sqrt{I_{16}\lambda_o^2} \left(\frac{n_e}{n_c}\right)^{1/2} \left(\frac{Z}{A}\right)^{1/4} \frac{1}{T_e^{1/4}}. \quad (14)$$

Here, $\omega_{pi} = \sqrt{4\pi n_i Z^2 e^2/m_i}$ is the ion plasma frequency, Z is the atomic charge and A is the atomic mass number.

In the following the simulation results are analysed with respect to the behaviour of the reflectivity of the incident laser beam, the induced activity of the parametric instabilities, and the phase space and Poynting vector. These issues are of course strongly interdependent.

3.2. Overall scenario

Two fundamental issues have to be addressed as far as the simulations are concerned.

- The relative importance of TPD and SRS. One would expect that the colder the plasma the stronger the TPD.
- Which mechanism is saturating the TPD and SRS activity at the quarter critical density $n_c/4$? Again this will be temperature dependent.

With respect to the first point it is indeed found that the higher the temperature the more pronounced SRS becomes, although it is a negligible energy loss mechanism as far as reflectivity is concerned. With respect to the second point one can observe a clear transition from LDI-induced saturation of Raman at 500 eV to saturation due to cavitation and density fluctuations for temperatures of 2 keV and above. For the low-temperature case c8 there is LDI and TPD activity but no SRS.

SBS is present for any temperature but decreases in importance the higher the temperature. The general conclusion of this set of simulations is that the plasma temperature plays a crucial role as far as the LPI scenario is concerned. Already a factor of two in the electron plasma temperature can significantly affect the relative importance of the parametric instabilities and their effect on the plasma dynamics and laser absorption. Realistic simulations and future experiments for SI require LPI to take place at the right temperature.

The overall energy balance is affected by the amount of the energy of the incident laser beam reflected due to SRS and SBS, Section 3.3, as well as by absorption into collective plasma modes and hot electrons, Sections 3.5 and 3.6. The time scales of the simulations allow study of the saturation of these instabilities related to electron modes, Section 3.4. It has been shown[19, 67] that the development and subsequent saturation of the electron-related instabilities takes place as a recurrent scenario.

3.3. Reflectivity

Figure 5 summarizes the reflectivities for the four temperatures. The reflectivities are evaluated using B_z integrated over the whole transverse direction. The use of B_z instead of E_y ensures that nothing is lost due to a possible opening angle of the backscattered light. There is basically no SRS activity for the cases c8 and h8. The substantial backscattering of the order of 10%–20% originates entirely from Brillouin. Increase of the temperatures strongly reduces SBS activity and increases SRS. However, the energy losses due to SRS even for the highest temperature case, h9, remain negligible. The bursty, spike-like nature of the SRS reflectivity is not related to inflationary SRS, as the corresponding frequency spectra relate the region of SRS activity to the vicinity of $n_c/4$, in accordance with standard absolute SRS excitation. In general, SRS is strongly Landau damped at temperatures of the order of a few keV. However, in the vicinity of the quarter critical density the wavevector of the backscattered light

$$k_1 \approx (1/c)\sqrt{\omega_1^2 - \omega_{\text{pe}}^2} = (\omega_o/c)\sqrt{1 - 2\sqrt{n/n_c}} \quad (15)$$

goes to zero, such that the momentum balance gives $k_{epw} \approx k_o$. The resulting Landau damping rate, see Equations (7) and (8), is of the order of $\gamma_L/\omega_o \approx 10^{-3}$ (for the highest temperature case h9, $T_e = 10$ keV). The threshold for SRS is independent of temperature, so one would expect that the SRS-induced part of the reflectivity will not change as the temperature is increased. However, locally TPD and SRS are competing for the pump and the threshold for TPD is strongly temperature dependent. As T_e increases, TPD activity is reduced and SRS can take more energy out of the pump, thereby increasing the reflectivity.

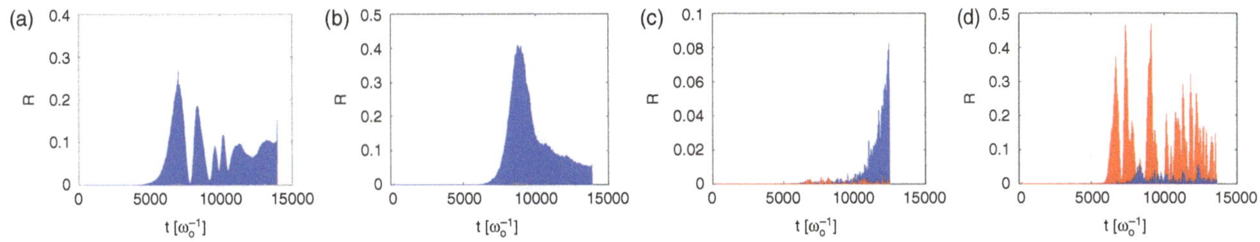

Figure 5. Reflectivities ($R = I/I_o$, i.e., reflected intensity over incident intensity at the centre of the speckle in the transverse direction) for the cases (a) c8, (b) h8, (c) h7 and (d) h9. The curves are 'filled' as the laser temporal period is resolved. The blue curve corresponds to SBS-like frequencies, summing the range $0.9\omega_o$–$1.1\omega_o$. The red curve corresponds to SRS-like frequencies, summing the range $0.0\omega_o$–$0.9\omega_o$. No frequencies are present in the interval $0.8\omega_o$–$0.9\omega_0$. Note: the time on the axis refers to the moment the reflected light crosses the boundary of the computational box; as the quarter critical density is located at $2200\,k_o^{-1}$, the light was actually refelected $\sim 2200\omega_o^{-1}$ earlier.

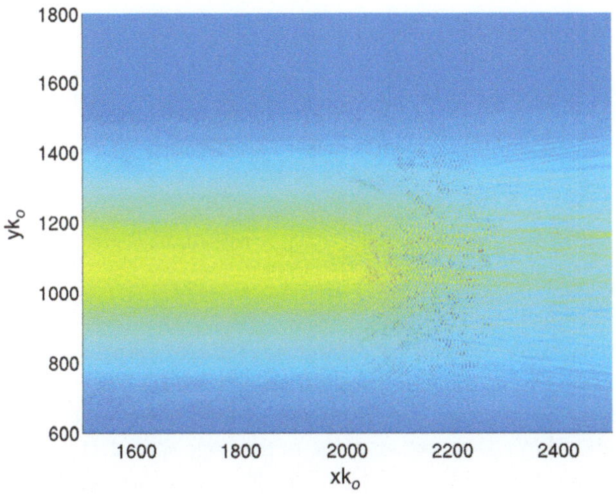

Figure 6. Poynting vector for the case h9 at $t\omega_o = 13\,600$.

The total reflectivities, i.e., the SRS and SBS contributions combined, reduce strongly as the temperature is increased. For the cases c8 and h8 the average reflectivity is of the order of 10%–15%, whereas for the high-temperature cases, h7 and h9, the overall reflectivity is reduced to 1%–2% only.

SBS develops everywhere in the profile, up to $n_c/4$. At $n_c/4$ SBS is strongly inhibited due to the fact that the laser beam is randomized due to the cavitation process and LDI, and the associated strong density fluctuations (see, e.g., Figure 6 for the case h9).

By contrast, SRS develops predominantly at $n_c/4$. The simulated time scales are short for the SBS evolution. It is therefore unclear what the saturation mechanisms for SBS are on longer time scales. The bursty behaviour of the reflectivities as visible in Figure 5 has been observed before and was discussed in the literature[19, 68]. In the case of SRS it is related to hot electron production and cavity creation.

Figure 7 shows the frequency spectra of the backscattered light, in agreement with the decomposition into SRS and SBS of the reflectivities. In the cold case, c8 (Figure 7(a)), the SRS signal is at $0.5\omega_o$ but is very weak in agreement with the reflectivity curves. In the hot case, h9 (Figure 7(b)), SRS is dominant. Clearly visible is a splitting of the SRS frequency around $0.5\omega_o$ with $\Delta k \approx 0.013\,k_o$. This most likely originates from density perturbations induced by TPD and/or cavitation. The localization of the SRS signal around $0.5\omega_o$ is a signature of the fact that SRS originates from the vicinity of the quarter critical density. The downshifted backscattered light from SRS has the correct frequency to be trapped locally in density cavities[19, 20].

Figure 8 shows the Fourier transforms of the electromagnetic field in the plasma for the two extreme cases, c8 and h9. In Figure 8(a) one can observe a feature at $|k| = 1.5\,k_o$ at large angle. This is a signature of TPD coupling with the laser and disappears at later time after saturation of TPD, as can be seen in Figure 8(b). The signal close to $0.86\,k_o$ stems from the incoming laser and the backscattered light due to SBS. The strong-signal opening angle increases in time due to side scattering and refraction from density perturbations. In the hot case, Figure 8(b, d), at later time two pronounced signals appear at $k_\parallel \approx 0.05\,k_o$ and $k_\parallel \approx 1.5\,k_o$. The first is the signature of SRS, as $k_1 \ll 1$ close to $n_c/4$ (see the discussion above), the latter one is the coupling of electron plasma waves at $\omega \approx 0.5\omega_o$ with the laser. As expected, the signal is mostly in the parallel direction as SRS has a small opening angle.

3.4. Electron-related mode activity: SRS, TPD and LDI

As already mentioned, there is almost no SRS for the cases c8 and h8. Strong SRS is present for higher temperatures above 2 keV. In general, it can be said that there is no SRS activity in the low-density part of the plasma corona. SRS is concentrated near the quarter critical density. The high-temperature case h9 has very strong absolute SRS activity but basically no TPD.

As can be seen from Table 2, the threshold for TPD is much lower then the threshold for SRS for the cases c8 and h8. As a consequence, TPD arises first and no or very little absolute SRS develops as TPD is initially depleting the

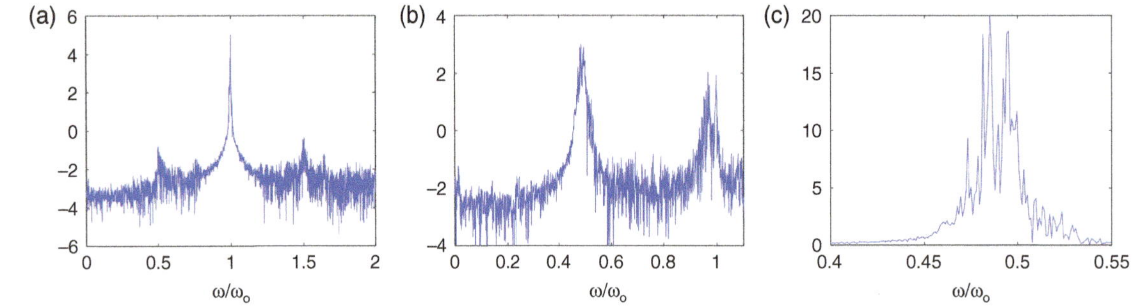

Figure 7. Frequency spectra for the cases (a) c8, (b) h9 and (c) a zoom of (b). Note: (a) and (b) are on log scale whereas (c) is on linear scale.

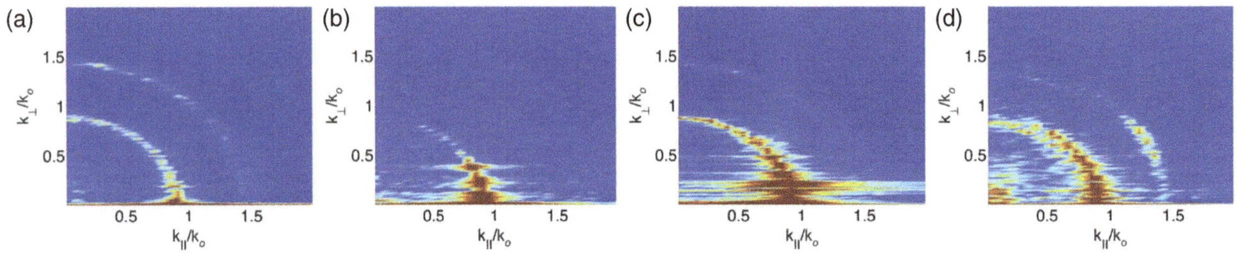

Figure 8. Two-dimensional Fourier spectra of the electromagnetic field B_z evaluated in the vicinity of $n_c/4$ for the cases c8 (a, c) and h9 (b, d) taken at times $t\omega_o = 4000$ (a, b) and $t\omega_o = 7000$ (c, d).

Table 2. Temperature-dependent occurrence of LPI phenomena. The number of stars gives a rough 'visual' interpretation of the strength of the process occurring, with ⋆⋆⋆⋆ strongest and ⋆ weakest. The numbers in the columns I_{16}^{TPD} and I_{16}^{SRS} are calculated from the corresponding Equations (5) and (6). The thresholds have to be compared with the laser intensity, which in units of 10^{16} W cm^{-2} is 1.2 for all cases. CAV = cavitation.

Case	T_e (keV)	CAV	LDI	SRS	SBS	TPD	I_{16}^{TPD}	I_{16}^{SRS}
c8	0.5	⋆	⋆⋆⋆⋆	⋆	⋆⋆⋆⋆	⋆⋆⋆⋆	0.014	0.11
h8	2	⋆⋆	⋆⋆⋆	⋆	⋆⋆⋆⋆	⋆⋆⋆⋆	0.056	0.11
h7	5	⋆⋆⋆⋆	⋆	⋆⋆⋆	⋆⋆	⋆⋆⋆	0.14	0.11
h9	10	⋆⋆⋆⋆	⋆	⋆⋆⋆⋆	⋆	⋆⋆	0.28	0.11

pump, as can be seen in Figure 9. The figure clearly shows the absence of any laser beam in the density profile behind the quarter critical density, which however is a transient feature due to TPD saturation.

As the SBS reflectivity data (see Figure 5(a)) show reflection of about 20%–30%, the remaining energy of order 70% at this time is absorbed in the plasma due to TPD.

At a later stage SRS still does not develop due to the strong, irregular small-scale density modulations induced by TPD. For the cold case c8, TPD is saturated by LDI which develops on the EPWs generated originally by TPD.

In the high-temperature cases (h7 and h9) the threshold for SRS is lower than for TPD and strong absolute SRS develops which induces cavitation and leads to saturation.

Figure 10(a) shows the Fourier transform of the ion density for the coldest case, c8, after TPD saturation. The large feature between 30° and 40° reproduces the IAWs which are the result of LDI decay of the plasmons excited by TPD and has k-vectors that are roughly twice the original k-vectors from the EPWs. The extent is due to the large spread of the k-vectors of the EPWs. The smaller feature with $k_{\|}$ close to zero results from the low-frequency beating of symmetrically excited plasmons, e.g., EPW$_1$ and EPW$'_1$ in Figure 4, summing up to purely perpendicular fluctuations[69]. The very small feature at $k_{\|} \approx 1.8\,k_o$ and in an almost perfect forward direction is the IAW signal generated by SBS (also visible in (b) for case h8).

The EPWs generated by TPD and the secondary waves generated by LDI evolve into turbulence, as observed in other recent simulations for SI[18, 70–72] (see the discussion of real-space figures in Section 3.5). As the temperature increases (Figure 10(b–d)), the LDI-induced signal disappears and the purely perpendicular signal from the EPW beating retreats to smaller wavevector (longer wavelength).

3.5. Plasma cavitation

The role of cavitation was clearly identified in previous work related to SI[19–21]. It is a fundamental mechanism of LPI which acts as a converter, transferring energy from the laser into kinetic energy of the plasma, and was originally identified in SBS activity[73]. Cavities present strong local plasma perturbations which can act as a dynamic random phase plate and induce coherence loss to the incident laser beam, i.e., they act as a means of plasma smoothing[68].

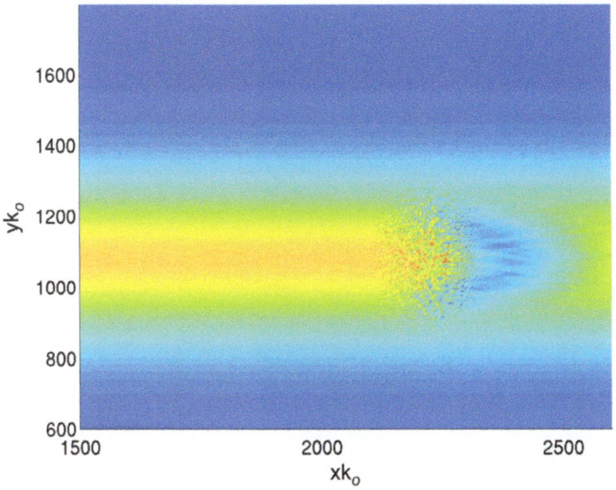

Figure 9. Poynting vector for the case c8 at $t\omega_o = 3600$. The 'hole' behind the density layer around $n_c/4$ is clearly visible.

They are also essential in saturating parametric instabilities at the quarter critical density. Finally, they are an important mechanism for laser energy absorption. Figure 11 shows the ion density cavities for the various simulation cases. As was shown before[19], the presence of the beating of the TPD waves facilitates the cavitation process. As the threshold for TPD is a linear function of T_e, see Equation (5), one would expect cavitation to occur later, the higher the temperature is. This is confirmed by the figure, which shows for each of the cases the time slice of maximum cavity activity. The cavities can be filled with either electrostatic[71, 72] or electromagnetic fields; for high laser intensity it was shown that they have a soliton-like structure[19, 20].

The high-temperature case h9, in particular, Figure 11(d), shows very large crater remnants from previous cavitations. These density perturbations strongly affect the incident laser light and act as a dynamic random phase plate, inducing fluctuating speckles for densities above $n_c/4$ (see Figure 6), thereby randomizing the laser beam that is transmitted across the perturbed density layer. In previous work it was established that SRS-induced cavitation and TPD might help the cavitating process by providing initial density modulation which serves as seeds for SRS-induced cavitation. In this particular case the characteristic size of the cavities is of the order of the laser wavelength, which is in contrast to electrostatic cavities having typical sizes of the order of some λ_D. The particular regimes where either one or the other is dominant are not yet fully clarified.

As discussed above (see Section 3.4), SRS activity increases with increasing temperature (also observed in Ref. [67]) at the quarter critical density. At the same time

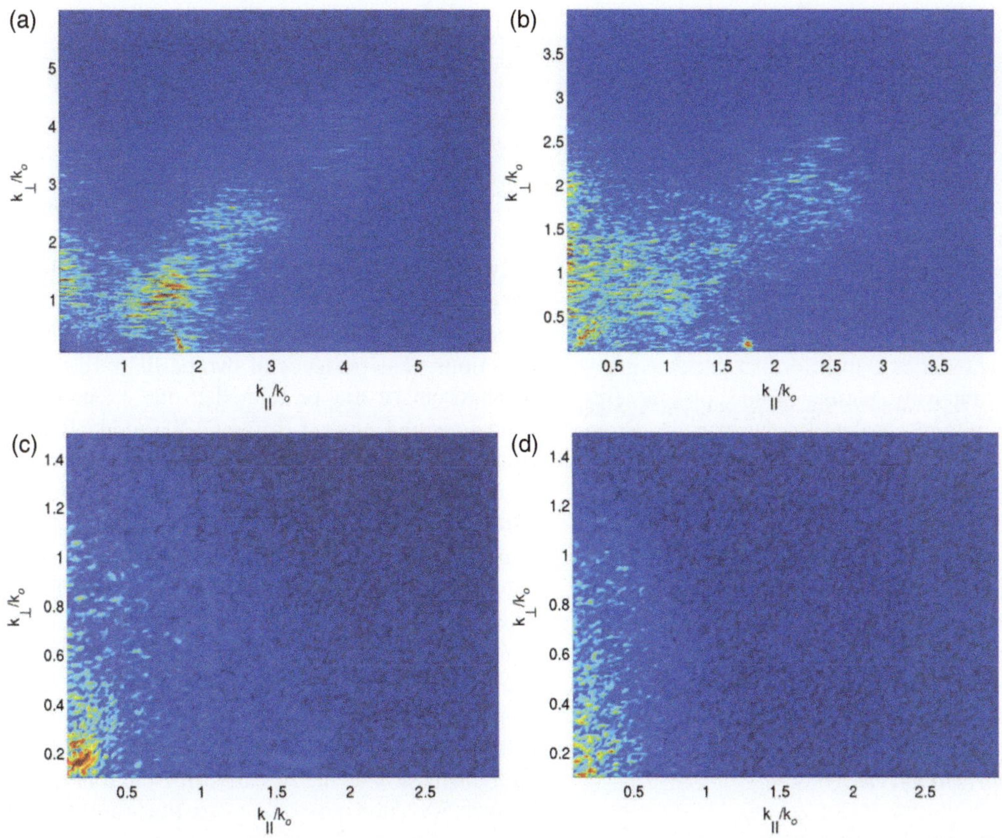

Figure 10. Fourier transform of the ion density corresponding to Figure 11. (a) Case c8 at $t\omega_o = 5600$, (b) case h8 at $t\omega_o = 7000$, (c) case h7 at $t\omega_o = 8300$ and (d) case h9 at $t\omega_o = 13\,600$. It should be noted that the axes for the various cases differ as the k-vectors become shorter as the temperature increases.

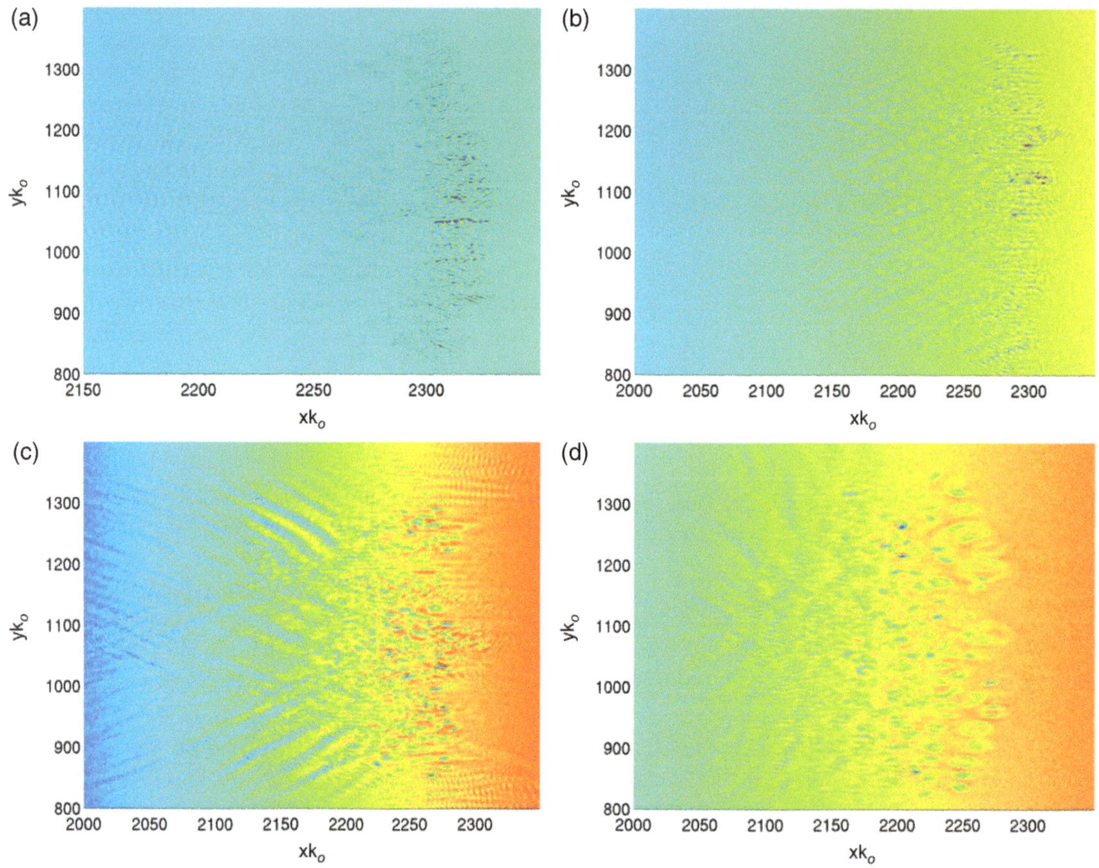

Figure 11. Ion density near the quarter critical density (located at $xk_o \approx 2300$. (a) Case c8 at $t\omega_o = 5600$, (b) case h8 at $t\omega_o = 7000$, (c) case h7 at $t\omega_o = 8300$ and (d) case h9 at $t\omega_o = 13\,600$. It should be noted that the colour scale used is not the same for each of the four sub-figures in order to enhance the visibility of the structures.

this also increases cavitation activity and size, as visible in Figure 11(a–d). In addition, a higher temperature implies a higher laser intensity such that the ponderomotive force can balance the plasma pressure so that cavities can be maintained.

The creation of the cavities is therefore an intricate interdependence of laser intensity, plasma profile, plasma temperature and growth rate of absolute Raman at $n_c/4$. From previous and present simulations it follows that the optimum is around 5 keV for cavity creation. Figure 11(c) has more cavities than the 10 keV case shown in Figure 11(d). The reason for this is that the TPD activity for the case h9 is reduced due to the higher threshold for exciting the instability, which, as said, facilitates the creation of cavities. The typical lifetime of these cavities is of the order of picoseconds, which implies that they are a recursive phenomenon.

3.6. Laser absorption into hot electrons

SRS, TPD and cavitation are all sources of hot electron production. However, the hot electrons differ as far as temperature and propagation direction are concerned[19, 20].

As discussed above, TPD is the dominant process in the cold case c8. As shown in Figure 4, the decay takes place in two directions, which is reflected in the phase space of the electrons, Figure 12(a). Clearly visible are the emerging clouds of accelerated electrons in the forward and backward directions. The presence of two partially separated features in phase space can be related to the excitation of plasma waves centred around different k-vectors, which has also been observed recently in other simulation work[74]. A corresponding signature is found in the p_y-space (not shown here). By contrast, the hot case h9, Figure 12(d), shows only a strong signature of SRS-accelerated electrons in the forward direction. No clear signal from TPD is visible. The intermediate cases, h7 and h8, show contributions from both processes, TPD and SRS. Most of the incident laser energy at this stage is absorbed in the vicinity of $n_c/4$ by either collective modes or hot electrons. For the case c8, Figure 9, depletion of the pump occurs after $2300\omega_o^{-1}$, i.e., no transmission at all. At early time the transmissivity is on average 50%, and it is 10% at later time (around $\sim 7000\omega_o^{-1}$). For case h8 one has 35% transmission at early time, and 15% at later time (same times as for the previous case c8). The case h7 is shown in detail in Figure 2(b) of Ref. [19]

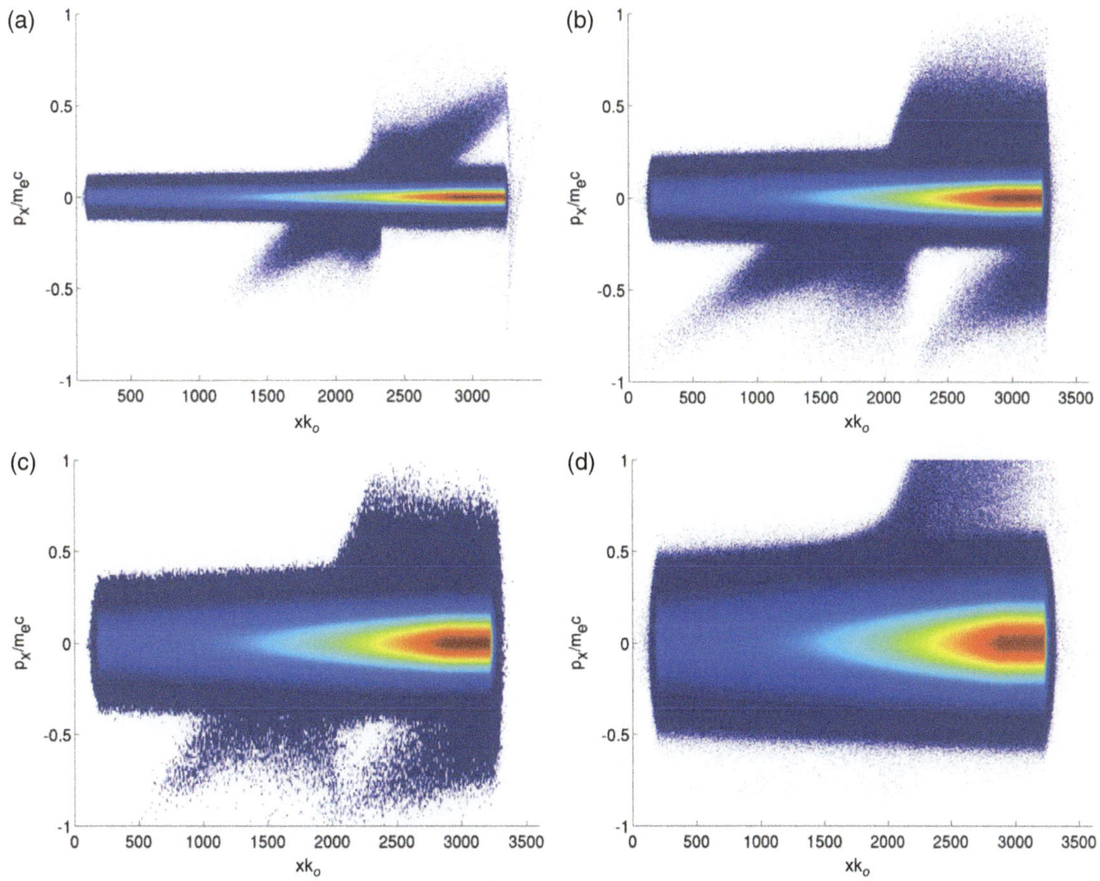

Figure 12. The transverse electron phase space as a function of the laser propagation direction for (a) case c8 at $t\omega_o = 5600$, (b) case h8 at $t\omega_o = 7000$, (c) case h7 at $t\omega_o = 7400$ and (d) case h9 at $t\omega_o = 5200$. The time slice for h9 is taken at an early time as the electrons start to recirculate quickly.

and has roughly 50% transmission. Finally, for the case h9 one has 25%–30% transmission increasing towards the end of the simulation. The hot electrons also act as a damping mechanism for EPWs and participate in the saturation of TPD and SRS. The distribution functions and temperatures are similar to those obtained previously[19, 20].

As was pointed out previously[10], the temperature characterizing the hot tail of the electron distribution function is less than 100 keV and is not dependent on the laser intensity but is determined by the resonant interaction of the electrons with the plasma waves generated by TPD and SRS. The laser intensity affects the number of hot electrons produced. The relevant parameters affecting the hot electron production are the phase velocity and the initial bulk electron temperature. Present and previous simulation work[19, 20] indicates that the hot electron temperature increases with the initial bulk temperature of the plasma. However, a more detailed analysis would be required to quantify this relationship.

4. Conclusions

The relative importance of the various phenomena (cavitation, LDI, SRS, SBS and TPD) is summarized in Table 2 in a qualitative way. The ratio between the thresholds for SRS and TPD is roughly $I^{\mathrm{SRS}}/I^{\mathrm{TPD}} \approx (24/T_e)(\lambda/L_n)^{1/3}$. Setting the ratio equal to 1 results in an electron plasma temperature of $\approx$4.3 keV, which separates the SRS-dominated regime from the TPD-dominated regime. This value agrees well with the observed switchover of the LPI scenario in the simulations. The laser intensities used in the simulations are above the threshold intensities for SRS and TPD for all temperatures used in the simulations. One should note that the above ratio depends strongly on the plasma temperature but only weakly on the gradient scale length of the plasma profile. Therefore, temperature should be the determining parameter for a given intensity. On increasing the intensity other instabilities might appear, such as iSRS in the low-density region, far below $n_c/4$, although saturation of SRS and TPD close to $n_c/4$ was observed. In addition, SBS will become even more dominant on the time scales considered in the simulations[19, 20]. Intensities higher than the ones considered here are not *a priori* excluded for the SI scenario and should motivate further studies in this respect.

An important conclusion of the simulation work presented is the necessity to have better absorption models for the laser beam in radiation hydrodynamic simulation for shock ignition. It is clear that a large fraction, or even most of it, dependent on the LPI parameters, is not absorbed via inverse Bremsstrahlung at the critical density, but rather by collective effects in the low-density corona. This has to be accounted for in hydrodynamic simulations of the implosion phase in a realistic way. Reliable integrated simulations for SI are necessary.

SI experiments and simulations therefore currently have some important caveats.

- The sensitivity of the participating LPI instabilities as far as the coronal plasma temperature is concerned limits the validity of experimental investigations; extrapolation to higher T_e than can be achieved on medium-size installations such as LULI, PALS, LIL or OMEGA requires particular care.
- The largest energy losses due to backscattering in the few-keV range come from SBS and not from SRS. Many SI-relevant present-day experiments concentrate on Raman rather than Brillouin.
- An open issue is the saturation mechanism for SBS which operates on time scales of at least tens of picoseconds.
- Longer time scales are required for kinetic LPI simulations. This was partially addressed by recent work[67].

Another important issue is how these instabilities are affected in the case of multiple overlapping beams as the driver[75].

Much more detailed experiments and simulations are needed to determine the presence and relative importance of the various participating instabilities.

A very interesting issue is to determine experimentally where exactly the laser energy is absorbed: in the low-density plasma corona, at the critical surface or at both locations (in which case the ratio would be important). Possible hot electrons have to be attributed clearly to either SRS or TPD. The distribution functions and directionality of the hot electrons will help in this respect. The simulations clearly show the importance of cavitation at the quarter critical density. The cavities are in general of the order of a few wavelengths. It should be possible to perform interferometry at $4\omega_o$ to image the presence of cavities.

Acknowledgements

Use of the computing center CCRT of the *Commissariat à l'énergie atomique et aux énergies alternatives* is acknowledged. C.R. acknowledges support from grant ANR-11-IDEX-0004-02 Plas@Par. S.W. benefitted from the support of the Czech Science Foundation (Project No. CZ.1.07/2.3.00/20.0279) and from ELI (Project No. CZ.1.05/1.1.00/02.0061).

References

1. G. von Guderley, Luftfahrt-Forsch. **9**, 302 (1942).
2. K. Brueckner and S. Jorna, Rev. Mod. Phys. **46**, 325 (1974).
3. V. Shcherbakov, Sov. J. Plasma Phys. **9**, 240 (1983).
4. R. Betti, C. Zhou, K. Anderson, L. Perkins, W. Theobald, and A. Solodov, Phys. Rev. Lett. **98**, 155001 (2007).
5. L. Perkins, R. Betti, K. LaFortune, and W. Williams, Phys. Rev. Lett. **103**, 045004 (2009).
6. X. Ribeyre, G. Schurtz, M. Lafon, S. Galera, and S. Weber, Plasma Phys. Control. Fusion **51**, 015013 (2009).
7. X. Ribeyre, M. Lafon, G. Schurtz, M. Olazabal-Loume, J. Breil, S. Galera, and S. Weber, Plasma Phys. Control. Fusion **51**, 124030 (2009).
8. X. Ribeyre, S. Guskov, J.-L. Feugeas, P. Nicolai, and V. Tikhonchuk, Phys. Plasmas **20**, 062705 (2013).
9. S. Atzeni, X. Ribeyre, G. Schurtz, A. Schmitt, B. Canaud, R. Betti, and L. Perkins, Nucl. Fusion **54**, 054008 (2014).
10. D. Batani, S. Baton, A. Casner, S. Depierreux, M. Hohenberger, O. Klimo, M. Koenig, C. Labaune, X. Ribeyre, C. Rousseaux, G. Schurtz, W. Theobald, and V. Tikhonchuk, Nucl. Fusion **54**, 054009 (2014).
11. M. Temporal, B. Canaud, W. Garbett, R. Ramis, and S. Weber, High Power Laser Sci. Eng. **2**, e8 (2014).
12. W. Theobald, K. Anderson, R. Betti, R. Craxton, J. Delettrez, J. Frenje, V. Glebov, O. Gotchev, J. Kelly, C. Li, A. Mackinnon, F. Marshall, R. McCrory, D. Meyerhofer, J. Myatt, P. Norreys, P. Nilson, P. Patel, R. Petrasso, P. Radha, C. Ren, T. Sangster, W. Seka, V. Smalyuk, A. Solodov, R. Stephens, C. Stoeckl, and B. Yaakobi, Plasma Phys. Control. Fusion **51**, 124052 (2009).
13. W. Theobald, R. Nora, M. Lafon, A. Casner, X. Ribeyre, K. Anderson, R. Betti, J. Delettrez, J. Frenje, V. Glebov, O. Gotchev, M. Hohenberger, S. Hu, F. Marshall, D. Meyerhofer, T. Sangster, G. Schurtz, W. Seka, V. Smalyuk, C. Stoeckl, and B. Yaakobi, Phys. Plasmas **19**, 102706 (2012).
14. S. Baton, M. Koenig, E. Brambrink, H. Schlenvoigt, C. Rousseaux, G. Debras, S. Laffite, P. Loiseau, F. Philippe, X. Ribeyre, and G. Schurtz, Phys. Rev. Lett. **108**, 195002 (2012).
15. M. Hohenberger, W. Theobald, S. Hu, K. Anderson, R. Betti, T. Boehly, A. Casner, D. Fratanduono, M. Lafon, D. Meyerhofer, R. Nora, X. Ribeyre, T. Sangster, G. Schurtz, W. Seka, C. Stoeckl, and B. Yaakobi, Phys. Plasmas **21**, 022702 (2014).
16. D. Batani, M. Koenig, S. Baton, F. Perez, L. Gizzi, P. Koester, L. Labate, J. Honrubia, L. Antonelli, A. Morace, L. Volpe, J. Santos, G. Schurtz, S. Hulin, X. Ribeyre, C. Fourment, P. Nicolai, B. Vauzour, L. Gremillet, W. Nazarov, J. Pasley, M. Richetta, K. Lancaster, C. Spindloe, M. Tolley, D. Neely, M. Kozlova, J. Nejdl, B. Rus, J. Wolowski, J. Badziak, and F. Dorchies, Plasma Phys. Control. Fusion **53**, 124041 (2011).
17. W. Kruer, *The Physics of Laser–Plasma Interaction* (Addison Wesley, New York, 1988).
18. H. Vu, D. DuBois, and B. Bezzerides, Phys. Rev. Lett. **86**, 4306 (2001).
19. C. Riconda, S. Weber, V. Tikhonchuk, and A. Heron, Phys. Plasmas **18**, 092701 (2011).
20. S. Weber, C. Riconda, O. Klimo, A. Heron, and V. Tikhonchuk, Phys. Rev. E **85**, 016403 (2012).
21. O. Klimo, S. Weber, V. Tikhonchuk, and J. Limpouch, Plasma Phys. Control. Fusion **52**, 055013 (2010).

22. O. Klimo, V. Tikhonchuk, X. Ribeyre, G. Schurtz, C. Riconda, S. Weber, and J. Limpouch, Phys. Plasmas **18**, 082709 (2011).
23. O. Klimo and V. Tikhonchuk, Plasma Phys. Control. Fusion **55**, 095002 (2013).
24. O. Klimo, J. Psikal, V. Tikhonchuk, and S. Weber, Plasma Phys. Control. Fusion **56**, 055010 (2014).
25. S. Depierreux, C. Labaune, D. Michel, C. Stenz, P. Nicolaï, M. Grech, G. Riazuelo, S. Weber, C. Riconda, V. Tikhonchuk, P. Loiseau, N. Borisenko, W. Nazarov, S. Hüller, D. Pesme, M. Casanova, J. Limpouch, C. Meyer, P. DiNicola, R. Wrobel, E. Alozy, P. Romary, G. Thiell, G. Souillié, C. Reverdin, and B. Villette, Phys. Rev. Lett. **102**, 195005 (2009).
26. J. Moody, B. MacGowan, D. Hinkel, W. Kruer, E. Williams, K. Estabrook, R. Berger, R. Kirkwood, D. Montgomery, and T. Shepard, Phys. Rev. Lett. **77**, 1294 (1996).
27. S. Depierreux, K. Lewis, C. Labaune, and C. Stenz, J. Phys. IV **133**, 317 (2006).
28. S. Depierreux, P. Loiseau, D. Michel, V. Tassin, C. Stenz, P.-E. Masson-Laborde, C. Goyon, V. Yahia, and C. Labaune, Phys. Plasmas. **19**, 012705 (2012).
29. C. Labaune, K. Lewis, H. Bandulet, S. Depierreux, S. Huller, P. Masson-Laborde, D. Pesme, and G. Riazuelo, J. Phys. IV **133**, 29 (2006).
30. H. Baldis, C. Labaune, J. Moody, T. Jalinaud, and V. Tikhonchuk, Phys. Rev. Lett. **80**, 1900 (1998).
31. R. Drake, R. Watt, and K. Estabrook, Phys. Rev. Lett. **77**, 79 (1996).
32. J. Fernandez, B. Bauer, K. Bradley, J. Cobble, D. Montgomery, R. Watt, B. Bezzerides, K. Estabrook, R. Focia, S. Goldman, D. Harris, E. Lindman, H. Rose, J. Wallace, and and B. Wilde, Phys. Rev. Lett. **81**, 2252 (1998).
33. H. Baldis, C. Labaune, E. Schifano, N. Renard, and A. Michard, Phys. Rev. Lett. **77**, 2957 (1996).
34. C. Goyon, S. Depierreux, D. Michel, G. Loisel, V. Yahia, P. Masson-Laborde, P. Loiseau, S. Huller, N. Borisenko, A. Orekhov, O. Rosmej, P. Nicolai, V. Tikhonchuk, and C. Labaune, EPJ Web Conf. **59**, 05006 (2013).
35. C. Goyon, S. Depierreux, V. Yahia, G. Loisel, C. Baccou, C. Courvoisier, N. Borisenko, A. Orekhov, O. Rosmej, and C. Labaune, Phys. Rev. Lett. **111**, 235006 (2013).
36. D. Michel, S. Depierreux, C. Stenz, V. Tassin, and C. Labaune, Phys. Rev. Lett. **104**, 255001 (2010).
37. L. Antonelli, D. Batani, A. Patria, O. Ciricosta, C. Cecchetti, P. Koester, L. Labate, A. Giulietti, L. Gizzi, A. Moretti, M. Richetta, L. Giuffrida, L. Torrisi, M. Kozlova, J. Nejdl, M. Sawicka, D. Margarone, B. Rus, G. Schurtz, X. Ribeyre, M. Lafon, C. Spindloe, and T. O'Dell, Acta Tecn. **56**, T57 (2011).
38. S. Depierreux, C. Goyon, K. Lewis, H. Bandulet, D. T. Michel, G. Loisel, V. Yahia, V. Tassin, C. Stenz, N. G. Borisenko, W. Nazarov, J. Limpouch, P. E. Masson Laborde, P. Loiseau, M. Casanova, Ph. Nicolaï, S. Hüller, D. Pesme, C. Riconda, V. T. Tikhonchuk, and C. Labaune, Plasma Phys. Control. Fusion **53**, 124034 (2011).
39. P. Koester, L. Antonelli, S. Atzeni, J. Badziak, F. Baffigi, D. Batani, C. A. Cecchetti, T. Chodukowski, F. Consoli, G. Cristoforetti, R. De Angelis, G. Folpini, L. A. Gizzi, Z. Kalinowska, E. Krousky, M. Kucharik, L. Labate, T. Levato, R. Liska, G. Malka, Y. Maheut, A. Marocchino, P. Nicolai, T. O'Dell, P. Parys, T. Pisarczyk, P. Raczka, O. Renner, Y. J. Rhe, X. Ribeyre, M. Richetta, M. Rosinski, L. Ryc, J. Skala, A. Schiavi, G. Schurtz, M. Smid, C. Spindloe, J. Ullschmied, J. Wolowski, and A. Zaras, Plasma Phys. Control. Fusion **55**, 124045 (2013).
40. H. Baldis and C. Walsh, Phys. Fluids **26**, 1364 (1983).
41. W. Seka, J.-F. Myatt, R. Short, D. Froula, J. Katz, V. Goncharov, and I. Iguminshchev, Phys. Rev. Lett. **112**, 145001 (2014).
42. Code developed by A. Heron and J.-C. Adam at CPHT. Ecole Polytechnique.
43. C. Liu and M. Rosenbluth, Phys. Fluids **19**, 967 (1976).
44. A. Simon, R. Short, E. Williams, and T. Dewandre, Phys. Fluids **26**, 3107 (1983).
45. C. Menyuk, N. El-Siragy, and W. Manheimer, Phys. Fluids **28**, 3409 (1985).
46. B. Afeyan and E. Williams, Phys. Rev. Lett. **75**, 4218 (1995).
47. B. Afeyan and E. Williams, Phys. Plasmas **4**, 3845 (1997).
48. D. Forslund, J. Kindel, and E. Lindman, Phys. Fluids **18**, 1002 (1975).
49. V. Oraevskii and R. Sagdeev, Sov. Phys. Tech. Phys. **7**, 955 (1963).
50. Y. Ichikawa, Phys. Fluids **9**, 1454 (1966).
51. S. Depierreux, J. Fuchs, C. Labaune, A. Michard, H. Baldis, D. Pesme, S. Hüller, and G. Laval, Phys. Rev. Lett. **84**, 2869 (2000).
52. S. Depierreux, C. Labaune, J. Fuchs, D. Pesme, V. Tikhonchuk, and H. Baldis, Phys. Rev. Lett. **89**, 045001 (2002).
53. J. Kline, D. Montgomery, B. Bezzerides, J. Cobble, D. DuBois, R. Johnson, H. Rose, L. Yin, and H. Vu, Phys. Rev. Lett. **94**, 0175003 (2005).
54. T. Obiki, R. Itatani, and Y. Otani, Phys. Rev. Lett. **20**, 184 (1968).
55. R. Drake and S. Batha, Phys. Fluids B **3**, 2936 (1991).
56. D. Villeneuve, K. Baker, R. Drake, B. Sleaford, B. LaFontaine, K. Estabrook, and M. Prasad, Phys. Rev. Lett. **71**, 368 (1993).
57. J. Fernandez, J. Cobble, B. Failor, D. Dubois, D. Montgomery, H. Rose, H. Vu, B. Wilde, M. Wilke, and R. Chrien, Phys. Rev. Lett. **77**, 2702 (1996).
58. R. Kirkwood, B. MacGowan, D. Montgomery, B. Afeyan, W. Kruer, J. Moody, K. Estabrook, C. Back, S. Glenzer, M. Blain, E. Williams, R. Berger, and B. Lasinski, Phys. Rev. Lett. **77**, 2706 (1996).
59. D. Montgomery, B. Afeyan, J. Cobble, J. Fernandez, M. Wilke, S. Glenzer, R. Kirkwood, B. MacGowan, J. Moody, E. Lindman, D. Munro, B. Wilde, H. Rose, D. Dubois, B. Bezzerides, and H. Vu, Phys. Plasmas **5**, 1973 (1998).
60. G. Bonnaud, D. Pesme, and R. Pellat, Phys. Fluids B **2**, 1618 (1990).
61. S. Karttunen, Phys. Rev. A **23**, 206 (1981).
62. J. Heikkinen and S. Karttunen, Phys. Fluids **29**, 1291 (1986).
63. B. Bezzerides, D. DuBois, and H. Rose, Phys. Rev. Lett. **70**, 2569 (1993).
64. B. Bezzerides, D. DuBois, H. Rose, and D. Russell, Phys. Scr. T **63**, 16 (1996).
65. R. Berger, C. Still, E. Williams, and A. Langdon, Phys. Plasmas **5**, 4337 (1998).
66. D. Russell, D. DuBois, and H. Rose, Phys. Plasmas **6**, 1294 (1999).

67. R. Yan, J. Li, and C. Ren, Phys. Plasmas **21**, 062705 (2014).
68. C. Riconda, S. Weber, V. Tikhonchuk, J.-C. Adam, and A. Heron, Phys. Plasmas **13**, 083103 (2006).
69. C. Riconda, S. Weber, O. Klimo, A. Heron, and V. Tikhonchuk, EPJ Web Conf. **59**, 05007 (2013).
70. H. Vu, D. DuBois, D. Russell, and J. Myatt, Phys. Plasmas **19**, 102708 (2012).
71. H. Vu, D. DuBois, J. Myatt, and D. Russell, Phys. Plasmas **19**, 102703 (2012).
72. H. Vu, D. DuBois, D. Russell, J. Myatt, and J. Zhang, Phys. Plasmas **21**, 042705 (2014).
73. S. Weber, C. Riconda, and V. Tikhonchuk, Phys. Rev. Lett. **94**, 055005 (2005).
74. R. Yan, C. Ren, J. Li, A. Maximov, W. Mori, Z.-M. Sheng, and F. Tsung, Phys. Rev. Lett. **108**, 175002 (2012).
75. J. Myatt, J. Zhang, R. Short, A. Maximov, W. Seka, D. Froula, D. Edgell, D. Michel, I. Igumenshchev, D. Hinkel, P. Michel, and J. Moody, Phys. Plasmas **21**, 055501 (2014).

4

Performance of rapid-grown KDP crystals with continuous filtration

Guohang Hu[1], Yueliang Wang[1,2], Junxiu Chang[1,2], Xiaoyi Xie[1,2], Yuanan Zhao[1], Hongji Qi[1], and Jianda Shao[1]

[1] *Key Laboratory of Materials for High Power Laser, Shanghai Institute of Optics and Fine Mechanics, Chinese Academy of Sciences, Shanghai 201800, China*

[2] *University of Chinese Academy of Sciences, Beijing 100049, China*

Abstract

Rapid growth processing of KDP crystals was improved by employing continuous filtration to eliminate bulk defects. The performances of the KDP crystals, including scattering defects, laser damage resistance and transmittance, were measured and analyzed. Compared with rapid-grown KDP without continuous filtration, the transmittance in the near-infrared was increased by at least 2%, almost all of 'micron size' defects were eliminated and 'sub-micron size' defects were decreased by approximately 90%. Laser damage testing revealed that the laser-induced damage thresholds (LIDTs), as well as the consistency of the LIDTs from sample to sample, were improved greatly. Moreover, it identified that 'micron size' defects were the precursors which initiated laser damage at relative lower laser fluence (4–6 J cm^{-2}), and there was a lower correlation between smaller size scattering defects and laser damage initiation. The improved consistency in the LIDTs, attributed to elimination of 'micron size' defects, and LIDT enhancement originated from the decreased absorption of the KDP crystals.

Keywords: continuous filtration; defect; KDP; laser damage; rapid growth

1. Introduction

KH_2PO_4 (KDP) crystals are important nonlinear optical materials in many applications, such as frequency converters and Pockel cells for high-power large-aperture laser systems[1]. In the 1990s, the rapid growth technique was developed to grow large-scale (40–50 cm) KDP boules at a rate of 10–20 mm/day[2, 3], which was one of the seven wonders during NIF laser construction[4]. Due to a growth rate an order of magnitude faster than conventional growth, a greater number of bulk defects were formed, degrading the performance of the crystals by causing pinpoint damage and optical losses, which adversely affected the quality of the downstream beam[1, 5]. It was found that typical damage pinpoints were caused by defects, which efficiently absorbed laser energy, leading to a temperature rise and subsequent 'micro-explosion'[5–7]. Various types of bulk defect, such as metallic impurities[8–10], electronic defects[11–13], organic materials, bacteria[14, 15], dislocation[16] and insoluble impurities, have been reported. In 1999, the continuous filtration technique was employed to improve the purity of the growth solution, and high-quality large-scale KDP crystal boules grown at a rate of 12–15 mm/day were demonstrated[17]. Although numerous studies have attempted to show the advantage of continuous filtration during rapid growth processing, no direct evidence has elucidated the reasons for the performance improvement. In the current study, the fundamental reason for the performance improvement was revealed via a comparison of the bulk defect properties, laser damage process and transmittance variation.

Correspondence to: Y. Zhao, No. 390, Qinghe Road, Jiading, Shanghai, China. Email: yazhao@siom.ac.cn

2. Experimental details

2.1. *Growth of KDP crystals*

KDP crystals were grown rapidly from an over-saturated aqueous solution of high-purity KDP powder in a 45 L growth vessel equipped with a continuous filtration unit, as shown in Figure 1. Both the growth vessel and the continuous filtration unit were placed in a thermostatic water bath. Two kinds of filter (the first-level filter was of the order of 0.1 μm, the second-level filter was of the order of 0.01 μm)

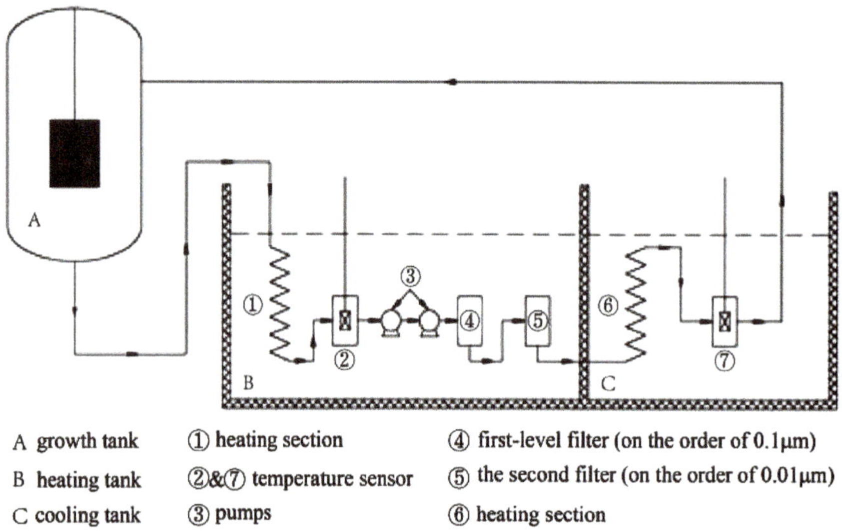

Figure 1. Schematic system for rapid growth of KDP boules using continuous filtration.

were connected inside the continuous filtration unit. A well-polished Z-cut crystal of size 8 mm × 8 mm × 4 mm was fixed in the center of the growth plate to initiate the crystallization. The growth rate was controlled to 10 mm/day by lowering the temperature from 318 to 305 K. The crystal was rotated in the 'forward–stop–backward' mode at a speed of 70 r.p.m. The temperature was controlled by a programmable Shimada controller (Model FP21) with an accuracy of 0.02 K. The KDP plates, labeled as CF-1 to CF-6, were cut in a Type I second harmonic generation (SHG) orientation from the same boule prepared by rapid growth with continuous filtration. A further six Type I KDP plates, labeled NCF-1 to NCF-6, prepared by rapid growth without continuous filtration, were chosen as a comparison. All KDP plates were cut to a size of 50 mm × 50 mm × 10 mm. They were annealed at 165°C and polished by diamond turning before taking measurements.

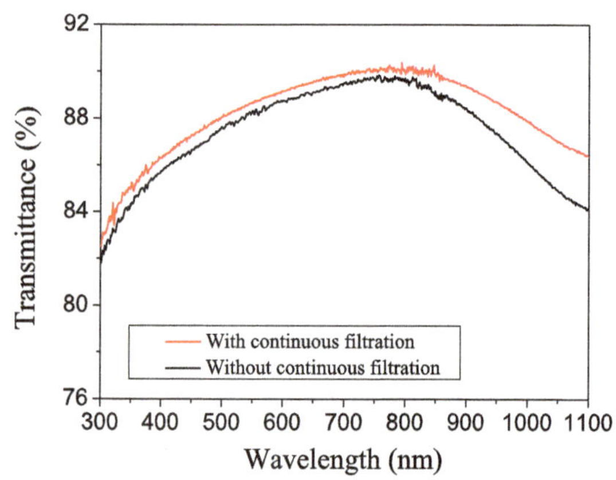

Figure 2. Transmittance spectra of KDP crystals.

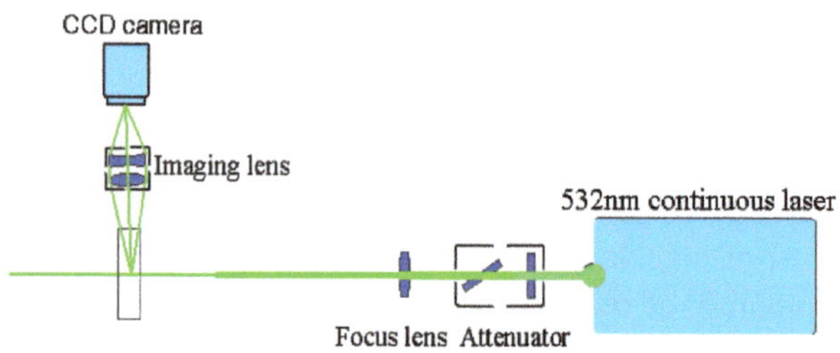

Figure 3. Schematic of the detection system for bulk defects in KDP crystals.

2.2. Measurement of transmittance spectra

The transmittance spectra of the samples were measured with a Lambda 1050 UV/Vis/NIR spectrophotometer in the wavelength range 300–1100 nm and are shown in Figure 2.

It was found that the transmittance of the CF series samples was slightly (approximately 0.2%–1%) higher than the NCF series samples in the wavelength range 300–850 nm, and at least 2% higher than the NCF series samples in the wavelength range 850–1100 nm.

2.3. Detection of bulk defects

Defects embedded in the crystals were detected by scattered light microscopy using high-intensity dark-field illumination (Figure 3). A CW 532 nm laser with a Gaussian spatial profile was focused and illuminated the bulk site at a power of 1.3 W. The focal length was 5000 mm, and the effective area of the spot on the sample was 0.8 mm[2]. The image of the illuminated region was captured, orthogonal to the direction of laser propagation, through the side of the sample.

Typical scattering defects of the KDP plates are shown in Figure 4. Because the scattering light magnified the

(a) (b) (c)

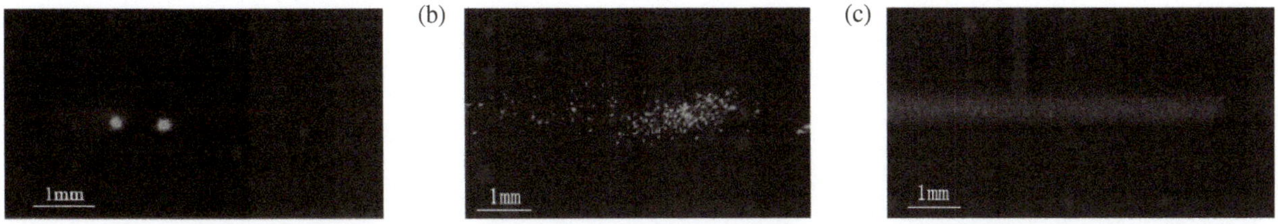

Figure 4. Typical scattering defects (a) 'micron size', (b) 'sub-micron size', (c) 'nano-scale cluster'.

defects and it was difficult to determine their actual size, the defects were categorized by their estimated size in this study. There were three kinds of scattering defect: sparsely distributed 'micron size' defects, high-density 'sub-micron size' defects and 'nano-scale' defect clusters. Statistical analysis of laser damage tests on 70-μm defects showed that all of the 'micron size' defects could generate laser damage at fluences higher than 4–6 $\mathrm{J\,cm^{-2}}$. 'Micron size' scattering defects were the most dangerous laser damage initiator. A fraction of 'sub-micron size' defects could lead to laser damage, with the probability of damage initiated being approximately 5%–40%. An increased local density of nano-scale defect clusters could also act as an initiator of laser damage.

The size of 'micron size' scattering defects was determined by microscope. The sample was wet-etched to bring the defects to the surface or near to the surface, where the size of the defects could be determined using a Nomarski microscope with 200× magnification. Figure 5 shows the morphology of a typical 'micron size' scattering defect. The color of 'micron size' defects was different from the surrounding KDP crystal. As is well known, the contrast in bright-field imaging depends on differences in light absorption and the refractive index. It is clear that the absorption or the refractive index of 'micron size' defects was different from the DKDP material. The sizes of the defects shown in Figures 4(b) and 4(c) were not determined in this way, and merely estimated from the scattered intensity, as it was very difficult to distinguish these defects after wet etching.

Due to the variations in the distribution of scattering defects, the density of defects was analyzed statistically by sampling 30 dark images randomly for each sample, as summarized in Figure 6. The density of bulk defects is the number of scattering sites ('micron size' and 'sub-micron size' defects) divided by the illuminated volume of the KDP crystal. For 'nano-scale clusters', it is more difficult to calculate the density than for the 'micron size' and 'sub-micron size' defects. The probability of capturing these defects from 30 dark images was calculated, and the difference of the brightness was not considered.

It was found that the 'micron size' defects were sparsely distributed, and the density of this kind of defect was approximately 0.008 $\mathrm{mm^{-3}}$ for the NCF series samples. However, they were eliminated after employing continuous

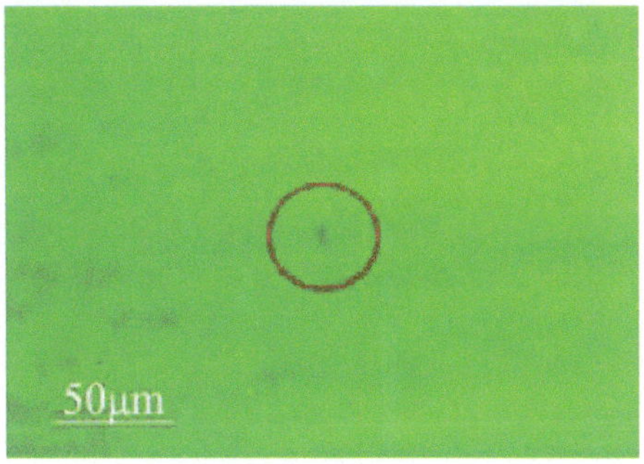

Figure 5. Image of a 'micron size' defect captured by the optical microscope.

filtration. Furthermore, the density of 'sub-micron size' defects in the CF samples was approximately one or two orders of magnitude lower than that of NCF samples. For the 'nano-scale clusters', although the probability of viewing this kind of defect in CF series samples was slightly lower than in NCF series samples, there was no obvious difference between the two series.

2.4. Laser damage testing

The damage test experiments were performed using a pulsed Nd:YAG laser operating at 30 Hz with an approximate output energy of 1.5 J at 1064 nm. The pulse duration was approximately 12 ns. The laser fluence on the target plane was adjusted by means of an energy attenuator, consisting of a half-wave plate and a polarizer. The laser beam was focused by the focusing lens onto the target plane. The focal length was 5000 mm, and the effective spot size on the sample was 0.6 $\mathrm{mm^2}$, as measured by a laser beam analyzer. The onset of laser damage was detected by an online imaging system with a resolution of the order of microns. A ramp (R-on-1) test was employed to analyze the laser damage resistance of the KDP witnesses, and at least 50 sites on each witness were sampled by laser pulses. The testing results are shown in Figure 7.

It was clear that the laser-induced damage probability shifted to higher fluences after employing continuous filtration. For the NCF series samples, some samples (NCF-3, NCF-4 and NCF-6) exhibited very low laser-induced damage thresholds (LIDTs), attributed to individual damage initiated

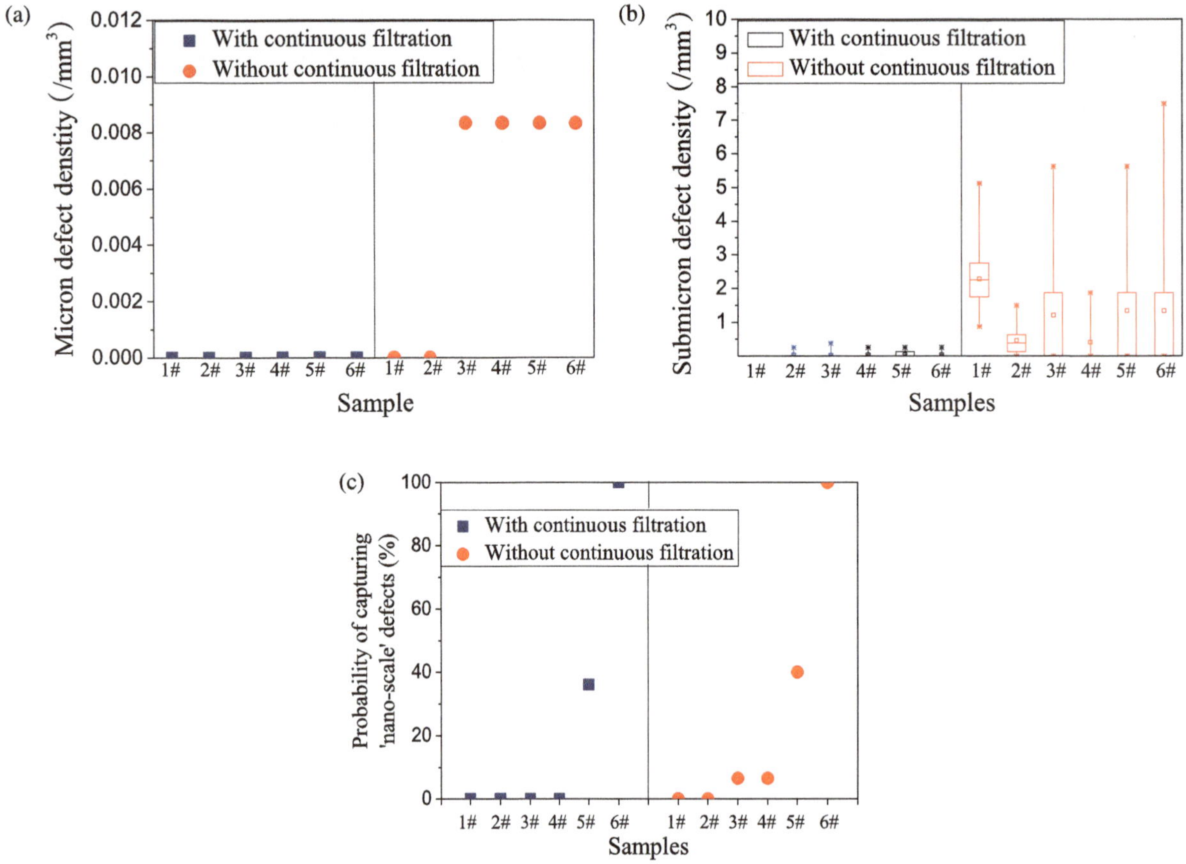

Figure 6. Comparison of (a) 'micron size' scattering defects, (b) 'sub-micron size' scattering defects and (c) 'nano-scale cluster' between the samples grown with and without continuous filtration.

at a very low laser fluence. The LIDTs of the NCF series samples ranged from 2.3 to 22.4 J cm^{-2}. For the CF series samples, the LIDTs range from 25.4 to 36.6 J cm^{-2}, and the consistency of the LIDTs was also improved.

3. Discussion

By using two-level filters in the continuous filtration system, the new growth method ruled out the probability of inclusions with sizes larger than the filter hole. The statistical data for the scattering defects showed that there were no 'micron size' defects after filtration and 'sub-micron size' defects decreased by 90%. This means that most of the defects with sizes larger than tens of microns were eliminated. However, there was no obvious change in 'nano-scale' defect clusters before and after employing continuous filtration, given its smaller size than the filter holes.

A decreased level of scattering defects reduced scattering losses, which was helpful in improving the transmittance of the KDP crystals. However, there was more noticeable improvement in transmittance for infrared wavelengths than in the visible and UV bands after employing continuous

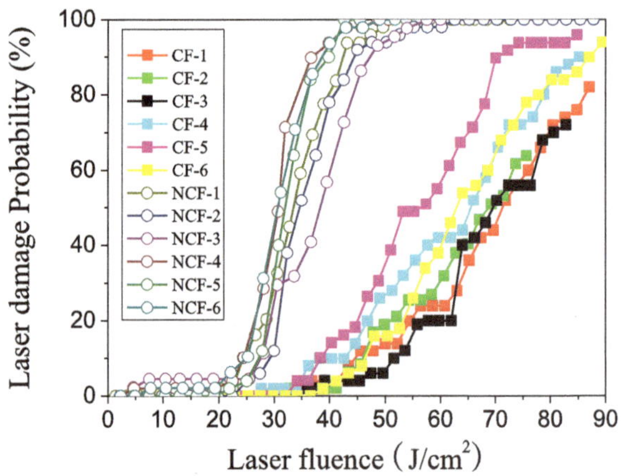

Figure 7. R-on-1 laser damage probability curves for the KDP samples.

filtration; this cannot be attributed to decreased scattering, because one would expect a higher scattering intensity at short wavelengths according to the scattering law. The higher transmittance of the CF series samples in the infrared band mainly comes from the decreased absorption of the KDP crystals.

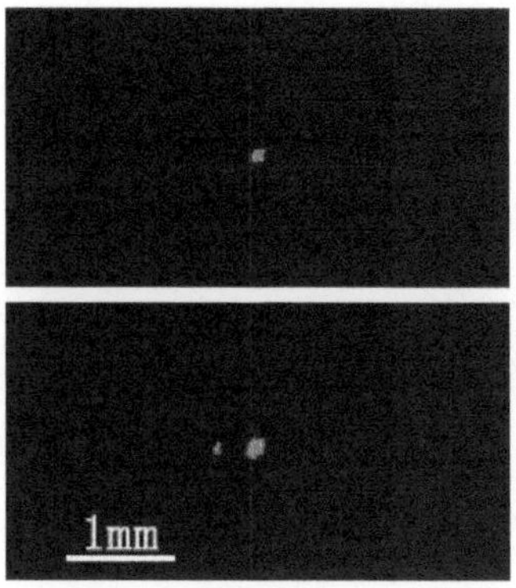

Figure 8. Laser damage initiated at 'micron size' defects in NCF series samples.

For KDP crystals, the prevailing consensus for laser damage initiation has been that the damage is dominated by precursor defects that efficiently absorb the laser energy, and that several kinds of defect may be responsible for damage initiation, such as impurities[18], interstitial atoms or vacancies[19]. The online monitoring technique, combining the light-scattering technique and LIDT testing, demonstrated that scattering defects were one of the damage precursors. Figure 8 showed that pinpoint damage initiated from the 'micron size' scattering defects.

The typical laser fluence for damage initiation from 'micron size' defects was approximately 4–6 J cm^{-2}, and could be defined as a low-fluence damage precursor. This was the reason for the low LIDT for three of the samples (NCF-3, NCF-4 and NCF-6) and the wide range of LIDTs of NCF series samples. Statistical analysis of the defect density showed that this kind of low-fluence damage precursor existed only in NCF series samples (see Figure 6), and most of the damage onset was correlated to 'micron size' scattering defects from the online monitor during laser damage testing. When continuous filtration was employed, there were no 'micron size' defects observed for the CF series samples. This was the reason for the improvements in LIDT consistency for the CF series samples.

Besides 'micron size' scattering defects, the density of 'sub-micron size' scattering defects in the CF series samples also decreased significantly compared with the NCF series samples. The density in the former was one or two orders of magnitude lower than in the latter. Online monitoring of damage initiation showed that there were a few damage sites ($<10\%$) initiated from 'sub-micron size' defects, and most of the damage was not initiated from pre-existing scattering defects (see Figure 9).

The effect of eliminating 'micron size' defects was to lower the damage probability at low fluences, and it did not change the probability of damage at high fluences. However, the curve of damage probability shifted to a higher laser fluence for the CF samples compared with the NCF samples, and 'sub-micron size' scattering defects were not the main cause for damage initiation. That means that there were high-fluence damage precursors other than 'sub-micron size' scattering defects. The pinpoint damage shown in Figure 9(b) showed that the high-fluence damage precursors were high-density distributed absorptive defects. Although there was no direct proof as to the high-fluence damage precursors from the experimental observations, we can get some information about them from the spectra curve. The transmission of the CF series samples was higher than that of the NCF series samples, with an increasing gap between the two curves on moving from short wavelengths to long wavelengths. The major difference was located in the infrared band, and was greater than 2%. The improvement of transmission may be attributed to decreased scattering losses and absorption losses. The scattering loss was certainly decreased for CF series samples compared with NCF series samples due to the change in scattering defect density shown in Figure 6. However, this was not the main reason for the increased transmission, as scattering is wavelength dependent and higher scattering losses should be expected at shorter wavelengths according to the scattering law. This means that the increasing transmission difference originated from decrease absorption losses, i.e., absorptive defects, giving rise to a shift of the damage probability curve to higher fluences. Therefore, continuous filtration reduced not only the scattering defects but also the absorptive defects, which was the reason for the improved quality of the rapid-grown KDP crystals.

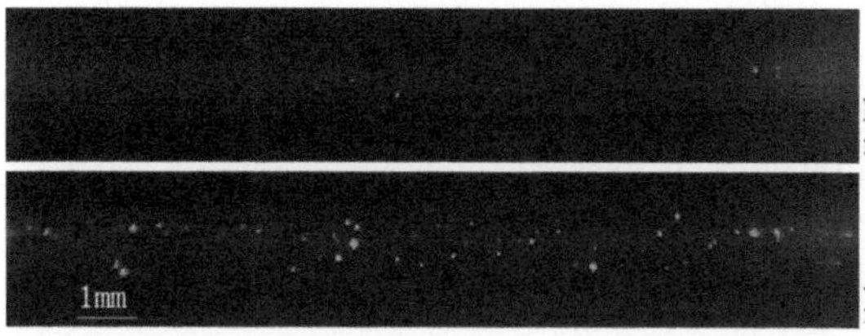

Figure 9. Typical damage initiation in CF series samples.

4. Conclusion

Rapid growth of KDP crystals with continuous filtration has been demonstrated. The transmittance of the crystals for the infrared band was at least 2% higher than that of samples grown without continuous filtration, which was attributed to the decrease of absorption. Micron size scattering defects were identified as low-fluence damage precursors, and can be eliminated by continuous filtration. The damage resistance of the crystals was greatly improved due to the elimination of low-fluence damage precursors and the decrease in absorption.

Acknowledgements

We are grateful for helpful discussions with Prof. Yuangen Yao of Fujian Institute of Research on the Structure of Matter, and Professors Xinguang Xu and Xun Sun of Shandong University.

References

1. J. J. De Yoreo, A. K. Burnham, and P. K. Whitman, Int. Mater. Rev. **47**, 113 (2002).
2. A. Chernov, N. Zaitseva, and L. Rashkovich, J. Cryst. Growth **102**, 793 (1990).
3. N. P. Zaitseva, J. J. DeYoreo, M. R. Dehaven, R. L. Vital, K. E. Montgomery, M. Richardson, and L. J. Atherton, J. Cryst. Growth **180**, 255 (1997).
4. N. Zaitseva and L. Carman, Prog. Cryst. Growth Ch. **43**, 1 (2001).
5. M. D. Feit and A. M. Rubenchik, Proc. SPIE **5273**, 74 (2004).
6. C. H. Chan, Appl. Phys. Lett. **26**, 628 (1975).
7. G. Duchateau, Opt. Express **17**, 10434 (2009).
8. R. A. Negres, P. DeMange, and S. G. Demos, Opt. Lett. **30**, 2766 (2005).
9. C. Marshall, S. Payne, M. Henesian, J. Speth, and H. Powell, J. Opt. Soc. Am. B **11**, 774 (1994).
10. G. Duchateau and A. Dyan, Opt. Express **15**, 4557 (2007).
11. S. Setzler and K. Stevens, Phys. Rev. B **57**, 2643 (1998).
12. C. S. Liu, Q. Zhang, and N. Kioussis, Phys. Rev. B **68**, 224107 (2003).
13. C. S. Liu, N. Kioussis, S. G. Demos, and H. B. Radousky, Phys. Rev. Lett. **91**, 015505 (2003).
14. K. E. Montgomery and F. P. Milanovich, J. Appl. Phys. **68**, 3979 (1990).
15. Y. Nishida, A. Yokotani, T. Sasaki, K. Yoshida, T. Yamanaka, and C. Yamanaka, Appl. Phys. Lett. **52**, 420 (1988).
16. B. Dam, P. Bennema, and W. J. P. van Enckevort, J. Cryst. Growth **74**, 118 (1986).
17. N. Zaitseva, J. Atherton, R. Rozsa, L. Carman, I. Smolsky, M. Runkel, R. Ryon, and L. James, J. Cryst. Growth **197**, 911 (1999).
18. A. Burnham, M. Runkel, M. Feit, A. Rubenchik, R. Floyd, T. Land, W. Siekhaus, and R. Hawley-Fedder, Appl. Opt. **42**, 5483 (2003).
19. G. Duchateaua, Proc. SPIE **7504**, 75041K (2009).

5

Inertial confinement fusion driven by long wavelength electromagnetic pulses

Baifei Shen, Xueyan Zhao, Longqing Yi, Wei Yu, and Zhizhan Xu

State Key Laboratory of High Field Laser Physics, Shanghai Institute of Optics and Fine Mechanics, Chinese Academy of Sciences, Shanghai 201800, China

Abstract

A method for inertial confinement fusion driven by powerful long wavelength electromagnetic pulses (EMPs), such as CO_2 laser pulses or high power microwave pulses, is proposed. Due to the high efficiency of generating such long wavelength electromagnetic pulses, this method is especially important for the future fusion electricity power. Special fuel targets are designed to overcome the shortcomings of the long wavelength electromagnetic pulses.

Keywords: CO_2 laser; Inertial confinement fusion; laser acceleration

1. Introduction

After the invention of laser, scientists began to study laser fusion for simulation of hydrogen bombs and as a long-term option to generate electric energy. Preliminary calculations suggested that a laser pulse of at least 100 kJ of energy and at most 10 ns duration would be required. However, recent experiments with NIF in United States implied that even 2 MJ might be still not enough by using the method of indirect drive[1].

One of the main problems is the hydrodynamic instabilities during the shell implosion. For inertial confinement fusion, in order to burn a substantial part of the fuel in a very short time, extremely high plasma densities have to be realized to achieve high reaction rates, since the confinement time is very short, typically 0.1 ns. To achieve such a high plasma density, one has to compress the fuel with very high pressure and design special targets as illustrated in Figure 1a. The targets consist of a hollow shell capsule with an outer ablator layer and a fuel layer of cryogenic DT.

Laser pulses, x-rays, or particle beams were proposed to heat the ablator to generate the required high pressure. The ignition threshold factor (ITF) is[1]

$$\mathrm{ITF} \bullet I_0 \left(\frac{M_{\mathrm{DT}}}{M_0}\right)\left(\frac{v}{v_0}\right)^8 \left(\frac{\alpha}{\alpha_0}\right)^{\bullet 4} \left(1 \bullet 1.2\frac{\Delta R^{\text{K-wtd}}_{\text{hotstspot}}}{R_{\text{hotstspot}}}\right)^4 \bullet \left(\frac{M_{\mathrm{clean}}}{M_{\mathrm{DT}}}\right)^{0.5} (1 \bullet P_{\mathrm{HS}}),$$

where I_0 is the overall multiplier, M_{DT} is the fuel mass, v is the implosion velocity, α is the adiabat, $\Delta R^{\text{K-wtd}}_{\text{hotstspot}}$ is the hot spot perturbation, M_{clean} is the clean fuel mass and P_{HS} is the hot spot purity factor. From this, we know that the fuel implosion velocity v is a critical parameter if the shell is not destroyed during the compression and the fuel mixture is not serious. The strong dependence of the minimum energy on the achievable implosion velocity means that only a factor of about 4 increase in the implosion velocity is required to go from a minimum energy of about 1 MJ to a minimum energy of 1 kJ[1]. However hydrodynamic instabilities prevent high implosion velocity while low entropy is kept.

The growth rate of the instabilities σ is represented by[1,2]

$$\sigma \bullet \sqrt{\frac{ak}{1 \bullet kL}} \bullet \beta k u_a,$$

where L is a characteristic density scale length at the ablation front, β is the coefficient in the range 1–3, k is the wave number of the perturbation, a is the acceleration, and u_a is the ablation velocity that is given by the mass ablation rate per unit surface area m divided by the ablation density,

$$u_a \bullet m/\rho_a.$$

It is clear that large u_a means small σ. In the indirect-drive approach[1], once laser energy is converted into x-ray radiation in a hohlraum, the mass ablation rate is generally much larger than that in the direct-drive approach. So, the requirement on laser-beam uniformity is relaxed and sensitivity to hydrodynamic instabilities is reduced. Therefore, the ICF programs around the world have concentrated most of their effort on the x-ray or indirect-drive approach.

Correspondence to: Baifei Shen, State Key Laboratory of High Field Laser Physics, Shanghai Institute of Optics and Fine Mechanics, Chinese Academy of Sciences, Shanghai 201800, China. Email: bfshen@mail.shcnc.ac.cn

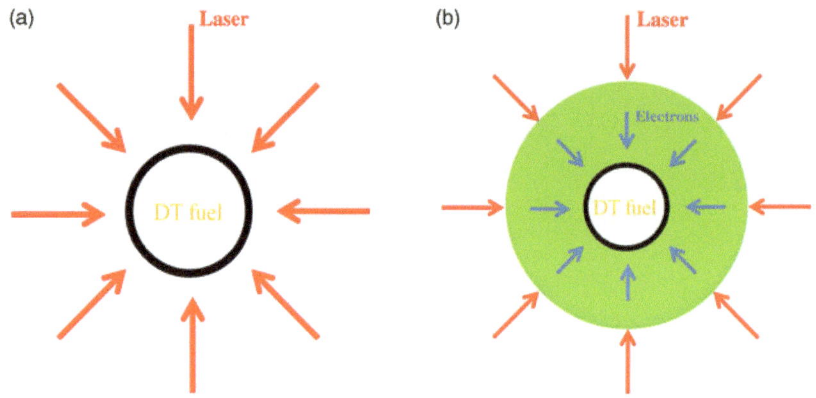

Figure 1. (Color online) Sketch map of fusion targets designed for (a) normal ICF Programs and (b) in our scheme.

But for inertial fusion energy (IFE), after the demonstration of ignition and energy gain, the overall efficiency from input electricity to output electricity must be considered[3]. At present, the efficiency from electricity to driving glass laser is only about 1%, although the efficiency may be increased to more than 10% in the future. Indirect-drive targets can further reduce the coupling efficiency from laser energy to the DT fuel.

Based on the above discussions, here we propose a new concept of inertial fusion by using powerful long wavelength electromagnetic pulses (EMPs), such as CO_2 laser pulses or high power microwave pulses, which can be efficiently generated. Special targets are designed to suppress the hydrodynamic instabilities.

2. Main concept

The main concept is illustrated in Figure 1b. A normal fusion target is surrounded by a gas layer of specially designed density and size. Many long wavelength EMPs illuminate it symmetrically from all directions.

The long wavelength EMPs could be CO_2 laser pulses or high power microwave pulses. The main reason to choose such pulses is that they can be very efficiently generated. For electricity power, it is very important, because the requirement for the energy gain from driver to fusion energy can be greatly reduced.

The problem is that the hydrodynamic instabilities can be easily developed, if we simply replace the 1 μm laser pulses (or their third harmonics) with long wavelength EMPs. Normal indirect drive with long wavelength EMPs is also not a good choice, since larger holes are needed to permit the pulses to penetrate into the hohlraum and the low density plasma in the hohlraum would prevent the transportation of the pulses of long wavelength (the critical density n_c becomes lower). Our solution is to use two-step driving, something like indirect drive but without a hohlraum. First, EMPs accelerate (not heat) electrons to about 2 MeV or less inside the low density gas layer just outside the normal fusion target. Second, the energetic electrons heat the ablator to produce high pressure to compress the fuel to high density and form a hot spot in the center of the fusion capsule.

In order to confirm the validity of the method, several main problems have to be solved. First, we should make sure electrons can truly be efficiently accelerated forward to about 1 MeV. The electron energy should not be too large, or too small. Second, since the current of the electron bunches is much larger than the Alfven limit in vacuum, we should make sure that such energetic beams of intense current can transport in plasmas. As a primarily concept design, we will discuss the two problems in Sections 3–5, although there are certainly many other problems.

3. Electron acceleration

Both x-rays and particle beams can penetrate into plasma of high density to produce large ablation velocity u_a. For indirect-drive fusion, x-rays of black body are used. For simpler direct-drive fusion, x-rays can also be generated if a high Z material is added at the outside of the capsule shell[4]. We may discuss this option in other publications. One defect of this method is that the x-rays go in all directions so that only half of the x-rays go inward, leading to low efficiency. Here we consider converting the energy of EMPs to inward moving energetic electron beams.

From the research of laser acceleration, we know that electrons can not only be heated but also be efficiently accelerated with the plasma wake and other mechanism driven by 800 nm laser pulses[5]. The laser wavelength should certainly not be restricted to 800 nm. Electrons should also be efficiently accelerated by long wavelength EMPs such as CO_2 laser pulses or high power microwave pulses, if the normalized amplitude of the laser field remains about $a = 1$ (the intensity is much lower than in the case of 800 nm laser pulses) and the plasma density decreases accordingly.

For the research of laser acceleration, the emphasis is usually put on the high energy and high quality of the energetic electron beams. Here for inertial fusion, the electron energy cannot substantially exceed one MeV, as we will discuss in Section 4. Therefore, a high plasma density near

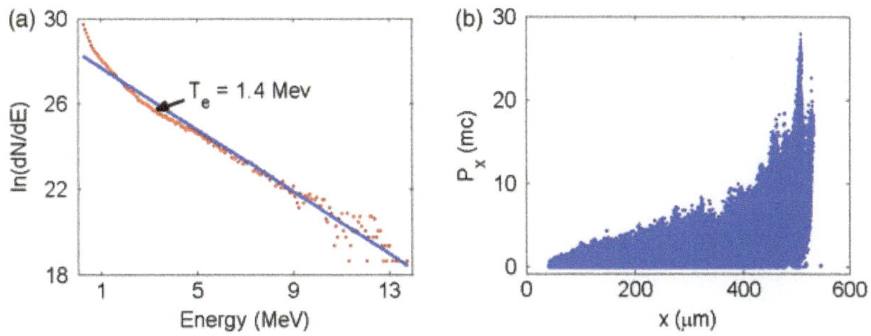

Figure 2. (Color online) Energy spectra (a) and longitudinal phase space distribution (b) of the energetic electrons generated by a CO_2 laser pulse (only electrons with energy larger than 500 keV are presented). The blue fitting line in (a) gives an electron temperature of 1.4 MeV.

to the critical density is used so that the acceleration length is limited. At the same time, in this way the number of energetic electrons increases greatly[6].

Here as an example, particle in cell (PIC) simulation is carried out to show the generation of energetic electrons driven by a CO_2 laser pulse of wavelength 10.6 μm, duration 200 fs, focal size 120 μm, intensity 7.6×10^{16} W/cm^2 and normalized amplitude $a = 2.5$. The total laser energy is about 29 J. The plasma density is 2×10^{18} cm^{-3}, which is $0.2\, n_c$.

After acceleration in a 0.5 mm long plasma, a bulk of energetic electrons is produced. Figure 2a indicates the energy spectrum exhibits typical property of Maxwell distribution with a temperature of 1.4 MeV. The longitudinal phase space distribution of the electrons with energy larger than 500 keV is shown in Figure 2b. The total charge of the hot electrons is 6 μC and the efficiency from laser to energetic electron in this case is 31.9%, which is sufficient for potential fusion applications in future. It should be mentioned that for the laser and plasma parameters used here acceleration mechanism is complicated. Stochastic acceleration, self modulated wake acceleration and direct laser acceleration may all play important roles[7] while other instabilities such as stimulated Raman scattering is not found to be important.

It should be noted that the method works as well for ns laser pulse required for actual fusion experiment which is much longer than in our simulation, which is limited by computer resource.

The initial part of the laser pulse should be of lower intensity than used in the above example so that the electron energy should be lower than 1 MeV. In this way the energetic electron can be stopped within a short distance, since in the initial phase of compression, the ρr of the shell is still very small.

If high power microwave pulses are used, lower gas density and larger gas size should be chosen, since the wavelength becomes even larger.

4. Electron beam driven fusion

Electron beam driven fusion has been studied since 1970s, because electrons can be efficiently accelerated with conventional accelerators[8,9]. It was estimated that an electron beam of energy 1 MJ, duration 10 ns and power 100 TW is required for inertial fusion[8,9]. Similar to x-rays, energetic electrons can easily penetrate into the fuel. It is good to increase the ablation rate to overcome the hydrodynamic instabilities. The electron energy cannot be too large, although the fuel shell in our approach should be thicker than the one in conventional target design. If only classical single-particle beam stopping is considered, electron energy cannot substantially exceed one MeV. The mechanisms responsible for energy loss by the beam electrons in the target plasma are generally binary coulomb collisions with the target electrons and ions, and excitation of plasma oscillations. In addition, some evidence of enhanced deposition due to beam-plasma instabilities has been reported[10], but the practical effect needs more investigation. The burn material may consist of two layers. The outer layer consists of middle Z ablators to help effectively stop the electrons while low Z ablators should be used for the inner layer to reduce the generation of bremsstrahlung which could preheat the cold fuel.

For electron driven fusion, similar to the well known fast ignition [11,12], one of the main difficulties is the propagation of the high current electron beams. For beam energy of about 1 MeV and beam power of 10^{14} W, the beam current would be 10^8 A. We know that the Alfven current in vacuum is $I_A = 17\beta\gamma$ kA. For 1 MeV electrons $\beta \sim 1$ and $\gamma \sim 3$, and one finds that $I_A \sim 50$ kA. This is far below the required current of 10^8 A. We know that NIF has 192 laser beams. Similarly, many electron beams bombarding the target from several sides can be used to significantly decrease the required current for each beam. But even for one hundred bunches, the current of each beam would be $\sim 10^6$ A and thus well above the Alfven limit. Anyway, it should be noted that this current is much lower than the current required for fast ignition.

Another problem with conventional accelerators is that the facility size is too large. There is always a long distance between the accelerators and the tiny fusion capsule. Since the current of each beam is still much larger than the Alfven limit in vacuum, the electron beam was hardly focused to the pellet due to the space-charge effect and that the current cannot propagate in vacuum. Fortunately, preformed

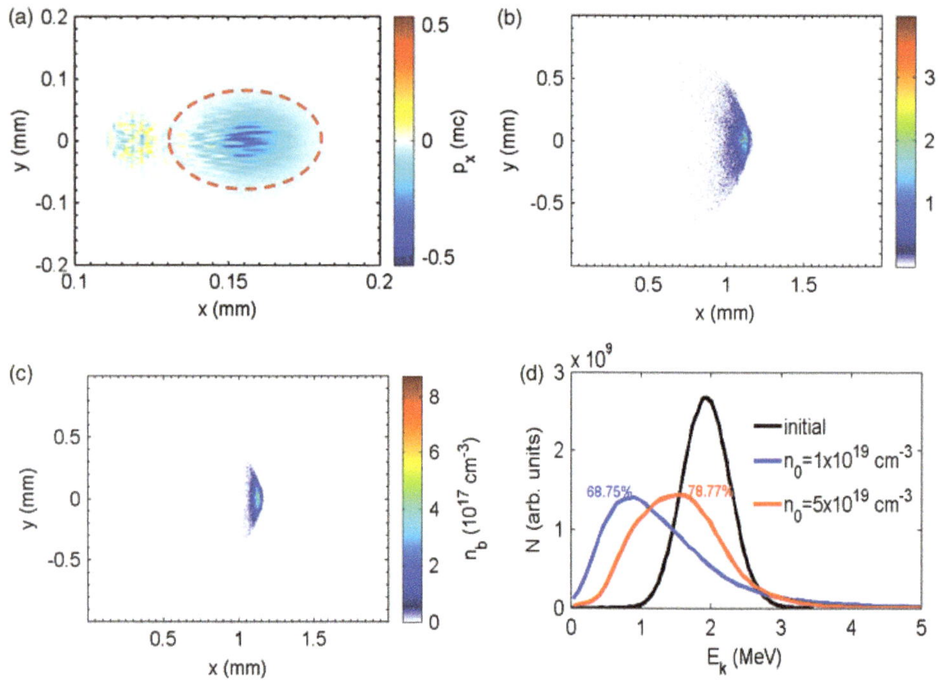

Figure 3. (Color online) Longitudinal momentum of background plasma electrons at the beam position (marked by the red dashed curve) (a). Density distribution of electron beam after propagating 1 mm in the plasma of density 1×10^{19} cm^{-3} (b) and 5×10^{19}cm^{-3} (c). The energy spectrum of electron beam at propagating distance $x = 0$ mm (black curve), $x = 1$ mm in the 1×10^{19} cm^{-3} plasma (blue curve), and $x = 1$ mm in the 5×10^{19} cm^{-3} plasma (red curve) (d).

current-carrying plasma discharge channels can be used to transport these beams. The efficiency of beam transportation and overlapping can be measured with the current density gain, which is the ratio of beam current density at the target surface to that of the individual beams. The problem of transportation is not completely solved so that light ion beams and heavy ion beams were proposed to replace the electron beams and laser pulses were chosen to be the most favorable driver for inertial fusion.

Now, in our case, many EMPs of long wavelength can be easily focused to the underdense plasma surrounding the fusion capsule, since photons are bosons. Inside the underdense plasma, electrons are accelerated inward directly to the capsule. There is no additional vacuum gap for the energetic electrons to go through.

In order to check how the energetic electron beams propagate in plasma, PIC simulations are performed. We suppose 100 electron beams are used to generate total power of 100 TW which is required for inertial fusion. If pulse duration is 5 ns and the efficiency from laser to electron is about 30% as estimated above, the total required laser energy is about 1.6 MJ, similar to the one provided by NIF. Therefore, the power of each beam is 1 TW. The intensity is 2×10^{16} W/cm^2 and the mean electron energy is 2 MeV. Therefore the current is 5×10^4 A, which is about 10 I_A. In the coronal area of fuel, the plasma density is usually larger than where the electrons are accelerated so that an energetic electron beam can more easily propagate. We consider two typical cases of background plasma density where the electron beam propagates $n_0 = 1 \times 10^{19}$ cm^{-3} and 5×10^{19} cm^{-3} for comparison. Due to the limitation of PIC codes, the beam length is 50 μm. Because collisions are not included in the PIC code, here we only consider the collective effects.

As shown in Figure 3a, when a dense electron beam with current larger than Alfven limit propagates in plasma, the beam charge induces collective shifting of background electrons, which forms a backward current within the incident electron beam. This induced current neutralizes the incident beam current and therefore permits its propagation in plasma. At the same time, comparing Figure 3a with Figure 3b–c, one can see that the electron beam is dispersed due to the repulsive space-charge force of all the electrons in the beam. This beam dispersion can be moderated by increasing the plasma density for the remaining plasma ions neutralize part of the space-charge force within the beam.

Apparently, the efficiency of energy transport is related to the density of background plasma as shown in Figure 3d. Usually in plasmas of higher density, more intense induced backward current is formed to ensure better transportation of the energetic electrons. As a result, about 68.7% of the total electron energy remains over after 1-mm-propagation in the plasma with density of 1×10^{19} cm^{-3}, while the proportion is 78.8% in the 5×10^{19} cm^{-3} plasma. So comparing to electron driven fusion scheme based on conventional accelerators, in our case, the electrons can propagate more efficiently through the dense plasmas in the coronal area of fuel to heat the ablator and generate the required high pressure. The beam energy is converted to implosion kinetic energy with efficiency typically of order 20%[9].

5. Discussion

There are certainly many other challenges to be overcome to realize this new fusion concept. One problem is that the stopping range of the electron is usually much larger than soft x-ray for indirect-drive fusion. For an electron of 2 MeV, the collisional stopping power is about 1.5 MeV cm^2/g. Therefore, a special fusion capsule of thick burn layer should be used. The electron energy of the front part of the electron beams can be designed to be smaller than the electron energy of the back part, since after compression, the ρr of the fusion target would become much larger.

For laser fusion, the laser wavelength is fixed during one shot. Only the laser power can change with time. For laser electron beam fusion, the electron energy is determined by the intensity of each laser pulse and the total beam energy is determined by the total laser power. Therefore, the time evolution of both the penetration depth and the burn pressure can be controlled.

Another problem is the filaments formed during the propagation which would easily excite the hydrodynamic instabilities. It was shown that the electron beam filaments are formed when $n_p/n_b = 10$[13]. Fortunately, the density of the energetic electron beams is about $10^{18}/cm^3$ in our case and the deposit plasma density for the beams is larger than $10^{22}/cm^3$. Therefore, the ratio $n_b/n_p < 10^{-4}$ at the deposition layer, where collision will dominate compared to self-collimation due to self generated magnetic field[12]. Based on the above discussion, we expect plasma heating at the deposition layer by the electron beams might be transversely smooth.

Ions can also be accelerated by a light pressure driven electrostatic field in the plasma of density slightly larger than critical density. The energetic electron beams in our proposal may be replaced by energetic light ions.

6. Conclusions

In conclusion, a new two-step scheme for inertial confinement fusion is proposed in this paper. Powerful long wavelength EMPs, which can be efficiently produced, accelerate large amounts of hot electrons in plasmas close to the critical density just outside the normal fusion target. Then energetic electrons heat the ablator to produce high pressure to compress the fuel to high density and form a hot spot in the center of the fusion capsule. The energy of the accelerated electrons follows Maxwell distribution. The temperature can be controlled to be around 1–2 MeV to ensure the electrons can easily penetrate into the fuel shell but without penetrating through it. PIC simulations indicate these hot electrons can be produced by CO_2 lasers with a relatively high efficiency of 31.9%. On the other hand, the problem that propagation of the high current (above Alfven limit) electron beams in our scheme can be overcome by special design of fuel targets. The high current electron beams can easily propagate in the dense plasmas in the coronal area of fuel because the current is offset by the collective backward shifting of the plasma electrons.

References

1. John Lindl, Phys. Plasmas **2**, 3933 (1995).
2. S. Atzeni, and J. Meyer-ter-Vehn, *Inertial Fusion-Beam Plasma Interaction, Hydrodynamic, Dense Plasma Physics* (Clarendon, Oxford, 2003),
3. S. Bodner, A. Schmitt, and J. Sethian, High Power Laser Sci. Eng. **1**, 2 (2013).
4. S. Fujioka, *et al.*, Phys. Rev. Lett. **92**, 195001 (2004).
5. E. Esarey, C. B. Schroeder, and W. P. Leemans, Rev. Mod. Phys. **81**, 1229 (2009).
6. B. Shen, Y. Wu, K. Dong, B. Zhu, Y. Gu, L. Ji, C. Jiao, J. Teng, W. Hong, Z. Zhao, L. Cao, X. Wang, and M. Y. Yu, Phys. Plasmas **19**, 033106 (2012).
7. C. Gahn, *et al.*, Phys. Rev. Lett. **83**, 4772 (1999); Z. M. Sheng et al., Phys. Rev. Lett. **88**, 055004 (2002).
8. G. Yonas, IEEE Trans. Nucl. Sci. NS **26**, 610 (1979).
9. M. Clauser, *et al.*, J. Phys. C **7**, 81 (1979).
10. S. Nakai, *et al.*, Nucl. Fusion Suppl. I, 207 (1977).
11. M. Tabak, *et al.*, Phys. Plasmas **1**, 1626 (1994).
12. C. Deutsch, H. Furukawa, K. Mima, M. Murakami, and K. Nishihara, Phys. Rev. Lett. **77**, 2483 (1996).
13. M. Honda, J. Meyer-ter-Vehn, and A. Pukhov, Phys. Rev. Lett. **85**, 2128 (2000).

6

Pump phase transfer and its control in hybrid seeded optical parametric amplifiers

Peng Yuan[1], Jingui Ma[1], Yongzhi Wang[2], Jing Wang[1], Daolong Tang[1], and Liejia Qian[1]

[1] *Key Laboratory for Laser Plasma (Ministry of Education) and Department of Physics and Astronomy, Shanghai Jiao Tong University, Shanghai 200240, China*

[2] *Department of Optical Science and Engineering, Fudan University, Shanghai 200433, China*

Abstract

Due to the existence of spatial walk-off and/or group-velocity mismatch effects, pump-to-signal phase transfer becomes inevitable during parametric amplification. We experimentally demonstrate that in hybrid seeded optical parametric amplifiers (OPAs) that include two OPA stages seeded by the signal and idler waves, respectively, the phase of the output signal can be restored to its initial value, although there are spatial and temporal phase fluctuations on the pump source. This method significantly relaxes the requirement for high pump beam quality, which is always very stringent in parametric amplification systems. With the introduction of this scheme into birefringent phase-matching OPAs or chirped-pulse OPAs, it should be promising to achieve intense femtosecond laser pulses that are close to the diffraction limit in space and ultra-high contrast in time, simultaneously.

Keywords: hybrid seeding; parametric amplifier; pump-to-signal phase transfer

1. Introduction

Optical parametric amplification (OPA) has demonstrated unique flexibility in producing femtosecond laser pulses tunable from 0.2 to 20 μm[1]. To increase the availability of ultra-short pulses with higher peak power, the scheme of chirped-pulse amplification (CPA) was introduced[2]. The integration of OPA and CPA into optical parametric chirped-pulse amplification (OPCPA) has rapidly pushed the peak power of amplified ultra-short laser pulses up to the petawatt level[3]. The development has benefited from some important advantages of OPCPA with respect to conventional CPA systems based on laser gain media: (i) the parametric gain within a single pass through a nonlinear crystal can be as high as many tens of decibels, minimizing the B-integral and allowing simple and compact setups[4]; (ii) parametric amplification supports much wider gain bandwidth (>100 nm), enabling the production of pulse durations in the few-cycle regime; (iii) there are few thermal loading effects during parametric amplification due to the weak parasitic absorption, ensuring good beam quality of the amplified signal. However, parametric amplification has a disadvantage that should not be ignored – its requirement for pump beam quality is much more stringent than that for conventional CPA[3]. During parametric amplification, any temporal modulations and random noise on the pump beam would be imparted to the signal and idler pulses due to the intensity-independent parametric gain. The contrast degradation of the amplified signal due to pump noise transfer has been theoretically and experimentally verified[5, 6]. However, the fact of pump-to-signal phase transfer, which could degrade the beam quality and temporal contrast of the amplified output severely, has been less appreciated.

During the parametric amplification process without an initial idler beam, the phases of the pump (at frequency ω_p), signal (ω_s) and idler (ω_i) waves obey[7]

$$\phi_s = \phi_s(0), \tag{1}$$

$$\phi_i = \pi/2 + \phi_p - \phi_s, \tag{2}$$

under the conditions of phase matching and no spatial walk-off or group-velocity mismatch (GVM) effects. At this point, the phase of the amplified signal, ϕ_s, is independent of the pump phase, ϕ_p, while the phase of the newly generated idler wave, ϕ_i, accommodates the phase difference between the pump and signal. This phase relation has been widely adopted as a nonlinear approach to control the carrier-envelope phase of ultra-fast light pulses[8]. However, when there is spatial walk-off and/or a temporal GVM effect, the phase relations are no longer valid, and both the signal and idler waves will partake of the pump phase in a very

Correspondence to: Jing Wang, Key Laboratory for Laser Plasma and Department of Physics and Astronomy, Shanghai Jiao Tong University, Shanghai 200240, China. Email: wangj1118@sjtu.edu.cn

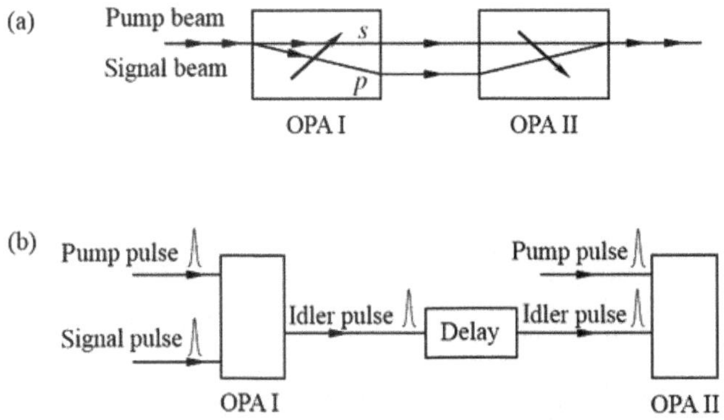

Figure 1. Schematic of optimized OPA designs: (a) walk-off-compensating configuration; (b) hybrid seeding configuration (the two-stage OPAs are seeded by signal and idler waves, respectively).

sophisticated manner. That is, the phase distortion will transfer from the pump laser to the amplified signal.

For parametric amplification based on quasi-phase matching (QPM) using a type-0 ($e_p \rightarrow e_s + e_i$) scheme, there is no spatial walk-off between interacting waves and, hence, no spatial phase transfer. A QPM parametric amplifier could thus use a multi-mode laser as the pump source[9, 10]. For parametric amplifiers adopting birefringent nonlinear crystal in which phase matching relies on the presence of both ordinary and extraordinary waves, spatial walk-off always exists and so does spatial phase transfer. This problem could be counteracted by using pairs of crystals arranged in a walk-off-compensated orientation[11], as shown in Figure 1(a), where the walk-off direction is reversed in the second crystal of each pair. However, there is no analogous method in the temporal domain that can eliminate the influence of GVM. Once the wavelengths of the interacting waves and the nonlinear crystal are chosen, the GVM relations (including its sign and value) are determined as well. Thus, reversal of the sign of GVM in a second OPA stage is much more difficult than reversal of the walk-off direction.

In this paper, we explore an approach toward controlling the pump-to-signal phase transfer in both the spatial and temporal domains. This scheme, as illustrated by Figure 1(b), is also known as hybrid seeded OPAs. It includes two OPA stages which are seeded by the signal and idler waves respectively. Hybrid seeding was previously adopted to eliminate the background of initial CW injection[12] or the undesired preceding and trailing oscillator pulses in the injection[13]. Here, we experimentally demonstrate that hybrid seeding can make it possible to restore the phase of the output signal to its initial value, regardless of the existence of GVM and walk-off effects in each OPA stage and the presence of both spatial and temporal phase distortion on the pump laser. Both the spatial beam quality and the temporal contrast of the amplified signal can thus be significantly improved as the pump phase transfer is blocked. Due to the tolerance of both spatial and temporal pump phase aberrations that normally accompany high energy pump lasers, the proposed method may provide a feasible route for the construction of high quality intense laser sources based on OPCPA.

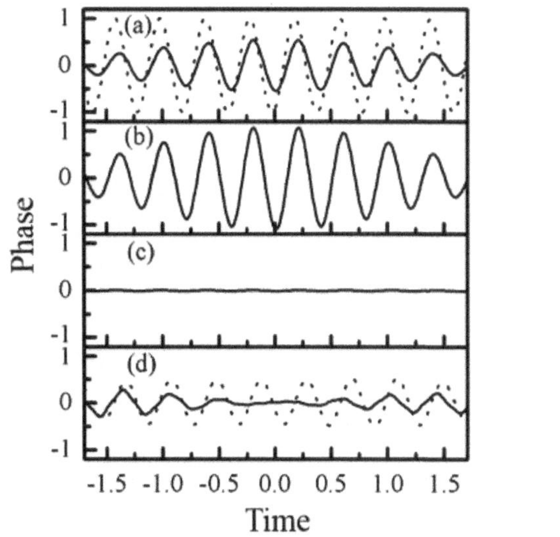

Figure 2. Calculated signal phase distortion and restoration in a two-stage OPA system. (a) The dotted line plots the initial phase modulation on the pump pulse. The solid line is the phase of the amplified signal from the first stage OPA. (b) The phase of the amplified signal from the second OPA stage which is seeded by the amplified signal from the first stage (conventional multistage OPA configuration). (c) The phase of the amplified signal from a hybrid seeded OPA with the condition $L_{\mathrm{GVM}(i-p)} = L_{\mathrm{GVM}(s-p)} = L_1 = L_2$. (d) The phase of the amplified signal (solid line) from a hybrid seeded OPA with the condition $L_{\mathrm{GVM}(i-p)} = 1.5L_{\mathrm{GVM}(s-p)} = 3L_1 = 2L_2$. The dotted line in (d) is the phase of the pump pulse after amplification. The nonlinear length L_{NL}[14], as an identification of the pump intensity, was fixed in the calculations to be $L_{\mathrm{NL}} = 0.3L_1$ for each stage.

2. Numerical simulations

To start with, we simulate the behavior of pump-to-signal phase transfer in a two-stage OPA using the standard coupled-wave equations[14]. Identical sinusoidal phase modulation, plotted as the dotted line in Figure 2(a), was introduced into the Gaussian pump pulse. The initial signal pulse having a uniform phase was also assumed to be Gaussian in shape with a duration of τ_s. Due to the pulse slippage caused by GVM, the pump phase modulation pattern was imprinted onto the amplified signal, as plotted by the solid line in Figure 2(a). When this distorted signal was injected into a second OPA stage, the phase fluctuation was further enhanced, as indicated by Figure 2(b). It is important to point out that the temporal phase transfer is governed only by the GVM between the pump and the idler (i.e., GVM_{i-p}). Only when $\mathrm{GVM}_{i-p} \neq 0$ will the pump phase transfer to the signal. However, the idler wave always inherits the pump phase pattern.

If we seeded the second OPA stage with the pump-phase-imposed idler output from the first stage, complete restoration of the uniform signal phase could be achieved, as shown in Figure 2(c). The restoration conditions used in the calculations included $\mathrm{GVM}_{i-p} = \mathrm{GVM}_{s-p}$ (the GVM between pump and idler equals the GVM between pump and signal), $L_1 = L_2$ (the two crystals have equal lengths) and the pump pulse overlapped with the idler well at the entrance of the second OPA stage. In a more general case, the establishment condition of phase restoration can extend to $L_1 \times \text{GVM}_{i-p} = L_2 \times \text{GVM}_{s-p}$, as illustrated by Figure 2(d). Even if this condition cannot be exactly satisfied in real systems, the signal phase can still be partially restored and the restoration degree relies on the amount of GVM_{s-i} (GVM between the signal and idler). That is, satisfactory control of pump-to-signal phase transfer necessitates a relatively small GVM_{s-i}. An analogous result was also found in the spatial domain. In type-I phase matching ($e_p \rightarrow o_s + o_i$) OPAs, both the signal and the idler are ordinary waves. Hence, there would not be spatial walk-off between the signal and the idler. Complete restoration of the spatial phase of the signal beam is thus easy to achieve.

3. Experimental setup and results

To experimentally demonstrate the capability of controlling phase transfer by hybrid seeding, a two-stage femtosecond OPA system was employed, as shown in Figure 1(b). Two 10 mm-thick uncoated $MgO:LiNbO_3$ crystals were used in type-I phase matching. A commercial femtosecond Ti:sapphire laser oscillator-regenerative amplifier (Spectra-Physics, Spitfire) producing 500 μJ, 70 fs laser pulses at 800 nm with a repetition rate of 1 kHz was adopted as the pump source. A 0.5 W 1064 nm diode-pumped CW $Nd:YVO_4$ laser operating in TEM_{00} mode served as the seed source. The pump beam was divided into two by a beam splitter with a reflectance of ~40%. These two beams were then telescoped and collimated to obtain equal intensities of ~90 GW cm^{-2} for pumping of the two OPA stages. The beam profile of the collimated pump is shown in Figure 3(a). In order to optimize the OPA gain and conversion efficiency, a negative chirp was deliberately introduced to the pump pulse to obtain a duration of ~150 fs for dispersion compensation[15]. This pulse duration as well as the 70 fs mentioned before was measured using a home-made self-correlator. The OPA was tuned to operate at a signal wavelength of ~1040 nm, where the signal and idler have approximately the same value of group velocity. Thus, control of the phase transfer in both the temporal and spatial domains is possible by using this hybrid seeding scheme.

In the spatial domain, the beam divergence caused by the Kerr effect was regarded as the spatial phase distortion of the pump laser in our experiment. Due to the presence of a significant accumulated Kerr nonlinear phase shift ($\Delta\varphi_{nl} \sim 6\pi$) in each OPA crystal, strong whole-beam self-focusing occurred across the pump beam, as conveyed by Figure 3(b). A small-size self-focusing effect was not observed, which is reasonable as the beam profile was very smooth. Since pump-to-signal phase transfer would occur only within the walk-off plane, the signal beam from the first OPA stage was found to be highly astigmatic, as shown in

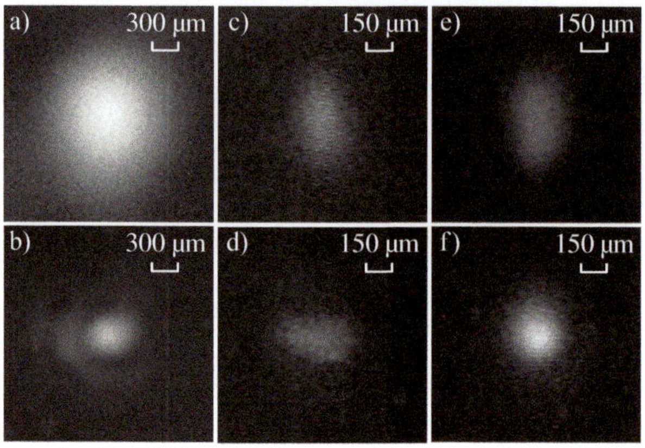

Figure 3. Beam patterns of the pump beam measured ~400 mm after the crystal (used for Kerr-effect-induced spatial phase distortion): (a) in the linear propagation regime; (b) at high intensity (90 GW cm^{-2}). Beam patterns of the amplified signal from the first OPA stage measured at the focus: (c) in the walk-off plane; (d) in the no-walk-off plane. Amplified signal from the second OPA stage measured at the focus in the no-walk-off plane: when the overlap between the pump and idler beams was not optimized (e) and when it was optimized (f).

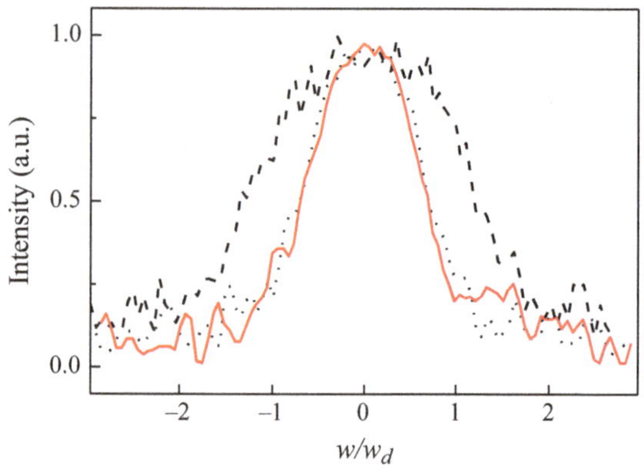

Figure 4. Measured beam profiles of the amplified signal from the hybrid seeded OPA. The dotted line plots the beam profile in the no-walk-off plane. The solid line is the beam profile in the walk-off plane after the spatial overlap between the pump and idler beams was optimized. The dashed line shows the beam profile in the walk-off plane when the pump–idler overlap was not optimized.

Figures 3(c) and (d). This astigmatism would become more severe if the signal were amplified further in a second OPA stage. By seeding the second stage with the idler from the first stage instead of the signal, although the output signal beam is also astigmatic in general (Figure 3(e)), a quite satisfactory compensation of the astigmatism (Figure 3(f)) can be obtained by adjusting the seeding idler beam to make it coincide with the pump beam in space at the entrance of the second OPA stage. As plotted in Figure 4, the astigmatism induced by pump phase transfer in the walk-off plane (solid line) was compensated close to the level of that in the anastigmatic no-walk-off dimension (dotted line). This indicates that the output signal beam of the hybrid seeded OPA could be very close to its diffraction limit. It should be pointed out that hybrid seeding would not reduce

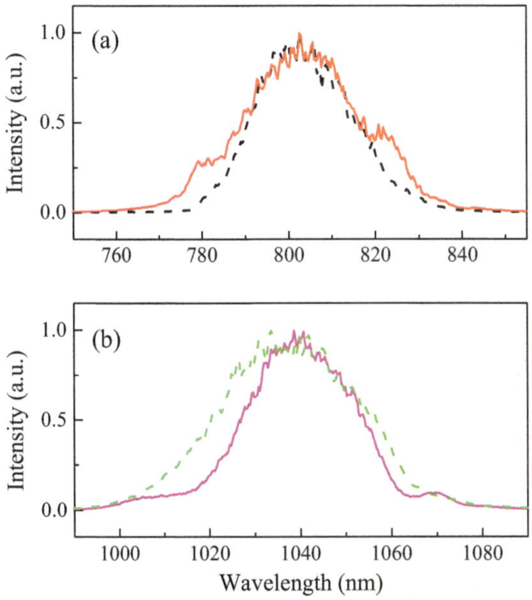

Figure 5. (a) Measured pump pulse spectra before (dashed line) and after (solid line) the crystal (where self-phase modulation occurred). (b) Measured signal pulse spectra when the time delay between the idler and the pump was ~150 fs (dashed line) and 0 (solid line), respectively.

the conversion efficiency. The obtained signal pulse energy of ~20 μJ was similar to that obtained by seeding both stages with the signal wave. The pump-to-signal conversion efficiency was ~5%. These experimental results clearly verify the capability of the hybrid seeded OPA scheme in signal phase restoration in the spatial domain.

To introduce some temporal phase distortion into the pump, the self-phase modulation effect when a high intensity pump laser passes through a nonlinear crystal was used. Figure 5(a) shows the pump spectrum before and after the crystal. Spectral broadening was observed, which indicated the occurrence of the desired self-phase modulation. During parametric amplification, the pump phase modulation was imparted to the signal pulse and led to signal spectrum broadening, as indicated by the dashed line in Figure 5(b). By seeding the second OPA stage with the idler pulse from the first stage and synchronizing the idler with the pump pulse carefully, a phase-restored signal pulse could be obtained, as indicated by the relatively narrower spectrum (solid line) in Figure 5(b).

4. Conclusion

We have demonstrated that by using a hybrid seeded OPA scheme, the phase distortion induced by pump phase transfer during parametric amplification can be well controlled in both the spatial and the temporal domain. To achieve higher output power, the two-stage configuration employed in our work could be further extended to several pairs of crystals, where the second crystal of each pair is seeded with the idler wave from the first crystal. In this way, the phase transfer in the second crystal could exactly cancel the phase transfer in the first crystal. That is, the pump-to-signal phase transfer can no longer accumulate with increase of the nonlinear interaction length. Moreover, the hybrid seeded OPA has almost the same conversion efficiency relative to classical multistage OPAs of the same overall length. Up to now, OPCPA systems have always resorted to various nonlinear approaches for pulse contrast enhancement, which have very low energy efficiency ~10%. By using the proposed method in OPCPA systems, the output temporal contrast can be enhanced while not sacrificing the amplified signal energy at all. We believe that this work provides new insight into the optimized design of further femtosecond OPA or OPCPA systems.

References

1. M. H. Dunn and M. Ebrahimzadeh, Science **286**, 1513 (1999).
2. A. Dubietis, G. Jonusauskas, and A. Piskarskas, Opt. Commun. **88**, 437 (1992).
3. E. A. Khazanov and A. M. Sergeev, Phys. Usp. **51**, 969 (2008).
4. Z. Xu, X. Yang, Y. Leng, H. Lu, L. Lin, Z. Zhang, R. Li, W. Zhang, D. Yin, S. Jin, J. Peng, B. Tang, and B. Zhao, Chin. Opt. Lett. **01**, 24 (2003).
5. N. Forget, A. Cotel, E. Brambrink, P. Audebert, C. L. Blanc, A. Jullien, O. Albert, and G. Chériaux, Opt. Lett. **30**, 2921 (2005).
6. I. N. Ross, G. H. C. New, and P. K. Bates, Opt. Commun. **273**, 510 (2007).
7. I. N. Ross, P. Matousek, G. H. C. New, and K. Osvay, J. Opt. Soc. Am. B **19**, 2945 (2002).
8. A. Baltuska, K. Fuji, and T. Kobayashi, Phys. Rev. Lett. **88**, 133901 (2002).
9. A. Galvanauskas, A. Hariharan, D. Harter, M. A. Arbore, and M. M. Fejer, Opt. Lett. **23**, 210 (1998).
10. I. Jovanovic, J. R. Schmidt, and C. A. Ebbers, Appl. Phys. Lett. **83**, 4125 (2003).
11. D. J. Armstrong, W. J. Alford, T. D. Raymond, A. V. Smith, and M. S. Bowers, J. Opt. Soc. Am. B **14**, 460 (1997).
12. H. Luo, L. Qian, P. Yuan, H. Zhu, and S. Wen, Opt. Express **13**, 9747 (2005).
13. I. Jovanovic, C. P. J. Barty, C. Haefner, and B. Wattellier, Opt. Lett. **31**, 787 (2006).
14. P. Zhu, L. Qian, S. Xue, and Z. Lin, Opt. Laser Technol. **35**, 13 (2003).
15. P. Yuan, L. Qian, H. Luo, H. Zhu, and S. Wen, IEEE J. Sel. Top. Quantum Electron. **12**, 181 (2006).

7

Inertial confinement fusion and prospects for power production

C.B. Edwards[1,2] and C.N. Danson[2]

[1] *HiPER Project, STFC Rutherford Appleton Laboratory, UK*

[2] *AWE plc, Aldermaston, UK*

Abstract
As our understanding of the environmental impact of fossil fuel based energy production increases, it is becoming clear that the world needs a new energy solution to meet the challenges of the future. A transformation is required in the energy market to meet the need for low carbon, sustainable, affordable generation matched with security of supply. In the short term, an increasing contribution from renewable sources may provide a solution in some locations. In the longer term, low carbon, sustainable solutions must be developed to meet base load energy demand, if the world is to avoid an ever increasing energy gap and the attendant political instabilities. Laser-driven inertial fusion energy (IFE) may offer such a solution.

Keywords: ICF; IFE; inertial confinement fusion; inertial fusion energy

1. Benefits of inertial fusion energy

Inertial fusion energy (IFE) has the potential to make a substantial contribution to meeting world energy needs in the second half of this century.

Security of supply and sustainability: IFE provides energy security and avoids geopolitical constraints because the key components of the fuel, deuterium and lithium, are abundant and widely distributed. Assessments carried out within the HiPER (European High Power Laser Energy Research Facility) project[1] show that sufficient materials are available for global power production at the 1 TWe (terawatt electrical) level for more than 1000 years.

Inherent safety: IFE is intrinsically very safe since it carries no risk of 'thermal runaway'. There is little stored energy within the system, no 'critical mass' issues and, under fault conditions, energy production would simply stop.

Low environmental impact: There are no carbon emissions from the fusion energy production process. With the use of suitable materials for the reaction vessel, the relatively small amount of radioactive waste generated from neutron activation will be short lived with the appropriate choice of materials and managed largely through recycling. Such materials are already available.

Affordable energy: Financial modelling based on reasonable assumptions of progress during the next phase of technology development and ignition physics shows that electricity derived from laser fusion may well be cost competitive with other environmentally acceptable sources[2], although the energy landscape in 30–50 years is uncertain and hence difficult to predict.

Separable technology: Physical separation of major systems, laser driver, reaction vessel, balance of plant, etc., allows accelerated development and reduced costs.

2. The D–T fusion reaction

The principle of fusion is simple, though its realization on an industrial scale suitable for commercial energy production is technologically extremely demanding. The underlying physics involves the use of powerful lasers to heat a mixture of two hydrogen isotopes, deuterium and tritium, to an extreme temperature of greater than 50 million degrees, whereupon the constituent nuclei fuse to form a helium ion (alpha particle) and a neutron, according to the reaction shown in Figure 1. In each fusion reaction, the helium ion

Correspondence to: C.N. Danson, AWE, Aldermaston, Reading, RG7 4PR, UK. Email: colin.danson@awe.co.uk

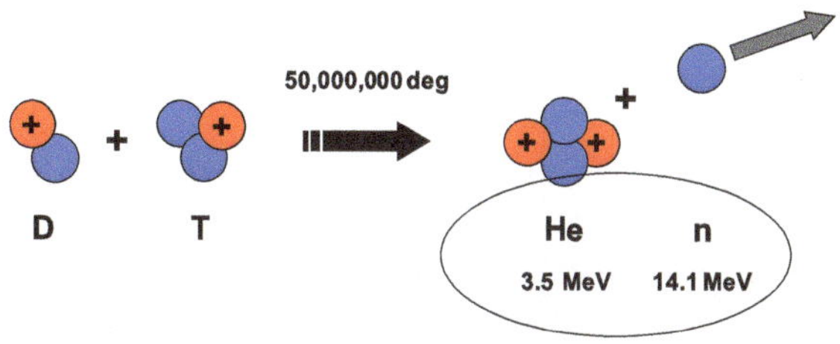

Figure 1. The D–T fusion reaction at the heart of laser energy.

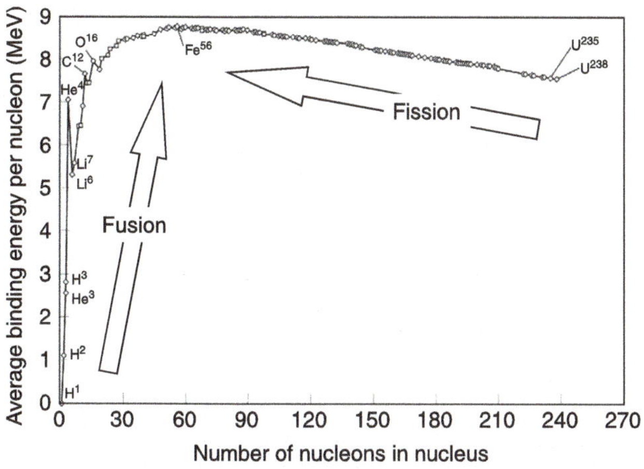

Figure 2. Binding energy variation with atomic number.

and the neutron carry excess energy totalling 17.6 MeV, millions of times greater than is liberated in a typical chemical reaction (~1 eV).

3. The source of nuclear energy

The energy released in nuclear fusion and fission reactions is derived from the same source. In both, the products of reaction have less mass than the constituents. The mass loss (m) is released as energy (E) according to Einstein's familiar equation $E = mc^2$, where c is the velocity of light. Since the velocity of light is very large, 3×10^8 m s^{-1}, a relatively small mass loss corresponds to a very large energy release.

This loss of mass is illustrated in graphic form in Figure 2, which plots the binding energy, or apparent loss of mass per nucleon, against atomic number. Energy release from fission corresponds to atoms of high atomic number, shown at the right-hand edge of the figure, splitting into lighter products. Fusion is represented by the joining of light elements, shown at the left-hand edge of the diagram, to produce a product of higher atomic number.

There are several potential fusion reactions, but the deuterium and tritium reaction has the highest cross section under the conditions attainable on Earth and is thus the most favourable for energy production in the foreseeable future.

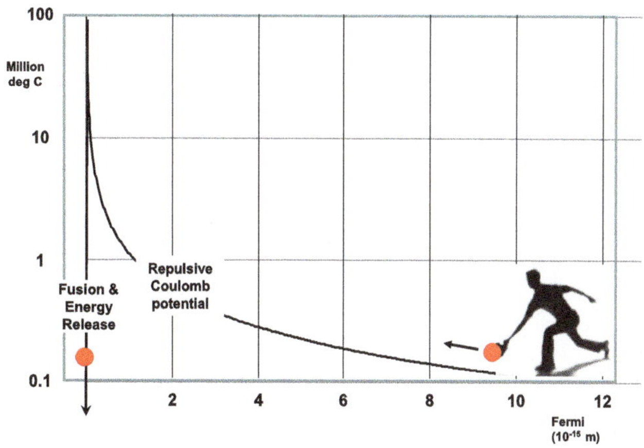

Figure 3. Potential energy schematic for fusion.

4. Physics of thermonuclear fusion

Although the liberated energy from both fission and fusion reactions has the same nuclear origins, there is an important difference between the physics of the two reactions which explains why power production from fusion is so technologically demanding. In the case of fission, some high atomic number nuclei are unstable and undergo spontaneous fission to produce lower atomic number products and energetic neutrons. These reactions occur at room temperature without the need to supply external energy to initiate or sustain them.

The physics of fusion is very different. When light nuclei approach to a separation comparable to their diameter, the strong nuclear force draws them even closer together until they fuse. However, this force only acts over very short distances, while at larger separation the nuclei are subjected to the repulsive Coulomb force which acts to push them apart. Only nuclei with sufficient kinetic energy to overcome the Coulomb barrier can approach closely enough to fuse. At room temperature an insignificant number of nuclei possess such energy, and external heating must be applied. It is the requirement to supply this heat energy that gives rise to the term 'thermonuclear'.

The height of the Coulomb barrier for deuterium and tritium is 1 MeV, corresponding to a temperature of 10 billion K. Fortunately, quantum mechanical tunnelling enables a significant number of neutrons to penetrate the barrier at lower energy, reducing the heating requirement to 5 keV (50 million K).

These two competing forces are shown in Figure 3, as a potential energy or 'bowling ball' diagram. The induction of fusion can be considered in terms of rolling a ball up the Coulomb potential with sufficient speed, or temperature, that it reaches the top of the barrier and falls into the potential well created by the strong nuclear force.

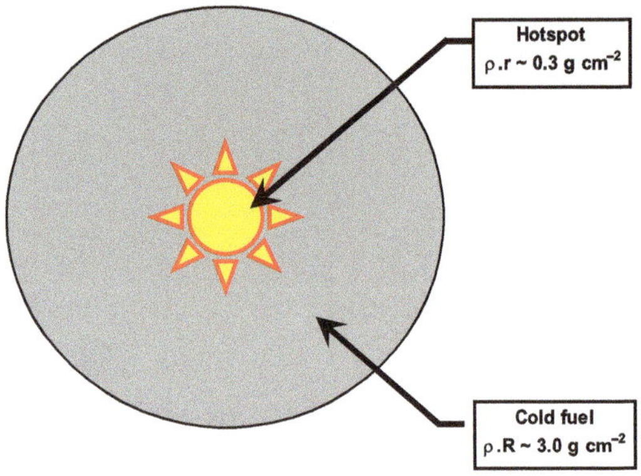

Figure 4. Assembly of the fuel at high density.

5. Fuel assembly for a self sustaining reaction

The fusing of nuclei by laser heating is a necessary first step in the energy production process, but this is not sufficient if the fusion process is to be efficient. The solution is to use the laser to heat a small volume of the D–T fuel, a 'hotspot', so that fusion reactions occur, and to arrange the geometry of the system so that the fusion energy released is deposited within the hotspot itself, raising the temperature further so that the fusion reactions proceed at a higher rate and heat the surrounding cold D–T fuel. This 'central hotspot ignition' approach[3, 4], shown schematically in Figure 4, can be likened to a match igniting a small amount of fuel, which then heats the surrounding material to ignition and produces a propagating burn wave.

The requirement for self heating of the hotspot region by the products of the laser-initiated reactions gives rise to a fundamental condition for efficient burning of the fuel. The 14 MeV neutron is small, highly energetic, and typically escapes from the central hotspot of the fuel without depositing significant energy. The alpha particle is larger, less energetic and much more likely to deposit its kinetic energy within the hotspot as required. The likelihood of trapping the alpha particle depends simply on the amount of material it must penetrate before leaving the hotspot region. This can be expressed as '$\rho \cdot r$' where ρ is the fuel density and r is the radius of the fuel in the hotspot, as shown in Figure 4. For 3.5 MeV alpha particles in a D–T mixture, the trapping condition is that $\rho \cdot r \sim 0.3$ g cm^{-2} (Ref. [5]).

We can now investigate the energy required to heat the hotspot region to the temperature at which fusion reactions occur. Consider the hotspot in Figure 4 with radius r and density ρ. The energy required to heat the hotspot to the temperature at which fusion reactions occur, T_i, is given by the product of the hotspot mass (volume × density) multiplied by the temperature rise T_i and the specific heat C_v.

Thus,

$$E = 4/3 \cdot \pi \cdot r^3 \cdot \rho \cdot T_i \cdot C_v,$$

which can be rearranged as

$$E = [(4/3 \cdot \pi \cdot T_i \cdot C_v) \cdot (\rho \cdot r)^3] \cdot \rho^{-2}.$$

However, since $\rho \cdot r$ is fixed at 0.3 g cm^{-2} by the alpha trapping condition, the energy can therefore be expressed as

$$E = \text{constant}/\rho^2.$$

Thus, by compressing the fuel to high density, the energy needed from the laser is reduced by the compression factor squared. For example, by compressing to a density of $\sim$500 g cm^{-3}, compared with the normal solid density of D–T ice ($\sim$0.2 g cm^{-3}), the energy required to heat the fuel is reduced by a factor of $(2500)^2$, or six million. It is this inverse square dependence of the ignition energy with density that enables inertial fusion to be brought within the reach of currently available laser driver technology.

In principle, the laser requirement could be reduced further by compressing to even higher densities. In practice, imperfections in the uniformity of the laser beams, the ability to control the laser pulse shape and imperfections in the manufacture of the fuel capsules themselves limit the achievable density to $\sim$500 g cm^{-3} (Ref. [5]).

In laser driven fusion, the laser must perform two distinct functions. First, the D–T fuel must be compressed to the density required to trap the fusion alpha particles; second, the hotspot must be heated to fusion temperature, producing a burn wave that propagates through the entire fuel mass.

Great care must be taken during the compression phase, as premature heating would cause the fuel pressure to increase, requiring increased drive energy from the laser. The solution is to shape the laser pulse so that the compression occurs adiabatically, in the absence of shocks, to keep the fuel temperature low. A further issue is the control of hydrodynamic and laser–plasma instabilities (LPI) during the compression phase, also requiring increased drive. Once the fuel capsule is fully compressed, the laser power is increased rapidly to launch a number of shock waves into the fuel, which coalesce at the centre of the hotspot and raise the temperature to the ignition point. The process of compression and heating by the laser is shown schematically in Figure 5.

6. Experimental proof of principle

The scientific proof of principle of the basis of inertial fusion was demonstrated in a series of underground nuclear tests in the 1980s. Since that time, vigorous experimental and computational programmes have been pursued in the US, France, Japan and elsewhere to demonstrate laser-driven ignition and burn of a D–T fuel capsule.

In 2010, the 192-beam, 1.8 MJ National Ignition Facility (NIF)[6, 7] was commissioned at LLNL in California with the combined mission of supporting the Stockpile

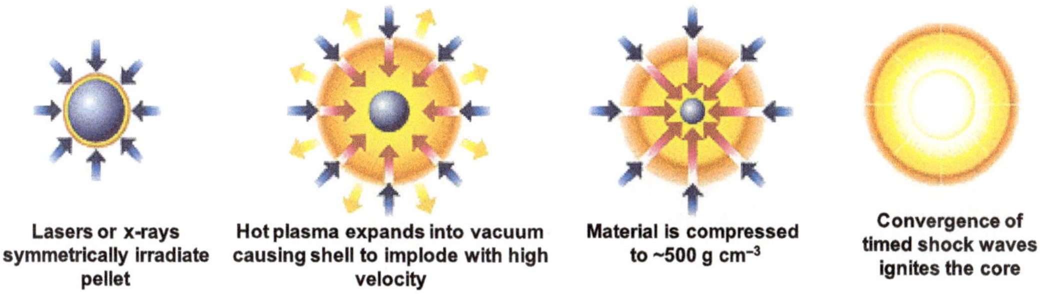

Figure 5. Compression and heating of the fuel capsule; ignition and burn of the D–T fuel.

Figure 6. The NIF facility at LLNL, California, USA (picture courtesy of LLNL).

Figure 7. One of the two NIF laser bays (picture courtesy of LLNL).

Figure 8. LMJ in Bordeaux, France (picture courtesy of CEA).

Stewardship Program, fundamental research in partnership with the academic community in the USA and elsewhere, and research into inertial confinement fusion through the National Ignition Campaign (NIC)[8]. The NIC experiments have made rapid progress towards achieving 'first ignition', and demonstration of net energy gain is anticipated within the next few years. The NIF facility building is shown in Figure 6, and Figure 7 shows one of the two NIF laser bays, each the size of a football field.

The achievement of first ignition at NIF would mark the culmination of 50 years of intensive research. It would be a full and sufficient demonstration that ICF has progressed from an elusive phenomenon of physics to a predictable, controllable technological process, ready to be harnessed for the benefit of mankind.

In 2016, the Laser Mégajoule (LMJ) facility[9] in Bordeaux, France will be completed for start up of operations by the CEA, and ignition-scale fusion experiments can begin in Europe. A photograph of the LMJ site during the later stages of construction is shown in Figure 8. LMJ beam time will be made available to the academic community, adding additional experimental capability to the international drive for fusion.

In Japan, the LFEX laser is in the final stages of commissioning and will be fully operational by 2015. LFEX is a high energy petawatt class short pulse facility which has been built alongside the Gekko XII facility at the Institute of Laser Engineering (ILE), Osaka University[10, 11], as shown in Figure 9. The combination of high energy short pulses coupled with the long pulse Gekko XII laser facility will enable experiments to be conducted into 'fast ignition', an advanced ICF scheme which it is hoped will produce high fusion energy gain at lower laser drive energy.

Figure 9. The Gekko XII and LFEX lasers at ILE, University of Osaka, Japan (picture courtesy of Osaka University).

Figure 10. The laser bay of AWE's Orion facility (picture courtesy of AWE).

In the UK, scientists have access to unique, large scale, world leading laser facilities – Vulcan[12] at Central Laser Facility (CLF), STFC Rutherford Appleton Laboratory and Orion[13] at AWE, as shown in Figure 10. These facilities are not powerful enough to reach the conditions for fusion gain, but they are well suited to conducting experiments designed to validate the numerical simulations on which inertial fusion depends and to investigate the underlying physics of the key processes.

UK facilities play two important roles in the laser driven fusion mission. First, the NIC in the USA has identified some areas in which the agreement between the latest experiments and simulations is poorer than expected[8]. Orion is particularly well suited to measuring the opacity and equations of state[14, 15] of materials at high density and temperature, and such measurements will help to improve the fidelity of the numerical simulations of importance to first ignition. Second, following first ignition, it will be necessary to tune the configuration of the fuel capsules and the laser parameters to optimize the energy gain of the system, to improve the robustness of ignition and to explore the sensitivity of the gain to imperfections. This will require a large number of experimental laser shots. With its high shot rate, Orion will be able to make an important contribution to this optimization process.

7. The HiPER project

HiPER[16, 17] is a pan-European ESFRI project (European Strategy Forum on Research Infrastructures), which seeks to demonstrate the production of secure, sustainable, safe and affordable energy with low environmental impact based on fusion driven by lasers and on a timescale relevant to meeting the challenges of the energy gap on the 2050 timescale.

Coordinated by the STFC (Science and Technology Research Council) at the Rutherford Appleton Laboratory, HiPER has created an exceptional opportunity for Europe to partner with the USA and other international partners in the approach to first ignition and to develop an international 'fast-track' approach to laser energy.

The HiPER consortium brings together 26 partners from 10 countries whose mission is to provide the scientific, technological and economic evidence to proceed to a demonstration power plant in a single step and at acceptably low commercial risk. A visualization of the HiPER concept is shown in Figure 11.

Studies of the economic viability of laser-driven fusion energy have been conducted for both HiPER and LIFE[18], an IFE project based at LLNL. Both studies are in broad agreement, although it is acknowledged that there are difficulties in predicting economic factors 20–50 years into the future. The major conclusions are that a future plant must run at a repetition rate of at least 10 Hz, that a fusion energy gain in excess of 60 is required and that the wall plug efficiency of the laser driver must exceed 10%. The HiPER strategy is to exploit the separability of the technology by demonstrating prototypes in each of the key areas during a phase of technology development. Once this is done, construction of the HiPER power production prototype facility will present a sufficiently low technological risk to attract investment funding.

The preparatory phase of the HiPER project ended in April 2013[19] and it has now entered its physics demonstration and technology development phase.

8. Technical and commercial challenges ahead

8.1. Physics demonstration

Demonstration of the ignition and burn of a fuel capsule, 'first ignition', at NIF, Laser Mégajoule or elsewhere is an essential precursor to major investment in the construction of a prototype power plant[20].

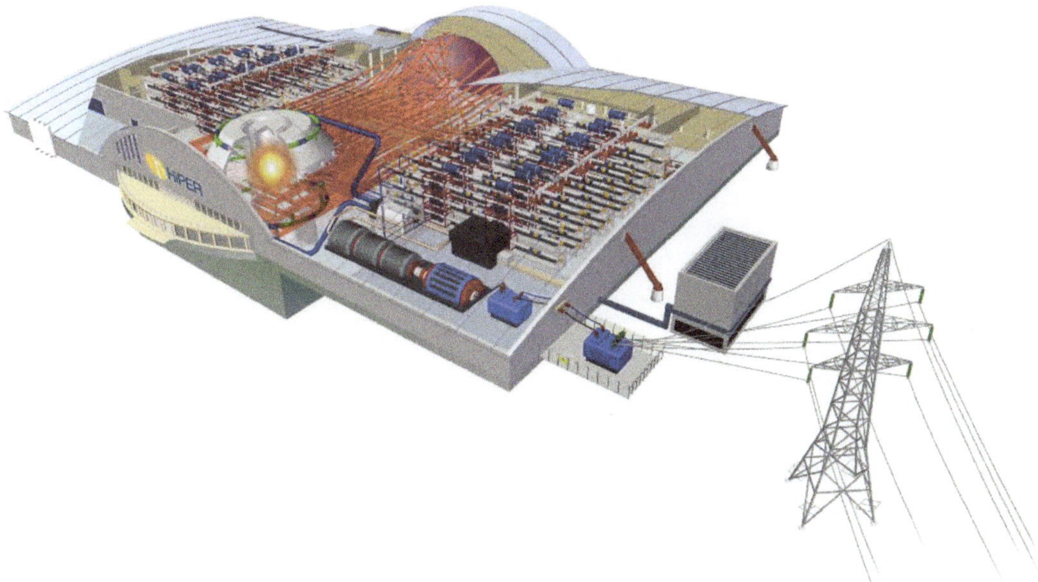

Figure 11. Visualization of the HiPER prototype facility.

The experimental programme at NIF has made exceptional progress towards first ignition in recent months, with net energy gain and indications of significant alpha particle heating of the fuel[21, 22]. Further improvements are still required, particularly in the control of the hydrodynamic and LPI instabilities, but progress is encouraging and the necessary enhancements are anticipated within the next few years.

8.2. Laser driver

For a commercial power plant, laser driver efficiencies of >10% and repetition rates of >10 Hz will be required. The NIF laser has been designed for scientific proof of principle of laser-driven fusion on a single shot basis. With an overall efficiency of 0.1% and a repetition rate of two full energy shots per day, the technology on which NIF is based is insufficient to meet the requirement for commercial power production! A dramatic step change in laser technology is therefore clearly required to meet the driver specification. This change is now available through the use of high efficiency laser diodes to replace flashlamps as the source of the optical pump for the laser driver.

Originally developed for use in the telecommunications industry, semiconductor laser diodes provide a highly efficient means of producing very specific, 'tailored' wavelengths of light which can be used for efficient pumping of high power laser gain media. This class of device, termed 'diode pumped solid state laser' (DPSSL)[23], offers high average power, high repetition rate operation and high efficiency. With cost reductions through mass production to meet rising demand, DPSSLs are ideal as drivers for IFE systems.

Within Europe, successful DPSSL development programmes are under way at the Central Laser Facility, STFC Rutherford Appleton Laboratory, UK[24], at Ecole Polytechnique, Paris, France[25], at the University of Jena, Germany[26] and at the HiLASE Project, Prague, Czech Republic[27, 28]. All seek to demonstrate 100 J/10 Hz operation at greater than 10% efficiency and all are making encouraging progress. The scaling of such systems to the 10 kJ level and the deployment of ~200 such beams around the fusion chamber to deliver multi-megajoule drive seems entirely feasible.

DPSSL technology has immediate exploitation potential in a wide range of industrial, medical and scientific applications. Furthermore, if the new technology can be delivered at acceptable cost, the high efficiency and low through-life operating costs will displace existing systems while making laser treatment economically viable in new sectors. The potential economic impact of such 'next generation' laser technology in the short and medium term provides a strong driver for investment in the technology, driving up the performance of the technology while driving down the cost of installed systems.

9. Energy capture and fuel cycle

Deuterium is an abundant constituent of seawater and can be extracted by chemical means. Tritium, however, is radioactive with a half-life of 12.3 years and must be generated '*in situ*' within the fusion fuel cycle. The process, which is based on neutron capture by lithium, is illustrated in Figure 12. Lithium is abundant and widely distributed in the Earth's crust. It can also be extracted from seawater.

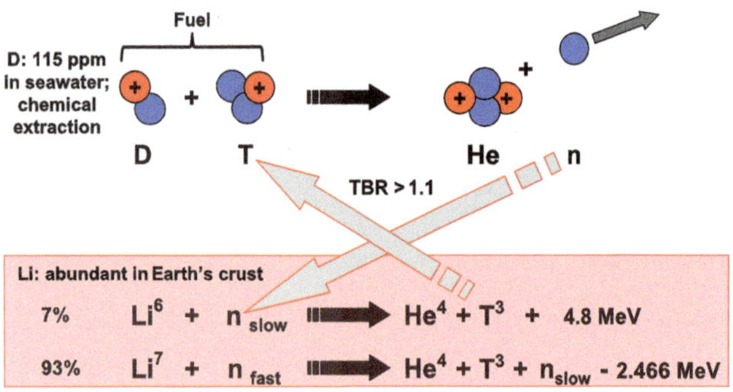

Figure 12. Neutron capture and tritium generation in lithium.

The reaction of a low energy neutron with lithium-6 is exothermic, adding to energy production, while the lithium-7 reaction is endothermic. Both reactions form tritium, and in the lithium-7 case the neutron is 'preserved' so that it can take part in a subsequent lithium-6 reaction. Addition of lithium-7 therefore enables excess tritium to be produced to make up losses and also to generate a surplus for fuelling other fusion power plants.

The LIFE project at LLNL has the most developed concept for energy extraction, based on a simple 'boiler tube' blanket design, with a mixture of liquid lithium-6 and lithium-7 isotopes flowing in tubes around the fusion chamber and through heat exchangers.

Engineering expertise and capability associated with liquid sodium cooling of fission reactor cores are readily transferable to fusion energy plants using lithium as the primary coolant. Extraction of tritium from the liquid lithium and its recovery from the waste stream of any unburnt fuel capsules remain to be demonstrated, but, again, substantial expertise exists in the nuclear industry.

10. Fuel capsule production

Possibly the greatest challenge facing energy production from laser-driven fusion is the mass production of fuel capsules to the required quality and cost. Each power plant will require approximately one million fuel pellet targets per day. Commercial modelling from the HiPER preparatory phase indicates a maximum acceptable unit cost for targets of approximately 0.5 Euro/Dollars. In comparison, the current cost of the most complex NIF ignition targets, which are made individually and require many manual assembly steps, exceeds this by more than four orders of magnitude, with production times of several days. Commercial viability of laser-driven fusion power plants clearly requires a completely different approach to manufacturing techniques to increase production rates and reduce cost.

Potential process solutions have been identified, but all need development to meet the systems requirement for commercial power production. As an incentive for development, the new technologies needed for mass production of fuel capsules have very significant commercial potential. Examples include the following[29]: advanced coating and polishing; microelectromechanical systems; microfluidics and dielectrophoresis; and the application of graphene based structures.

Microtarget design and manufacturing facilities are associated with high power laser programmes throughout the world. Together, these centres of excellence, in partnership with industry and academia, are ideally placed to make a significant contribution to the field, advancing IFE research worldwide and realizing economic impact in both the short and the long term.

11. Conclusion

Laser-driven IFE is based on the conversion of isotopes of hydrogen into helium through the process of fusion, using lasers as a driver. This technology could be producing energy on the 2050 timescale, with the potential to supply a significant proportion of world energy needs in the following decades.

IFE development is approaching a seminal moment, with proof-of-principle demonstration at the NIF in the USA expected within the next few years. Ignition at NIF or elsewhere will pave the way for a programme of technology development followed by construction of a prototype IFE plant to demonstrate power production on a commercially viable basis.

The HiPER project has defined a path to a commercially attractive prototype plant. The success of HiPER to date has created an exceptional opportunity for international partnerships in the run up to first ignition and thereby for the development of an international 'fast-track' approach to laser energy.

Global investment in this field over the next 5–10 years will secure an important stake in this high impact, strategic technology, providing stand-alone benefit to participating nations and positioning them as suppliers of IFE rather than customers of this highly promising technology. The net cost of such an investment would largely be offset by the value of the arising intellectual property and the economic impact of its exploitation in a diverse range of applications in the short and medium term.

Acknowledgements

The HiPER Preparatory Phase Project was supported by FP7-Infrastructures-2007-1 (Grant Agreement number 211737): 'The European High Power laser Energy Research facility Preparatory Phase Study'; the UK Science, Technology and Facilities Council; the Ministry of Education, Youth and Sports of the Czech Republic and many 'in-kind contributions' from the HiPER partners and their sponsoring agencies.

References

1. M. Dunne, N. Alexander, F. Amiranoff, P. Aguer, S. Atzeni, H. Azechi, V. Bagnoud, P. Balcou, J. Badziak, D. Batani, C. Bellei, D. Besnard, R. Bingham, J. Breil, M. Borghesi, S. Borneis, A. Caruso, J. C. Chanteloup, R. J. Clarke, J. L. Collier, J. R. Davies, J.-P. Dufour, P. Estraillier, R. G. Evans, M. Fajardo, R. Fedosejevs, G. Figueria, J. Fils, J. L. Feugeas, M. Galimberti, J.-C. Gauthier, A. Giulietti, L. A. Gizzi, D. Goodin, G. Gregori, S. Gus'kov, L. Hallo, C. Hernandez-Gomez, D. Hoffman, J. Honrubia, S. Jacquemot, M. Key, J. Kilkenny, R. Kingham, M. Koenig, F. Kovacs, K. Krushelnic, C. Labaune, K. Lancaster, C. Leblanc, P. H. Maire, M. Marklund, W. Martin, A. McEvoy, P. McKenna, J. T. Mendonça, J. Meyer-ter-Vehn, K. Mima, G. Mourou, S. Moustaizis, Z. Najmudin, P. Nickles, D. Neely, P. Norreys, M. Olazabal, A. Offenberger, N. Papadogianis, J.-P. Perin, J. M. Perlado, J. Ramirez, R. Ramis, Y. Rhee, X. Ribeyre, A. Robinson, K. Rohlena, S. J. Rose, M. Roth, C. Rouyer, C. Rulliere, B. Rus, W. Sandner, A. Schiavi, G. Schurtz, A. Sergeev, M. Sherlock, L. Silva, R. A. Smith, G. Sorasio, C. Strangio, H. Takabe, M. Tatarakis, V. Tikhonchuk, M. Tolley, M. Vaselli, P. Velarde, T. Winstone, K. Witte, J. Wolowski, N. Woolsey, B. Wyborn, M. Zepf, and J. Zhang, HiPER: The European High Power Laser Energy Research Facility, Technical Background and Conceptual Design Report 2007; RAL-TR-2007-008 (2007).
2. C. D. Orth, Nucl. Fusion **42**, 354 (2002).
3. D. Lindl, P. Amendt, R. L. Berger, S. G. Glendinning, S. H. Glenzer, S. W. Haan, R. L. Kauffman, O. L. Landen, and L. J. Suter, Phys. Plasmas **11**, 339 (2004).
4. S. H. Glenzer, B. J. MacGowan, P. Michel, N. B. Meezan, L. J. Suter, S. N. Dixit, J. L. Kline, G. A. Kyrala, D. K. Bradley, D. A. Callahan, E. L. Dewald, L. Divol, E. Dzenitis, M. J. Edwards, A. V. Hamza, C. A. Haynam, D. E. Hinkel, D. H. Kalantar, J. D. Kilkenny, O. L. Landen, J. D. Lindl, S. LePape, J. D. Moody, A. Nikroo, T. Parham, M. B. Schneider, R. P. J. Town, P. Wegner, K. Widmann, P. Whitman, B. K. F. Young, B. Van Wonterghem, L. J. Atherton, and E. I. Moses, Science **327**, 1228 (2010).
5. S. Atzeni and J. Meyer-ter-Vehn, *The Physics of Inertial Fusion* (Oxford University Press, 2004).
6. G. H. Miller, E. I. Moses, and C. R. West, Opt. Eng. **43**, 2841 (2004).
7. E. I. Moses, Nucl. Fusion **49**, 104022 (2009).
8. J. Lindl, O Landen, J. Edwards, and E. Moses, Phys. Plasmas **21**, 020501 (2014).
9. J. Ebrardt and J. M. Chaput, J. Phys.: Conf. Ser. **244**, 032017 (2010).
10. H. Shiraga, S. Fujioka, M. Nakai, T. Watari, H. Nakamura, Y. Arikawa, H. Hosoda, T. Nagai, M. Koga, H. Kikuchi, Y. Ishii, T. Sogo, K. Shigemori, H. Nishimura, Z. Zhang, M. Tanabe, S. Ohira, Y. Fujii, T. Namimoto, Y. Sakawa, O. Maegawa, T. Ozaki, K. A. Tanaka, H. Habara, T. Iwawaki, K. Shimada, H. Nagatomo, T. Johzaki, A. Sunahara, M. Murakami, H. Sakagami, T. Taguchi, T. Norimatsu, H. Homma, Y. Fujimoto, A. Iwamoto, N. Miyanaga, J. Kawanaka, T. Jitsuno, Y. Nakata, K. Tsubakimoto, K. Sueda, N. Morio, S. Matsuo, T. Kawasaki, K. Sawai, K. Tsuji, H. Murakami, T. Kanabe, K. Kondo, R. Kodama, N. Sarukura, T. Shimizu, K. Mima, and H. Azechi, High Energy Density Phys. **8**, 227 (2012).
11. H. Azechi, K. Mima, S. Shiraga, S. Fujioka, H. Nagatomo, T. Johzaki, T. Jitsuno, M. Key, R. Kodama, M. Koga, K. Kondo, J. Kawanaka, N. Miyanaga, M. Murakami, K. Nagai, M. Nakai, H. Nakamura, T. Nakamura, T. Nakazato, Y. Nakao, K. Nishihara, H. Nishimura, T. Norimatsu, P. Norreys, T. Ozaki, J. Pasley, H. Sakagami, Y. Sakawa, N. Sarukura, K. Shigemori, T. Shimizu, A. Sunahara, T. Taguchi, K. Tanaka, K. Tsubakimoto, Y. Fujimoto, H. Homma, and A. Iwamoto, Nucl. Fusion **53**, 104021 (2013).
12. C. N. Danson, P. A. Brummitt, R. J. Clarke, J. L. Collier, B. Fell, A. J. Frackiewicz, S. Hancock, S. Hawkes, C. Hernandez-Gomez, P. Holligan, M. H. R. Hutchinson, A. Kidd, W. J. Lester, I. O. Musgrave, D. Neely, D. R. Neville, P. A. Norreys, D. A. Pepler, C. J. Reason, W. Shaikh, T. B. Winstone, R. W. W. Wyatt, and B. E. Wyborn, Nucl. Fusion **44**, S239 (2004).
13. N. Hopps, C. Danson, S. Duffield, D. Egan, S. Elsmere, M. Girling, E. Harvey, D. Hillier, M. Norman, S. Parker, P. Treadwell, D. Winter, and T. Bett, Appl. Opt. **52**, 3597 (2013).
14. D. J. Hoarty, P. Allan, S. F. James, C. R. D. Brown, L. M. R. Hobbs, M. P. Hill, J. W. O. Harris, J. Morton, M. G. Brookes, R. Shepherd, J. Dunn, H. Chen, E. Von Marley, P. Beiersdorfer, H. K. Chung, R. W. Lee, G. Brown, and J. Emig, High Energy Density Phys. **9**, 661 (2013).
15. D. J. Hoarty, P. Allan, S. F. James, C. R. D. Brown, L. M. R. Hobbs, M. P. Hill, J. W. O. Harris, J. Morton, M. G. Brookes, R. Shepherd, J. Dunn, H. Chen, E. Von Marley, P. Beiersdorfer, H. K. Chung, R. W. Lee, G. Brown, and J. Emig, Phys. Rev. Lett. **110**, 265003 (2013).
16. J. Hien, L. O. Silva, G. Korn, L. A. Gizzi, and C. B. Edwards, (Eds) *SPIE Volume 8080, Proceedings of 'Diode-Pumped High Energy and High Power Lasers; ELI: Ultra-relativistic Laser–Matter Interactions and Petawatt Photonics; and HiPER: the European Pathway to Laser Energy', Prague, Czech Republic, April 18, 2011.*
17. http://www.hiperlaser.org.
18. M. Dunne, Fusion Sci. Technol. **60**, 19 (2011).
19. http://www.hiper.org/Resources/HiPER_Preparatory_Phase_Completion_Report.pdf.
20. http://www.nap.edu/catalog.php?record_id=18289.
21. https://www.llnl.gov/news/newsreleases/2013/Feb/NR-13-02-07.html.
22. http://www.bbc.co.uk/news/science-environment-24429621.
23. A. Tunnermann, H. Zellmer, W. Schome, A. Giesen, and K. Contag, Top. Appl. Phys. **78**, 369 (2000).
24. S. Banerjee, K. Ertel, P. D. Mason, P. J. Phillips, M. Siebold, M. Loeser, C. Hernandez-Gomez, and J. L. Collier, Opt. Lett. **37**, 2175 (2012).
25. J.-C. Chanteloup, A. Lucianetti, D. Albach, and T. Gonçalvès-Novo, Plasma Fusion Res. **8**, 3404043 (2013).
26. M. Hornung, H. Liebetrau, A. Seidel, S. Keppler, A. Kessler, J. Korner, M. Hellwing, F. Schorcht, D. Klopfel, A. K. Arunachalam, G. A. Becker, A. Savert, J. Polz, J. Hein, and C. Kaluza, High Power Laser Sci. Eng. **2**, e20 (2014).
27. M. Divoky, M. Smrz, M. Chyla, P. Sikocinski, P. Severova, O. Novak, J. Huynh, S. S. Nagisetty, T. Miura, J. Pilar, O. Slezak, M. Sawicka, V. Jambunathan, J. Vanda, A. Endo, A. Lucianetti, D. Rostohar, P. D. Mason, P. J. Phillips, K. Ertel, S. Banerjee, C. Hernandez-Gomez, J. L. Collier, and T. Mocek, High Power Laser Sci. Eng. **2**, e14 (2014).
28. A. Lucianetti, M. Sawicka, O. Slezak, M. Divoky, J. Pilar, V. Jambunathan, S. Bonora, R. Antipenkov, and T. Mocek, High Power Laser Sci. Eng. **2**, e13 (2014).
29. M. Tolley and C. Spindloe, in *Laser–Plasma Interactions and Applications*, P. McKenna, D. Neely, R. Bingham and D. A. Jaroszynski, eds. Ch. 17, (Springer, 2013), p. 431, ISBN 978-3-319-00037-4.

8

Numerical simulation of debris-removal trajectories on the transport mirrors in high-power laser systems

Yangshuai Li[1,2], Jianqiang Zhu[1], Xiangyang Pang[1], Hua Tao[1,2], Xiang Jiao[1,2], and Yongzhong Wu[1,2]

[1]*National Laboratory on High Power Laser and Physics, Shanghai Institute of Optics and Fine Mechanics, Chinese Academy of Sciences, Shanghai 201800, China*

[2]*University of Chinese Academy of Sciences, Beijing 100049, China*

Abstract

In high-power laser systems (HPLSs), understanding debris-removal trajectories is important in eliminating debris from the surfaces of transport mirrors online and keeping other optical components free from contamination. NS equations, the RNG k–ε model and the discrete phase model of the Euler–Lagrange method are used to conduct numerical simulations on the trajectories of contaminant particles of different sizes and types on the mirror surface using Fluent commercial software. A useful device is fabricated based on the simulation results. This device can capture and collect debris from the mirror surface online. Consequently, the effect of debris contamination on other optical components is avoided, cleaning time is shortened, and ultimately, the cleanliness of the mirrors in HPLSs is ensured.

Keywords: cleanliness of the mirror; debris trajectory; Fluent; high-power laser systems; numerical simulations

1. Introduction

High-power laser systems (HPLSs) are large optical instruments for high-energy physics and inertial confinement fusion. The transport mirrors of HPLSs direct light from the main amplifiers to the target and impact remarkably upon the quality of laser beams arriving at the target[1]. Current mirrors work at a fluence of several joules per square centimeter, but future operating conditions will require these mirrors to sustain high fluences of up to tens of joules per square centimeter or even higher. Therefore, rigorous cleanliness is critical in preventing damage to the mirror surfaces and prolonging the lifetime of the optical devices[2].

The debris on mirrors includes natural dust falls, metal fragments, and organic fibers left during operations. Numerous methods have been used to keep the optical devices clean, but to date, few experimental results have demonstrated the optical cleanliness of the whole laser facility[3]. Wiping mirrors with a clean cloth is the commonly practiced technique. However, hard debris often scratches transport mirrors because uneven forces are used in wiping away debris. Thus, the Lawrence Livermore National Laboratory designed an *in situ* surface debris-removal system for the upward-facing transport mirrors of the National Ignition Facility. This system effectively removes debris with particle sizes larger than 10 μm using high-speed airflow generated by an air knife[2,4,5]. However, most of the debris returns to the environment, which easily leads to the contamination of other optical components and re-contamination of the mirrors. Therefore, understanding debris-removal trajectories and realizing the capture and collection of the debris during the blowing process are necessary for efficient debris removal.

In the present study, the flow field and the removal trajectories of debris with varied types and sizes on the transport mirror are numerically investigated. A device is fabricated based on the simulation results. This device can capture and collect debris from the mirror surface online.

2. The numerical model

Computational fluid dynamics (CFD) is an important technique for investigating flow fields. With computer numerical calculation and image display, CFD can conduct numerical simulations on a system containing relevant physical phenomena, such as fluid flow and heat. Numerical analysis of the flow field with basic physical quantities (such as speed, pressure, temperature and concentration), as well as the distribution of fluid trajectory, is now a popular topic in fluid studies.

Correspondence to: Y. Li, No. 390, Qinghe Road, Jiading, Shanghai, CN 201800, China. Email: yshli@siom.ac.cn

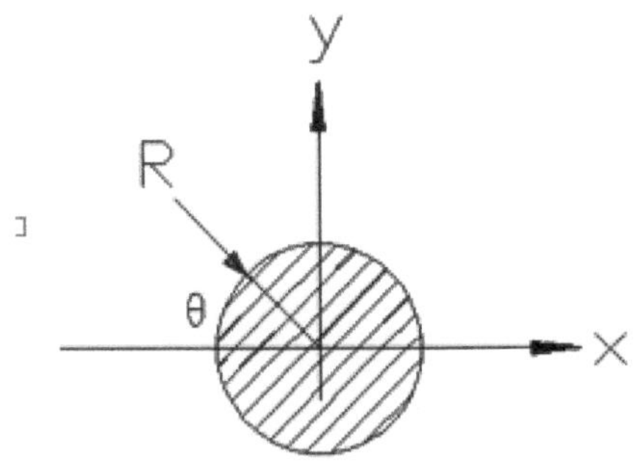

Figure 1. Interaction force between the gas and a spherical particle.

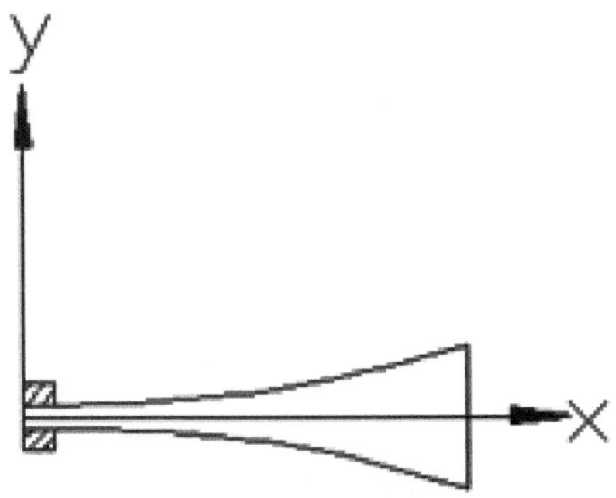

Figure 2. Schematic diagram of air ejection.

2.1. The model to be calculated

A two-dimensional (2D) model satisfies the requirements of understanding the fluid field and the trajectories of contaminant particles. The 2D model in this study included one air knife, one transport mirror and one cover (Figure 1). The air knife was used to provide high-speed airflow. The transmission mirror surface, covered with particulate pollutants, was the purged surface. The cover connecting to a vacuum cleaner was used to provide a low pressure area, change the flow field and realize the capture of contaminant particles. Several assumptions were considered to simplify the model. First, the fluid was compressible nitrogen, which provided a stable inlet pressure. Second, the loaded solid particles were spheres with smooth surfaces. Similar solid particles had similar densities. No energy was lost during the particle and wall collisions, and the rebounds were symmetrical. Third, only the steady aerodynamic resistances affecting the solid particles caused by the fluid flow were considered, and the Basset and Saffman forces were ignored.

The 2D model was implemented using the business software Gambit 2.4.6, used for geometry and mesh generation, and the 20 640 quadrilateral elements were divided by quadrilateral mesh structures[6, 7]. The coupled implicit solver, steady-state calculations and second-order upwind discretization were adopted. The Courant number was 1. The simulation of the gas–solid flow field was calculated using the discrete phase model. The convergence criteria of the mass residuals and energy residuals were less than 10^{-5}.

2.2. Boundary conditions and associated settings

The best way to remove a particle is to blow in the direction of the x-axis, where the particle can be subjected to maximum thrust R (Figure 2). The airflow ejecting from the nozzle is similar to a trapezoidal distribution (Figure 3). The maximum speed of the cross-section is at the center line (x-axis). Thus, the particle is subjected to the maximum force and easily removed when the blowing surface is situated on the x-axis. Therefore, a suitable distance S between the up-surfaces of the transport mirror and the air knife was chosen to make the transport mirror surface close to, or coinciding with, the centerline of the airflow (Figure 1). Thus, a suitable air pressure should be selected to obtain a suitable flow velocity. Figure 1 shows that after ejecting from the air knife, the airflow initially flowed along the curved edge of the air knife, then the flow direction changed to 90°. Finally, the airflow entered the up-surface L of the transport mirror and moved along the up-surface. After a series of experiments, a good particle removal effect was obtained when the air pressure was 9 atm and S was 3 mm.

Boundary conditions were set as follows. The outlet of the air knife was the pressure inlet. The gage total pressure was 9 atm, the supersonic/initial gage total pressure was 7 atm, the inlet temperature was 300 K and the turbulent energy and default dissipation rates were 1.

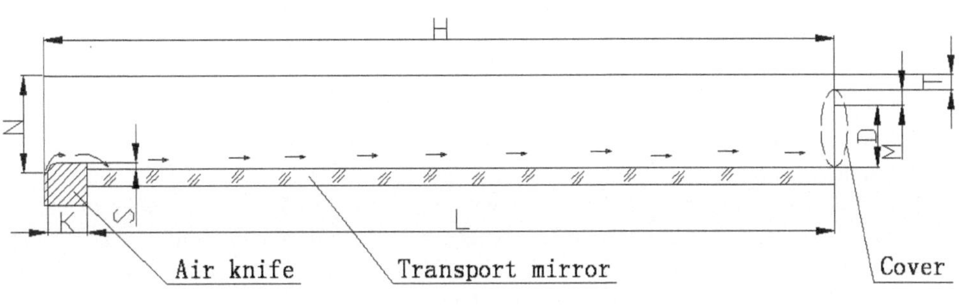

Figure 3. Model of the air knife blowing.

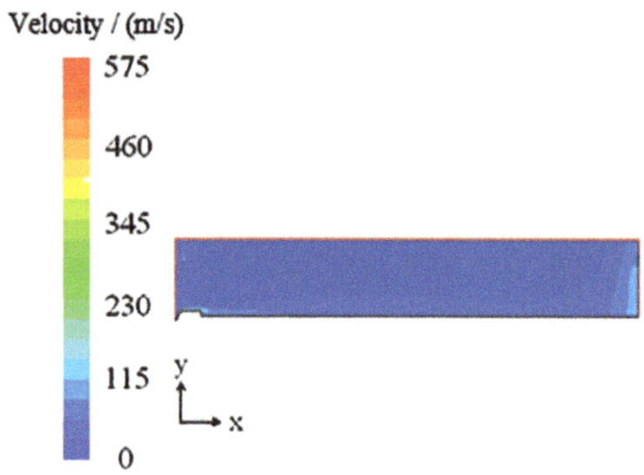

Figure 4. Simulation results of the flow.

Figure 5. Velocity vector of the flow near the air knife.

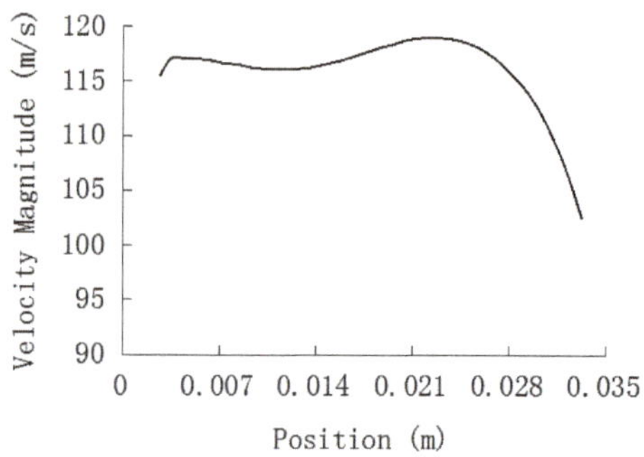

Figure 6. Speed of the flow at the pressure outlet D in the y-axis.

Boundaries H, N, D, and T were the pressure outlets. $H = 423.05$ mm, $N = 50$ mm, $D = 32$ mm, and $T = 8$ mm. The outlet temperature was 300 K. The turbulent energy and default dissipation rate were 1. The gage total pressures of boundaries H, N, D and T were 1 atm. The gage total pressure of boundary D was 0.8 atm.

The remaining boundaries were walls, which were non-slip walls. $K = 21.05$ mm, $T = 8$ mm, and the length L of the up-surface of the transport mirror was 400 mm. Dust, aluminum particles and stainless steel particles, with sizes of 10, 20, 40, and 80 μm, were used as debris, and they were uniformly distributed on the surface of the transport mirror. The densities of dust, aluminum (Al) particles and stainless steel (ss) particles were 1000, 2700 and 7930 kg m^{-3}, respectively.

2.3. Governing equations

The basic equations of fluid analysis, turbulent flow and particle motion were used as the control equations[8]. These basic equations included the equations for the conservation of mass, momentum and energy. The turbulent flow model was the RNG k–ε model, which is an improved model of the standard turbulent model of k–ε. The particle motion equation was the force-balance equation in Cartesian coordinates.

3. Simulation results and discussion

3.1. Flow field

Figures 4 and 5 are the simulation results of the whole flow and the vector of the flow near the air knife. The maximum speed of the air flow coming from the air knife reaches 575 m s^{-1} in the region near the outlet of the air knife (Figure 4). Air velocity decreases along the wall from the outlet of the air knife to the far end of the transport mirror with the length of 400 mm. Figure 6 is the velocity of pressure outlet D. The negative pressure of outlet D can steer flow. The air speed is 115 m s^{-1}, which approximately decreases along the y-axis. These high speeds blow the particles away from the transport mirror.

3.2. Debris trajectories

Dust particles with four different sizes were considered. For each size, 200 dust particles were evenly placed on the transport mirror up-surface L from the left to the right, and a numerical simulation was performed based on the numerical model and settings stated in Section 2. The removal trajectories of dust particles were obtained (Figure 7). Figure 7(a)–(d) correspond to the particles with sizes of 10, 20, 40 and 80 μm, respectively. Each curve represents one type of movement tendency of some particles. There are a total of three kinds of movement trends: type 1, type 2 and type 3. Besides this, curves of type 1 in Figure 7(a) and (b), and curves of type 3 in Figure 7(c) and (d), are also corresponding in respect of the maximum trajectories of particles in four sizes.

The simulation results demonstrate that the motion displacements of the majority of the particles are small in the vertical direction, almost slipping along the upper surface of the mirror (type 2). The y-coordinate of the upper mirror surface is 2 mm; thus the minimum value of Y in Figure 7 is 2 mm. Only 10 particles (type 1 and type 3) have different trajectories compared with the other particles because the static pressure in most of the mirror surface area is greater

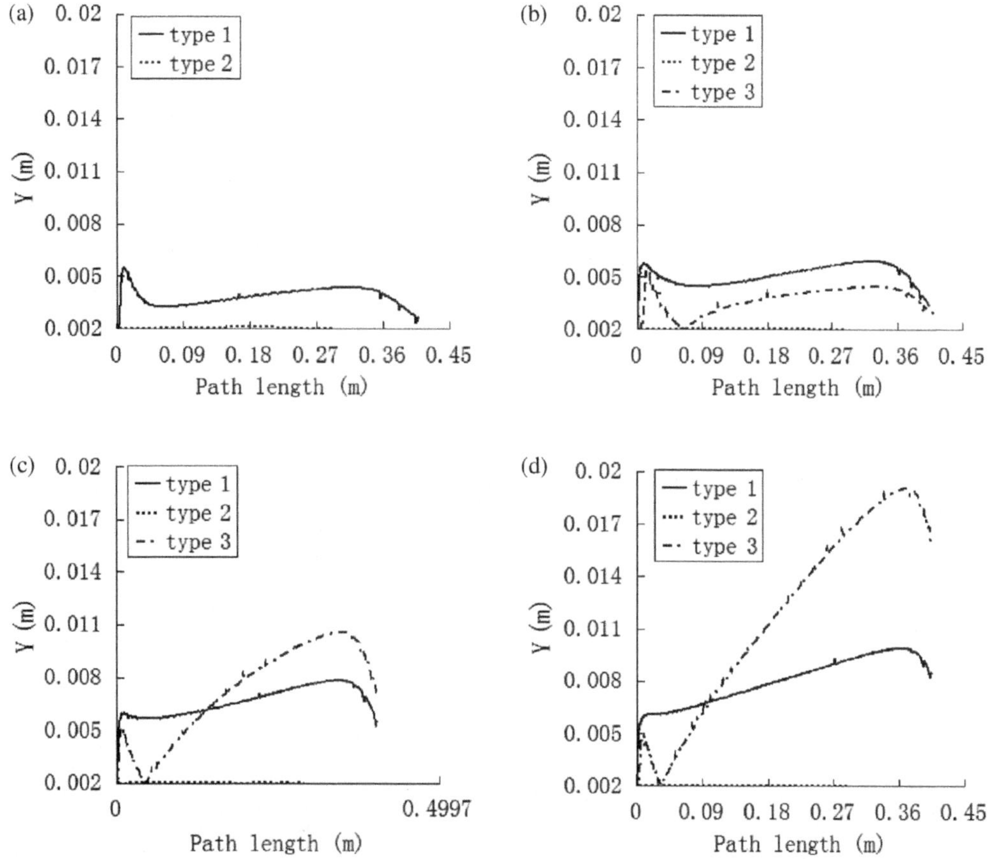

Figure 7. Trajectories of dust particles in four sizes. The sizes of the dust particles in (a)–(d) correspond to 10, 20, 40 and 80 μm.

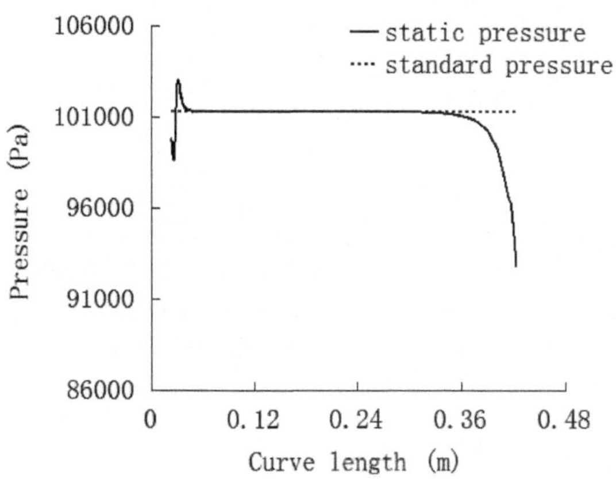

Figure 8. Hydrostatic transmission curve on the mirror surface.

than or equal to the standard atmospheric pressure, except for the area that is 0 to 6 mm away from the mirror edge closing to the air knife (Figure 8). The pressure is negative in this area, and 10 particles labeled 0 to 9 are distributed in this area.

The Y values of particles increase rapidly, decrease, then increase along the path, and finally decrease (type 1). This event is due to the fact that the region of the 10 particles instantly becomes a negative pressure zone when the air knife is blowing. Then, turbulences (Figure 5) occur and drive the particles to instantly generate upward movements. With the increase in the Y values, the pneumatic forces weaken, gravity gradually dominates, and the velocities decrease to zero. During this period, the Y values reach maximum. Then, the particles move along the negative direction of the y-axis, the Y values decrease, and the vortices are strengthened. The velocities of the particles along the negative direction of the y-axis are gradually reduced to zero when the pneumatic forces are large enough to overcome gravity. Then, the velocity direction reverses, and the Y values increase. Gradually, the Y values decrease because the negative pressure of the outlet D can steer flow.

With the increase in particle sizes, the numbers of particles of type 3 that begin to reach the maximum Y values, hit the mirror surface and finally rebound off the surface also increase. This phenomenon is due to insufficient pneumatic forces to overcome the additional gravity caused by the increase in the particle sizes in the falling phase, and more particles crash into the mirror surface and rebound off. The maximum Y value is 19.1 mm and the size of the dust is 80 μm at the moment when the four sizes of particles are present.

Similar results are also obtained when numerical simulations are conducted for stainless steel and aluminum particles

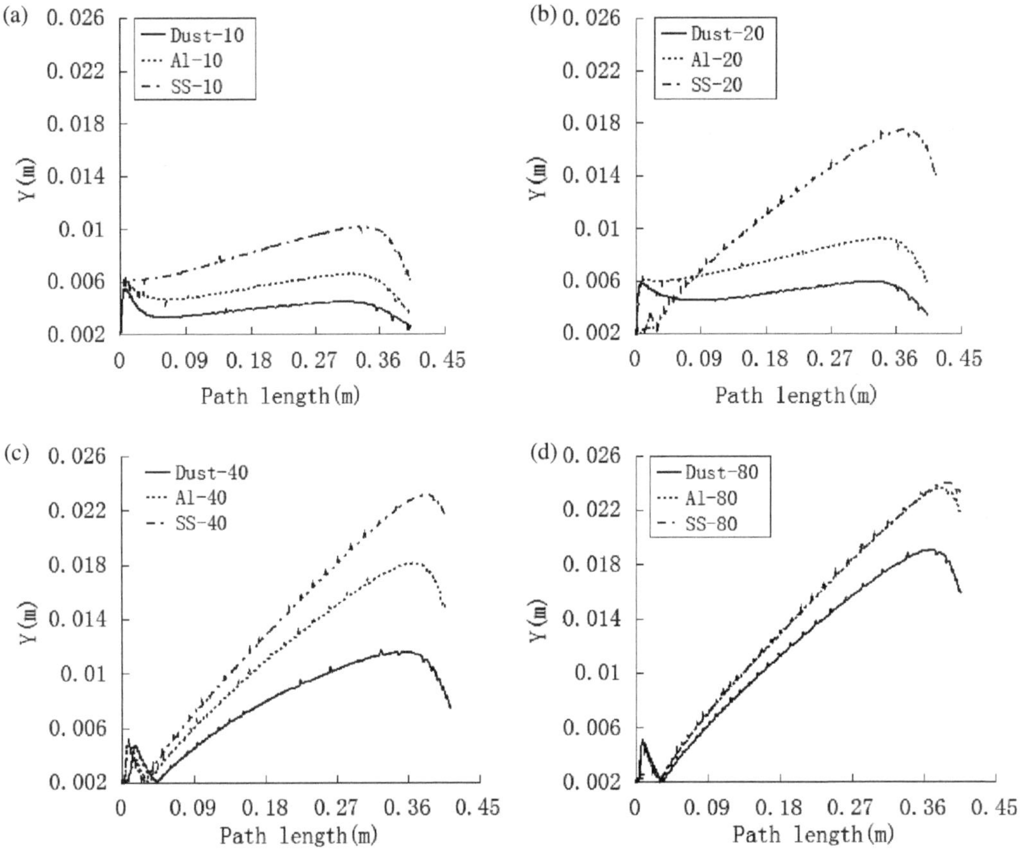

Figure 9. Maximum Y values of three particle types in four sizes. The sizes of the particles in (a)–(d) correspond to 10, 20, 40, and 80 μm.

in four sizes. The corresponding maximum Y values are 23.6 and 24 mm, and their sizes are both 80 μm.

Comparing the maximum Y values of the dust, the aluminum particles and the stainless steel particles, the greater the particle density, the greater the maximum Y value when the particles are of the same size (Figure 9).

4. The device

A device was designed based on the above model of the air knife (Figure 1). The device had the same boundary conditions and associated settings as those mentioned above. The device could capture and collect contaminant particles (Figure 10). The cover was connected to a vacuum cleaner, which provided a negative pressure. The device performance was tested using dust particles with sizes that ranged from 25 to 50 μm and 50 to 100 μm as the contaminants, because most contaminant particles on the transport mirror are dust.

Ordinary glass with the size of 300 mm × 400 mm was used as a substitute for the transport mirrors. The glass was divided into four pieces to facilitate the measurement of the number of dust particles using a microscope. Each piece had the same size of 150 mm × 200 mm. Circular measurement areas of 4×4 with a diameter of 10 mm were set. Positioning centers represented by '+' were distributed on the opposites

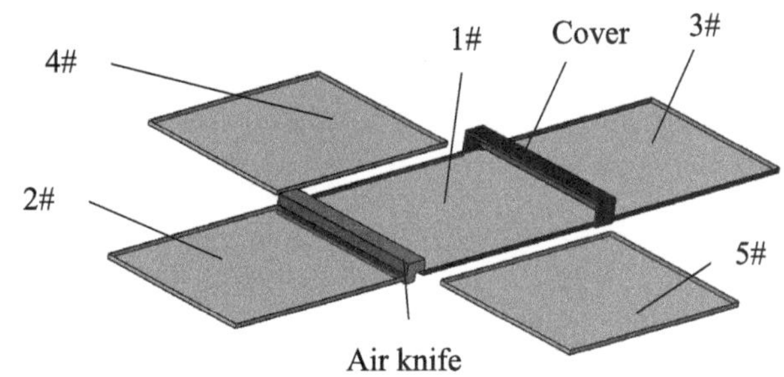

Figure 10. Layout of the device. #1, #2, #3, #4, and #5 are transport mirrors.

of the measurement surfaces (Figure 11). The total number of particles measured in the 16 regions represented the number of particles on the mirror surface to reduce experimental errors. Four clean mirrors surrounded mirror #1 to monitor and judge the moving trend of the dust particles during air blowing. The air knife used was a commercial air knife type 12014s[9]. The distance S from the upper mirror surface to the upper air knife surface was 3 mm (Figures 1 and 10). The inlet pressure was 9 atm. The changes in the number of dust particles on the surfaces of the mirrors before and after using the device were determined. The cover provided a negative pressure of 0.8 atm during blowing. Experiments were

Table 1. The numbers of particles before and after blowing.

No. of the mirrors	Size of the particles (μm)	Without debris collector: Number of particles before blowing	Without debris collector: Number of particles after blowing	With debris collector: Number of particles before blowing	With debris collector: Number of particles after blowing
#1	25–50	612	97	495	79
	50–100	632	55	668	43
#2	25–50	0	12	0	15
	50–100	0	7	0	4
#3	25–50	0	20	0	15
	50–100	0	8	0	2
#4	25–50	0	15	0	9
	50–100	0	4	0	2
#5	25–50	0	29	0	6
	50–100	0	9	0	2

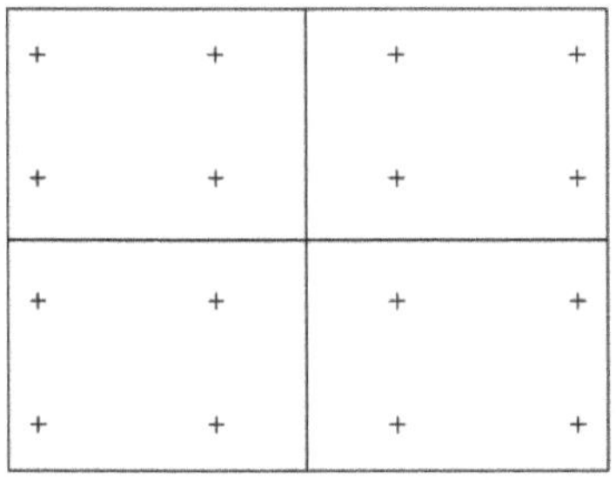

Figure 11. Sample of the transport mirror.

conducted in an ultra-clean environment with a cleanliness level of class 1000.

The experimental results are shown in Table 1. When a debris collector is not used, the numbers of particles in the particle size ranges of 25–50 μm and 50–100 μm are reduced from 612 to 97 and from 632 to 55 respectively after blowing. The efficiency of the air knife clearance of dust particles in the two ranges is 84% and 91%, respectively. These percentages indicate that the air knife can efficiently remove stainless steel particles. At this time, 12% and 4% of the dust particles in the two size ranges are transferred to the surfaces of #2, #3, #4 and #5 transport mirrors. Many particles on the surface of transport mirror #1 are transferred to the surrounding environment in the blowing process.

However, the numbers of particles in the particle size ranges of 25–50 μm and 50–100 μm are reduced from 495 to 79 and from 668 to 43 respectively after blowing when a debris collector is used. This result indicates that the removal efficiency of dust particles in the two ranges is 84% and 94%, which is better than the case without a debris collector. Therefore, using a debris collector can improve removal efficiency. Additionally, 9% and 1% of the dust particles in the two size ranges transfer to the surfaces of #2, #3, #4 and #5 transport mirrors, and these percentages are lower than the case without the debris collector. This result suggests that the debris collector can collect dust particulates and reduce the probability that particles transfer to the surrounding environment, and partly validates the simulation results of the debris-removal trajectories.

5. Conclusions

In the present study, numerical simulations of debris-removal trajectories on the transport mirrors of HPLSs are conducted using Fluent commercial software. The trajectories of contaminant particles of different sizes and types on the transport mirror surface are determined. A device is built based on the simulation results to efficiently capture and collect debris from the surface of the mirror online. Ultimately, debris contamination of other optical components is avoided, cleaning time is shortened and the cleanliness of the mirrors is ensured. Only mirrors laid horizontally have been considered in the present work. Although many transport mirrors are laid non-horizontally in practical applications, more studies are required to understand the debris-removal trajectories of horizontal mirrors. Meanwhile, the cover should also be optimized to capture and efficiently collect debris particles, and further verify the simulation results of the debris trajectories.

Acknowledgements

The authors would like to acknowledge Dr Mingying Sun for modifying and proofreading the manuscript. Grants from the Chinese and Israeli cooperation project on high-power laser technology (2010DFB70490) supported this study.

References

1. M. A. Norton, C. J. Stolz, E. E. Donohue, W. G. H. K. Listiyo, J. A. Pryatel, and R. P. Hackel, Proc. SPIE **5991** (2005).
2. W. H. Gourdin, E. Dzenitis, D. Martin, K. Listiyo, G. Sherman, W. Kent, R. Butlin, C. J. Stolz, and J. Pryatel, Proc. SPIE **5647**, 107 (2004).

3. B. Wang, M. Wang, M. Zhu, X. Chen, and W. Wu, Adv. Mater. Res. **765–767**, 2288 (2013).
4. C. J. Stolz, Proc. SPIE **6834**, 2 (2007).
5. C. J. Stolz, LLNL-CONF-406214 (2008).
6. W. Ruijin, Z. Kai, and W. Gang, *The Base and Application Examples of Fluent Technology* (Tsinghua University Press, Beijing, 2007).
7. H. Zhanzhong, *Fluent—Simulation Examples and Analysis of Fluid Engineering* (Beijing Institute of Press, Beijing, 2009).
8. W. Fujun, *Computational Fluid Dynamics Analysis—CFD Software Principles and Applications* (Tsinghua University Press, Beijing, 2004).
9. http://item.taobao.com/item.htm?spm=a230r.1.14.41.BQzBJJ&id=38566407374&ns=1&abbucket=13#detail.

51.5 W monolithic single frequency 1.97 μm Tm-doped fiber amplifier

Xiong Wang, Pu Zhou, Xiaolin Wang, Hu Xiao, and Lei Si

College of Optoelectronic Science and Engineering, National University of Defense Technology, Changsha 410073, China

Abstract

We demonstrate a monolithic single frequency Tm-doped fiber amplifier with output power of 51.5 W. A single frequency fiber laser at 1.97 μm is amplified by a cascaded master oscillator power amplifier (MOPA) system with all-fiber configuration. The optical-to-optical conversion efficiency of the main fiber amplifier is 45%. No amplified spontaneous emission (ASE) or stimulated Brillouin scattering (SBS) effect is observed in the fiber amplifier. The output power could be further scaled by launching more pump power.

Keywords: Single frequency; Tm-doped fiber laser; fiber MOPA; monolithic configuration.

1. Introduction

Single frequency lasers near 2 μm, which play a significant role in many applications such as coherent LIDAR, coherent beam combining, nonlinear frequency conversion, medicine, atmospheric sensing and spectroscopy, have attracted intense interest and have been widely investigated[1–12]. In recent years, 2 μm single frequency fiber lasers have undergone rapid development[11–22] since fiber lasers have outstanding advantages including high efficiency, considerable compactness, convenience for thermal management, high beam quality and excellent stability. To realize single frequency fiber laser near 2 μm, distributed Bragg reflector (DBR)[15] and distributed-feedback (DFB)[12,19] techniques have been employed, the output power is limited and should be further amplified for practical applications. At present, most of the reported high power single frequency fiber lasers near 2 μm are achieved based on bulk configuration[3,4]. Meanwhile, monolithic fiber lasers based on all-fiber configuration possess the advantages of being more stable and practical, so it is really attractive to investigate and develop single frequency fiber lasers near 2 μm employing monolithic fiber configuration[20]. An admirable previous work has been done by Gapontsev *et al.*, who demonstrated a 20 W monolithic single frequency fiber amplifier at 1.93 μm in 2007[18].

In this paper, we demonstrate a monolithic single frequency fiber amplifier at 1.97 μm based on master oscillator power amplifier (MOPA) configuration. The maximum output power has reached 51.5 W with slope efficiency of 45%. There was no amplified spontaneous emission (ASE) or stimulated Brillouin scattering (SBS) effect observed in the amplifier, which means the output power can be improved further if the pump power is increased.

Correspondence to: Xialong Wang, College of Optoelectronic Science and Engineering, National University of Defense Technology, Changsha 410073, China. Email: wangxiong23@gmail.com

2. Experimental setup

In the MOPA system, the single frequency seed fiber laser based on ultrashort cavity[16] operates at 1.97 μm and has an output power of 35 mW, and the linewidth is less than 100 kHz. Figure 1 shows the two-stage Tm-doped fiber pre-amplifier. The first stage of the pre-amplifier consists of a 1550 nm pump laser, a 1550/2000 wavelength division multiplexer (WDM), an isolator (ISO) and a piece of 2.5 m long single cladding Tm-doped fiber (NA is 0.15) with core diameter of 9 μm and cladding diameter of 125 μm. The second stage of the pre-amplifier employs a $(2 + 1) \times 1$ signal–pump combiner and two 793 nm laser diodes (LDs). The amplified 2 μm laser signal was injected into a piece of 8 m long double cladding Tm-doped fiber (DC-TDF) with core diameter of 10 μm and inner cladding diameter of 125 μm. The Tm-doped fiber's NA is 0.15 and the absorption coefficient at 793 nm is 3 dB/m. The fiber pre-amplifier finally amplified the seed laser's power to 3 W.

The main fiber amplifier's configuration is shown in Figure 2. A $(6+1) \times 1$ signal–pump combiner was employed to combine the pre-amplified 2 μm signal laser and the 793 nm pump laser together. The 3 m long large mode area (LMA) DC-TDF with core and inner cladding diameters of

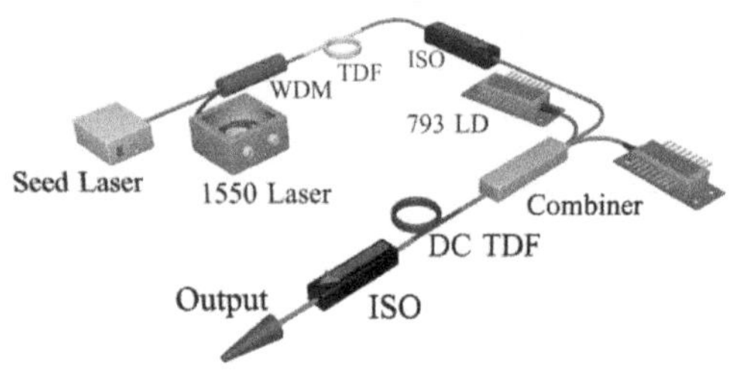

Figure 1. Schematic diagram of the Tm-doped fiber pre-amplifier. WDM, wavelength division multiplexer; TDF, Tm-doped fiber; ISO, isolator; DC-TDF, double cladding Tm-doped fiber.

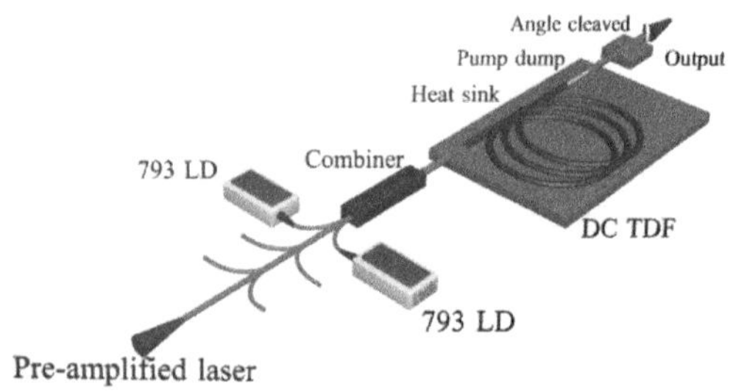

Figure 2. Schematic diagram of the main fiber amplifier.

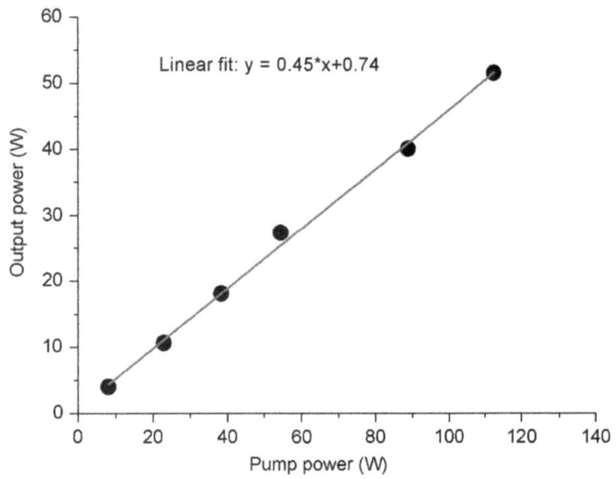

Figure 3. Single frequency fiber amplifier output with linear fit to slope efficiency of 45%.

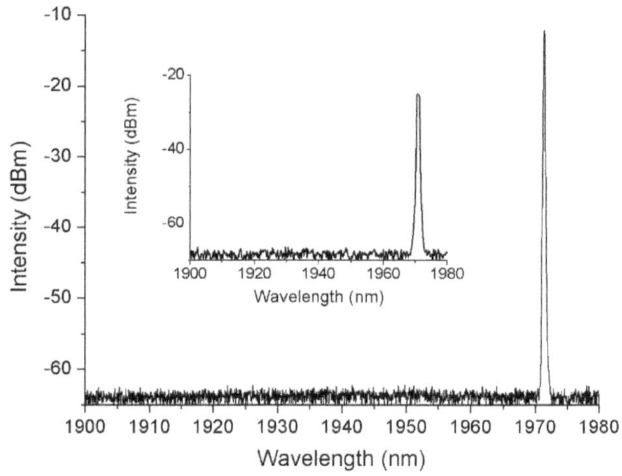

Figure 4. Spectrum of the single frequency fiber amplifier. Inset: spectrum of the seed laser.

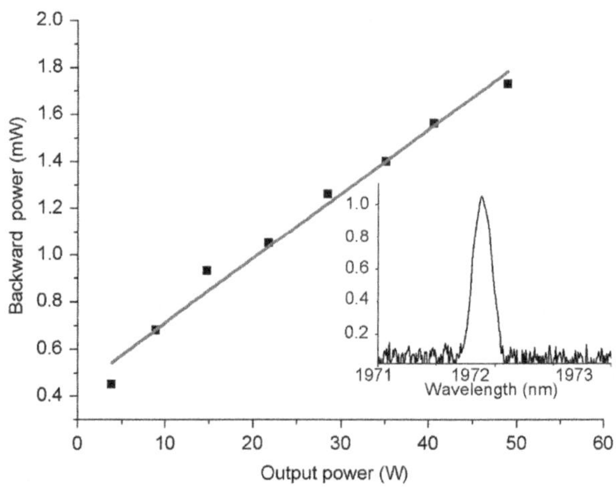

Figure 5. Backward power of the single frequency fiber amplifier. Inset: spectrum of the backward power.

25 and 250 μm acted as the gain medium. The absorption coefficient of the LMA DC-TDF at 793 nm is about 8 dB/m. The output end of the combiner was fusion spliced to one end of the DC-TDF and coated with low refractive index polymer. A piece of 0.5 m 25/250 passive fiber was fusion spliced to the other end of the active fiber. The fusion splice joint of the active and passive lasers was covered by gel with high refractive index to dump the unabsorbed pump power. Water-cooled heat sinks were employed to protect the fusion splice joint and active fiber from thermally-induced damage. The active fiber is coiled within a diameter of less than 12 cm to ensure good beam quality[23,24]. The output end of the passive laser was angle cleaved to avoid the unwanted reflection. The fusion splice joints in the MOPA system make it a really monolithic fiber configuration. The output power was measured at the angle cleaved end and the spectrum was recorded by an optical spectrum analyzer (OSA, YOKOGAWA) with resolution of 0.05 nm.

3. Results and discussion

The output of the single frequency fiber amplifier is shown in Figure 3. Only two of the six ports of the combiner were employed to launch 793 nm pump power into the active fiber. The maximum output power reaches 51.5 W when the pump power is about 112 W. The slope efficiency is 45% with respect to the launched pump power. One can find out that the output of the fiber amplifier is far from saturated, and more output power is realizable if the pump power is increased.

Figure 4 depicts the spectrum of the single frequency fiber amplifier with output power of 51.5 W. The single frequency fiber amplifier's central wavelength is 1972.13 nm. No ASE or parasitic lasing is observed from the spectrum. The spectrum of the single frequency seed laser was inserted into Figure 4. There are no evident differences observed between the spectra of the seed laser and the amplifier.

The backward power of the single frequency MOPA system was measured at a spare port of the signal–pump combiner in Figure 2[25,26]. The results are shown in Figure 5, from which we can find out that the backward power increases regularly as the output power mounts up, and the maximum backward power does not exceed 1.8 mW when the output power reaches 51.5 W. The spectrum of the backward power shows that there is no SBS existing in the amplifier. These results mean that the laser power is below the SBS threshold in the system, and the output power is not limited by the nonlinear effects in the fiber amplifier. Actually, the fiber length and core diameter in this experiment are nearly the same with those of Goodno *et al.*[3], so we can coarsely estimate that the SBS threshold of our system is above several hundred watts. Much more parameters, such as bend radius, temperature gradient and Brillouin gain spectrum, should be considered to precisely estimate the SBS threshold in theoretical analysis.

4. Conclusion

In conclusion, we experimentally demonstrate a monolithic single frequency fiber amplifier at 1.97 μm. The system has all-fiber configuration and the maximum output power has reached 51.5 W. The slope efficiency with respect to the launched pump power is 45%. The spectrum of the fiber amplifier was measured and no ASE was observed. From the backward power results we have verified that there was no SBS effect in the amplifier. Higher output power of monolithic single frequency fiber amplifier near 2 μm is realizable if more pump power is available.

References

1. D. Y. Shen, W. A. Clarkson, L. J. Cooper, and R. B. Williams, Opt. Lett. **29**, 2396 (2004).
2. B. Yao, X. Duan, D. Fang, Y. Zhang, L. Ke, Y. Ju, Y. Wang, and G. Zhao, Opt. Lett. **33**, 2161 (2008).
3. G. D. Goodno, L. D. Book, and J. E. Rothenberg, Opt. Lett. **34**, 1204 (2009).
4. L. Pearson, J. W. Kim, Z. Zhang, M. Ibsen, J. K. Sahu, and W. A. Clarkson, Opt. Express **18**, 1607 (2010).
5. Y. Liu, J. Liu, and W. Chen, Chin. Opt. Lett. **9**, 090604 (2011).
6. H. Lv, P. Zhou, H. Xiao, X. Wang, and Z. Jiang, Chin. Opt. Lett. **10**, 051403 (2012).
7. J. Li, S. H. Yang, H. Y. Zhang, W. Xie, and C. M. Zhao, Laser Phys. **20**, 1806 (2010).
8. C. Gao, M. Gao, Y. Zhang, Z. Lin, and L. Zhu, Opt. Lett. **34**, 3029 (2009).
9. C. Gao, Z. Lin, M. Gao, Y. Zhang, L. Zhu, R. Wang, and Y. Zheng, Appl. Opt. **49**, 2841 (2010).
10. P. Zhou, X. Wang, Y. Ma, K. Han, and Z. Liu, Laser Phys. **21**, 184 (2011).
11. W. Shi, E. B. Petersen, D. T. Nguyen, Z. Yao, A. Chavez-Pirson, N. Peyghambarian, and J. Yu, Opt. Lett. **36**, 3575 (2011).
12. J. Geng, Q. Wang, T. Luo, S. Jiang, and F. Amzajerdian, Opt. Lett. **34**, 3493 (2009).
13. S. Agger, J. H. Povlsen, and P. Varming, Opt. Lett. **29**, 1503 (2004).
14. J. Wu, Z. Yao, J. Zong, and A. Chavez-Pirson, Proc. SPIE **7195**, 71951K (2009).
15. Z. Zhang, A. J. Boyland, J. K. Sahu, W. A. Clarkson, and M. Ibsen, IEEE Photon. Technol. Lett. **23**, 417 (2011).
16. J. Geng, J. Wu, and S. Jiang, Opt. Lett. **32**, 355 (2007).
17. M. Meleshkevich, A. Drozhzhin, N. Platonov, D. Gapontsev, and D. Starodubov, Proc. SPIE **5709**, 117 (2005).
18. D. Gapontsev, N. Platonov, M. Meleshkevich, M. Shkurikhin, S. Agger, P. Varming, and J. H. Poylsen, *Lasers and Electro-Optics, CLEO 2007. Conference on. IEEE***1–2** (2007).
19. N. Y. Voo, J. K. Sahu, and M. Ibsen, IEEE Photon. Technol. Lett. **17**, 2550 (2005).
20. Q. Fang, W. Shi, K. Kieu, E. Petersen, A. Chavez-Pirson, and N. Peyghambarian, Opt. Express **20**, 16410 (2012).
21. H. Lv, P. Zhou, H. Xiao, X. Wang, and Z. Jiang, Chin. Opt. Lett. **10**, 051403 (2012).
22. W. Guo, D. Shen, J. Long, and F. Wang, Chin. Opt. Lett. **10**, 091406 (2012).
23. J. P. Koplow, D. A. V. Kliner, and L. Goldberg, Opt. Lett. **25**, 442 (2000).
24. L. Li, Q. Lou, J. Zhou, J. Dong, Y. Wei, and J. Li, Chinese J. Lasers **34**, 323 (2007).
25. D. Wenbo, W. Xiaolin, H. Kai, Z. Pu, X. Xiaojun, and S. Bohong, Chinese J. Lasers **38**, 1105009 (2011).
26. D. Yunfeng, Z. Peng, H. Bangcai, P. Rong, and N. Ding, Chinese J. Lasers **36**, 640 (2009).

10

Novel transparent ceramics for solid-state lasers

Hao Yang[1], Jian Zhang[1], Dewei Luo[2], Hui Lin[2], Deyuan Shen[1], and Dingyuan Tang[1,2]

[1] *School of Physics and Electronic Engineering, Jiangsu Normal University, Xuzhou 221116, China*

[2] *School of Electrical and Electronic Engineering, Nanyang Technological University, Singapore 639798, Singapore*

Abstract

Recent progress on rare-earth doped polycrystalline YAG transparent ceramics has made them an alternative novel solid-state laser gain material. In this paper, we present results of our research on polycrystalline RE:YAG transparent ceramics. High optical quality YAG ceramics doped with various rare-earth (RE) ions such as Nd^{3+}, Yb^{3+}, Er^{3+}, Tm^{3+}, and Ho^{3+} have been successfully fabricated using the solid-state reactive sintering method. Highly efficient laser oscillations of the fabricated ceramics are demonstrated.

Keywords: RE:YAG; Solid State Lasers; Transparent Ceramics

1. Introduction

Recently, there has been a resurgence of interest in rare-earth (RE) doped YAG (RE:YAG) transparent ceramics as laser gain media, because the optical quality of the ceramics has been greatly improved. It is now comparable to that of single-crystal materials[1,2]. Compared with single-crystal-based laser materials, apart from possessing exactly the same spectroscopic properties, ceramic laser materials also have the advantages of ease to fabricate large size and complicated structures, short fabrication time, and the possibility of mass production.

Polycrystalline ceramics became available for optical applications in 1962, when Coble first prepared translucent Al_2O_3 ceramics[3]. Since then, various kinds of transparent ceramics have been developed for optical and related applications. In 1995, Ikesue *et al.* achieved highly efficient continuous-wave (CW) laser emission in a Nd:YAG transparent ceramic[4–6]. They fabricated transparent polycrystalline Nd:YAG ceramics by a solid-state reactive sintering method using high-purity powders as the starting materials. A laser output of about 70 mW was obtained with a slope efficiency of 28%. Their result clearly demonstrated the potential high functionality and flexibility of the laser ceramics. Since 1999, Yagi's group in Konoshima Chemical and Ueda's group in the University of Electro-communications, Japan, have developed highly transparent ceramic laser materials in different ways: using nanopowder technology and using a vacuum sintering method. Scattering losses were greatly reduced in Konoshima's YAG transparent laser ceramics[7,8]. In 2000 and 2001, Lu *et al.* reported the operation of diode pump Nd:YAG ceramic lasers with ceramic samples made by Konoshima Chemical. The output power was gradually increased from 31 W to 110 W[8–11]. Extensive studies have shown that the thermal and mechanical properties of the ceramics are better than those of YAG single crystals[12,13]. In 2006, the Lawrence Livermore National Laboratory (LLNL) succeeded in achieving a world record high power of 67 kW emission from its solid-state heat capacity laser system using a large-scale transparent Nd:YAG ceramic with samarium edge cladding supplied by Konoshima Chemical[14]. Very recently, the emission of a Nd:YAG ceramic laser system with output power beyond 100 kW was further demonstrated[15–17].

In the past years, most of the research work on transparent laser ceramics has focused on Nd:YAG ceramics. Recently, some research groups have also paid attention to developing other rare-earth doped YAG ceramics, such as Yb:YAG, Er:YAG, Ho:YAG, the sesquioxide ceramics (Y_2O_3, Lu_2O_3, Sc_2O_3), and composite ceramics. Dong *et al.* reported continuous-wave and passively Q-switched microchip lasers made of Yb:YAG ceramics[18]. Heavily doped Yb:YAG ceramics are more suitable for thin disk lasers than Yb:YAG single crystal. Sesquioxides possess higher thermal conductivities than YAG. For high-power lasers, thermal conductivity is an important parameter of the laser gain media. Kim *et al.* reported the fabrication of Yb:Lu_2O_3 transparent ceramic by a hot isostatic pressing (HIP) approach[19]. They successfully demonstrated efficient lasing with fabricated Yb:Lu_2O_3 ceramic. An output power of more than 16 W at 1080 nm was obtained with a slope efficiency of 74%. Ikesue *et al.* demonstrated laser-

Correspondence to: Dingyuan Tang, 101 Shanghai Road, Tongshan District, Xuzhou 221116, Jiangsu, China. Email: edytang@ntu.edu.sg

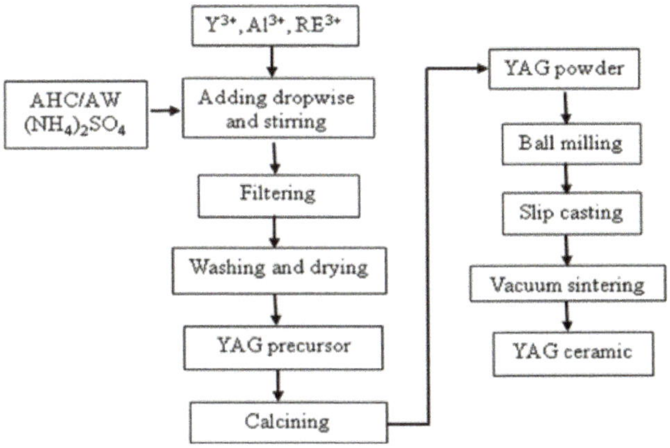

Figure 1. Fabrication process of YAG ceramic by co-precipitation and vacuum sintering method.

quality composite transparent ceramics by using a layer by layer approach[2]. Messing *et al.* developed a colloidal tape casting process to make composite laser ceramics. A ceramic rod with segments of undoped YAG, 0.25% Er:YAG, and 0.5% Er:YAG was fabricated. With such a composite ceramic rod of length 62 mm, lasing with a slope efficiency of 27% was achieved, compared with the 39% slope efficiency achieved on a commercial 45 mm long 0.5% Er:YAG single crystal rod[20].

2. Fabrication of YAG laser ceramics

Suitable fabrication methods are required to obtain YAG transparent ceramics of >99.99% density. So far, two different approaches have been successfully implemented to produce high-quality laser ceramics. Figure 1 shows the fabrication process for YAG transparent ceramics by the co-precipitation and vacuum sintering method[21]. Konoshima Chemical succeeded in fabricating high-quality transparent YAG-based laser ceramics using this method. The method is based on co-precipitation of a metal nitrate solution with NH_4HCO_3. Aqueous solutions of aluminum, yttrium and rare-earth nitrates are mixed together. The mixed aqueous solution is added dropwise and mixed with an aqueous solution of NH_4HCO_3. Steps of filtration and washing with water are repeated several times, and the resulting material is then dried for 2 days at ~120°C. The precursor thus obtained is calcined at 1200°C, and then the YAG powder is obtained. This powder is milled with solvent, binder, and dispersion medium for 24 hours. The milled slurry is put in a gypsum mold and dried to obtain the desired form, or the slip casting forming method is used to obtain a high-density and uniformly green body. After removing the organic component by calcining, such material is vacuum sintered at 1700°C; then highly transparent Nd:YAG ceramics can be obtained.

Ikesue reported laser-quality polycrystalline Nd:YAG transparent ceramics made using a reactive sintering method. Figure 2 shows the fabrication process of the method[4].

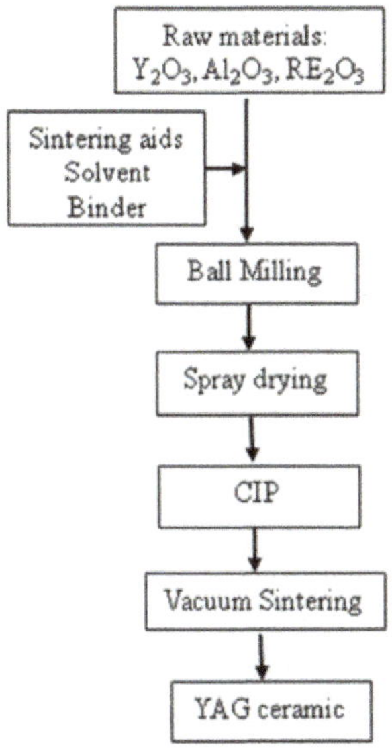

Figure 2. Fabrication process of YAG ceramic by reactive sintering method.

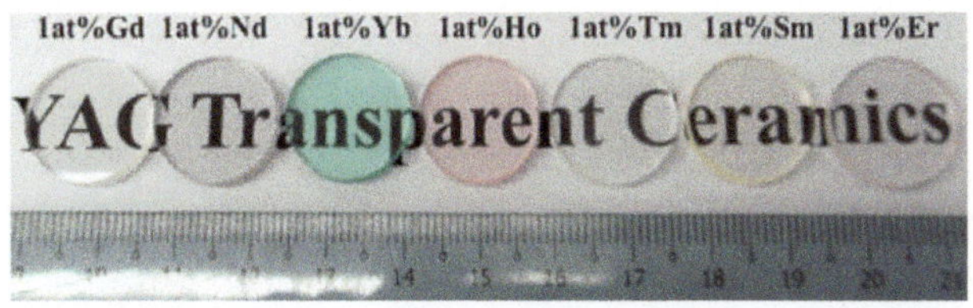

Figure 3. Photograph of different rare-earth doped YAG transparent ceramics fabricated in-house.

Commercial α-Al_2O_3, Y_2O_3, and RE_2O_3 powders are selected as the starting materials. The powders are weighed precisely according to the chemical stoichiometric composition and mixed with 99.99% ethanol. The mixed slurry is then ball milled using a planetary milling machine for 15 hours. TEOS is introduced at 0.5 wt% relative to the total powder mass to introduce SiO_2 as a sintering aid. After milling, the powder mixture is dried at 120°C for 24 hours in an oven and then sieved through a 200-mesh screen. After removing organic components by calcining at 800°C for 3 hours, the powders are dry pressed in a stainless steel die at 15 MPa. The green body pellets are further cold isostatically pressed (CIP) at 200 MPa. After CIP, the relative density of the green body is ~53%. The green bodies are sintered in a high temperature vacuum sintering furnace. The sintered pellets are then annealed in air at 1400°C for 15 hours to completely remove internal stress and eliminate the oxygen vacancies. After polishing, the specimen can be used as a laser medium.

3. Properties of the RE:YAG ceramics

Using the reactive sintering method, we have succeeded in fabricating high-quality transparent YAG ceramics doped with different rare-earth ions such as Nd^{3+}, Yb^{3+}, Er^{3+},

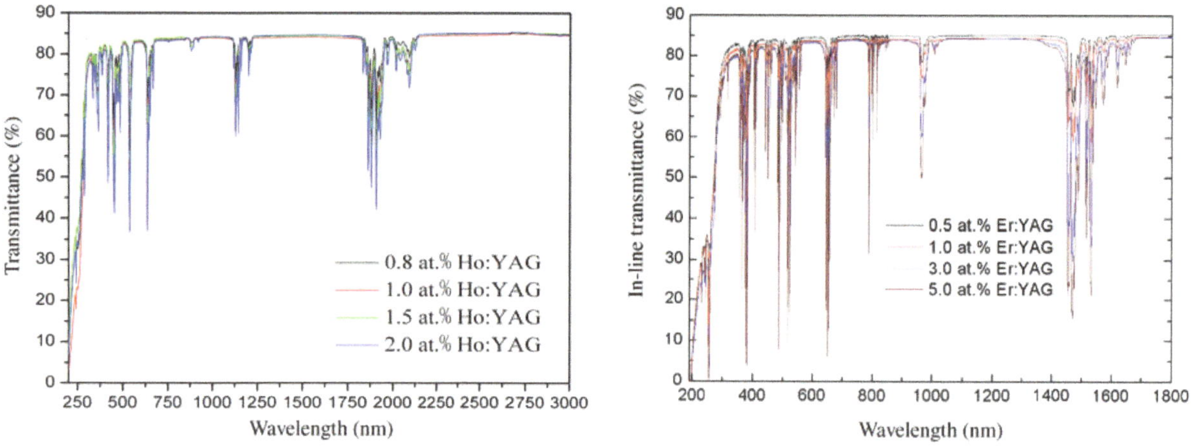

Figure 4. The in-line transmittance of Ho:YAG and Er:YAG transparent ceramics with different doping concentrations.

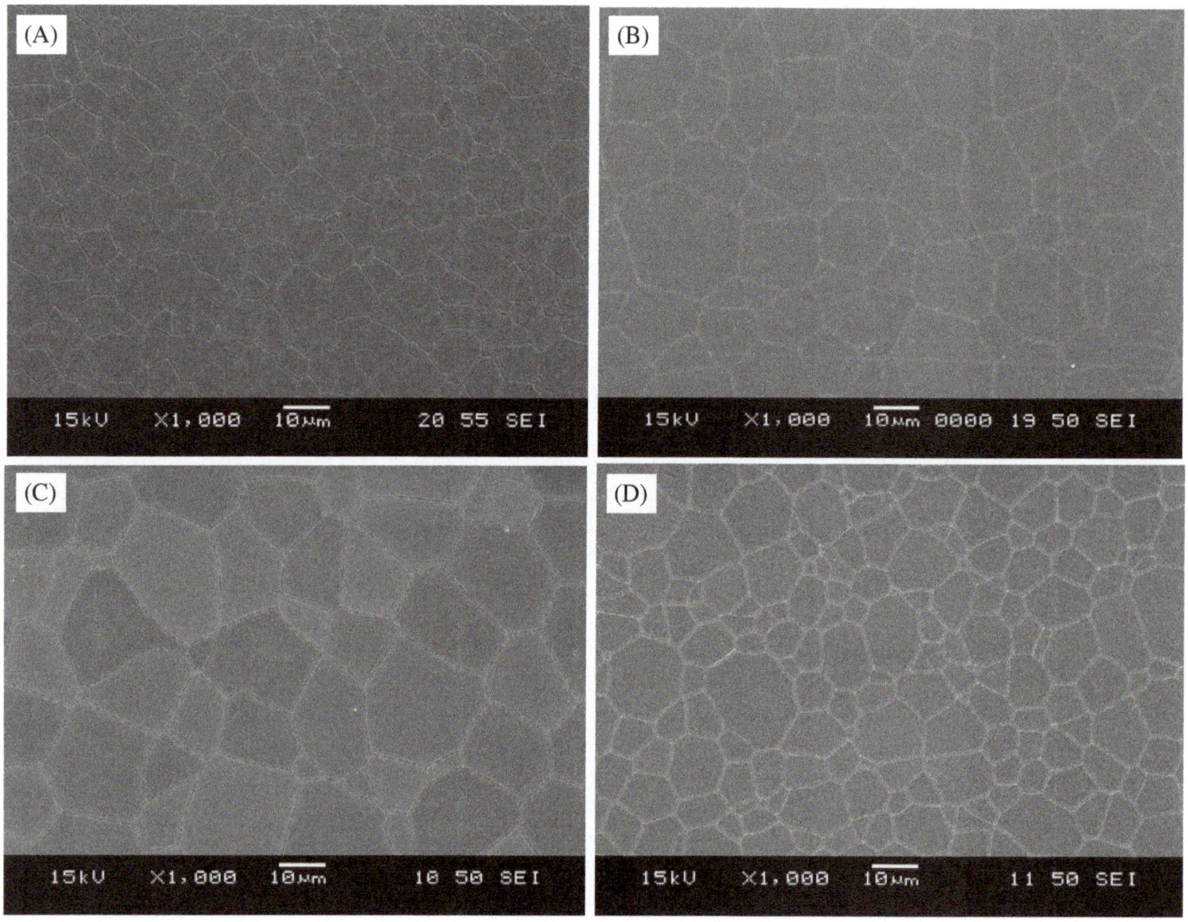

Figure 5. The microstructures of the RE:YAG ceramics fabricated in-house: (A) Er:YAG, (B) Ho:YAG, (C) Tm:YAG, and (D) Nd:YAG.

Ho^{3+} and Tm^{3+} [22–28]. Figure 3 shows a photograph of various rare-earth-ion doped YAG ceramics fabricated in-house. To illustrate the high optical quality of the fabricated YAG ceramics, as examples, we have shown in Figure 4 the optical transmission of Er^{3+} and Ho^{3+} doped YAG under different doping concentrations. The transmittance of all the ceramics at 1100 nm is above 84%, which is very close to the theoretic transmittance.

Grain sizes, grain-boundary phases, and pores are usually considered as the main factors affecting the optical quality of transparent ceramics. Figure 5 shows the microstructures of YAG ceramics doped with different types of rare-earth ion after thermal etching at 1500°C for 1 hour. The transparent

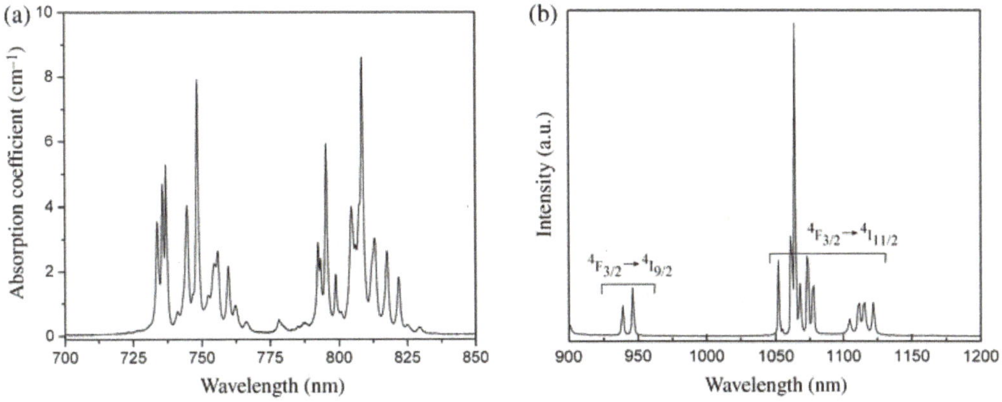

Figure 6. The absorption and emission spectra of the fabricated Nd:YAG ceramics.

ceramics are composed of grains with an average size ~10 µm. It can be seen that the grain size changed very little with the different types of rare-earth-ion doping. No grain-boundary phases and pores can be observed in the microstructures of the RE:YAG ceramics.

4. Laser performance of the fabricated ceramics

There has been extensive research on the lasing performance of Nd:YAG and Yb:YAG ceramics, not only because high optical quality Nd:YAG ceramics were successfully produced, but also because their single-crystal counterparts have been widely used in commercial solid-state laser systems. Nd^{3+} or Yb^{3+} doped lasers emit in the 1.0 µm wavelength regime. High-power solid-state lasers operating in this wavelength regime have found widespread applications in various fields ranging from industrial materials processing to medical devices. Compared with Nd^{3+} and Yb^{3+} ion doped YAG ceramics, Er^{3+}, Tm^{3+}, and Ho^{3+} doped YAG ceramics emit in the 1.5–2.0 µm wavelength regime. 1.5–2.0 µm laser emission is considered eye-safe, as the cornea absorbs wavelengths greater than 1.4 µm. Light with shorter wavelengths can reach the more sensitive retina. Therefore, the potential of eye damage by the scattered laser beam is much lower for lasers operating at wavelengths above 1.4 µm. Eye-safe solid-state lasers have applications such as in the range finders, remote sensing, and medical devices.

Using the in-house fabricated RE:YAG transparent ceramics, we designed ceramic lasers operating at different wavelengths. Our main purpose is to study the laser performance of the fabricated laser ceramics.

4.1. 1.0 µm *laser oscillation*

Figure 6 shows the absorption and emission spectra of a 1.0 at.% Nd:YAG ceramic fabricated in-house. Compared with those measured with a Nd:YAG single crystal, no measurable difference could be identified. The absorption peak of the ceramic is located at 808 nm, and the peak absorption coefficient is measured as 8.64 cm^{-1}, which corresponds

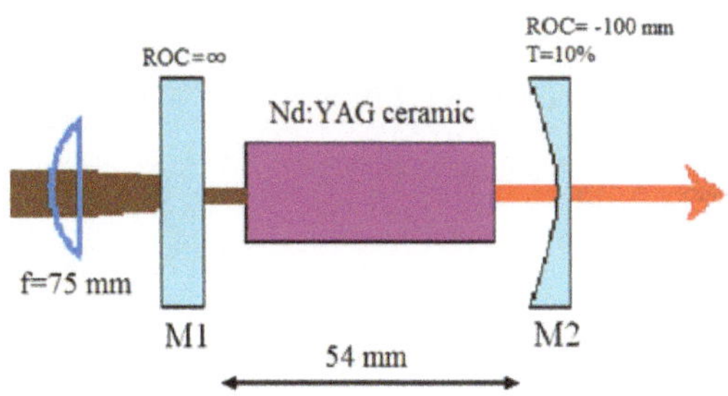

Figure 7. A schematic of the Nd:YAG ceramic laser setup.

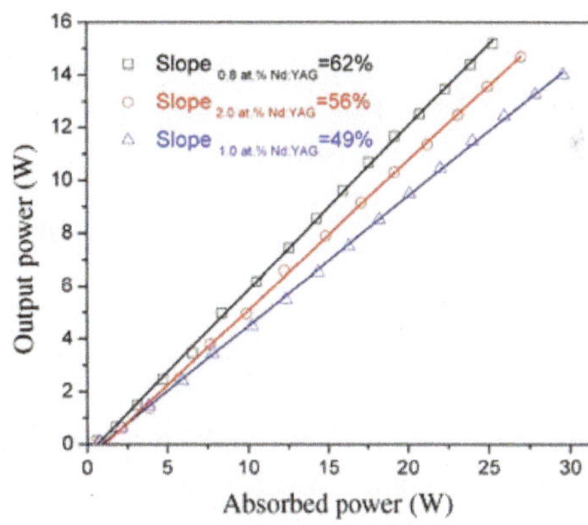

Figure 8. Laser performance of the Nd:YAG ceramic lasers.

to an absorption cross-section of 6.26×10^{-20} cm^2. From 900 to 1200 nm, there are two main emission bands centered at 946 and 1064 nm, which correspond to the $^4F_{3/2} \rightarrow {}^4I_{9/2}$ and $^4F_{3/2} \rightarrow {}^4I_{11/2}$ transitions of the Nd^{3+} ions, respectively. The strongest emission peak is at 1064 nm, which is same as that of Nd:YAG single crystals. Figure 7 shows a schematic of our diode-pumped Nd:YAG ceramic laser setup. The Nd:YAG ceramics were cut into pieces of dimension $3 \times 3 \times 5$ mm^3. Nd:YAG ceramics with three Nd^{3+} doping concentrations (0.8 at.%, 1.0 at.%, and 2.0 at.%) were used. The laser performance of the Nd:YAG ceramics is shown in Figure 8. The best performance was achieved with the 0.8 at.% Nd:YAG ceramic. A maximum output power of

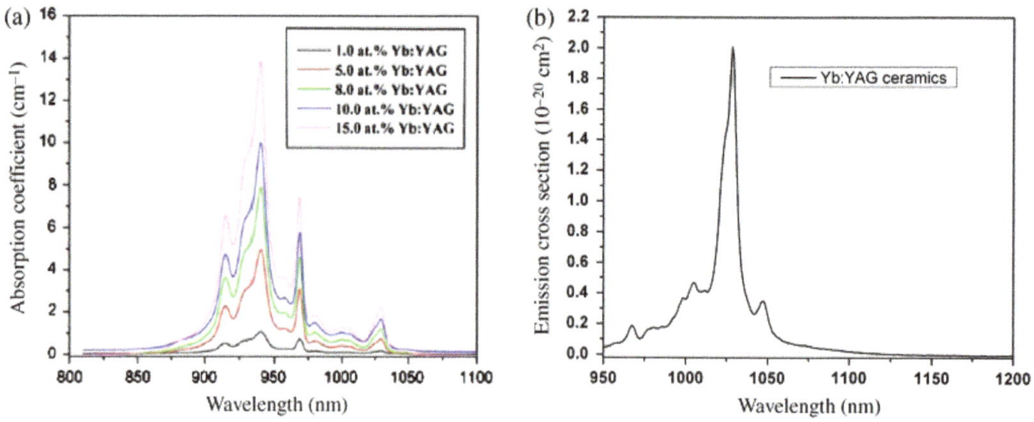

Figure 9. The absorption and emission spectra of the fabricated Yb:YAG ceramics.

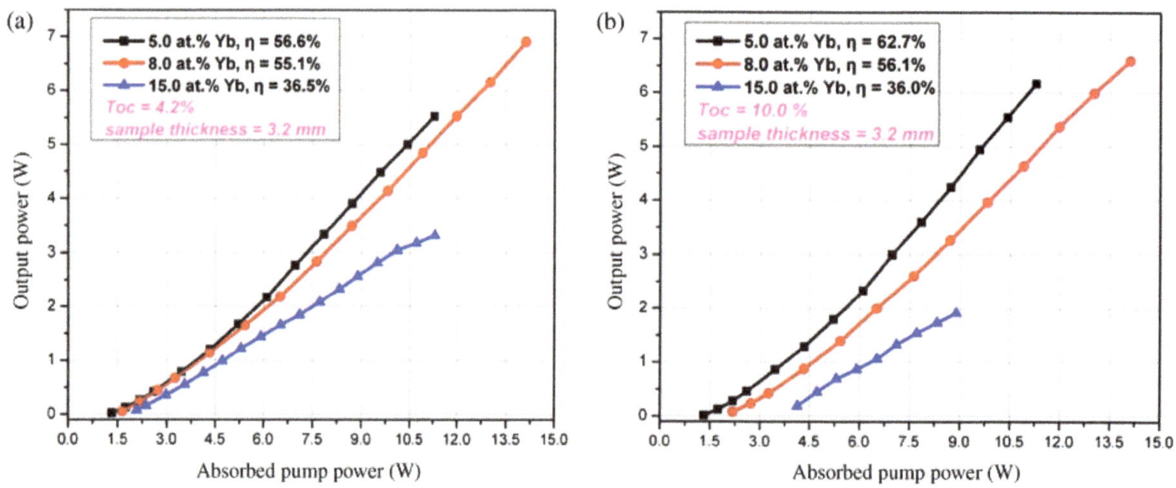

Figure 10. Laser performance of the Yb:YAG ceramic lasers.

15.2 W at 1064 nm was obtained under an absorbed pump power of 25 W. The slope efficiency is 62%, which is very close to that of Nd:YAG single crystals.

The absorption spectra of the fabricated Yb:YAG ceramics under different Yb^{3+} doping concentrations are plotted in Figure 9(a). All the samples have a very strong absorption peak located at 940 nm and two moderate absorption peaks located at 915 and 969 nm, respectively, corresponding to the $^2F_{7/2} \rightarrow {}^2F_{5/2}$ transition of Yb^{3+}. The absorption coefficient at 940 nm increased from 1.06 cm^{-1} under the 1.0 at.% Yb^{3+} doping concentration to 13.87 cm^{-1} under the 15.0 at.% doping concentration. Absorption at 1030 nm is clearly observable, which is attributed to the quasi-three-level nature of the Yb doped gain media. Figure 9(b) shows the emission cross-section of the Yb:YAG ceramics. The peak is at 1028 nm, and it has a value of 2.01×10^{-20} cm^2. The laser performance of the Yb:YAG ceramics under two different output couplings (T_{oc}) was experimentally studied. Yb:YAG ceramics with different doping concentrations (5.0 at.%, 8.0 at.%, and 15.0 at.%) were cut into pieces of dimension $3 \times 3 \times 3.2$ mm^3. The CW laser emission was at 1030 nm. Under $T_{oc} = 4.2\%$, the laser threshold for the 5.0 at.%, 8.0 at.%, and 15.0 at.% Yb:YAG ceramics was 1.31 W, 1.64 W, and 2.08 W, respectively. Under $T_{oc} = 10\%$, it was 1.3 W, 2.16 W, and 4.21 W, respectively, as shown in Figure 10(a) and (b). Due to the reabsorption of the laser, the output power did not increase linearly with the absorbed pump power, but has the tendency that the larger the absorbed pump power, the higher the slope efficiency. With the experimentally available pump power, a maximum output power of 6.9 W was achieved on the 8.0 at.% sample under $T_{oc} = 4.2\%$, while the highest slope efficiency of 62.7% was obtained with the 5.0 at.% Yb:YAG ceramic sample under $T_{oc} = 10\%$. No saturation was observed on all the Yb:YAG ceramic lasers. We believe that, with further optimization of the doping concentration and experimental conditions, even higher output power and slope efficiency could be obtained.

As there is no concentration quenching for the Yb^{3+} doped laser materials, high doping is preferred to achieve high power and high efficiency of Yb^{3+} doped lasers. However, a problem associated with Yb^{3+} doped YAG is that, as the Yb^{3+} doping concentration is increased, the thermal conductivity of the doped YAG drops significantly. To solve the problem, the compound material $Lu_3Al_5O_{12}$

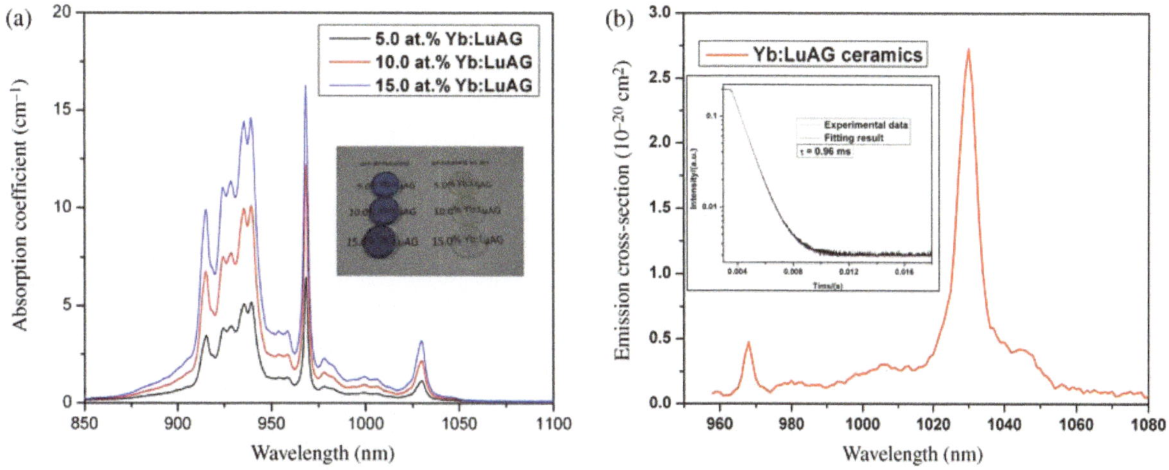

Figure 11. The absorption and emission spectra of the fabricated Yb:LuAG ceramics.

was proposed as an alternative to YAG. $Lu_3Al_5O_{12}$ has high density (6.72 g cm^{-3}), high thermal conductivity (9.6 w/m · K), and a cubic structure. Although the thermal conductivity of LuAG is a little bit smaller than that of YAG, its conductivity decreases only slightly under high Yb doping, which is of benefit to high-power laser operation. We have also fabricated high optical quality Yb:LuAG ceramics with different Yb^{3+} doping concentrations[29] Figure 11(a) shows the absorption spectra of the as-fabricated Yb:LuAG ceramic samples with different Yb^{3+} doping concentrations. A photograph of the Yb:LuAG samples is also shown in the inset of the figure. The samples appear blue in color before annealing. After being annealed in air they become colorless and fully transparent. The absorption spectra of the Yb:LuAG ceramics are similar in profile to those of the Yb:YAG ceramics. However, the highest absorption peak is located at 968.4 nm. For the samples with 5%, 10%, and 15% at.% Yb^{3+} concentrations, the peak absorption coefficients were measured as 6.4 cm^{-1}, 12.2 cm^{-1}, and 16.3 cm^{-1}, respectively. The highest absorption peak of Yb:LuAG is very sharp, with a full width at half maximum (FWHM) of 2.5 nm. The second highest absorption peak is at 940 nm; it is a band consisting of double peaks, centered at 935.6 nm and 939.6 nm, respectively. The two peaks have nearly the same absorption coefficient. The FWHM of the absorption band is 21.6 nm, which favors direct diode pumping with InGaAs laser diodes. The absorption cross-section at the 939 and 968 nm absorption peaks is calculated to be 0.73×10^{-20} cm^2 and 0.91×10^{-20} cm^2, respectively. In addition, there is an absorption peak located at 1030 nm, which is clearly observable on the spectrum. Figure 11(b) shows the emission cross-section and the fluorescent decay curve (inset) for a 5.0 at.% Yb:LuAG ceramic. The emission peak of the Yb:LuAG ceramic is also located at the 1030 nm wavelength, whose emission cross-section is calculated to be 2.7×10^{-20} cm^2 using the Füchtbauer–Ladenburg (F–L) equation. The value is almost 30% higher than that of Yb:YAG (~2.0×10^{-20} cm^2). The fluorescence decay of the Yb^{3+} ions in the LuAG ceramic sample was measured; the results are shown in the inset of Figure 11(b). A single exponential decay curve was observed. It gives a lifetime of 0.96 ms. The CW laser performance of a 5.0 at.% Yb:LuAG ceramic was experimentally investigated under different output couplings. The Yb:LuAG samples were cut into pieces of dimension $3 \times 3 \times 3.2$ mm^3. The result is shown in Figure 12. Under Toc = 50%, a maximum output power of 7.2 W was obtained at an absorbed pump power of 12.4 W, corresponding to a slope efficiency of 65%.

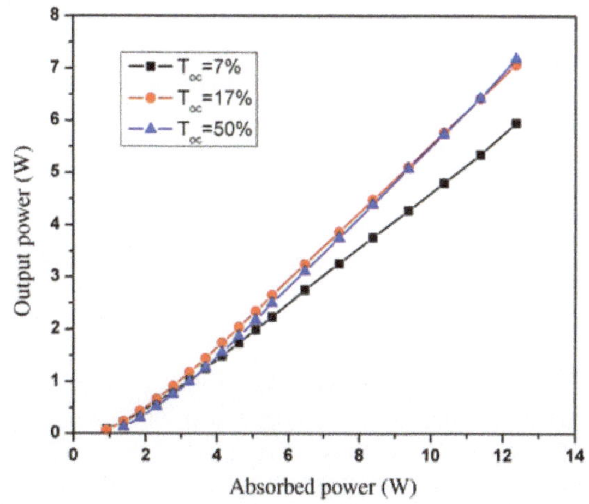

Figure 12. Laser performance of the Yb:LuAG ceramic lasers.

4.2. 1.6 μm *laser oscillation*

The laser performance of the Er^{3+} doped YAG ceramics was also experimentally investigated. Figure 13(a) shows the absorption spectra of the Er:YAG ceramics under different doping concentrations[25]. The absorption coefficient of the Er:YAG ceramics increased with the increase of the Er^{3+} concentration. At 1532 nm, the absorption coefficient increased from 0.64 to 5.06 cm^{-1} as the Er^{3+} concentration increased from 0.5 at.% to 5.0 at.%. Figure 13(b) shows the

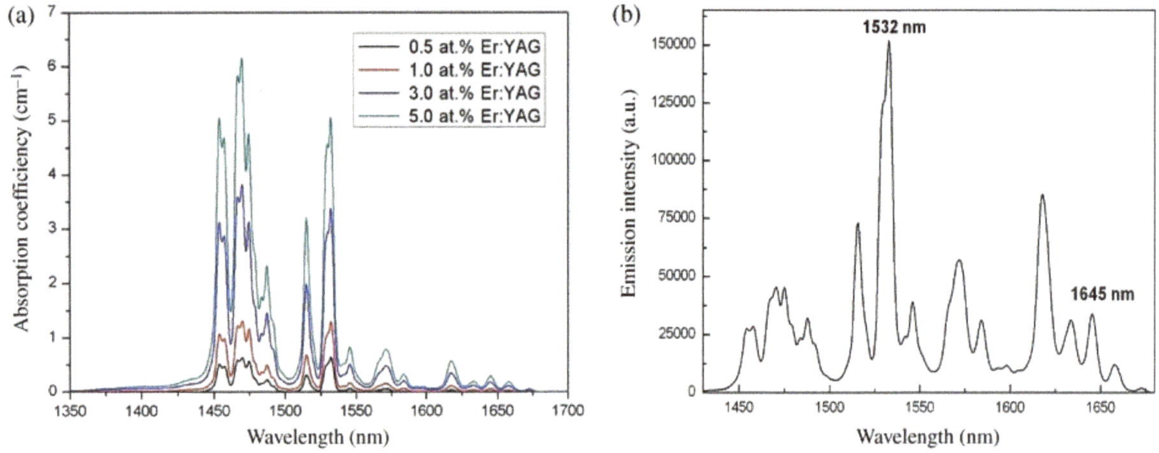

Figure 13. The absorption and emission spectra of the fabricated Er:YAG ceramics.

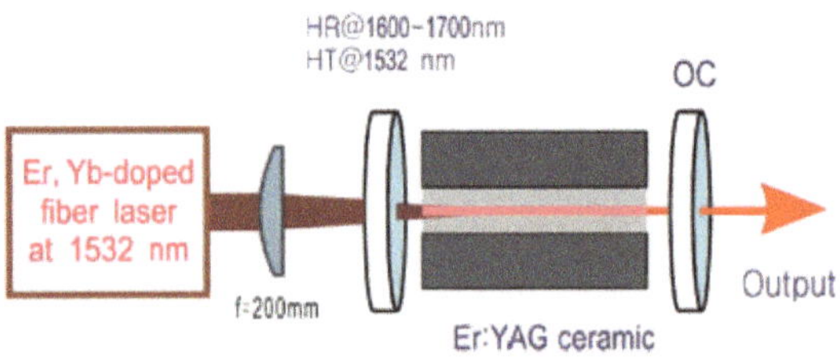

Figure 14. A schematic of the Er:YAG ceramic laser setup.

emission spectrum of the Er:YAG ceramic with a doping concentration of 0.5 at.% in the range 1400–1700 nm. The emission is due to the $^4I_{13/2} \rightarrow {}^4I_{15/2}$ transition of the Er^{3+} ions. The highest emission peak is at 1532 nm, with an emission cross-section of 0.99×10^{-20} cm^2. Figure 14 shows the schematic of the Er:YAG ceramic laser setup used in our experiment. The laser is in-band pumped by an Er–Yb fiber laser at 1532 nm. A 1.0 at.% Er:YAG ceramic sample, cut into pieces of dimension $2 \times 4 \times 14$ mm^3, was used as the gain medium. An output coupler with 10% transmission at 1645 nm was used. The performance of the Er:YAG laser is shown in Figure 15. The laser has a pump threshold of ~1.3 W. It reached an output power of 13.8 W at 1645 nm under 27.3 W of incident pump power. A slope efficiency with respect to the incident pump power of 54.5% and an optical–optical conversion efficiency of 50.5% were achieved. No power saturation was observed on the maximum available pump power, suggesting that further power scaling of the laser is possible by simply increasing the input pumping power.

4.3. 2.0 μm *laser oscillation*

Figure 16 shows the absorption and emission spectra of the Ho:YAG ceramics fabricated in-house[18]. As shown, the strongest absorption peak is at 1907 nm, which is the same as that measured with the Ho:YAG single crystals. As the Ho^{3+} doping concentration increased from 0.8 to 2.0 at.%, the absorption coefficient at 1907 nm increased linearly from 0.72 to 1.78 cm^{-1}. Based on the measured

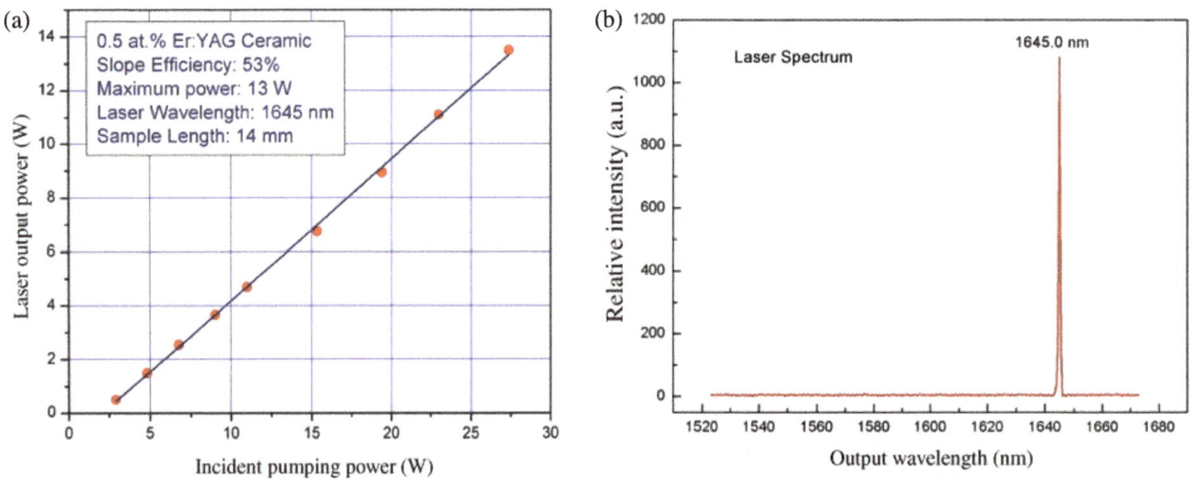

Figure 15. Laser performance of the Er:YAG ceramic laser.

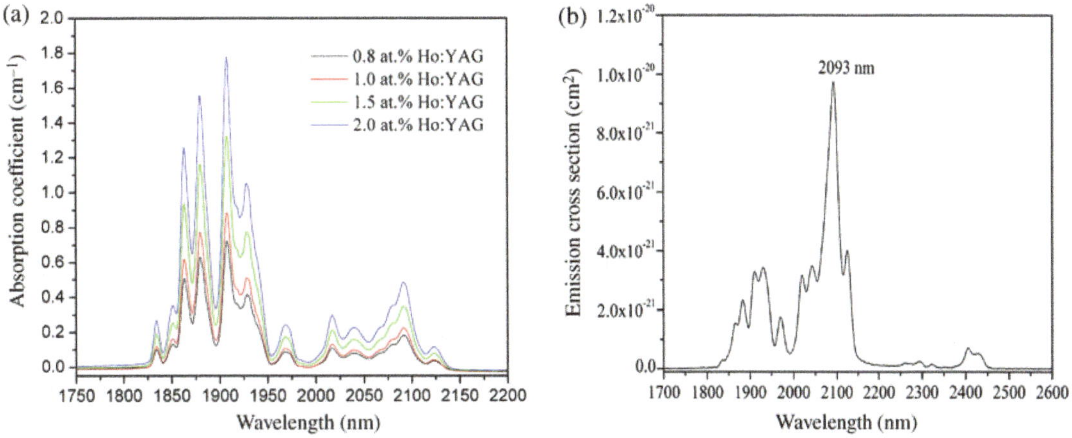

Figure 16. The absorption and emission spectra of the fabricated Ho:YAG ceramics.

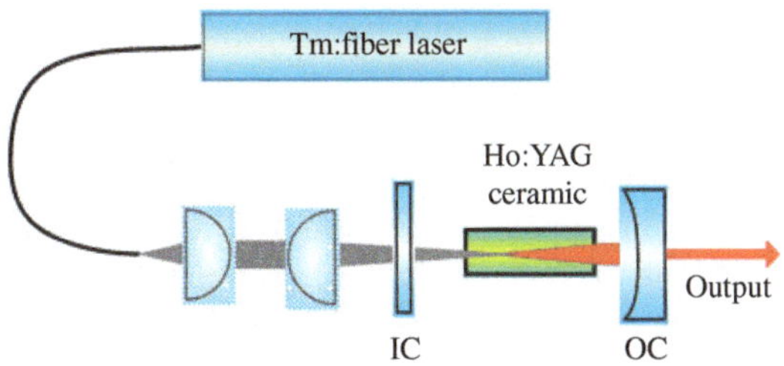

Figure 17. A schematic of the Ho:YAG ceramics laser setup.

absorption coefficients, the absorption cross-section of the Ho:YAG ceramic is calculated as 0.645×10^{-20} cm^2. The emission spectrum from 1700 to 2600 nm of a 1.0 at.% Ho:YAG ceramic measured at room temperature is shown in Figure 16(b). The main emission band is centered at 2093 nm, which corresponds to the $^5I_7 \to {}^5I_8$ transition of Ho^{3+}. The emission cross-section at 2093 nm is calculated to be 0.97×10^{-20} cm^2 by using the F–L equation. Figure 17 shows a schematic of the Ho:YAG ceramic laser experimentally used. The Ho:YAG ceramic laser is in-band pumped by a home-made Tm:YAG fiber laser at 1907 nm. The dimensions of the Ho:YAG ceramic used were 2 mm × 3 mm × 14 mm. A 1.5 at.% Ho:YAG ceramic was used as the laser gain medium. With 35 W incident pump power, a maximum output power of 20.6 W at 2097 nm was obtained, corresponding to a slope efficiency of 60.9% and an optical-to-optical conversion efficiency of 58.8%, as shown in Figure 18.

Besides Ho:YAG ceramics, Tm:YAG ceramics also emit in the 2 μm wavelength regime. Figure 19 shows the transmittance and emission spectrum of the in-house fabricated Tm:YAG ceramics. There are six strong absorption bands centered at 356, 458, 684, 781, 1202, and 1626 nm, which correspond to the transitions of Tm^{3+} from the ground state $^3H_6$ to its excited state, $^1D_2$, $^1G_4$, $^3F_3$, $^3H_4$, $^3H_5$, and $^3F_4$, respectively. The strong emission peaks are at wavelengths of 1626 nm, 1702 nm, 1746 nm, 1785 nm, 1882 nm, 1960 nm, and 2016 nm, respectively. Tm:YAG is a quasi-three-level system. Hence, laser emission with wavelength

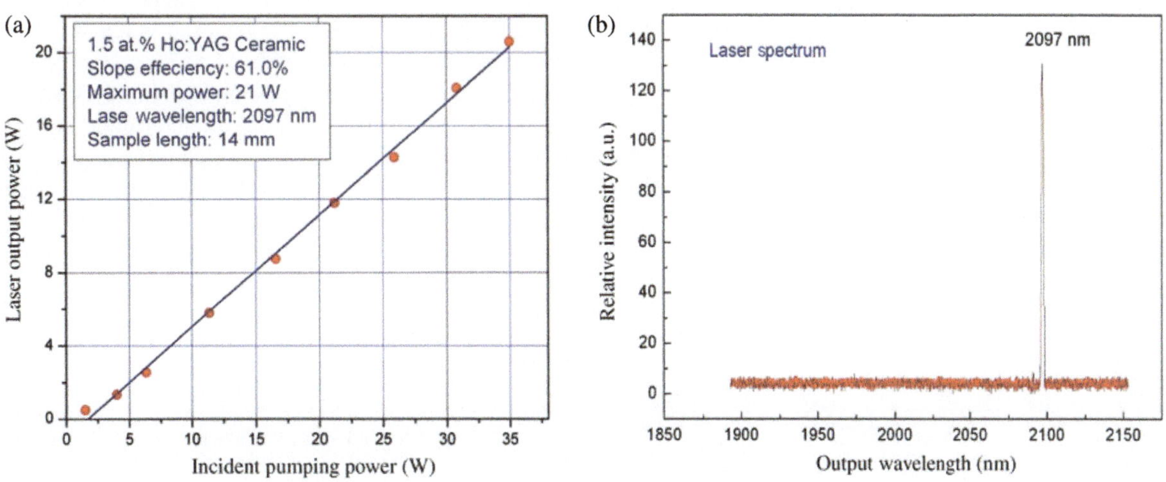

Figure 18. Laser performance of the Ho:YAG ceramic laser.

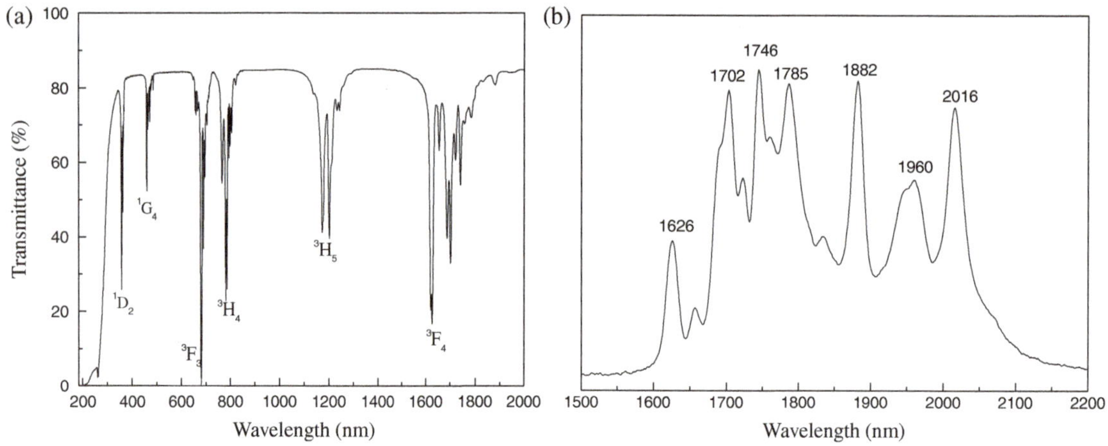

Figure 19. The transmittance and emission spectra of a fabricated Tm:YAG ceramic.

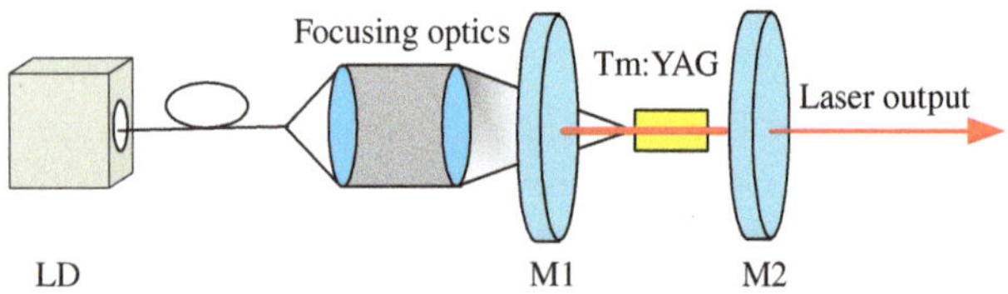

Figure 20. A schematic of the diode-pumped Tm:YAG ceramic laser setup.

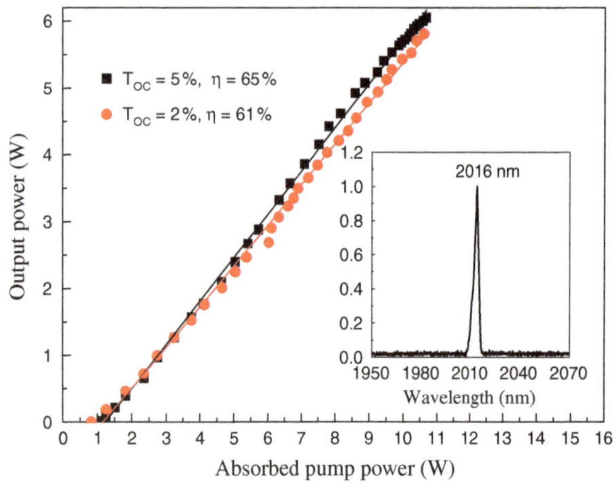

Figure 21. Laser performance of the Tm:YAG ceramic lasers.

below 1.9 μm is difficult to achieve, due to the reabsorption by the Tm^{3+} ions.

Tm:YAG lasers can be directly pumped by commercially available high-power GaAlAs diodes at ~800 nm (the $^3H_6 \rightarrow {}^3H_4$ transition). Benefitting from a fortuitous 'two-for-one' cross-relaxation process, pumping quantum efficiencies approaching 2 could be achieved with pumping under high Tm^{3+} doping concentrations. Figure 20 shows the schematic of the diode-pumped Tm:YAG ceramic laser used to study the lasing performance of the fabricated Tm:YAG ceramics. The diode laser is operating at 786 nm. Tm:YAG ceramics with different doping concentrations were cut into pieces of dimension $3 \times 3 \times 3.2$ mm^3. The laser emitted at 2016 nm. With an output coupler of 5% coupling, a maximum output power of 6.05 W was obtained under an absorbed pump power of 10.68 W, giving a slope efficiency of 65%.

In Table 1, we have summarized the laser performance of our fabricated RE:YAG ceramics.

5. Conclusions

High optical quality polycrystalline YAG transparent ceramics doped with various rare-earth ions were successfully fabricated by the reactive sintering method. The laser performance of the fabricated RE:YAG ceramics were studied experimentally. The main results achieved in our works can be summarized as follows.

(1) High-quality RE:YAG transparent ceramics were fabricated using the reactive sintering method under vacuum conditions. Irrespective of the different doped rare-earth ions and their doping levels, defect-free microstructures were observed for all the samples produced.

(2) Highly efficient CW laser oscillations in the 1.0 μm regime were obtained in the Nd:YAG and Yb:YAG ceramics.

Table 1. Summary of the results of the ceramic laser experiments

	Nd:YAG	Yb:YAG	Yb:LuAG	Er:YAG	Tm:YAG	Ho:YAG
Pumping wavelength	808 nm	940 nm	980 nm	1532 nm	786 nm	1907 nm
Output wavelength	1064 nm	1030 nm	1030 nm	1645 nm	2016 nm	2097 nm
Maximum output power	15 W	6.9 W	7.2 W	13.8 W	6.05 W	21 W
Slope efficiency	62.0%	62.7%	65.0%	54.5%	65.0%	61.0%

The highest slope efficiency achieved on the Nd:YAG ceramics is 62 %, which is very close to that of Nd:YAG single crystal.

(3) Er:YAG ceramic is suitable for achieving ~1.6 μm laser emission. When end pumped by an Er,Yb-fiber laser at 1532 nm, a maximum output power of 13.8 W at 1645 nm was obtained with a slope efficiency of 54.5%.

(4) By using a 1.5 at.% Ho:YAG ceramic as the laser medium, 20.6 W laser output at 2097 nm was achieved under 35 W of incident pump power, and a slope efficiency of 61% has been achieved.

(5) High optical quality Tm:YAG laser ceramics have been fabricated. Pumped by a laser diode at 768 nm, high-efficiency CW laser operation has been demonstrated.

Acknowledgements

This project is supported by the Priority Academic Program Development of Jiangsu Higher Education Institutions (PAPD) and the National Research Foundation of Singapore under contract NRF-G-CRP-2007-01.

References

1. G. L. Messing, and A. J. Stevenson, Science **322**, 383 (2008).
2. A. Ikesue, and Y. L. Aung, Nature Photonics **2**, 721 (2008).
3. R. L. Coble, J. Amer. Ceram. Soc. **45**, 123 (1962).
4. A. Ikesue, T. Kinoshita, K. Kamata, and K. Yoshida, J. Am. Ceram. Soc. **78**, 1033 (1995).
5. A. Ikesue, I. Furusato, and K. Kamata, J. Am. Ceram. Soc. **78**, 225 (1995).
6. A. Ikesue, K. Kamata, and K. Yoshida, J. Am. Ceram. Soc. **78**, 2545 (1995).
7. J. R. Lu, M. Prabhu, J. Song, *et al.*, Appl. Phys. B **71**, 469 (2000).
8. J. R. Lu, J. Song, M. Prabhu, *et al.*, Jpn. J. Appl. Phys **39**, L1048 (2000).
9. J. Lu, K. Ueda, H. Yagi, T. Yanagitani, Y. Akiyama, and A. A. Kaminskii, J. Alloy. Compd. **341**, 220 (2002).
10. J. Lu, T. Murai, K. Takaichi, *et al.*, Appl. Phys. Lett. **78**, 3586 (2001).
11. J. Lu, H. Yagi, K. Takaichi, *et al.*, Appl. Phys. B **79**, 25 (2004).
12. L. Maziex, and D. J. Green, Int. J. Appl. Ceram. Technol. **3**, 166 (2006).
13. H. Yagi, K. Takaichi, K. Ueda, Y. Yamasaki, T. Yanagitani, and A. A. Kaminskii, Laser Phys. **15**, 1338 (2005).
14. R. M. Yamanoto, *Proc. Adv. Solid State Photon., Nara, Japan, WC5* (2008).
15. B Bishop, "Northrop Grumman scales new heights in electric laser power, achieves 100 kW from a solid-state laser", http://www.irconnect.com/noc/press/pages/news_releases.html (March 18, 2009).
16. A. Mandl, and D. E. Klimek, in *Conference on Lasers and Electro-Optics*, OSA Technical Digest (CD) (Optical Society of America 2010), paper JThH2.
17. R. Guoguang, and H. Yunian, Laser & Infrared **36**, 617 (2006).
18. J. Dong, G. Z. Xu, J. Ma, *et al.*, Opt. Mater. **34**, 959 (2012).
19. W. Kim, C. Baker, G. Villalobos, *et al.*, J. Am. Ceram. Soc. **94**, 3001 (2011).
20. E. R. Kupp, G. L. Messing, J. M. Anderson, and V. Gopalan, J. Mater. Res. **25**, 476 (2010).
21. H. Yagi, T. Yanagitani, K. Takaichi, K. Ueda, and A. A. Kaminskii, Opt. Mater. **29**, 1258 (2007).
22. H. Gong, D. Y. Tang, H. Huang, and J. Ma, J. Am. Ceram. Soc. **92**, 812 (2009).
23. H. Yang, X. P. Qin, J. Zhang, *et al.*, Opt. Mater. **34**, 940 (2012).
24. D. W. Luo, J. Zhang, C. W. Xu, X. P. Qin, D. Y. Tang, and J. Ma, Opt. Mater. **34**, 936 (2012).
25. X. P. Qin, H. Yang, G. H. Zhou, *et al.*, Opt. Mater. **34**, 973 (2011).
26. Y. Wang, D. Y. Shen, H. Chen, *et al.*, Opt. Lett. **36**, 4485 (2011).
27. H. Yang, J. Zhang, X. P. Qin, *et al.*, J. Am. Ceram. Soc. **95**, 52 (2012).
28. W. L. Gao, J. Ma, G. Q. Xie, *et al.*, Opt. Lett. **37**, 1076 (2012).
29. C. W. Xu, D. W. Luo, J. Zhang, *et al.*, Laser Phys. Lett. **9**, 30 (2012).

11

SG-II-Up prototype final optics assembly: optical damage and clean-gas control

Dongfeng Zhao[1,2], Li Wan[1], Zunqi Lin[1], Pin Shao[1], and Jianqiang Zhu[1]

[1] *Shanghai Institute of Optics and Fine Mechanics, Chinese Academy of Sciences, No. 390, Qinghe Road, Jiading District, Shanghai 201800, China*

[2] *University of Chinese Academy of Sciences, No. 19A Yuquan Road, Beijing 100049, China*

Abstract

The Shenguang-II Upgrade (SG-II Up) facility is an under-construction high-power laser driver with eight beams, 24 kJ energy, 3 ns pulse duration and ultraviolet laser output, in the Shanghai Institute of Optics and Fine Mechanics, China. The prototype design and experimental research of the prototype final optics assembly (FOA), which is one of the most important parts of the SG-II Up facility, have been completed on the ninth beam of the SG-II facility. Thirty-three shots were fired using 1-ω energy from 1000 to 4500 J and 3-ω energy from 500 to 2403 J with a 3 ns square pulse. During the experiments, emphasis was given to the process of optical damage and to the effects of clean-gas control. A numerical model of the FOA generated by the Integrated Computer Engineering and Manufacturing code for Computational Fluid Dynamics (ICEMCFD) demonstrated that a flux within 1–5 l s^{-1} and a 180 s period is effectual to avoid contaminant sputtering to the optics. The presence of surface 'mooning' damage and surface spots located outside the clear aperture are induced by contaminants such as wire, silica gel and millimeter order fiber and metal.

Keywords: Final Optics Assembly; optical damage; SG-II-Up

1. Introduction

The Shenguang-II Upgrade (SG-II Up) is a kilojoule-class solid-state laser and targeting facility under construction by the National Laboratory on High Power Lasers and Physics. The amplified 1053 nm beams from the Nd:glass driver are transported (at 40 kJ for 3 ns) in 2 × 2 quads with eight beams with aperture size 310 mm × 310 mm to the 2.4 m-diameter target chamber where eight final optics assemblies (FOAs) convert each beam to the third harmonic, separate the residual 1053 and 527 nm beams, and finally focus the 351 nm beam onto the target.

The prototype design and manufacture of the FOA have been finished, and it is different from the NIF'FOA and the SG-II'FOA[1–6]. The FOA consists of fixed interfaces to the target chamber and five modules that house eight full-aperture optics, as shown in Figure 1. The aperture sizes of these optics are as large as 370 mm × 370 mm. The corresponding functions of the various optics, as shown in the aforementioned figure, are as follows.

Correspondence to: D. Zhao, Shanghai Institute of Optics and Fine Mechanics, Chinese Academy of Sciences, No. 390, Qinghe Road, Jiading District, Shanghai 201800, China. Email: dfzhao@siom.ac.cn

1. Phase plate – conditions the beam phases to form an ~750 μm flat-top profile.
2. Vacuum window – provides a near-vacuum FOA environment.
3. Frequency conversion crystals – convert the 1053 nm beam to a 351 nm beam.
4. Wedged plate – separates the residual 1053 and 527 nm light with 351 nm light, and the refractive low-energy 351 nm light to a power sensor on the output surface.
5. Focus lens – focuses the 351 nm beam onto the target; the focal length is 2234 mm.
6. Beam sampling grating (BSG) and main debris shield (MDS) – providing a low-efficiency grating to the input surface for energy sampling (calorimetry) and large-object protection for upstream (more expensive) optics.
7. Disposable debris shield (DDS) – thin, inexpensive optics for primary debris protection.

The frequency converter is a cascade sum-frequency generation design consisting of a 12.5 mm-thick Type-I KDP

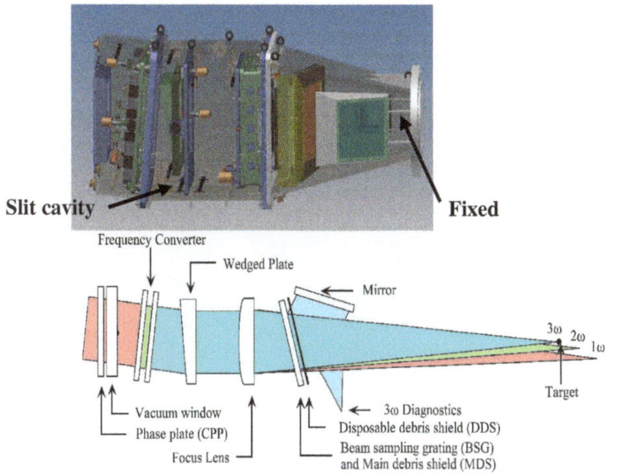

Figure 1. The prototype FOA is composed of fixed interfaces to the target chamber and five modules that house eight full-aperture optics. The clear aperture size is 310 mm × 310 mm.

doubler and a 10.5 mm-thick Type-II KDP tripler, optimized to achieve >70% peak power conversion efficiency to the third harmonic at a $1\omega_0$ driver irradiance of 3 GW cm^{-2} (Ref. [7]). The wedged plate, which is a prism with an edge angle of 11.22°, combines with the focus lens to realize a 2 mm separating distance of the fundamental and second harmonic away from the third harmonic target. The focus lens has a 45 mm-thick meniscus-aspheric surface for avoiding optical damage induced by ghost images of the fourth order for 1ω, 2ω, and 3ω. The BSG and MDS are made with 8 mm-thick fused silica possessing a 0.2% low-efficiency grating on the input surface for energy sampling. This setup provides transitions from the near-vacuum FOA environment to the hard-vacuum target chamber environment. The disposable debris shield is 2 mm-thick borosilicate glass. Here, the angle of the normal direction of the BSG and MDS to the DDS and the direction of the incident beam is 13.5°, which is used to deviate ghosts below the fourth order for $1\omega_0$, $2\omega_0$ and $3\omega_0$ from the direction of the incident beam. Inside the prototype FOA, the pressure is 10 Torr with 10 SLPM before the laser shot and 50 Torr with 40 SLPM after the laser shot using clean dry nitrogen to purge.

2. Clean-gas control

The optics in the prototype FOA are coated with a sol-gel anti-reflection film, but residuary (0.5%–1%) reflection also exists. Hence, these eight optics (including 16 surfaces) could form thousands of ghost images within the fourth-order reflection. If these ghost images are located on the optics, the optics would be directly damaged; if these ghost images move to other parts such as the mechanism, wire and silica gel (the paste around the optical elements, to guarantee the stability of the stress and deformation), the optics would be indirectly damaged because of the contaminants. As shown in the last section, the BSG, MDS and DDS are tilted by 13.5°, and most of the ghost images are deflected from the main beam path to protect the optics. These ghost images lead to another problem, i.e., the presence of contaminants. To avoid contaminant sputtering to the optics, the following procedures are performed. First, the mechanism surfaces are treated with anodic oxidation. Second, clean gas is used to blow the optics surface to eliminate contaminants. In this section, a numerical simulation is carried out based on the FOA construction and a blowpipe arrangement, and the experimental results are presented.

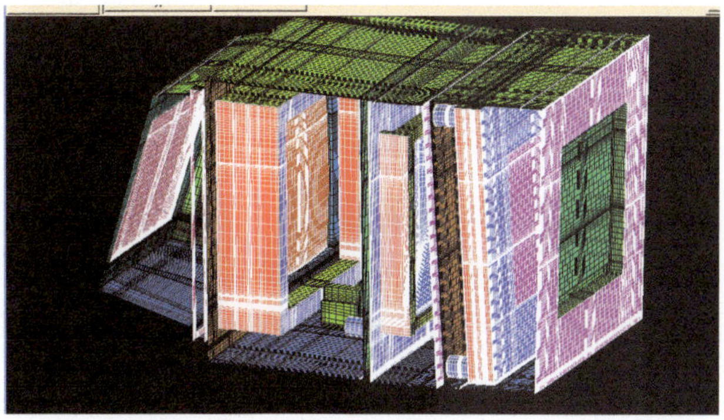

Figure 2. Numerical model of the FOA generated by ICEMCFD.

2.1. Simulation analysis

The main physics process involved in the prototype FOA is the convection–diffusion process in the slit cavity (as indicated in Figure 1). The governing equations for this process can be written as

$$\frac{\partial \rho}{\partial t} + \frac{\partial}{\partial x_k}(\rho V_k) = 0, \tag{1}$$

$$\frac{DV_j}{Dt} = -\frac{1}{\rho_0}\frac{\partial (P - P_0)}{\partial x_j} + \nu \frac{\partial^2 V_j}{\partial x_k \partial x_k}, \tag{2}$$

$$\frac{Dc}{t} = \Gamma \frac{\partial^2 c}{\partial x_k \partial x_k}, \tag{3}$$

where V_i is the velocity vector ($i = 1, 2, 3$), t is time, ρ is the density of N_2, υ is the kinematic viscosity of N_2, c is the density of the aerosol in N_2, p is the pressure of N_2, p_0 is the reference pressure, and ρ_0 is the density in the reference point. Equation (1) is the continuity equation, Equation (2) is the momentum equation, and Equation (3) is the composition equation.

The standard *k–e* model is used in the numerical simulation. This model is based on the hexahedron, and the gridding number is 2.9 million, as shown in Figure 2. The gridding is optimized with the Smoother tools of the Integrated Computer Engineering and Manufacturing code

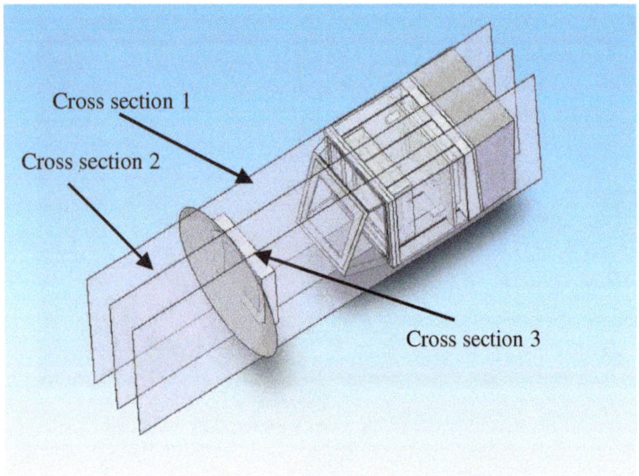

Figure 3. Schematic of the observed cross-section distributions in the prototype FOA.

Figure 4. Density distribution of the contaminants in cross-section 2 as the flux ranges within 1–5 l s^{-1} after 30 s.

for Computational Fluid Dynamics (ICEMCFD) based on the condition that the quality of the deformation rate above 0.8 is >90% and that all deformation rates are >0.3. The deformation rate is defined as the ratio of the diameter of the excircle to that of the incircle. A larger deformation rate corresponds to higher grinding quality.

Based on the above model, three cross-sections are analyzed, as shown in Figure 3. Flow situations of the flow field in the cavities of the prototype FOA at 100 Torr and constant pressure flux of 5 or 50 l s^{-1} are simulated. Then, the ventilation efficiency on the optics surfaces and the density distribution of contaminants in the FOA are investigated for the above two fluxes. An entrance and an exit also exist in each cavity of the FOA.

The simulation results show that both 5 and 50 l s^{-1} fluxes can maintain the relative density of the contaminants on the optics surfaces below 20% when the designed inlet pipe, exhaust pipe and spatial distribution are used. About 50 s after the laser launch, the contaminants decrease to 0.1% when 5 l s^{-1} is used and 6% when 50 l s^{-1} flux is used, as shown in Figure 4. Therefore, considering the contamination-elimination time, the flux is chosen within 1–5 l s^{-1} and a 180 s period.

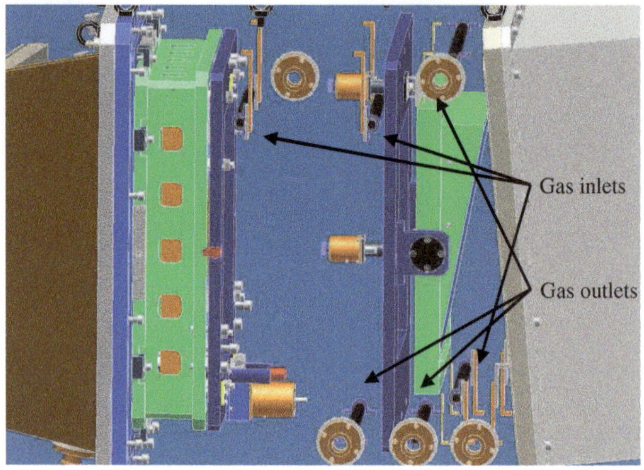

Figure 5. Distribution of the gas inlets and outlets in the FOA.

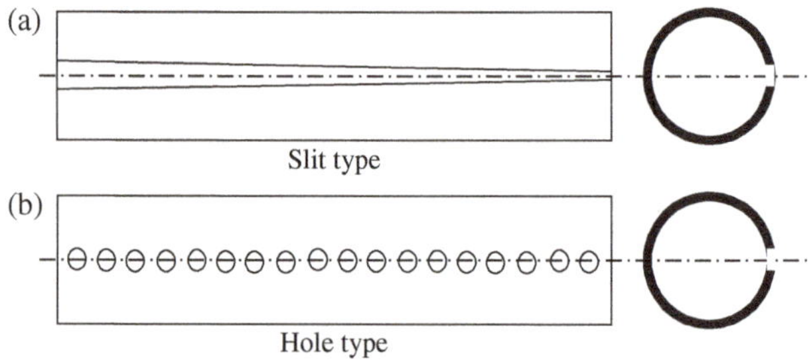

Figure 6. Two types of inlet and exhaust pipes.

2.2. *Gas-blowing scheme*

In this section, the sizes, quantities and positions of the entrance and the exit as well as the gas-flow rate and hold time required for the contaminant elimination are given, based on numerical analysis of the gas-flow characteristics in the prototype FOA. Figure 5 shows the distributions of the inlet pipes and the corresponding flanges. The FOA includes nine entrances and five exits.

The N_2 used is 99.99% pure, and the H_2O content is less than 3 ppm. Gas is transported to the location where the aerosol needs to be blown by the inlet pipe. Figure 6 presents two types of inlet and exhaust pipes. To ensure uniform speed of the gas flow along the axial direction, the slit width is gradually increased from the point near the entrance to that near the exit for the slit-type pipe, as shown in Figure 6(a). For the small hole type (Figure 6(b)), the hole diameter is increased and the distance between the adjacent holes is decreased gradually from the point near the entrance to that near the exit.

The configurations of the exhaust pipes are similar to those of the inlet pipes. To ensure uniform speed of the exhaust flow along the axial direction, the slit width is gradually increased from the point near the exit to the far end for the slit-type pipe. For the small hole type, the hole diameter is increased and the distance between the adjacent holes

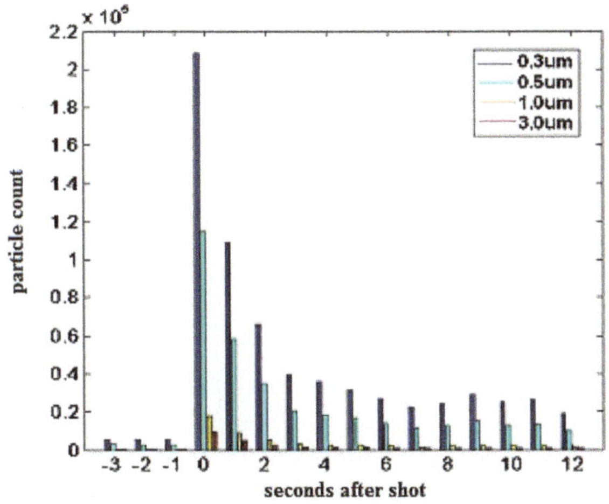

Figure 7. Graph showing the real-time contaminant quantity.

is gradually decreased from the point near the exit to the far end.

2.3. Experimental results

In the experiment, the flow controller is triggered at the same time as the laser fires. The flow flux is 40 SLPM. The particle counter is used to detect contamination. Figure 7 shows that the contaminant quantity increases from 3000 particles before the laser fires to 140 000–210 000 particles after the firing. Thus, the particle density contrast is about 1:70. The optics surfaces is spotted after 33 shots. During the experiment, results obtained by the particle counter are taken from one of the exits. Hence, these results represent the relative density variation of the contaminants rather than the absolute quantity.

3. Optical damage

3.1. Damage morphology induced by the contaminants

Using an Msu25D high-resolution microscope, damage morphology is detected. Figure 8 shows the presence of surface 'mooning' damage and a surface spot located outside the clear aperture and induced by contaminants apart from the common damage morphologies, such as surface pinpoint damage, block peeling, long scratching, and bulk filament[7–9]. Here, surface mooning damage is serious in the center and changes the color of the film, induced by metal contaminants.

The components and morphologies of the contaminants are different; thus, the damage morphologies are also

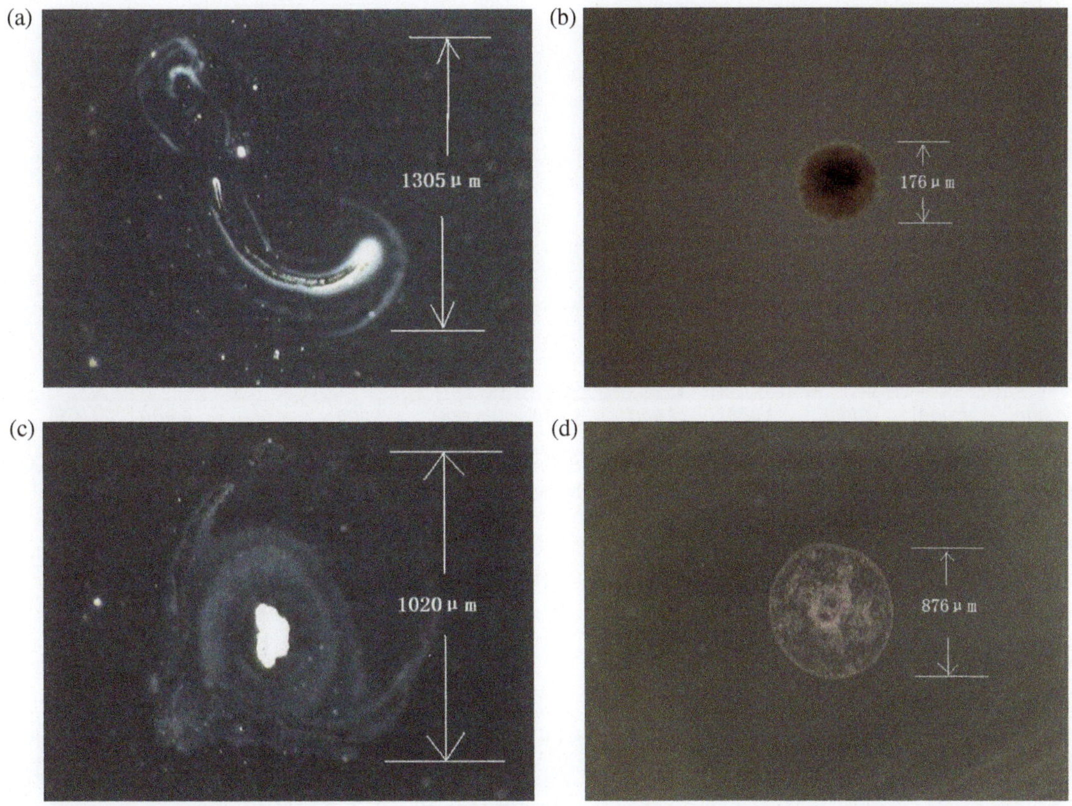

Figure 8. Surface damage morphologies induced by contaminants: (a) film damage located on the clear aperture brim, (b) surface spot outside the clear aperture, (c) clear aperture brim and (d) surface 'mooning' damage outside the clear aperture.

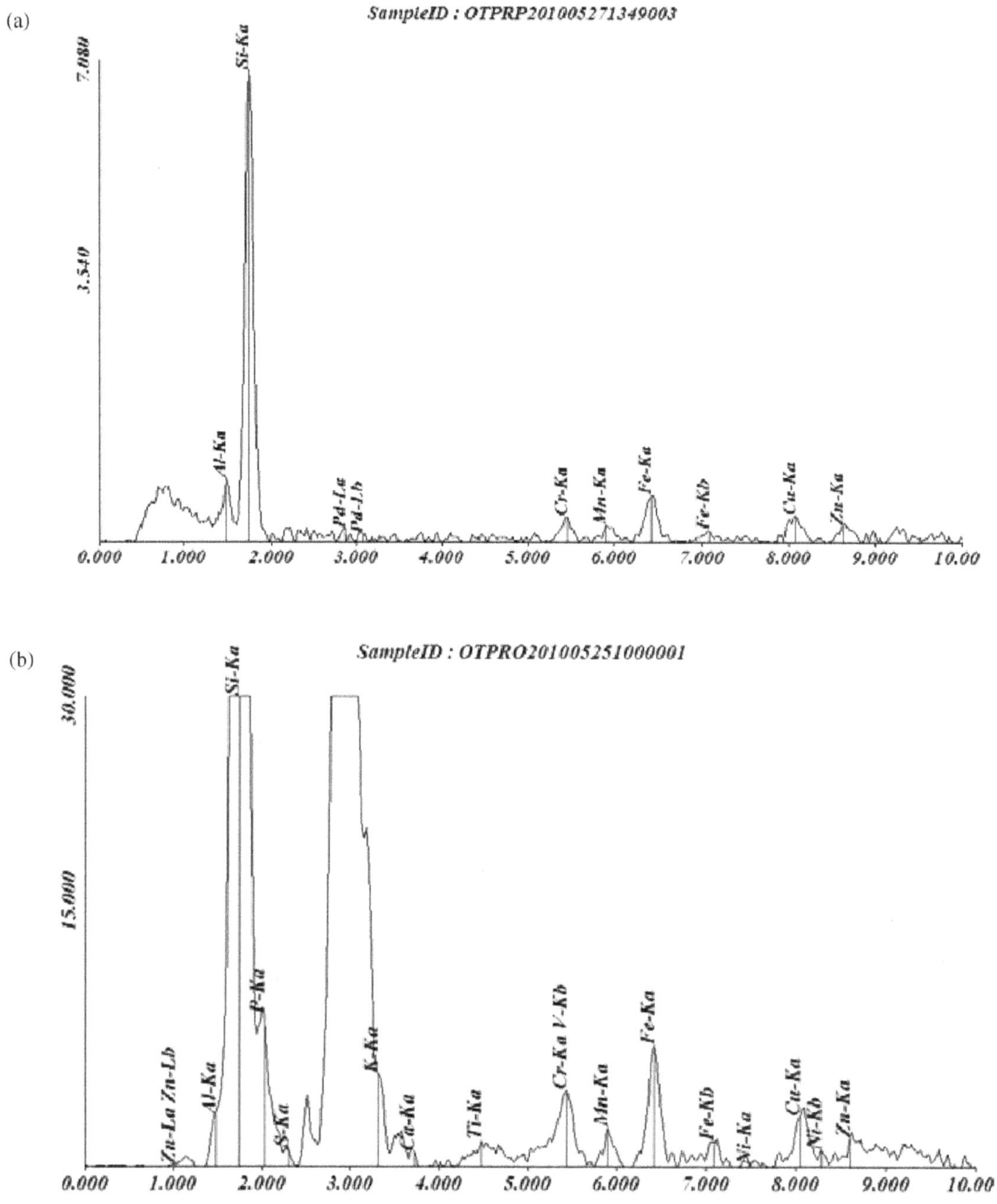

Figure 9. Results obtained by x-ray fluorescence spectrometry: (a) unused silica coated by sol-gel film (including Al, Si, Pd, Cr, Mn, Fe, Cu and Zn) and (b) used silica coated by sol-gel film (including Zn, Al, Si, P, S, K, Ca, Ti, Cr, Mn, Fe, Ni and Cu; because the Pb amount is negligible, it is not labeled).

different. The different figure subparts are explained as follows.

(a) Figure 8(a) is induced by millimeter order fibers absorbing on the optics surfaces. After less than 0.1 J cm^{-2} order laser irradiation, the fiber is ablated and induces film damage. Suffering from multiple irradiations, the damage size increases and induces substrate damage. Given that the heat absorption coefficient of the fiber is low, this kind of damage cannot induce catastrophic bulk peeling.

(b) Figure 8(b) is induced by wire and silica gel. Occurrences of this kind of morphology are numerous on the optics surfaces. If the wire and silica gel are irradiated by a laser, parts of them volatilize and are then absorbed on the optics surfaces. After several irradiations by stray light, they appear ablated but do not damage the sol-gel film.

(c) Figure 8(c) is induced by millimeter order metal contaminants. The substrate is damaged, and the surface 'mooning' damage of the sol-gel film appears. This kind of damage is due to the optics surface being affected by metal particle sputtering. After less than 0.1 J cm^{-2} order laser irradiation, the damage size increases and the sub-millimeter order damage points are uniformly distributed in the clear aperture because of the high absorption coefficient of the metal.

(d) Figure 8(d) is induced by the same millimeter order metal contaminants. The substrate is seriously damaged in the center, and surface 'mooning' damage of the sol-gel film appears in the periphery. This phenomenon is also induced by the metal contaminants but occurs outside the clear aperture and is irradiated by the stray light.

3.2. Component analysis of the contamination

Parameters such as the components of the laser-induced damage point, the irradiated point without damage and the un-irradiated point, as well as the mechanical material, wire, silica gel, unused silica and chemical films, are analyzed by energy-dispersive x-ray fluorescence spectrometry. Figure 9 shows that the new components S, P, K, Ca, Mn, Al and Pd appear and that the contents of the Fe and Cu increase in the used optics compared with the unused optics. Al and Pd come from the sol-gel film; S, P and K are from the silica gel used to fix the optics; Ti and Cu come from the metal material sputtering of the prototype FOA. Molecular contaminants were not analyzed by Raman spectrometry because of the large optics size (360 mm × 370 mm).

4. Conclusions

The uniform distribution of hundred-micron-order punctiform damage in the clear aperture and the uniform distribution of 'mooning' damage or surface spots outside the clear aperture suggest that the contaminants induced by stray light because of the residual reflection from the mechanical metals, silica gel and wire are the main reasons for the damage increase. The gas flow used in the experiment eliminates certain contaminants, but ineffectively. Surface treatment of the metallic material is not completely effective for elimination of contaminants. Stray light absorbing traps are required in the prototype.

References

1. S. H. Glenzer, B. J. MacGowan, P. Michel, N. B. Meezan, L. J. Suter, S. N. Dixit, J. L. Kline, G. A. Kyrala, D. K. Bradley, D. A. Callahan, E. L. Dewald, L. Divol, E. Dzenitis, M. J. Edwards, A. V. Hamza, C. A. Haynam, D. E. Hinkel, D. H. Kalantar, J. D. Kilkenny, O. L. Landen, J. D. Lindl, S. LePape, J. D. Moody, A. Nikroo, T. Parham, M. B. Schneider, R. P. J. Town, P. Wegner, K. Widmann, P. Whitman, B. K. F. Young, B. Van Wonterghem, L. J. Atherton, and E. I. Moses, Science **327**, 1228 (2010).
2. B. M. Van Wonterghem, S. C. Burkhart, C. A. Haynam, K. R. Manes, C. D. Marshall, J. E. Murray, M. L. Spaeth, D. R. Speck, S. B. Sutton, and P. J. Wegner, Proc. SPIE **5341**, 55 (2004).
3. Z. Lin, Chin. J. Lasers **37**, 2202 (2010).
4. D. Zhao, G. Zhen, P. Shao, N. Hua, Z. Qiao, Y. Dai, and J. Zhu, Chin. J. Lasers **34**, 637 (2007).
5. C. A. Haynam, P. J. Wegner, J. M. Auerbach, M. W. Bowers, S. N. Dixit, G. V. Erbert, G. M. Heestand, M. A. Henesian, M. R. Hermann, K. S. Jancaitis, K. R. Manes, C. D. Marshall, N. C. Mehta, J. Menapace, E. Moses, J. R. Murray, M. C. Nostrand, C. D. Orth, R. Patterson, R. A. Sacks, M. J. Shaw, S. B. Sutton, W. H. Williams, C. C. Widmayer, R. K. White, S. T. Yang, and B. M. Van Wonterghem, Appl. Opt. **46**, 3276 (2007).
6. Z. Lin, S. Wang, and D. Fan, Chin. J. Lasers **B10**, IV6 (2001).
7. D. Zhao, L. Wang, Z. Lin, P. Shao, L. Ji, Z. Cai, R. Wu, Y. Dai, and J. Zhu, Chin. J. Lasers **38**, 0702001 (2011).
8. S. Papemov and A. W. Schmid, Proc. SPIE **7132**, 71321J (2008).
9. T. I. Suratwala, P. E. Miller, J. D. Bude, R. A. Steele, N. Shen, M. V. Monticelli, M. D. Feit, T. A. Laurence, M. A. Norton, C. W. Carr, and L. L. Wong, J. Am. Ceram. Soc. **94**, 416 (2011).

12

Application of image processing based on multiple filters in an alignment system

Peiying Zeng, Baoqiang Zhu, De'an Liu, and Jianqiang Zhu

National Laboratory on High Power Laser and Physics, Shanghai Institute of Optics and Fine Mechanics, Chinese Academy of Sciences, Shanghai 201800, China

Abstract

Beam alignment depends on CCD real-time image analysis and processing. In order to improve the quality of the alignment, multiple filters are used in far-field and near-field image processings. These multiple filters are constituted of an average filter and a median filter in different connection sequences, so that they can deal with different kinds of noise. To reduce the effect of the unknown nonlinear relationship between motor running steps and deviation pixels, a feasible methodology is offered to improve this phenomenon and a fuzzy algorithm is applied to the motor feedback control process. Because of the fuzzy control it is not necessary to establish an accurate mathematical model, so the impact of the nonlinear relationship will be reduced.

Keywords: alignment; fuzzy control; image processing; multiple filters

1. Introduction

The alignment system of a high power laser facility is used for deviation adjustment of beams. For an already running facility, the error caused by the vibrations will be magnified through beam transmitting and lead to beam deviation[1]. The alignment process mainly depends on the CCD real-time optical image analysis and processing; in general, these images contain a lot of noise. The noise mainly include (1) noise caused by electromagnetic interference, (2) noise caused by effects of stray light, (3) Gaussian noise that is generated by the internal resistive components, and (4) salt and pepper noise that is generated by image cutting or the photoelectric conversion process[2, 3]. When the noise is serious, it not only affects spot identification and segmentation but also affects the calculation of center coordinates. The common approach is to remove this noise by an average filter or median filter, and then to use the threshold method to calculate the center of the spot for beam auto-alignment. A median filter can smooth the noise point but will cause image blur, while a median filter can avoid image blur, but will lose image detail such as corner and line information[4, 5].

In this paper, an improved multiple filter method for image denoising is proposed to replace traditional single filter methods. These multiple filters are constituted of an average filter and a median filter in different connection sequences, so that all advantages of both filters can be combined. Then, an adaptive step feedback adjustment mechanism is built to improve the alignment quality and processing time.

Correspondence to: Peiying Zeng, PO Box 800-211, Shanghai 201800, China. Email: zeng_peiying@126.com

2. Multiple filter denoising

For the average filter algorithm, also called the neighborhood linear average method, the basic idea is use the average gray value of several neighboring pixels instead of the gray value of each pixel. The neighborhood is usually constituted of four or eight pixels that are around one pixel, as shown in Figure 1. An average filter is a kind of linear filter, using the smoothing feature of a linear filter to restrain the Gaussian noise and impulse noise. However, it will also cause a spot image in which especially the edges become blurred.

For a given image, we can use a two-dimensional function $f(x, y)$ to describe all pixels of this image. For a pixel point (x, y), supposing that the neighborhood is $r * r$ (r is even), the output of the average filter $\overline{f}(x, y)$ is calculated by

$$\overline{f}(x, y) = \frac{1}{r * r} \sum_{i=-\frac{r}{2}}^{\frac{r}{2}} \sum_{j=-\frac{r}{2}}^{\frac{r}{2}} f(x + i, y + j). \qquad (1)$$

The basic idea of a median filter is to sort all gray values of each pixel in one neighborhood first, and then to use the middle position gray value instead of the gray value of

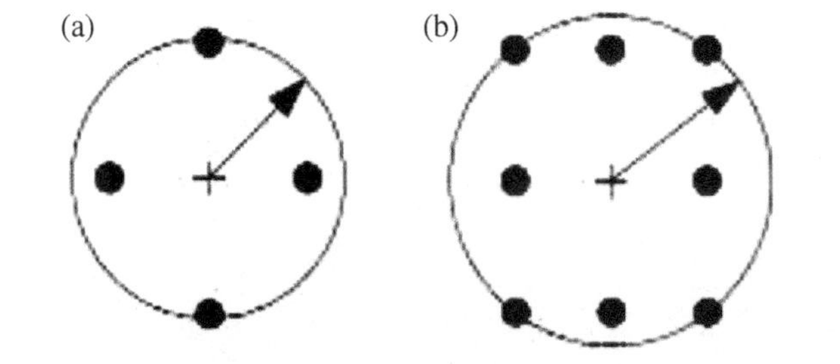

Figure 1. The neighborhood of an average filter; the point '+' is the selected pixel. (a) The neighborhood is constituted of four pixels; this is the simplest mode. (b) The neighborhood is constituted of eight pixels; the precision will be better than with four pixels.

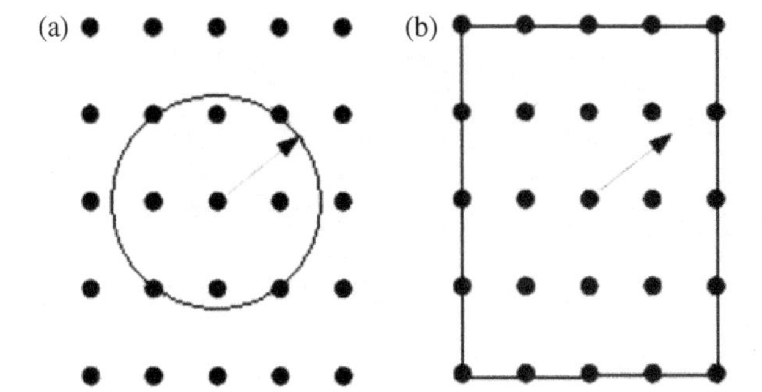

Figure 2. The neighborhood of a median filter; the center point is the selected pixel. (a) The neighborhood is constituted of nine pixels; this is the most commonly used mode. (b) The neighborhood is constituted of 25 pixels; the precision will be better than for the mode with nine pixels but requires more processing time.

the center pixel of the neighborhood. The neighborhood is usually constituted of nine or 25 pixels that are around one pixel, as shown in Figure 2. Because salt and pepper noise often appears in the form of isolated points, and each point is constituted by only a few pixels, a median filter is effective in restraining salt and pepper noise and has almost no influence on the image edges. However, for large noise, and especially for Gaussian noise, the restraining ability of a median filter is weaker than that of an average filter.

For a given image, a two-dimensional function $f(x, y)$ can describe all pixels. For a pixel point (x, y), supposing that the neighborhood is $a*a$ (a is odd), so that there are $n = a*a$ gray values $\{a_1, a_2, \ldots, a_n\}$ in this neighborhood, the value in the middle position will output as the selected pixel gray value after sorting, so the output of the median filter $\overline{f}(x, y)$ can be expressed by

$$\overline{f}(x, y) = \mathrm{med}\{a_1, a_2, \ldots, a_n\}. \tag{2}$$

In order to combine the advantages of these filters, multiple filters are built to perform the image processing; these multiple filters are constituted of an average filter and a median filter in different connection sequencess. According to the sequence of connection, there are two types of multiple filter, the median–linear filter and the linear–median filter.

For the median–linear filter, referred to as ML, the output is calculated by

$$\overline{f}(x, y) = \frac{1}{r*r} \sum_{i=-\frac{r}{2}}^{\frac{r}{2}} \sum_{j=-\frac{r}{2}}^{\frac{r}{2}} \mathrm{med}(f(x+i, y+j)). \tag{3}$$

For the linear–median filter, referred to as LM, the output is calculated by

$$\overline{f}(x, y) = \mathrm{med}\left(\frac{1}{r*r} \sum_{i=-\frac{r}{2}}^{\frac{r}{2}} \sum_{j=-\frac{r}{2}}^{\frac{r}{2}} (f(x+i, y+j))\right). \tag{4}$$

For the median–linear filter, median filtering will be performed first. This procedure can remove the salt and pepper noise without losing image data. It also has the ability to weaken the effect of Gaussian noise. Then the average filter is used to filter most of Gaussian noise. Since the median filtering has the ability to distinguish the edge, the boundary of the spot image and the noise will have a significantly improved result. Under these circumstances, if the spot area of the image is large enough and the gray values of the spot and the noise have an obvious difference, then the effect of blurring that is produced by the median filter can be ignored.

If the noise is in the form of isolated points appearing in the image and rarely affects multiple pixels, and the gray values of the spot and the noise are closer, then the linear–median filter will be better.

After the performance of image denoising, the threshold gravity method[6, 7] is normally used to calculate the spot center position. This method usually sets a threshold value and then converts the images into two-dimensional value images. During this process, if the size of some image is $A*B$, $f(i, j)$ is the gray value of one pixel and the gray threshold value is set as T, then the threshold procedure can be expressed as

$$F(x, y) = \begin{cases} f(x, y), & f(x, y) \geqslant T, \\ 0, & f(x, y) \leqslant T. \end{cases} \tag{5}$$

When the center of gravity method is used to calculate centrality, a binarized matrix is actually calculated to obtain the center coordinates (x_0, y_0), as follows:

$$x_o = \frac{\sum_{x=1}^{A} \sum_{y=1}^{B} F(x, y)x}{\sum_{x=1}^{A} \sum_{y=1}^{B} F(x, y)},$$
$$y_o = \frac{\sum_{x=1}^{A} \sum_{y=1}^{B} F(x, y)y}{\sum_{x=1}^{A} \sum_{y=1}^{B} F(x, y)}. \tag{6}$$

The original beam images that are captured by digital CCDs are shown in Figure 3. Figures 3(a) and (b) are near-field images. From these images we can see that the near

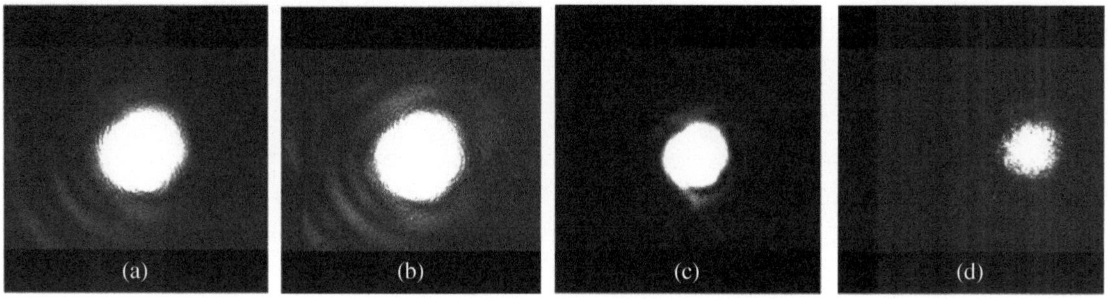

Figure 3. Original images of the near and far fields. (a), (b) Near-field images. (c), (d) Far-field images.

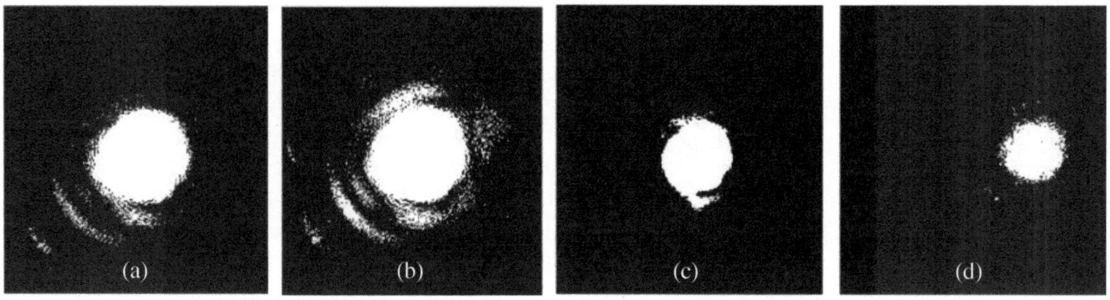

Figure 4. Images of threshold processing without a filter. (a), (b) Near-field images. (c), (d) Far-field images.

field has a clear image and a larger spot; Gaussian white noise is the main noise. Therefore, the main purpose of the near-field image is to remove Gaussian white noise, so the median–linear filter is appropriate for this situation. Figures 3(c) and (d) are far-field images. For these images, the imaging is not clear and the spot area is smaller than for the near-field ones. The main noise is salt and pepper noise, so the linear–median filter is better here.

If we directly perform threshold processing on these images without denoising, the results are shown in Figure 4. Gaussian noise is very obvious in Figures 4(a) and (b); salt and pepper noise is very obvious in Figures 4(c) and (d). The presence of these types of noise will have a serious effect on the calculation of the center.

For the near-field images, using the median–linear filter to denoise first and then performing the threshold processing, the detailed procedure is as follows:

(i) execute median filtering to remove salt and pepper noise;

(ii) execute average filtering to weaken Gaussian noise;

(iii) execute threshold processing and obtain the gray-scale map;

(iv) obtain the spot center by Equation (6) and calculate the deviation from the reference position;

(v) adjust the mirrors through the motor feedback control system.

The results of processing the image are shown in Figure 5.

The detailed procedure for the far field is similar to the above steps; the only difference is that the average filtering is performed first, and then the median filter is used to remove salt and pepper noise as well as to restore the clarity of the images. The results are shown in Figure 6.

3. Motor feedback control

The processed image captured by digital CCD is used to calculate the deviation, and the feedback control is used to align the beam path, which is accomplished by converting the deviation into motor rotation steps. The image center calculation and motor rotation are key factors, which directly affect the reflector adjustment. To finish the alignment in the shortest time, parallel alignment is conducted in the multi-path laser system. Therefore, the speed of image processing for a single path determines the alignment efficiency of the whole system[8, 9].

Ideally, the motor rotation step number is a fixed value corresponding to a pixel, according to the step number; it is only necessary to align once to adjust the beam center back to the reference position. However, as it is affected by interference factors, the relation between the pixel and the motor rotation step number is nonlinear in real time. The alignment feedback adjustment procedure adopts multiple iterations to adjust the beam path until the deviation of the beam center position is acceptable. The number of iterations is a random function of the pixel and the motor step number, which theoretically depends on the goodness of the fit

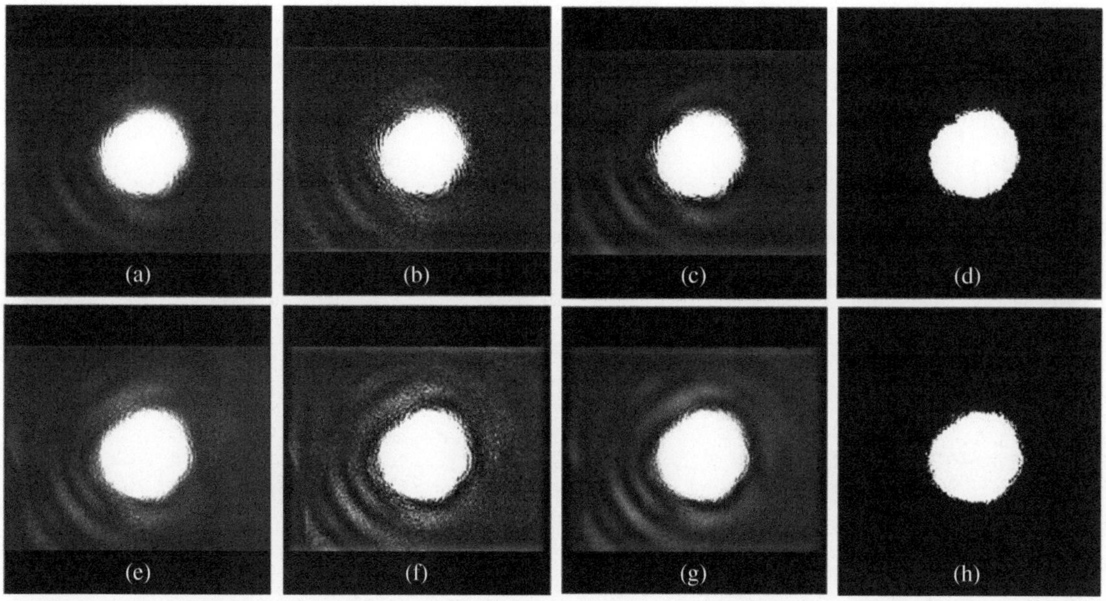

Figure 5. Near-field images after multiple filters. (a), (e) The original images captured by near-field CCD. (b), (f) After use of a median–linear filter with 25 pixels. (c), (g) After use of an average filter with eight pixels. (d), (h) The images after threshold processing has been performed.

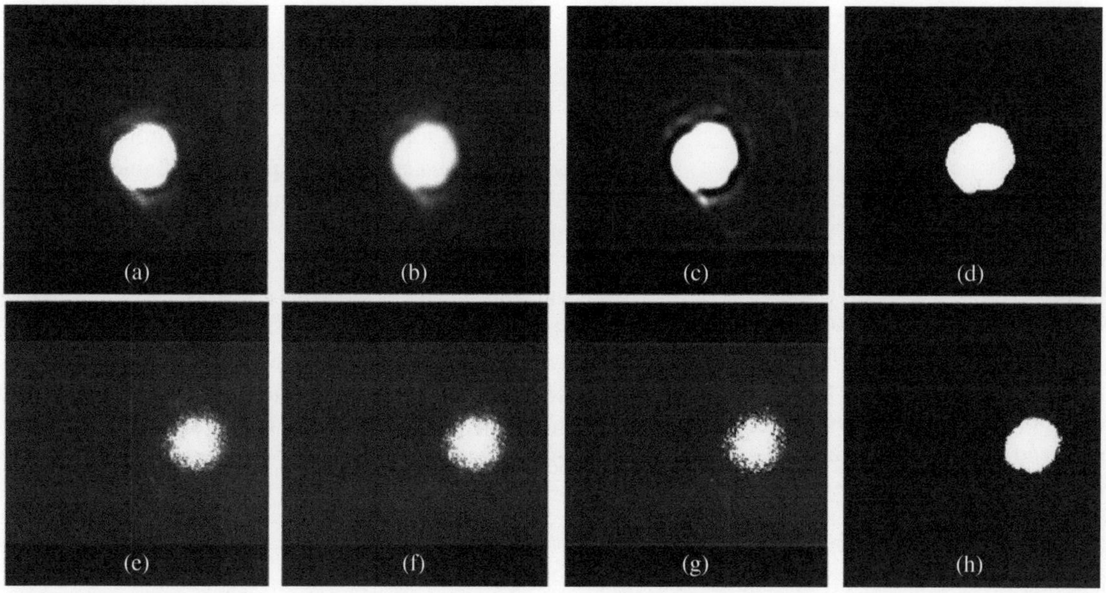

Figure 6. Far-field images after multiple filters. (a), (e) The original images captured by far-field CCD. (b), (f) After use of an average filter with eight pixels. (c), (g) After use of a median–linear filter with 25 pixels. (d), (h) The images after threshold processing has been performed.

between the random function and the nonlinear relationship. Therefore, the adjustment time fluctuates violently.

In order to eliminate the influence of the nonlinear relation between the pixel and the motor rotation step number, and moreover to reduce the number of iterations of mirror adjustment, we introduce a fuzzy control algorithm to rapidly align the beam with the specified center. The fuzzy control algorithm can realize universal approximation performance of the unknown control structure and obtain an adaptive variable related to the pixel and motor rotation step number. In other words, the real-time adaptive variable step control is determined by the actual deviation distance, which is determined by a fuzzy decision in accordance with the fuzzy rule. This ensures the fast and accurate automatic alignment, and enhances the robustness of the system [10, 11].

First, fuzzy linguistic rules are used to change the experience and knowledge of the facility into fuzzy control rules. Because this process does not depend on an accurate mathematical model of the controlled object, we can ignore the effect of non-linear quantitative distance adjustment when

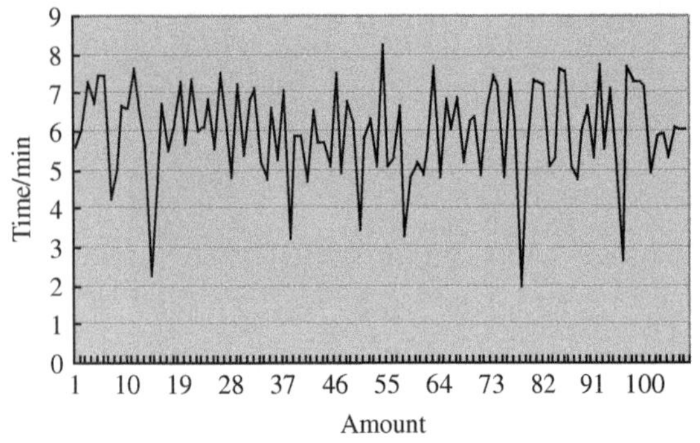

Figure 7. The results of the iteration method without a filter.

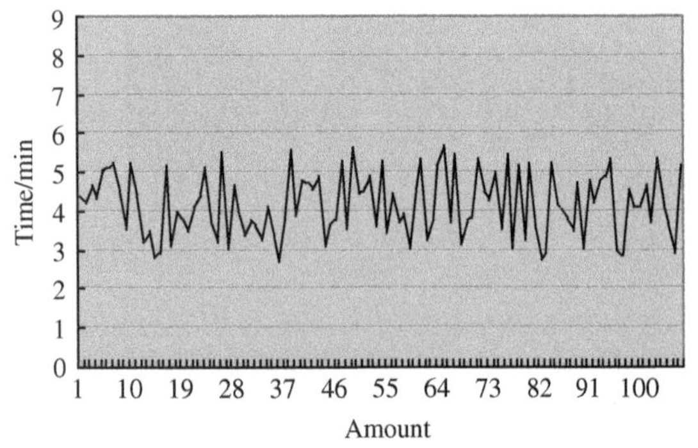

Figure 8. The result of the adaptive variable step method with filter denoising.

the deviation is large. The real-time adaptive variable step will reduce when the spot center approaches the reference position; when it is close to the reference position, the adaptive variable step will become small. The adaptive variable step value is determined by the specific evaluation function and is quantified by fuzzy rules.

The fuzzy control uses a two-dimensional fuzzy controller. The deviation e and the deviation change e_c are inputs, their corresponding linguistic variables are set as E and E_C, the output is the control quantity u and the corresponding linguistic variable is set as U.

The linguistic variable value set of E_C and U can be expressed as

$$E_C = U = \{NB, NM, NS, O, PS, PM, PB\}. \tag{7}$$

The corresponding meanings are negative-big/negative-medium/negative-small/zero/positive-small/positive-medium /positive-big/.

The linguistic variable value set of E can be expressed as

$$E = \{NB, NM, NS, NO, PO, PS, PM, PB\}. \tag{8}$$

The corresponding meanings are negative-big/negative-medium/negative-small/negative-zero/positive-zero/positive -small/positive-medium/positive-big.

Table 1. Fuzzy control rule.

	NB	NM	NS	O	PS	PM	PB
NB	PB	PB	PB	PB	PM	O	O
NM	PB	PB	PB	PB	PM	O	O
NS	PM	PM	PM	PM	O	NS	NS
NO	PM	PM	PS	O	NS	NM	NM
PO	PM	PM	PS	O	NS	NM	NM
PS	PS	PS	O	NM	NM	NM	NM
PM	O	O	NM	NB	NB	NB	NB
PB	O	O	NM	NB	NB	NB	NB

The fuzzy rule of e and e_c is

$$E = \begin{cases} 0, & e \leqslant 2; \\ INT(e/25), & 2 < e. \end{cases}$$

$$E_c = \begin{cases} 0, & e_c \leqslant 2; \\ INT(e_c/20), & 2 < e_c. \end{cases} \tag{9}$$

The conditional statement of the fuzzy control rules is

$$\text{if } E = A_i \quad \text{then} \quad \text{if} \quad E_c = B_j \quad \text{then } U = C_{ij} \\ (i = 1, 2, \ldots, 8;\ j = 1, 2, \ldots, 7\}. \tag{10}$$

The A_i, B_j, and C_{ij} respectively belong to deviation set E, deviation change E_C, and control quantity set U. Equation (10) can also be expressed as Table 1.

When the alignment is begun, CCDs will capture beam images for processing to obtain the deviation e and the amount of deviation change e_c. Then, according to Equations (7) and (8) we can get the values of E and E_C; thus, the output of the control quantity U can be obtained from Equation (10) by inquiry, Table 1; the value of U determines the variable step size and number of rotation steps. The above procedures should be repeated until the center of the beam reaches the specified location.

4. Experimental results

In order to verify the validity of the image processing and fuzzy control in this paper, the following experiment is conducted. A small simulation platform is built in accordance with the SG II ninth beam path, which is driven by a motor to move 14 reflectors. The platform is monitored by seven sets of optical imaging systems composed of CCDs and optical lenses. In the experiment, the reflectors in a path are moved by some random steps to deflect the beam center. Then, the beam path is adjusted to make the beam center return to the reference position. The experiment is repeated 100 times, and the corresponding alignment times are recorded to be plotted as shown in Figures 7 and 8.

Figure 7 is the time curve of the original scheme. The original scheme first calculates the center error by a center of gravity method for the threshold value, and then adjusts the beam to the path center by multiple iterations. Figure 8 is the time curve of the new scheme. The new scheme first calculates the center error after filtering by the hybrid filter. Then, the optical path is adjusted by an adaptive variable step method. It can be seen that the time consumed and stability of the new scheme have a very significant improvement. The reflector adjustment process is completed in less than 4 min and the overall time for the system is less than 5 min.

In these 100 experiments, the fastest image processing time is 9 s (in SF1 far-field CCD) and the slowest image processing time is 57 s (in SF7 near-field CCD); the main processing time distribution is 27–46 s. The far-field image processing time is faster than the near-field image processing time. Because the image is processed three times, the time cost is increased, but the fuzzy control can effectively reduce the amount of iteration adjustment and save time. By using the fuzzy control, the adjustment can be finished in 3–5 steps and the accuracy is up to ± 1 pixel; the old scheme needs 6–10 adjustment steps and the accuracy is only up to ± 1 pixel.

5. Conclusion

This method provides improved image processing of a laser alignment system based on mean filtering and median filtering, which is combined with the operating characteristics of a large amount of Gaussian noise and impulse noise in the imaging process of the laser alignment system. The correction speed of the path center and the computational accuracy are improved by different hybrid filters in accordance with the near- and far-field spot features. In addition, an adaptive step feedback adjustment mechanism based on a fuzzy algorithm is built to lower the impact of the nonlinear random relation between the unit pixel and motor adjustment steps. It improves the fast automatic alignment capability and the robustness of the system.

Acknowledgements

This paper was supported by grants from the Chinese and Israeli cooperation project on high power laser technology (2010DFB70490).

References

1. F. Liu, Z. Liu, L. Zheng, H. Huang, and J. Zhu, High Power Laser **1**, 29 (2013).
2. D. R. Mendoza, *International Congress on Instrumentation in Aerospace Simulation Facilities* (Pacific Grove, California, 1997), p. 22.
3. H. Faraji and W. J. MacLean, Image Proc. IEEE **15**, 2676 (2006).
4. T. Bose, *Digital Signal and Image Processing* (John Wiley & Sons Inc, New York, 2003).
5. V. Gupta and M. Shandilya, *International Conference on Engineering* (Academic, Ahmedabad, 2012), p. 1.
6. E. R. Davies, IEE Proc. Vis. Image Signal Process. **146**, 279 (1999).
7. L. Liu and G. Y. Kuang, J. Image Grap. **14**, 622 (2009).
8. F. N. Lv, D. Z. Liu, R. F. Xu, J. Z. Cao, and D. Y. Fan, Opt. Technol. **31**, 335 (2005).
9. D. Liu, L. Ding, and Y. Gao, Chin. J. Lasers **36**, 1101 (2009).
10. C. C. Lee, Syst. Man Cyber. **20**, 404 (1990).
11. Z. Kovacic and S. Bogdan, *Fuzzy Controller Design Theory and Applications* Y. Hu, L. Zhang, Y. Liu, and Y. Chen (trans.) (China Machine Press, Beijing, 2010).

13

Splicing technology of Ti:sapphire crystals for a high-energy chirped pulse amplifier laser system

Yanqi Liu, Yuxin Leng, Xiaoming Lu, Yi Xu, and Cheng Wang

State Key Laboratory of High Field Laser Physics, Shanghai Institute of Optics and Fine Mechanics, Chinese Academy of Sciences, Shanghai 201800, China

Abstract

We develop a splicing technology of Ti:sapphire crystals for a high-energy chirped pulse amplifier laser system that can suppress the parasitic lasing to improve the amplification efficiency compared to a large-size single Ti:sapphire crystal amplifier. Theoretical investigations on the characteristics of the amplifier with four splicing Ti:sapphire crystals, such as parasitic-lasing suppression and amplification efficiencies, are carried out. Some possible issues resulting from this splicing technology, including spectral modulation, stretching or splitting of the temporal profile, and the sidelobe generation in the spatial domain (near field and far field), are also investigated. Moreover, the feasibility of the splicing technology is preliminarily demonstrated in an experiment with a small splicing Ti:sapphire crystals amplifier. The temporal profile and spatial distribution of the output pulse from the splicing Ti:sapphire crystal amplifier are discussed in relation to the output pulse from a single Ti:sapphire crystal amplifier.

Keywords: chirped pulse amplification; splicing technology; Ti:sapphire crystal amplifier

1. Introduction

Several high-power femtosecond laser systems have been recently built because of the urgency in developing ultra-intense and ultra-short lasers for high-field physics[1–5]. A petawatt (PW) level peak power laser pulse has even been achieved based on the Ti:sapphire chirped pulse amplifier (CPA) concept.

Although Ti:sapphire crystals have been used widely in PW-level ultra-intense and ultra-short lasers because of their advantages, such as broad gain bandwidth for supporting 10 fs-level pulse duration and excellent optical and mechanical properties, Ti:sapphire crystals and the related CPA scheme have not been considered as a suitable gain medium and laser amplification scheme for the even higher 10 PW-level peak power laser. The critical problem is that parasitic lasing (PL)[6], which results from transverse amplified spontaneous emission (ASE), exists in larger-sized Ti:sapphire crystal amplifiers for higher energy laser pulse amplification. When the chirped laser pulses are amplified to near 100 J energy, further energy amplification would be difficult, and the chirped pulse beam quality and energy gain decrease quickly because of the parasitic lasing in a Ti:sapphire CPA laser system[1]. However, large-diameter Ti:sapphire crystal growth is complex and expensive, and will take a long period. Thus, optical parametric chirped pulse amplification (OPCPA)[5] based on a large-size nonlinear crystal has to be used in developing a 10 PW even EW laser, because this technology can avoid the influence of the PL[6], although there is higher efficiency and more reliability and stability for a CPA laser systems based on Ti:sapphire. The possibility of using many pieces of Ti:sapphire crystals to compose a larger splicing Ti:sapphire crystal for a high-energy CPA amplifier was considered several years ago. However, further research has not been carried out to our knowledge[2].

In this paper, we develop a Ti:sapphire crystal splicing technology, and demonstrate its feasibility in a CPA laser. The theoretical investigation demonstrates that PL can be suppressed efficiently with a larger aperture Ti:sapphire CPA amplifier. Moreover, compared with a single Ti:sapphire crystal amplifier, the splicing Ti:sapphire crystal amplifier have the similar results of the theoretical and experimental investigations, such as, the spectrum, the pulse duration, and the beam spatial distribution in the far field.

2. Theory

For a high-energy CPA system, a large aperture of the laser crystal is necessary for saturable pumping or damage flux existing. Unfortunately, PL will be obtained more readily with the enlarging of the aperture, thereby reducing the amplification efficiency of the system. In this section, the

Mailing address: Yuxin Leng: No. 390, Qinghe road, Jiading, Shanghai, CN 201800, China. Correspondence to: Email: lengyuxin@siom.ac.cn principal condition of parasitic lasing is discussed, and it is shown that splicing technology can suppress the parasitic lasing theoretically. The possible issues resulting from the splicing Ti:sapphire crystal technology, and the influences of these issues, such as spectral modulation, the temporal profile of the compressed pulse, and the spatial distribution, are also discussed.

The threshold of parasitic lasing depends on the gain of the crystal. The gain of the crystal based on the amplifier (which in this case is Ti:sapphire) can be calculated with inverted population density, which can be obtained by using the Frantz–Nodvik (F–N) equation[7]:

$$G = \exp\left\{\frac{W_0}{2}[\exp(-\alpha_p \cdot z) + \exp(-\alpha_p \cdot (L_z - z))] \times \left(\frac{\alpha_p}{\hbar \cdot \omega_p}\right) \cdot 2D \cdot 3e^{-19}\right\}, \tag{1}$$

where G is the gain of the crystal, W_0 is the density of the pump energy, α_p is the absorption coefficient, L_z is the thickness of the Ti:sapphire, $\hbar$, ω_p are the Planck constant and the angular frequency, D is the clear aperture, and z is the distance from the surface to the inner of the crystal. The gain of the transverse ASE is the highest on the surface, and the parasitic lasing is also the strongest; thus the situation of parasitic lasing on the surface is used to represent the crystal[8]; in this model, $z = 0$. Equation (1) indicates that a stronger pump generates higher G, and the threshold of the parasitic lasing can be more easily reached. The equation $R_1 \cdot R_2 \cdot G_{\text{ASE}} = 1$ can be used as the criterion for parasitic lasing, where R_1 and R_2 are the reflectivities of the crystal side surfaces. G_{ASE} is the gain at the crystal surface, which also indicates the threshold of the parasitic lasing previously mentioned. Then Equation (1) can be expressed as follows:

$$R_1 \cdot R_2 \cdot \exp\left\{\frac{W_0}{2}[\exp(-\alpha_p \cdot z) + \exp(-\alpha_p \cdot (L_z - z))] \times \left(\frac{\alpha_p}{\hbar \cdot \omega_p}\right) \cdot 2 \cdot (2 \cdot r) \cdot 3e^{-19}\right\} = 1, \tag{2}$$

where $2r = D$. Equation (2) indicates that there should be a threshold of pump energy with a special diameter.

The result is shown by the dashed curve in the Figure 1.

The dashed curve from Figure 1 indicates that the density of the pump energy should be decreased to avoid the occurrence of parasitic lasing, even if the aperture is increased. A common method is cladding with refractive-index-matched liquid or direct absorber coating on the side surfaces[1] of the crystals to decrease the reflectivity. This method has been used widely in PW-level systems. However, in a system with higher peak power output, the aperture of the crystal is enlarged to avoid damage. Equation (2) shows that the threshold of PL would be decreased[4].

Meanwhile, the transverse gain can be decreased and the threshold can be increased significantly when the optical path length of the ASE can be cut off, which can be achieved with splicing technology. Figure 2 shows that the larger crystal can be composed of 2 × 2, 4 × 4, or more crystals. The gap can be filled with an index-matched material, liquid or solid; and the material has high absorption coefficient about ASE avoiding the reflection of ASE at the side face of Ti:sapphire crystal. In this way, the threshold can be increased, and the result is shown in Figure 1, which is calculated from Equation (2). The 2 × 2 splicing crystals (the dot curve) and 4 × 4 splicing crystals (the dashed curve) have a higher threshold than the single crystal (the solid curve). For example, when the radius is 5 cm, the threshold is 6.68 J cm^{-2} for the 4 × 4 splicing crystals; the threshold is 4.72 J cm^{-2} for the 2 × 2 splicing crystals, and the threshold is 2.36 J cm^{-2} for the single crystal.

Figure 1 also indicates that this technology has the advantage of short growing period and low cost. For example,

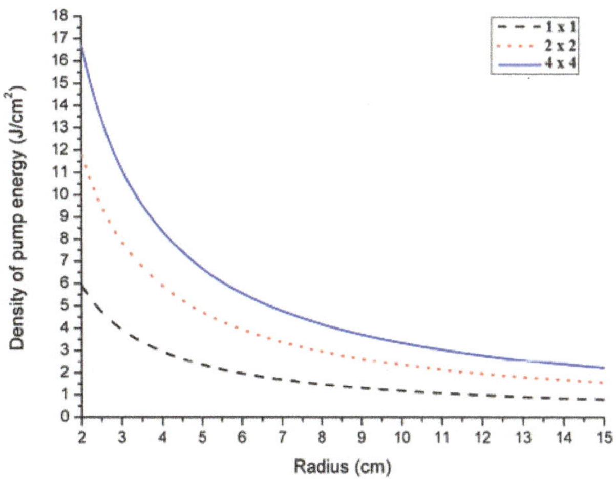

Figure 1. Theoretic relationship between the radius and the threshold of pump energy with a single crystal (dashed line), 2 × 2 splicing crystals (dot line), and 4 × 4 splicing crystals (solid line). The absorption coefficient is 0.94.

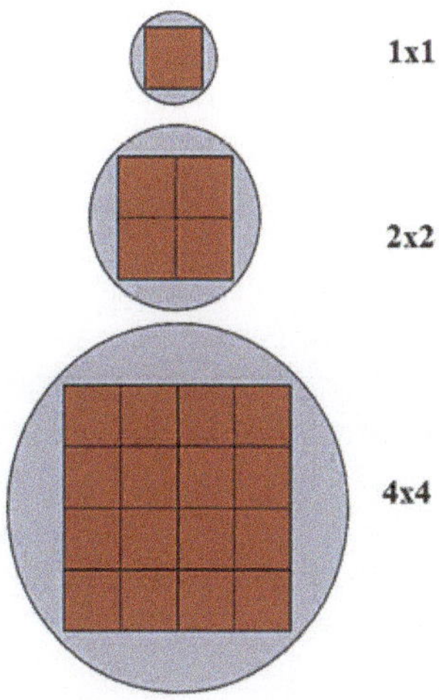

Figure 2. Scheme of splicing crystals.

in a 10 PW CPA system, when the pulse width of the extraction laser is 30 fs, the output energy would be 300 J. Considering the efficiency of the compressor and the conversion efficiency from pump to extraction, the pump energy of the final amplifier will be more than 850 J. With the ordinary index-matched material having a reflectivity index of 0.048%, the aperture is much more than for the largest Ti:sapphire that has been reported. Even when better index-matched materials are used, the aperture would still be more than 300 mm. However, using the splicing technology of the Ti:sapphire crystal, only four 120 mm × 120 mm crystals are needed to compose a large-aperture splicing crystal with clear aperture of 240 mm.

In the period of the CPA system running, the accurate threshold of the PL, in our experience, is always lower than the results previously shown, and the calculation process is also more complicated because of the influence of some uncertain factors, such as the beam homogeneity, and the delay jitter between the pump laser and the extraction laser. However, these factors can hardly affect the conclusion of this section, because the splicing crystals have similar effects to those of the single crystal. For example, the threshold of PL with the single crystal decreased from 2.36 J cm^{-2} to 1.18 J cm^{-2}, while the threshold of the 2×2 splicing crystals decreased from 4.72 J cm^{-2} to 2.36 J cm^{-2} under the same conditions.

According to the analysis above, crystal splicing technology has the potential to increase the threshold of PL in a CPA system. However, as previously mentioned, possible issues should be considered, including spectral modulation, stretching or splitting of temporal profile, and the sidelobe in spatial domain which may worsen the consistency of the focusing spot in the spatial domain or time domain compared with the single crystal amplifier. Furthermore, the solving of these issues is limited to inevitable errors in the process of crystal splicing, such as the different thicknesses of the crystals, angle errors of the optical axis, and the gap of the splicing crystal. Thus, details of these issues will be discussed theoretically in what follows.

(i) Spectral modulation and the temporal profile of the compressed pulse.

Unlike the single crystal amplifier, the errors mentioned before can be hardly corrected by manual adjustment during the period of the system running. For example, as the 2 × 2 splicing crystals have four crystals and four optical axes correspondingly, if one of the crystals is adjusted and its optical axis is corrected, the errors of the other crystals may be worsened, which would generate spectral modulation and a sidelobe pulse in the time domain as well as thickness differences. For these reasons, these kinds of influence should be calculated.

The Ti:sapphire crystal is a uniaxial crystal, and the grating in the compressor system is sensitive to the laser pulse polarization[9]. Therefore, a Ti:sapphire crystal and grating are equivalent to a birefringent plate and a polarizer, and the

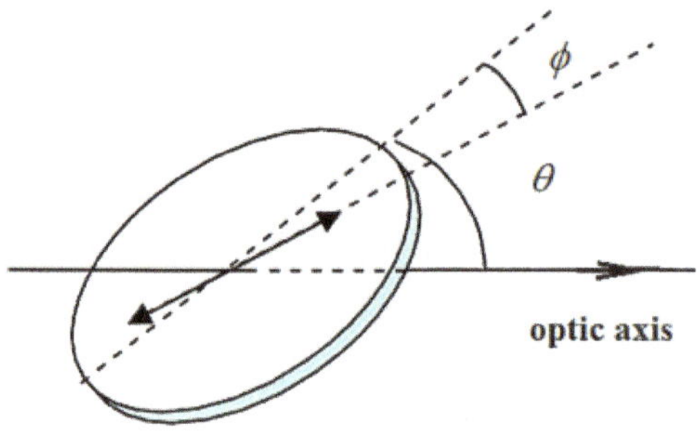

Figure 3. Schematic of θ and φ.

angle errors (Figure 3) would modulate the spectra of the laser[10, 11]. The equation for the spectral modulation is as follows:

$$T(\lambda) = 1 - \sin^2(2\phi)\frac{n_o^4 - n_o^2\cos^2(\theta)}{(n_o^2 - \cos^2\phi\cos^2\theta)^2} \times \sin^2\left\{\frac{\pi t n_e[1 + \cos^2\theta\cos^2\phi/n_e^2 - \cos^2\theta\cos^2\phi/n_o^2]}{\lambda[1 - \cos^2\theta\sin^2\phi/n_e^2 - \cos^2\theta\cos^2\phi/n_o^2]^{1/2}} - \frac{\pi t n_o}{\lambda[1 - \cos^2\theta/n_o^2]^{1/2}}\right\}, \quad (3)$$

where $T(\lambda)$ is the transmittance related to the wavelength (λ); n_o, n_e are the principal refractive indices; and θ, φ are the angle errors.

Equation (3) shows that the spectral modulation is influenced by three factors: the thickness of the crystal, and the angle errors θ and φ; different values of these factors would generate different situations of spectral modulation. Figure 4a shows the situation of spectral modulation at $\theta = 10°$, $\varphi = 18.5°$; the thickness of the crystal is 10 mm, and the beam single-passes the crystal.

As we know, the spectral modulation changes the temporal profile of the compressed pulse. Thus, using Fourier transformation, the temporal profile from the spectral modulation can be calculated, as shown in Figure 4a shows. Figure 4b is then obtained, indicating that two sidelobes are generated and the intensity is 10% of the main pulse[12].

Another inevitable error that can influence the temporal profile of the compressed pulse is the thickness difference, show in Figure 5. Different thicknesses would stretch or even split the pulse duration. For example, in the splicing crystal CPA system with a duration of 30 fs, only about 5–6 mm of thickness differences (the influence of the GDD are ignored) can stretch the pulse duration about 60 fs, or even cause it to be split. This phenomenon is a very serious problem for ultra-short lasers, in which the pulse duration is of fs order. Thus, this error should be given special attention in the process of splicing the crystals.

(ii) The spatial distribution of the focused point.

The existence of a gap for the compression of parasitic lasing is inevitable, which may affect the spatial distribution. This phenomenon can be explained easily by diffraction theory[13]. In fact, the result can be easily obtained through

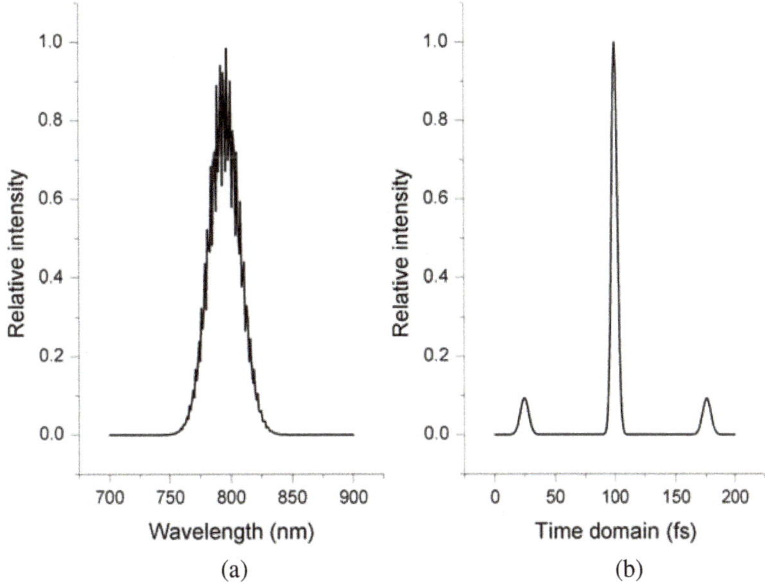

Figure 4. (a) Situation of spectral modulation (single-pass) at $\theta = 10$, $\varphi = 18.5$; the thickness of the crystal is 10 mm. (b) Sidelobes generated by spectral modulation.

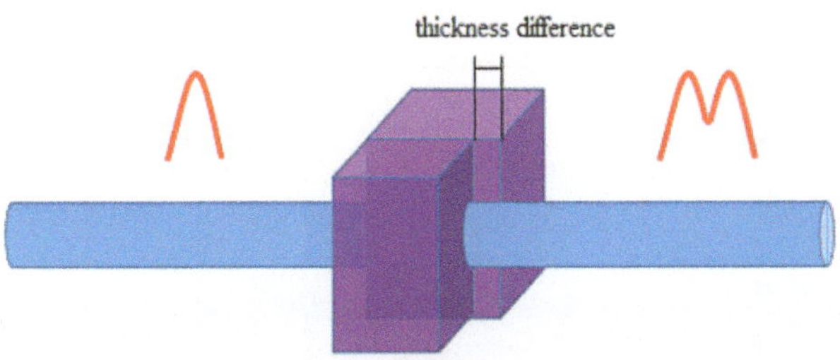

Figure 5. Schematic of the influence of thickness difference on the temporal profile.

FFT (fast Fourier transformation) using Matlab. Figure 6 shows the simulated result, which indicates that the gap would generate a series of sidelobes in the frequency domain. In this case, the second intensity sidelobe is about 10% of the main peak; the width of the gap is 10% of the beam diameter.

3. Experiment

As we discussed before, some inevitable errors in the processing of crystal splicing generate possible issues, and the issues are solved in two aspects [(i) and (ii)]. However, this study focuses on the influence of these possible issues, that is, whether these inevitable errors can be limited to be small enough to be ignored using existing crystal processing technology. Thus, in this study, some experiments have been carried out to test the influence of these issues with the two aspects in theory, and the parameters used below are processed with ordinary accuracy. Notably, the experiment is limited to the aperture of the splicing crystals (10 mm); an experiment about suppressing PL has not been carried out.

The experiment was carried out in a typical CPA system based on Ti:sapphire. The scheme is shown in Figure 7a–c. The experiment shows (a) measurement of the energy, spot in the near and the far fields, and the spectra; (b) measurement of the autocorrelator trace; and (c) a photo of 2×2 spliced Ti:sapphire 7.5×7.5 mm^2 in size, 10 mm in thickness, and an absorptivity of 80% for 532 nm. The extraction beam passed through the splicing crystals twice when the crystal was pumped. The wavelength of the pump beam is 532 nm, and that of the extraction beam is from 760 nm to 830 nm. A spectrometer was used to measure the situation of the spectral modulation; a calorimeter was used to measure the energy of the amplified beam, and a CCD was used to observe the spatial distribution. Finally, a Glan prism was used to replace the gating to measure the temporal profile of the compressed pulse with autocorrelation and CCD.

The diameters of the extraction and pump beams are 10 mm; the extraction beam is 0.5 mJ and the pump is 1.7 J; we obtained 1.1 mJ output for a single pass and 3.1 mJ for a double pass. Considering that the absorptivity of the crystal without coating is 80% for the pump, this result is in accordance with the theoretically expected result.

(i) Experiment of the spectral modulation and the temporal profile.

The calculation result shows that the angle errors generate spectral modulation. However, when the angle errors are less than 1° (ordinary accuracy), this kind of spectral modulation can be ignored. Figure 8 shows the spectra of the whole beam measured by the spectrometer, and Figure 9 shows the spectra of each beam passing through the four smaller crystals of the splicing crystals.

In Figure 8, the dashed curve indicates the original spectrum of the extraction beam, the dotted curve indicates the

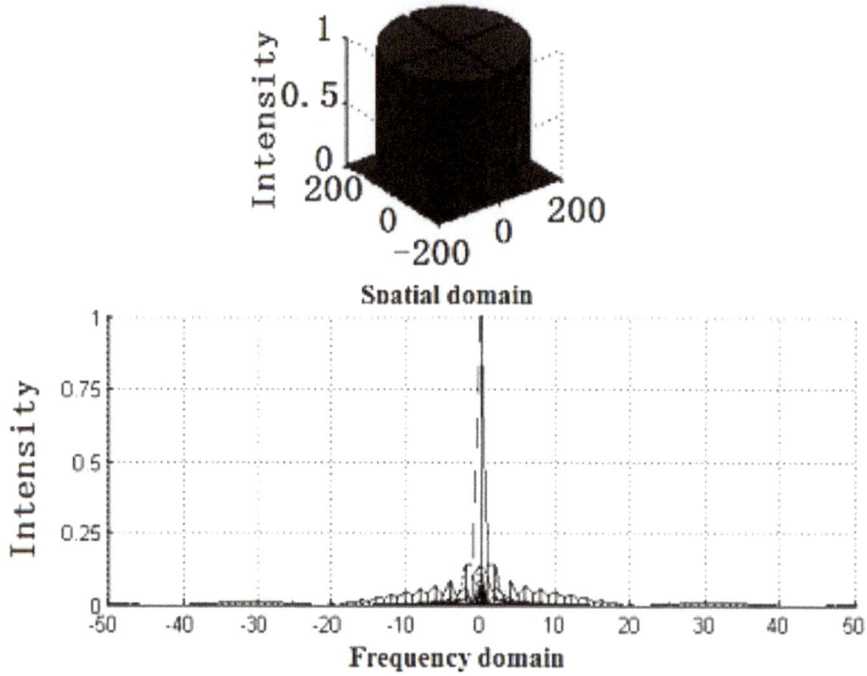

Figure 6. Theoretical relationship between the spot in the near field and the spot in the far field.

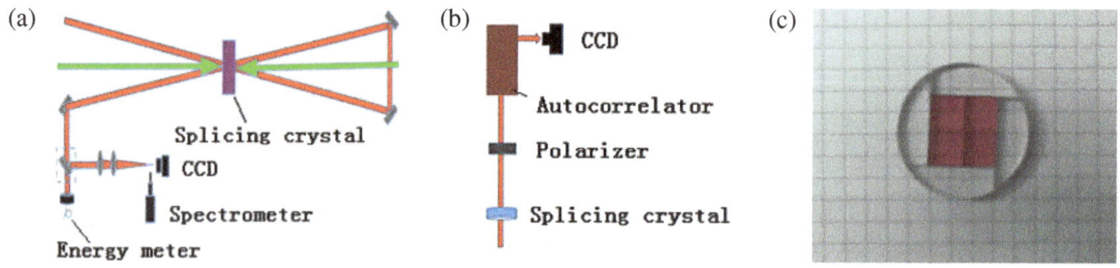

Figure 7. Scheme of experiment: (a) measurement of the energy, spot in the near and the far fields, and spectra; (b) measurement of the autocorrelator trace; (c) photo of the 2 × 2 splicing crystals.

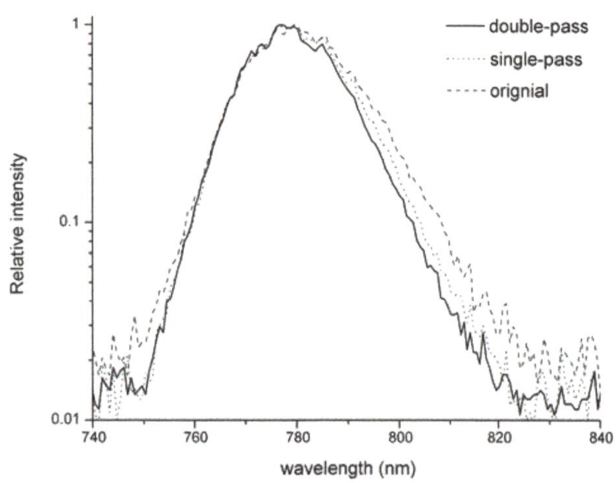

Figure 8. Spectrum of the seed: the dashed curve is the original spectrum; the dotted curve is the spectrum amplified by a single pass; the solid curve is the spectrum amplified by a single pass.

extraction beam single passed through the crystal and the solid curve is that for a double pass. The spectra did not have any obvious change, and the spectra of each beam that passed through the four smaller crystals of the splicing crystals in Figure 8 also support this point. The solid line in Figure 9 indicates a single pass, and the dotted line indicates a double pass.

The temporal profile of the compressed pulse can also be measured in the experiment, and the result is shown in Figure 10.

The result in Figure 10 was measured by autocorrelation and CCD. The width of the curve respects the pulse duration[14]: (a) is the autocorrelation trace of the beam that did not pass through splicing crystals, and (b) is the autocorrelation trace of the beam that passed through the splicing crystals. The pulse duration is 63 fs in (a) and 62 fs in (b), respectively. The results show that the pulse duration did not stretch.

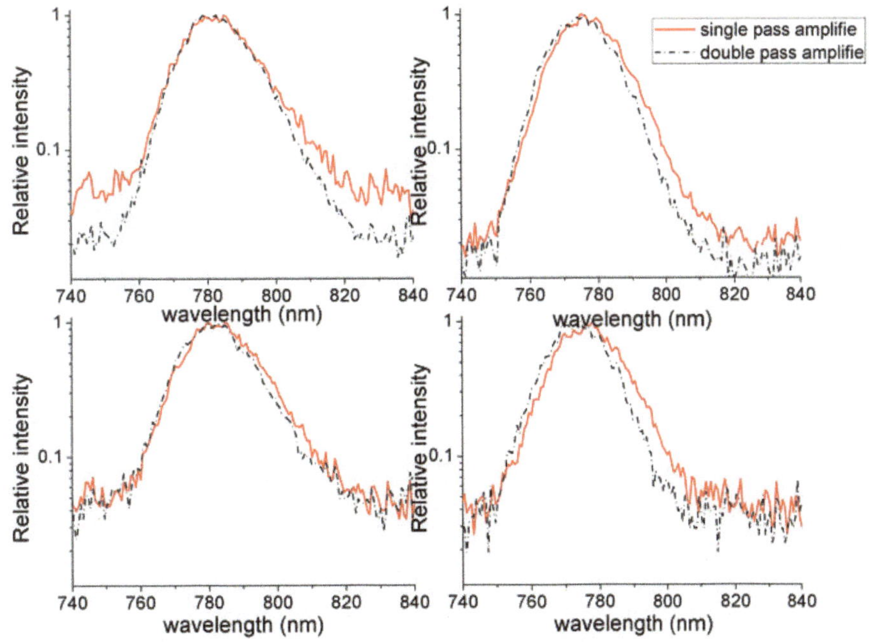

Figure 9. Spectrum of the beam passed through the smaller crystals of the 2 × 2 splicing crystals.

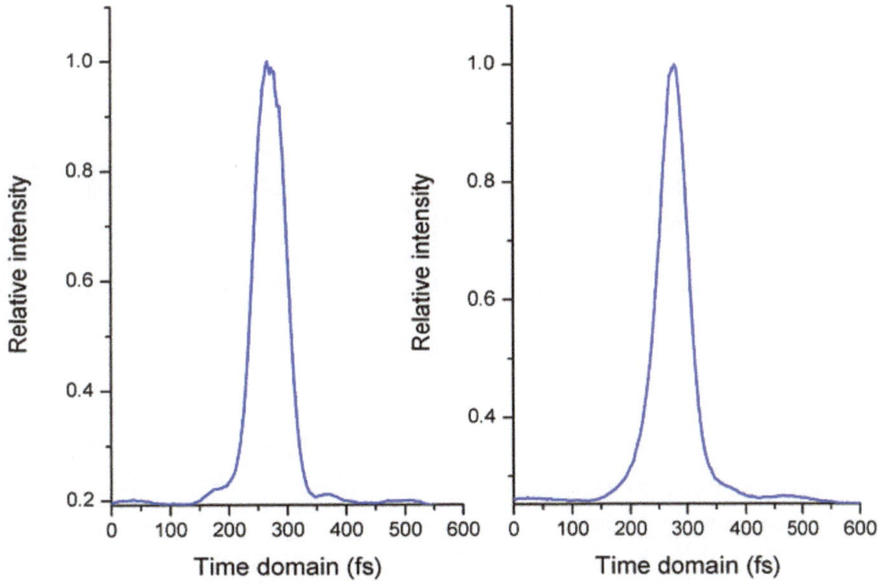

Figure 10. Autocorrelation trace in the experiment: (a) original trace; (b) the trace of the splicing crystals.

(ii) Experiment of the spatial distribution of the focused point.

There are two problems about the spatial distribution: the diffraction effect in the near field and the spatial sidelobes in the far field. Figure 11a shows the diffraction of the extraction beam that passed through the splicing crystals in the near field, and the width of the gap is about 10% of the aperture. The diffraction effect is intense, which may damage the optical elements. However, a soft-edge aperture can solve this problem. Figure 11b shows the spot with the soft-edge aperture, and the shape is +, as Figure 11c shows. This figure indicates that the diffraction effect significantly decreased.

The spatial distribution in the far field is also measured. Figure 12a shows the original spot in the far field in which the beam did not pass through the crystal, (b) shows the spot single-passed through the splicing crystal without the aperture, and (c) shows the spot single-passed through the splicing crystal with a soft-edge aperture. These results show that, in the far field, the spot that passed through the crystal

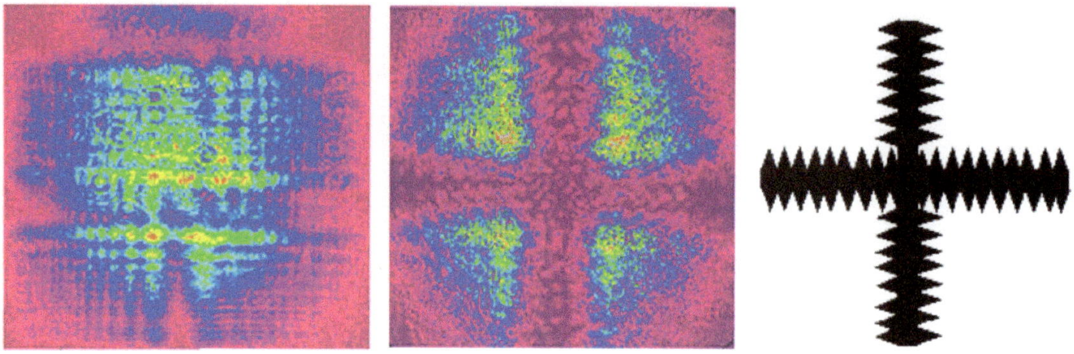

(a) (b) (c)

Figure 11. Spot of the extraction beam in the near field: (a) with soft-edge aperture; (b) without edge aperture, and (c) schematic of the soft-edge aperture.

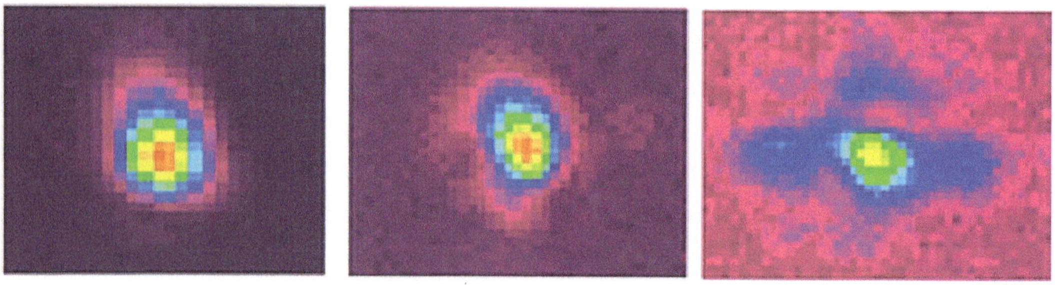

(a) (b) (c)

Figure 12. Spot of the extraction beam in the far field: (a) did not pass through the crystal, (b) passed through the crystal without the soft-edge aperture, and (c) passed through the crystal with the soft-edge aperture.

(Figure 12b) has similar spot distribution to the original spot distribution (Figure 12a), where the gap is about the 10% of the width of the clear aperture. Figure 12c shows that the sidelobes are generated because the soft-edge aperture is too wide (about 20% of the clear aperture).

4. Conclusion

We have developed a splicing technology of Ti:sapphire crystals for chirped pulse amplifier (CPA) laser systems. Theoretical and experimental investigations on the amplifier with four splicing Ti:sapphire crystals, such as amplification efficiency, spectral modulation, temporal profile of the compressed pulse, and the beam spatial distribution (near field and far field), have been carried out. The theoretical result shows that splicing technology of the Ti:sapphire crystal can be a possible scheme to enlarge the aperture of the crystal and suppress PL in the CPA systems. Some experiments were carried out to demonstrate that the possible issues generated by the inevitable errors can be ignored with ordinary accuracy.

Acknowledgements

This work is partially supported by the National Basic Research Program of China (Grant No. 2011CB808101), the National Natural Science Foundation of China (NSFC) (Grant Nos. 61221064, 61078037, 11127901, 11134010), and the International S&T Cooperation Program of China (Grant No. 2011DFA11300).

References

1. X. Liang, Y. Leng, C. Wang, C. Li, L. Lin, B. Zhao, Y. Jiang, X. Lu, M. Hu, C. Zhang, H. Lu, D. Yin, Y. Jiang, X. Lu, H. Wei, J. Zhu, R. Li, and Z. Xu, Opt. Express **15**, 15335 (2007).
2. T. Tajima and G. Mourou, Phys. Rev. Special Topics-Accelerators and Beams **5**, 031301 (2002).
3. Y. Xu, J. Wang, Y. Huang, Y. Li, X. Lu, and Y. Leng, High Power Laser Science and Engineering **1**, 98 (2013).
4. Y. Chu, X. Liang, L. Yu, Y. Xu, L. Xu, L. Ma, X. Lu, Y. Liu, Y. Leng, R. Li, and Z. Xu, Opt. Express **21**, 29231 (2013).
5. A. Dubietis, G. Jonusauskas, and A. Piskarskas, Opt. Commun. **88**, 473 (1992).
6. Gilles Cheriaux, *et al.*, *Light at Extreme Intensities 2011*. Vol. 1462 (AIP Publishing, 2012).
7. L. M. Frantz and J. S. Nodvik, J. Appl. Phys. **34**, 2346 (1963).
8. K. Ertel, C. Hooker, S. J. Hawkes, B. T. Parry, and J. L. Collier, Optics Express **16**, 8039 (2008).
9. F. Kong, Y. Jin, S. Liu, S. Chen, H. Guan, K. He, Y. Du, and H. He, Chin. Optics Lett. **11**, 102302 (2013).

10. X. Lu, C. Li, Y. Leng, C. Wang, C. Zhang, X. Liang, R. Li, and Z. Xu, Chin. Optics Lett. **5**, 493 (2007).
11. Y. Leng, L. Lin, W. Wang, Y. Jiang, B. Tang, and Z. Xu, Optics & Laser Technology **35**, 425 (2003).
12. D. R. Preuuss and J. L. Gole, Appl. Optics **19**, 702 (1980).
13. M. Born and E. Wolf, *Principles of Optics: Electromagnetic Theory of Propagation, Interference and Diffraction of Light* (Cambridge University, United Kingdom, 1999).
14. F. Salin, P. Georges, G. Roger, and A. brun, Appl. Optics **26**, 4528 (1987).

14

Effects of target pre-heating and expansion on terahertz radiation production from intense laser-solid interactions

X.H. Yuan[1,2], Y. Fang[1], D.C. Carroll[3,4], D.A. MacLellan[3], F. Du[2], N. Booth[4], M. Burza[5], M. Chen[1], R.J. Gray[3], Y.F. Jin[1], Y.T. Li[2], Y. Liu[1], D. Neely[4], H. Powell[3], G. Scott[3,4], C.-G. Wahlström[5], J. Zhang[1,2], P. McKenna[3], and Z.M. Sheng[1,3]

[1] *Key Laboratory for Laser Plasmas (Ministry of Education) and Department of Physics and Astronomy, Shanghai Jiao Tong University, Shanghai 200240, China*

[2] *Beijing National Laboratory of Condensed Matter Physics, Institute of Physics, Chinese Academy of Sciences, Beijing 100190, China*

[3] *SUPA Department of Physics, University of Strathclyde, Glasgow G4 0NG, UK*

[4] *Central Laser Facility, Rutherford Appleton Laboratory, Oxfordshire OX11 0QX, UK*

[5] *Department of Physics, Lund University, S-22100 Lund, Sweden*

Abstract

The first experimental measurements of intense ($\sim 7 \times 10^{19}$ W cm^{-2}) laser-driven terahertz (THz) radiation from a solid target which is preheated by an intense pulse of laser-accelerated protons is reported. The total energy of the THz radiation is found to decrease by approximately a factor of 2 compared to a cold target reference. This is attributed to an increase in the scale length of the preformed plasma, driven by proton heating, at the front surface of the target, where the THz radiation is generated. The results show the importance of controlling the preplasma scale length for THz production.

Keywords: target heating; terahertz radiation

1. Introduction

Investigation of terahertz (THz) radiation from intense laser-produced plasmas is important both from the perspective of exploring new fundamental physics and because of the potential to develop compact sources for application. High-power laser-produced plasma provides a unique environment for exploring matter under extreme temperature and density conditions, in which ultra-high electric and magnetic fields are generated and evolve on ultrafast timescales. Measurement of the radiation emitted from such plasmas provides a powerful diagnostic tool to explore their evolution. For example, THz radiation emission can be used to diagnose the movement of fast electrons along the target front surface[1–3] and to diagnose the evolution of plasma at the target rear surface[4]. In addition, the development of relatively compact high-power laser-driven THz sources could find applications, for example, in inducing phase transitions in material science[5–7] and for imaging in biology and medicine[8, 9].

Correspondence to: Email: paul.mckenna@strath.ac.uk; Y.T. Li, Beijing National Laboratory of Condensed Matter Physics, Institute of Physics, Chinese Academy of Sciences, Beijing, China. Email: ytli@iphy.ac.cn

Previous studies have shown that THz emission is particularly strong when foil targets are used at the focus of a high-power laser pulse. Strong THz emission is induced at both the front[3] and rear surfaces[4]. The sensitivity of THz generation to laser pulse and interaction parameters (such as the pulse energy, polarization, and incident angle) were reported in both 100-femtosecond-level (or even shorter) duration regimes and in the weakly relativistic or non-relativistic intensity regimes[3, 10–13]. The influence of the preplasma generated at the front surface of the target foil has also been explored[10]. On increasing the preplasma generation by lowering the laser peak-to-ASE (amplified spontaneous emission) intensity contrast ratio, Li *et al.*[10] measured reduced THz emission (a factor of 10 decrease for a similar reduction in contrast ratio, from 3×10^{-8} down to 5×10^{-6}). However, working with similar laser pulse parameters to Li *et al.*, Hamster *et al.*[1] report an opposite trend: a factor of 10 increase in THz emission when the laser pulse intensity contrast was reduced from 2.5×10^{-6} to 5×10^{-4}. In another experimental study, reported in Sagisaka *et al.*[2], THz emission was shown to be rather insensitive to the scale length of the preformed plasma. Although three

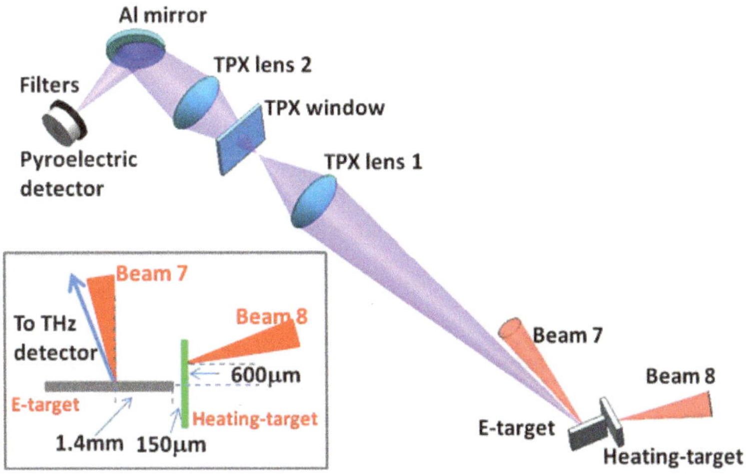

Figure 1. (Color online) Schematic of the experiment layout showing the positions and orientations of the targets, the laser beams, and the THz measurement optics and detector. TPX lens 1, and both targets and beams are in the vacuum sealed by the TPX window. The inset shows a top view of the orientation of the laser beams and targets in closer detail.

different THz generation mechanisms are invoked to explain the dramatically different observations, further investigation of the influence of preformed plasma is necessary for a fuller understanding of the physics underpinning THz generation.

In this article, we report on experimental and theoretical investigations on THz generation from relativistically intense ($>7 \times 10^{19}$ W cm^{-2}) laser pulses irradiating solid targets which have been preheated with an intense pulse of multi-MeV protons to induce preplasma formation. A factor of 2 decrease in the total energy of the THz emission is measured compared to equivalent measurements with an initially cold solid.

2. Experiment and results

The experiment was performed using the Vulcan laser at the Rutherford Appleton Laboratory in the UK. Figure 1 shows a schematic of the experimental arrangement. Two picosecond synchronized laser beams, referred to as beams B7 and B8 (beams 1–6 are the long pulse beams of the Vulcan laser facility and are not used in this study), are used to drive THz generation and proton acceleration for heating, respectively. These beams are hereafter referred to as B7 and B8. B7 delivers pulses of 60 ± 7 J on the target with a duration equal to 1 ps, resulting in a calculated peak intensity of $(7.6 \pm 0.9) \times 10^{19}$ W cm^{-2} when focused to a spot of 10 μm diameter (FWHM) at an incident angle of 8° (with respect to the target normal). Note that this relatively small angle of incidence was chosen to minimize the production of transient surface currents of fast electrons[14, 15], which can be a dominant driver of THz generation. B8 pulses have energy and duration equal to 200 ± 15 J and 12 ps, respectively. They are focused to a 12 μm diameter (FWHM) spot, at an incident angle of 15° (again with respect to the target normal), resulting in a calculated peak intensity of $(1.5 \pm 0.1) \times 10^{19}$ W cm^{-2}. Both beams were controlled independently to enable either beam to be used separately or both used together with variable temporal separation of the pulses.

The THz generation target was a 200 μm thick amorphous carbon foil (3 mm × 3 mm); it is labeled E-target in Figure 1. The proton source target was a 20 μm thick gold foil (2 mm × 2 mm); it is labeled Heating-target. The inset of Figure 1 shows a top view of the experimental arrangement when both laser beams and targets were employed. The targets were orientated at a slight angle (not illustrated in the figure) such that the expanding proton beam from the Heating-target irradiated the E-target in the region of the B7 focus. The distance between the two laser foci was 1.7 mm. The temporal separation between the pulses was set such that B8 arrived on target (85 ± 7) ps prior to B7. This enables sufficient time for the propagation of the multi-MeV proton beam from the Heating-target to the E-target, and an additional several tens of picoseconds for preplasma expansion.

The main diagnostic is a broadband pyroelectric THz radiometer (Model SPI-A-62 from Spectrum Detector Inc.) coupled to a Tektronix oscilloscope. The detector has a flat response coefficient from 0.3 to 21 THz, with a calibrated responsivity of $\sim 10^4$ V W^{-1}. Two 50 mm diameter polymethylpentene (TPX) lenses were used to collect and refocus the THz radiation emitted in the backward direction at 20° to the B7 laser axis. The focal lengths of these two optics are $f = 200$ mm and $f = 100$ mm. The electromagnetic (EM) radiation was transmitted out of the vacuum chamber through a 2 mm thick TPX window. The collection angle is limited to 8×10^{-4} sr by the aperture size of the second lens. To prevent the detector being irradiated by high-energy particles, the sensor is oriented at an angle to the target and the measurement axis is turned by an aluminium mirror so that only the EM radiation is reflected onto the detector. The

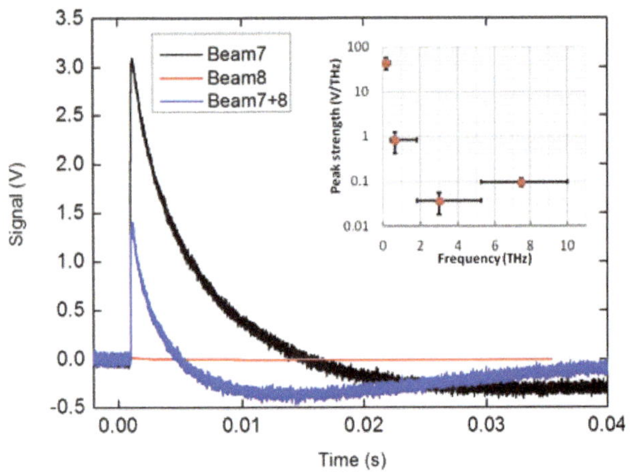

Figure 2. (Color online) Typical THz waveforms (from an oscilloscope) as measured with a pyroelectric detector operating in the range 0.3–1 THz. Black: B7 only; red: B8 only; blue: both B7 and B8. The inset shows a reconstructed THz spectrum from a B7-only shot.

whole detector was housed in a 5 mm thick aluminum cylindrical cavity, which was further shielded by 50 mm thick lead bricks, leaving only a 5 mm diameter entrance hole. Several types of filter were used to ensure that the measured signal was in the THz frequency regime. A 1 mm thick piece of high-resistivity silicon (HR-Si) wafer was tightly fixed around the entrance hole to ensure that the arrangement was light-tight. Copper and stainless steel meshes of period 3 mm and 1 mm were stacked to remove EM noise below 0.3 THz. Different low-pass filters (QMMF mesh filter from QMC Instruments Ltd) were further employed to limit the upper cutoff frequency measured, to 10 THz, 5 THz, 1 THz, and 0.3 THz, enabling a crude THz spectrum to be determined.

Example waveforms measured with the THz detector are shown in Figure 2, for three cases: (1) B7 only, (2) B8 only, and (3) both B7 and B8. The filters were chosen to transmit EM radiation in the range 0.3–1 THz. For the B7-only case, the measured signal peak voltage was 3.2 V and the duration (FWHM) was 4 ms. The signal decay time is set by the detector response, and is much longer than the actual THz pulse duration. The pulse energy detected is determined from the calibration to be 0.184 μJ (equivalent to 230 μJ sr^{-1} if assuming isotropic emission and taking account of the diagnostic acceptance solid angle and the attenuations of the lens, window, and filters). The shot-to-shot reproducibility of the signal was checked over a series of laser shots on the same target type, and it showed a variation of $<10\%$. The inset of Figure 2 shows a crude THz emission spectrum for the B7-only case. The central frequency for each point is the average of adjacent cutoff frequencies of the low-pass edge filters, and the horizontal error bars are obtained from their differentiation. The amplitude of the signal is plotted at the detector voltage per unit frequency (in THz). The vertical error bars reflect the degree of shot-to-shot fluctuation and account for systematic errors. The result indicates that the measured THz emission is mainly at relatively low (up to 1 THz) frequencies.

To determine if the B8 interaction with the proton source target contributes to the measured THz signal, measurements were made with B8 only. The energy of the THz pulse in this case was only 1.4 μJ sr^{-1}, which is more than two orders of magnitude smaller than the B7 interaction, indicating that the THz emission is dominated by the B7 interaction. In shots for which both B7 and B8 are used, i.e., corresponding to proton pre-heating of the main target (E-target), the measured THz energy decreases, compared to the equivalent case without proton heating, by more than a factor of 2, to 106 μJ sr^{-1}.

We now consider the changes to the main target induced by the proton beam. The interaction of B8 with the thin gold target accelerates protons via the target normal sheath acceleration (TNSA) mechanism[16]. About 5% of the laser pulse energy is converted to produce a proton beam with an exponential spectral distribution, with temperature ~2 MeV and cutoff energy equal to 14 MeV. The spectrum is measured, using a stack of dosimetry film, with the E-target in position so that the missing fraction, i.e., the protons absorbed by the E-target, can be determined.

Proton stopping within the target is governed by the Bragg peak energy deposition profile. The resulting target heating and hydrodynamic expansion are modelled using the 1D HELIOS-CR radiation-hydrodynamic code[17] and the PROPACEOS equation of state (EoS). The temporal separation of the arrival of B7 and B8 at their respective targets is 85 ps. Protons with energy of 8 MeV, for example, will transverse the 1.7 mm distance from the source to the focal position of B7 in 43 ps, with the larger number of 2 MeV protons arriving after 85 ps. The majority of the protons thus arrive and start heating the E-target in the temporal window from around 40 ps prior to the arrival of the peak of B7. This sets the range of expansion times to be modelled in the HELIOS calculations. Note that, as with similar studies involving target heating driven by TNSA protons from a secondary laser-foil interaction (e.g., [18]), we assume that, compared to proton heating, x-ray and fast-electron emission from the secondary foil contributes relatively little to the heating of the front surface of the primary target, due to the Bragg peak energy deposition profile and relatively high directionality of the proton beam. Another factor which must be considered is the amplified spontaneous emission (ASE) of B7 which arrives at the target prior to the protons. Contrast measurements made on the Vulcan laser indicate that the level of the ASE intensity is of the order of 10^{10} W cm^{-2} at 0.1–1 ns, and 10^{11} W cm^{-2} at 50–100 ps prior to the peak of the laser pulse.

3. Simulations and discussion

HELIOS simulations were performed for two cases: (1) B7 ASE only (i.e., no proton heating) and (2) B7 ASE plus the

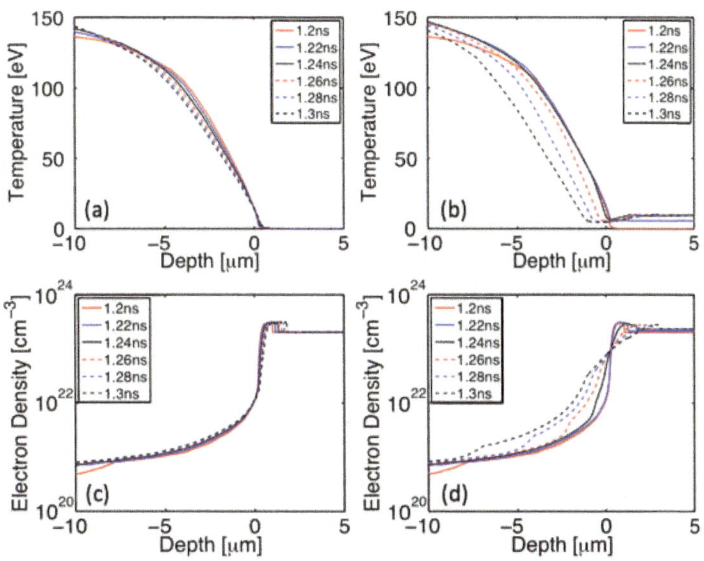

Figure 3. (Color online) HELIOS simulation results. (a) and (c) Temperature and electron density profiles at the front surface of a carbon target at given simulation times, driven by the laser (B7) ASE only. (b) and (d) Corresponding temperature and density profiles for the same target driven by the laser (B7) ASE plus proton heating driven by B8. See the main text for details.

proton beam. The simulation starts at $t = 0$ with the arrival of the ASE on target and runs until $t = 1.3$ ns. The ASE intensity is 10^{10} W cm^{-2} from $t = 0$–1 ns, and 10^{11} W cm^{-2} from $t = 1$–1.3 ns. For case (2), 8 MeV protons arrive at the target at $t = 1.2$ ns, while lower-energy protons arrive later: 1 MeV protons arrive at $t = 1.28$ ns. The peak of B7 also arrives at $t = 1.28$ ns.

The resulting target heating and front surface plasma expansion profiles at given simulation times are shown in Figure 3. The ASE-only simulations (Figures 3(a) and 3(c)) show that the primary effect of the ASE pedestal is to heat the underdense region of the target and drive a low-temperature shock wave into the target, resulting in compression of the first micron depth of the target by ~25%. The target heating and expansion driven by the ASE is relatively small. For the case in which the proton beam is added, we find that the first few tens of microns depth from the front surface of the target is heated to temperatures of ~10 eV, as shown in Figure 3(b), with an exponential decrease in temperature to about 0.3 eV at a depth of 100 μm. Similar laser-driven proton heating profiles have been reported in other work[18]. Importantly, we find that the target front side expands faster into vacuum in response to this heating, resulting in slightly larger plasma scale length, compared to the case of ASE-only heating.

There have been a few mechanisms proposed for THz emission from relativistic laser interaction with solid targets. One of them is associated with the space-charge field or laser wakefield excitation at the vacuum–plasma interface driven by the laser ponderomotive force[1, 19–21]. This mechanism requires a large volume of underdense plasma at densities below 10^{19} cm^{-3}, with scale length longer than the length of the driving laser pulse; otherwise, the generation efficiency is low. Our fluid simulations of preplasma formation reveal that the additional proton heating, which strongly affects THz emission, mainly changes the density scale length in a localized region at densities of the order of n_c (~10^{21} cm^{-3}). It therefore appears that the contribution of THz emission due to this mechanism is less important. As discussed above, other THz generation mechanisms are associated with energetic electron production, such as the surface transient current at the front[3] or rear[22], antenna emission[2], and transient radiation[23–25], which occur when the preplasma has moderate-to-short density scale lengths. Generally, all of these mechanisms depend upon the preplasma conditions in two ways, by changing (1) the absorption of laser energy to fast electrons, and (2) the injection properties of the fast-electron distribution (see reference [26] for an example study of the sensitivity to preplasma scale length). A relatively large preplasma usually produces more hot electrons with higher temperature, which often leads to higher transient currents near the target surface when the laser is incident at relatively large angles with respect to the target normal. Two features distinguish the present work from previous studies: the laser (B7) is incident at a relatively small angle (8°), and our approach using proton heating over a large area of the target surface drives quasi-1D preplasma expansion. The results which show that THz emission is reduced with increasing preplasma expansion, which are in broad agreement with a previous study under different conditions[10], are explained partially by the fact that transient currents formed deeper into the expanded preplasma cannot emit THz waves out of it because of their low frequency. These effects will be investigated in more detail in further work.

4. Conclusion

In summary, we have compared THz radiation emission measurements from the interaction of an intense laser pulse with a solid target with and without front surface expansion driven by heating with laser-accelerated protons. The experiment demonstrates that pre-heating and expansion of the target significantly reduces the level of THz emission, even though the laser is incident at a small angle with respect to the target normal direction. The results suggest that, under these conditions, hot-electron generation and transport effects in the preplasma play an important role in THz emission.

Acknowledgements

The authors acknowledge the expert support of the staff at the Central Laser Facility at Rutherford Appleton Laboratory, UK. This work was supported by National Basic Research Program of China (grant nos. 2013CBA01500 and 2014CB339801), National Natural Science Foundation of China (grant nos. 11121504, 11205100, 11220101002 and 11135012), the EPSRC (grant nos. EP/J003832/1 and EP/L001357/1), and the Swedish Research Council.

References

1. H. Hamster, A. Sullivan, S. Gordon, and R. W. Falcone, Phys. Rev. Lett. **71**, 2725 (1993).
2. A. Sagisaka, H. Daido, S. Nashima, S. Orimo, K. Ogura, M. Mori, A. Yogo, J. Ma, I. Daito, A. S. Pirozhkov, S. V. Bulanov, T. Zh. Esirkepov, K. Shimizu, and M. Hosoda, Appl. Phys. B **90**, 373 (2008).
3. Y. T. Li, C. Li, M. L. Zhou, W. M. Wang, F. Du, W. J. Ding, X. X. Lin, F. Liu, Z. M. Sheng, X. Y. Peng, L. M. Chen, J. L. Ma, X. Lu, Z. H. Wang, Z. Y. Wei, and J. Zhang, Appl. Phys. Lett. **100**, 254101 (2012).
4. A. Gopal, T. May, A. S. Herzer, A. Reinhard, S. Minardi, M. Schubert, U. Dillner, B. Pradarutti, J. Polz, T. Gaumnitz, M. C. Kaluza, O. Jackel, S. Riehemann, W. Ziegler, H. P. Gemuend, H. G. Meyer, and G. G. Paulus, New J. Phys. **14**, 083012 (2012).
5. M. K. Liu, H. Y. Hwang, H. Tao, A. C. Strikwerda, K. Fan, G. R. Keiser, A. J. Sternbach, K. G. West, S. Kittiwatanakul, J. Lu, S. A. Wolf, F. G. Omenetto, X. Zhang, K. A. Nelson, and R. D. Averitt, Nature (2012)
6. M. Chen, X. H. Yuan, and Z. M. Sheng, Appl. Phys. Lett. **101**, 161908 (2012).
7. M. Chen, A. Pukhov, X. Y. Peng, and O. Willi, Phys. Rev. E **78**, 046406 (2008).
8. B. Ferguson and X. C. Zhang, Nature Mat. **1**, 26 (2002).
9. H. T. Chen, W. J. Padilla, J. M. O. Zide, A. C. Gossard, A. J. Taylor, and R. D. Averitt, Nature **444**, 597 (2006).
10. C. Li, M. L. Zhou, W. J. Ding, F. Du, F. Liu, Y.-T. Li, W.-M. Wang, Z.-M. Sheng, J.-L. Ma, L.-M. Chen, X. Lu, Q.-L. Dong, Z.-H. Wang, Z. Lou, S.-C. Shi, Z.-Y. Wei, and J. Zhang, Phys. Rev. E **84**, 036405 (2011).
11. F. Du, C. Li, M. L. Zhou, W. M. Wang, L. N. Su, Y. Zheng, X. L. Ge, Y. T. Li, J. L. Ma, X. L. Liu, L. Zhang, Z. M. Sheng, L. M. Chen, X. Lu, Q. L. Dong, Z. H. Wang, Z. Y. Wei, and J. Zhang, Sci. China Inf. Sci. **55**, 43 (2012).
12. F. Du, C. Li, M. L. Zhou, W. M. Wang, L. N. Su, Y. Zheng, Y. T. Li, J. L. Ma, Z. M. Sheng, L. M. Chen, X. Lu, Z. H. Wang, Z. Y. Wei, and J. Zhang, Sci. China-Phys. Mech. Astron. **55**, 589 (2012).
13. Y. Gao, T. Drake, Z. Y. Chen, and M. F. DeCamp, Opt. Lett. **33**, 2776 (2008).
14. P. McKenna, D. C. Carroll, R. J. Clarke, R. G. Evans, K. W. D. Ledingham, F. Lindau, O. Lundh, T. McCanny, D. Neely, A. P. L. Robinson, L. Robson, P. T. Simpson, C.-G. Wahlstrom, and M. Zepf, Phys. Rev. Lett. **111**, 074802 (2007).
15. R. J. Gray, X. H. Yuan, D. C. Carroll, C. M. Brenner, M. Coury, M. N. Quinn, O. Tresca, B. Zielbauer, B. Aurand, V. Bagnoud, J. Fils, T. Kuhl, X. X. Lin, C. Li, Y. T. Li, M. Roth, D. Neely, and P. McKenna, Appl. Phys. Lett. **99**, 171502 (2011).
16. S. C. Wilks, A. B. Langdon, T. E. Cowan, M. Roth, M. Singh, S. Hatchett, M. H. Key, D. Pennington, A. MacKinnon, and R. A. Snavely, Phys. Plasmas **8**, 542 (2001).
17. J. J. MacFarlane, I. E. Golovkin, and P. R. Woodruff, J. Quant. Spectrosc. Radiat. **99**, 381 (2006).
18. A. Mancic, J. Robiche, P. Antici, P. Audebert, C. Blancard, P. Combis, F. Dorchies, G. Faussurier, S. Fourmaux, M. Harmand, R. Kodama, L. Lancia, S. Mazevet, M. Nakatsutsumi, O. Peyrusse, V. Recoules, P. Renaudin, R. Shepherd, and J. Fuchs, High Energy Density Physics **6**, 21 (2010).
19. Z. M. Sheng, H. C. Wu, K. Li, and J. Zhang, Phys. Rev. E **69**, 025401(R) (2004).
20. Z. M. Sheng, K. Mima, J. Zhang, and H. Sanuki, Phys. Rev. Lett. **94**, 095003 (2005).
21. L. M. Gorbunov and A. A. Frolov, J. Exp. and Theoret. Phys. **102**, 894 (2006).
22. A. Gopal, S. Herzer, A. Schmidt, P. Singh, A. Reinhard, W. Ziegler, D. Brommel, A. Karmakar, P. Gibbon, U. Dillner, T. May, H.-G. Meyer, and G. G. Paulus, Phys. Rev. Lett. **111**, 074802 (2013).
23. J. Zheng, K. A. Tanaka, T. Miyakoshi, Y. Kitagawa, R. Kodama, T. Kurahashi, and T. Yamanaka, Phys. Plasmas **10**, 2994 (2003).
24. C. B. Schroeder, E. Esarey, J. van Tilborg, and W. P. Leemans, Phys. Rev. E **69**, 016501 (2004).
25. W. J. Ding, Z. M. Sheng, and W. S. Koh, Appl. Phys. Lett. **103**, 204107 (2013).
26. P. McKenna, D. C. Carroll, O. Lundh, F. Nurnberg, K. Markey, S. Bandyopadhyay, D. Batani, R. G. Evans, R. Jafer, S. Kar, D. Neely, D. Pepler, M. N. Quinn, R. Redaelli, M. Roth, C.-G. Wahlstrom, X. H. Yuan, and M. Zepf, Laser Part. Beams **26**, 591 (2008).

15

Coherent X-ray mirage: discovery and possible applications

Tatiana Pikuz[1,2], Anatoly Faenov[1,2,5], Sergey Magnitskiy[3], Nikolay Nagorskiy[2], Momoko Tanaka[1], Masahiko Ishino[1], Masaharu Nishikino[1], Yuji Fukuda[1], Masaki Kando[1], Yoshiaki Kato[4], and Tetsuya Kawachi[1]

[1]*Quantum Beam Science Directorate, Japan Atomic Energy Agency, Kizugawa, Kyoto 619-0215, Japan*

[2]*Joint Institute for High Temperatures, Russian Academy of Sciences, Moscow 125412, Russia*

[3]*International Laser Center of M.V. Lomonosov Moscow State University, Moscow, Russia*

[4]*The Graduate School for the Creation of New Photonics Industries, Hamamatsu, Shizuoka 431-1202, Japan*

[5]*Institute for Academic Initiatives, Osaka University, Suita, Osaka, 565-0871, Japan*

Abstract

In the far field of the intensity distribution of the beam delivered by a two-stage transient–collisional excitation X-ray laser (XRL), a non-expected interference pattern that is stable from shot to shot has been discovered. It is demonstrated that the interference is caused by the emergence of an imaginary source in the amplifying plasma, which is phase matched to the radiation of the generator. The observed phenomenon is called an X-ray coherent mirage. To explain the obtained results, a new theoretical approach is developed. The basic essential conditions for formation of the X-ray mirage are formulated, and possible applications are discussed. This paper details the experiments, including the formulation of the necessary and sufficient conditions for formation of the X-ray mirage, and possible applications are discussed.

Keywords: coherent seeded beams; mirage; phase-matching; X-ray lasers; X-ray plasma optics

1. Introduction

X-rays are a form of electromagnetic radiation with wavelengths shorter than the visible part of the spectrum. This means that practically all the phenomena existing in the optical, visible range should have their analogues in X-rays. However, while more than 100 years have passed since the discovery of X-rays, and despite the impressive achievements in X-ray physics since then, optical mirages[1], an amazing and mysterious phenomenon of optical refraction, have never been observed in the X-ray band. Indeed, the appearance of a mirage in the X-ray is extremely difficult because for all conventional materials the refractive indexes are close to 1, and gradients of the refractive indexes are low in the X-ray frequencies. Obviously, such properties practically eliminate refraction and the possibility of mirage formation in the X-ray spectral range. Nevertheless, recently a new unique medium has been created: laser-induced plasma, which has a noticeable X-ray refraction and is effective for X-ray laser (XRL) generation[2–5]. After the first realization of an XRL, significant interest emerged in developing a coherent table-top XRL. In particular, a plasma-based transient collisional excitation XRL[6–9] allowed one to effectively realize a compact coherent soft X-ray source. However, the coherence of the output radiation of a modern XRL is far from perfect, and the efforts of many research groups are aimed at improving both the temporal and the spatial coherence of XRLs.

Recently, using radiation from a plasma-based transient collisional excitation XRL[10], the observation of an unknown interference structure in the spatial profile of the output beam of a two-stage plasma media XRL has been reported[11]. Such a result led to founding and, after scrupulous theoretical analysis, to understanding this new mechanism, which could have a drastic influence on the spatial characteristics of XRL beams — the formation of X-ray coherent mirages[11], which are unique phenomena for the X-ray spectral range. In this article, we present some new results and provide more detailed discussions about the X-ray coherent mirage phenomenon and about its possible applications in science and technology.

Correspondence to: Tatiana Pikuz, Quantum Beam Science Directorate, Japan Atomic Energy Agency, Kizugawa, Kyoto 619-0215, Japan. Email: tapikuz@yahoo.com

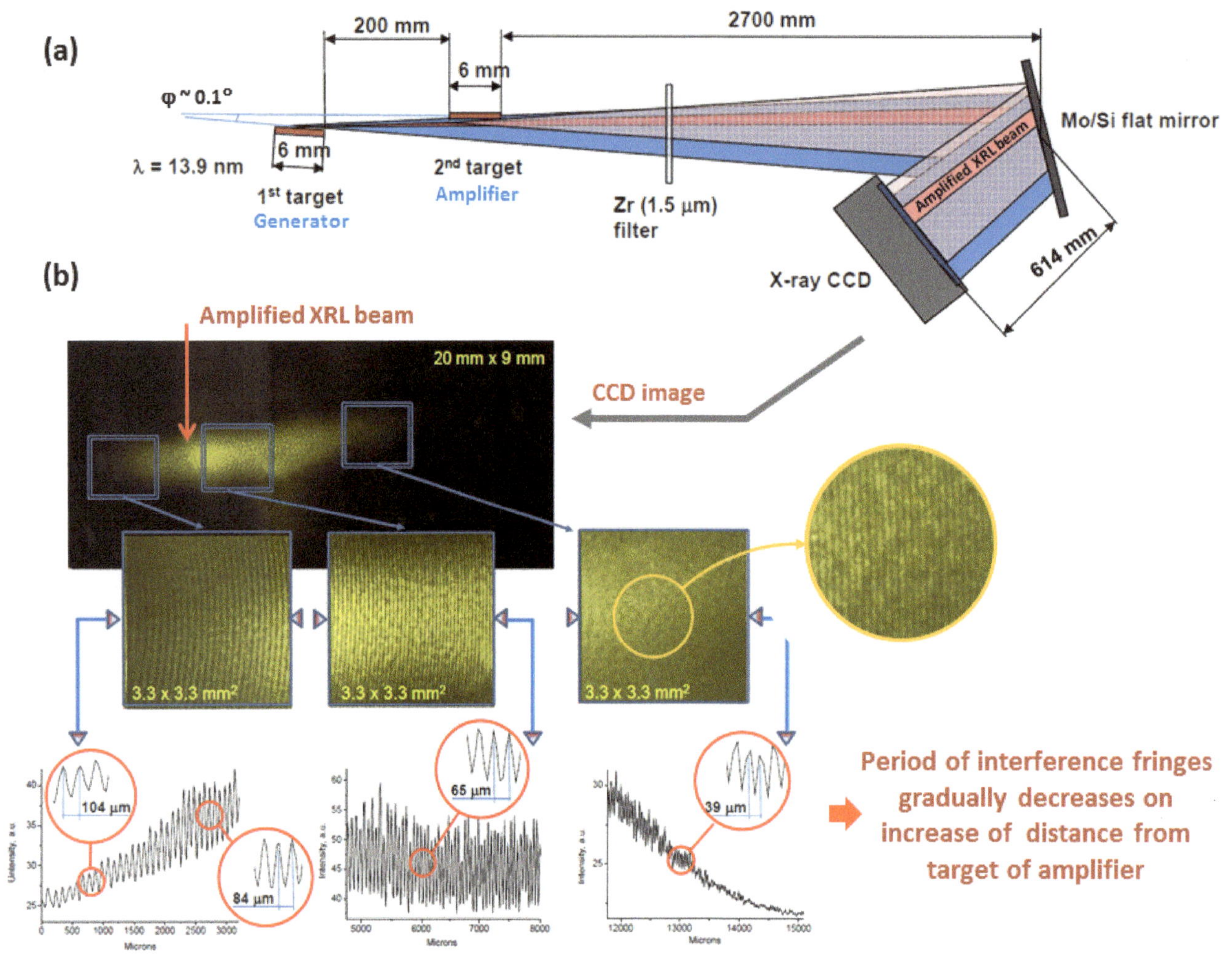

Figure 1. (a) Scheme of the experiment. (b) Typical image of an XRL far-field intensity distribution, obtained in one shot.

2. Experimental set-up and results

The results presented here were obtained during an experimental study of the spatial characteristics of the far-field wave front of an XRL, built on a two-stage transient-collisional excitation oscillator–amplifier Ni-like Ag plasma (λ =13.9 nm) scheme[10], at Kansai Photon Science Institute of the Japan Atomic Energy Agency (Figure 1(a)). Thin silver foils for the oscillator and the amplifier were pumped, respectively, with the line-focused travelling wave laser pulses adjusted to the proper timing for effective amplification. The XRL pulse after the amplifier had energy of ~300 nJ, bandwidth narrower than 10^{-4}, and pulse duration of ~ 7 ps.

At specific conditions of XRL operation (delay between generator and amplifier pumping pulses < 10 ps and relative angle between oscillator and amplifier targets < 0.6°), we found the appearance of an interference pattern (a sequence of dark and light rings) in the form of concentric rings in the output of the XRL beam. A typical image obtained by a back-illuminated X-ray charge-coupled device (CCD) with 2048 × 2048 pixels (the size of a pixel is 13.5 μm) at a distance of 3314 mm from the output of the XRL in one shot is shown in Figure 1(b). In the image, the bright spot elongated in the horizontal direction is a footprint of amplified XRL radiation. Concentric interference fringes are present over practically all of the beam profile. As can be clearly seen from the expanded insets of the image in Figure 1(b), the period of the fringes is gradually decreasing from the center to the peripheries. Such a structure was observed only in the case when two-stage XRL lasers were operated. This means that, in this case, the generator and the amplifier were turned on simultaneously. Special experiments, when the generator or amplifier was turned off, were also performed.

In the case of the output intensity distribution of the beam, such an interference pattern was never observed. Moreover, we checked the stability of such phenomenon appearing during XRL operation, and discovered that, under certain pumping parameters of the XRL, the ring structures were recorded consistently from shot to shot.

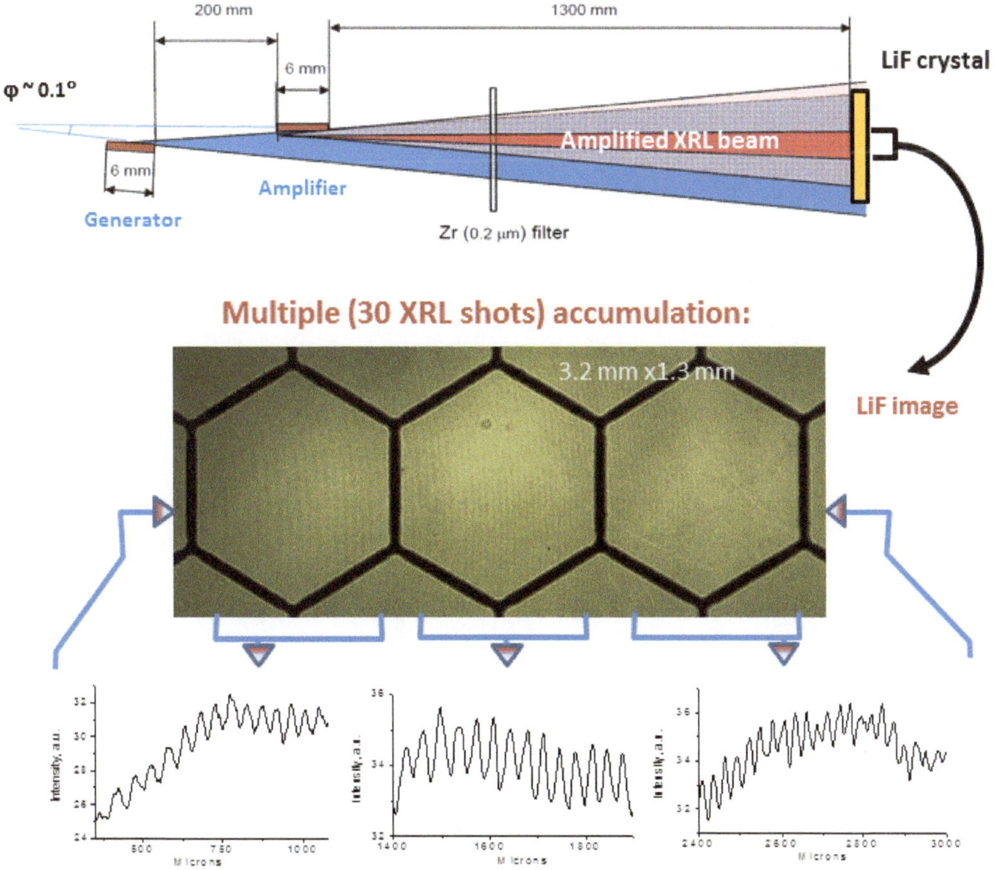

Figure 2. Experimental observation of the interference pattern in the intensity distribution of the output of an oscillator–amplifier XRL beam after 30 accumulated shots.

Such high stability of generation of interference fringes allowed us to provide an additional series of experiments. A LiF crystal detector[12–14] with submicron spatial resolution was used for measurements of the spatial-intensity distribution of the XRL beam, which were averaged over 30 laser shots. The scheme of the experiment and a typical image are presented in Figure 2. Our measurement indicates high stability of the interference patterns, which appears despite the accumulation of a large number of XRL shots. This gives us the possibility to conclude that the observed phenomenon has a fundamental character; both sources are rigidly phased and do not significantly depend on fluctuating plasma parameters.

3. Discussion

The appearance of such a concentric interference pattern in the spatial profile of the output beam is a rather unexpected observation, because there are no apparent physical causes that can lead to it. Indeed, according to optical interferometry fundamentals, the observation of circular concentric fringes indicates an existence of two point coherent sources, separated in space, located on the axis, passing through the center of the circles (see Figure 3). This means that detected rings are produced by an interference of radiation emitted by two coherent point sources, which were produced in our experimental conditions. Obviously the first source is located in the XRL generator (see Figure 1(a) and Figure 2). To determine the position of the second source, we calculated interference patterns, and by matching their period with the period of experimentally observed interference patterns (see Figure 4) we found that second source must be located at the distance of 203 $\pm$ 2 mm from the first one. From the layout of the XRL system (see Figure 1(a) and Figure 2) it follows that the second source should be located somewhere in the middle of the amplifying plasma. In fact, the distance between the generator and the amplifier targets is 200 mm, and the length of both targets is 6 mm. So the distance 203 mm is the distance from the exit of generator to the center of the amplifier. Although this geometrical consideration is quite simple, it is not easy to find physical reasons for the emergence of the second point source in the amplifying plasma. Moreover, to be able to interfere with the source from the generator, the second source should be phase matched to the first one.

Based on our observations, we undertook theoretical analyses, which showed that the user of the standard system of Maxwell–Bloch equations, typically describing amplification of seeded beams in plasma media, could not explain

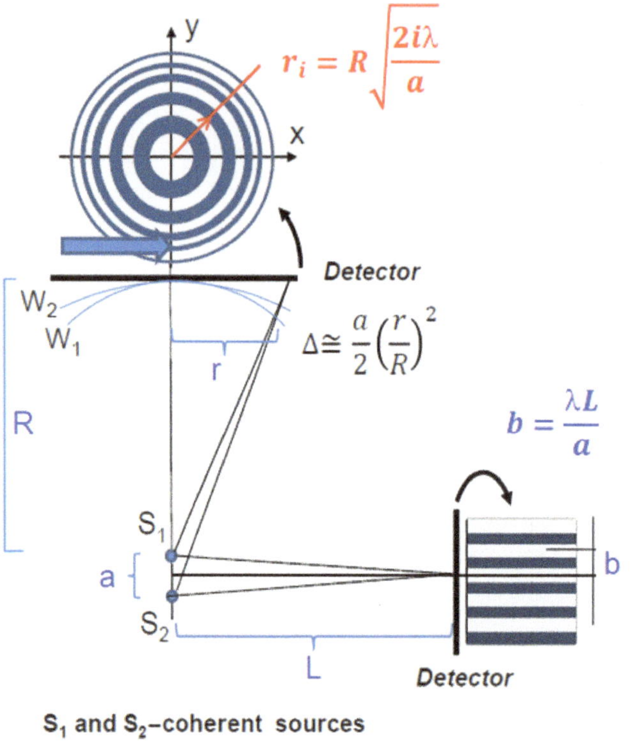

Figure 3. Optical scheme of fringe formation by two point coherent sources.

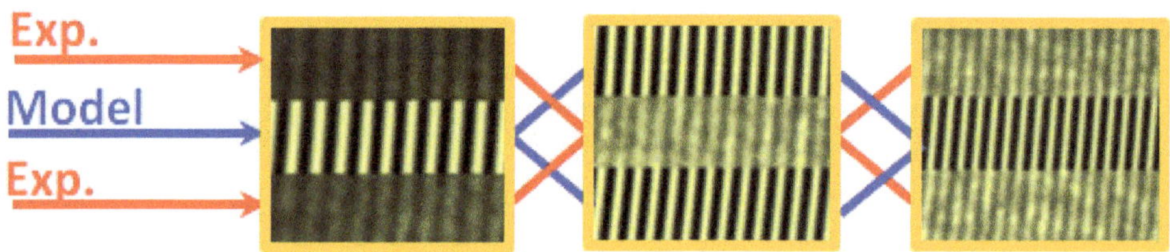

Figure 4. Comparisons of experimental and modeled interference patterns for the case of the distance between two phase-matched X-ray point sources being 203 mm.

the emergence of the second source and the observed phenomenon. Indeed, from our experiments we could see that the phase front of the stimulated emission should repeat the phase front of an initial wave. This means that the amplified seeded X-ray beam could not create a spherical wave with a phase front of another curvature. This led to the necessity of developing a new theoretical approach. Such an approach is proposed in[11] and is based on the fact that in our experiments (due to the high density and strong localization of laser-produced plasma media in space) the X-ray beam propagation and transformation are similar to the phenomenon of classical mirages in a strongly inhomogeneous atmosphere.

It is supposed in[11] that, mathematically, a mirage is a part of the total radiation, which remains after the initial radiation is subtracted from it, under the assumption that the initial radiation propagates through a medium without distortion in the medium. We took to the account that the complex amplitude of the output XRL beam $A(x, y, z)$ consisted of two terms: the amplitude of the incident beam propagating in empty space $A_{inc}(x, y, z)$ and the amplitude of mirage radiation emerging in the plasma $A_M(x, y, z)$:

$$A(x, y, z) = A_{inc}(x, y, z) + A_M(x, y, z).$$

To determine which imaginary source corresponds to the output mirage radiation $A_M^{im}(x; y; z)$, it is enough to solve the equation $A_M(x, y, z) = A(x, y, z) - A_{inc}(x, y, z)$ for the output radiation and to run the complex amplitude $A_M(x, y, z)$ backward in empty space. Schematically, the algorithm is shown in Figure 5.

Suppose that a wide divergent X-ray beam is incident on a small piece of plasma medium. Several factors will influence the formation of a mirage. First of all, the spatial distribution of the refractive index, but also the gain of the medium, will play a significant role in the process. Besides, since in typical XRL the gain coefficient is very high, it is considered to be the effects of saturation in signal amplification. Under

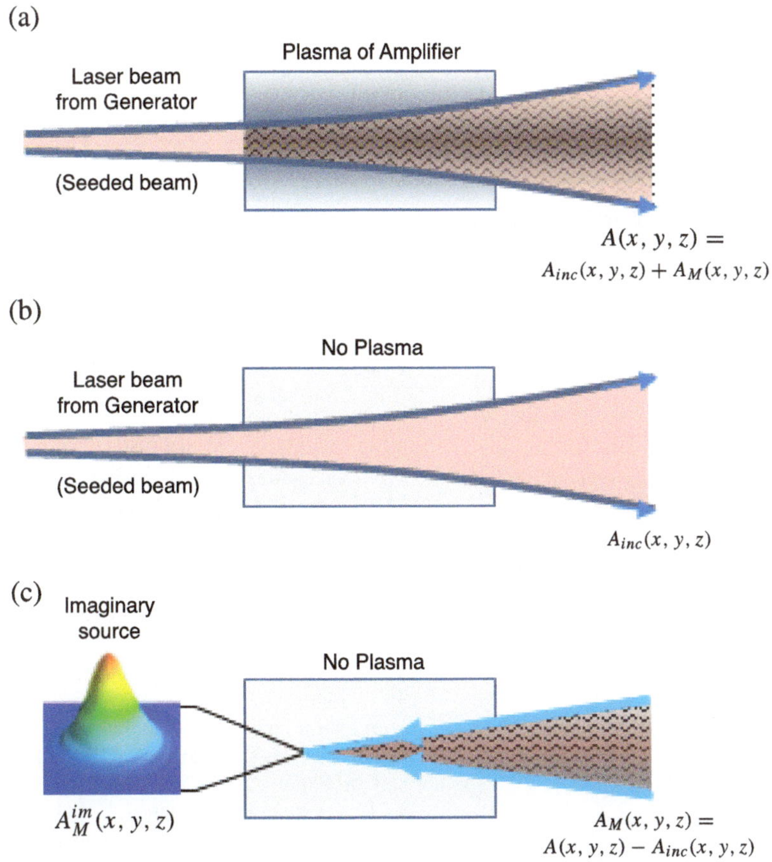

Figure 5. Schematic explanation of the theoretical algorithm for X-ray mirage formation. (a) XRL radiation from the generator (seeded beam) propagated in empty space with complex amplitude $A_{inc}(x, y, z)$; (b) after propagation of the seeded beam through plasma of the amplifier, the complex amplitude of the distorted radiation $A(x, y, z)$ consists of two terms: the complex amplitude $A_{inc}(x, y, z)$ and the complex amplitude of the mirage $A_M(x, y, z)$; (c) formation of imaginary source $A_{Mim}(x, y, z)$ by running the complex amplitude $A_M(x, y, z)$ backward in empty space.

these assumptions, the optical properties of the plasma will be similar to the optical properties of some effective inhomogeneous lens, where the thickness of the lens imitates the refractive index of the plasma. This means that the lower the refractive index, the thicker the lens. In the case when a seeded beam passes through such an optical system, it will consist of a set of outgoing waves generated by either a point or an extended imaginary source located along the axis of symmetry. If the plasma has a gain, the central part of the beam will be predominantly amplified. Since the electron density $n_e(x; y)$ surface is approximated by a paraboloid, the central part of the plasma works as an ideal aberration-free lens. From such a consideration it follows that the higher the gain, the narrower the axial region of the plasma which contributes to the beam amplification. This leads to the conclusion that an emerged virtual source will be closer to a point one and, evidently, will be phased with the incident radiation. The numerical simulations showed that, for the typical parameters of XRL plasma media, a coherent mirage in the form of an imaginary point source arises. It is found that the position of this point source in space is in accordance with our obtained experimental data.

The development of the theory and carrying out of the modelling allowed us to formulate the necessary and sufficient conditions for formation of an X-ray mirage in plasma media: (1) there should be a local maximum in the $N_e(x; y)$ distribution, (2) the spatial location of small-signal of gain distribution $G(x, y)$ in the plane x, y, which is perpendicular to the XRL beam propagation should coincide with the position of the local maximum of $N_e(x; y)$, the (3) spatial size of the gain region should be equal to or less than the width of the $N_e(x; y)$ local maximum, and (4) the uniformity of the plasma along the direction of beam propagation should be sufficiently high.

It is clear that these four conditions are difficult to reach in experiments. On the other hand, each of these conditions in itself is not unique to a plasma induced by short laser pulses on the surface of solid targets and correspondent to the conditions, which is usually necessary for optimal transient–collisional XRL generation. This means that X-ray radiation amplification in mirage formation is the key issue. If there is no gain, the whole area of plasma filled with radiation would participate in the process of beam formation. As a result, output radiation with non-uniform intensity and with a sufficiently inhomogeneous phase front would be formed. But this is not a mirage. As not in every inhomogeneous atmosphere does an optical illusion occur, similarly not in every inhomogeneous plasma does an X-ray mirage occur.

4. Conclusions

The experimental observation of an X-ray coherent mirage in a two-stage XRL system, and a theoretical approach, which explained a mechanism of that phenomenon, are discussed. It is shown that, in XRL plasma media, both refraction and diffraction accompany the process of amplification and play significant roles in the emergence of the virtual source of the mirage. There are various possibilities of how the X-ray mirage phenomenon could be useful for further improvement of XRL systems and in developing new advanced methods of plasma media diagnostics. The discovery of X-ray mirages could be applied to further develop modern X-ray technology, for example, for generation of adaptively controlled phased X-ray beams and plasma X-ray optics.

Acknowledgement

This work was partly supported by RFBR-JSPS collaboration program (RFBR Grant 14-02-92107).

References

1. A. A. Knowlton, Science **50**, 328 (1919).
2. R. C. Elton, *X-ray Lasers* (Academic, 1990).
3. H. Daido, Rep. Progr. Phys. **65**, 1513 (2002).
4. P. Jaegle, *Coherent Sources of XUV Radiation* (Springer, 2006).
5. S. Suckewer and P. Jaegle, Laser Phys. Lett. **6**, 411 (2009).
6. P. V. Nickles, V. N. Shlyaptsev, M. Kalachnikov, M. Schnürer, I. Will, and W. Sandner, Phys. Rev. Lett. **78**, 2748 (1997).
7. J. Dunn, A. L. Osterheld, R. Shepherd, W. E. White, V. N. Shlyaptsev, and R. E. Stewart, Phys. Rev. Lett. **80**, 2825 (1998).
8. T. Kawachi, M. Kado, M. Tanaka, A. Sasaki, N. Hasegawa, A. V. Kilpio, S. Namba, K. Nagashima, P. Lu, K. Takahashi, H. Tang, R. Tai, M. Kishimoto, M. Koike, H. Daido, and Y. Kato, Phys. Rev. A. **66**, 033815 (2002).
9. B. A. Reagan, W. Li, L. Urbanski, K. A. Wernsing, C. Salsbury, C. Baumgarten, M. C. Marconi, C. S. Menoni, and J. J. Rocca, Opt. Express **21**, 28380 (2013).
10. M. Nishikino, N. Hasegawa, T. Kawachi, H. Yamatani, K. Sukegawa, and K. Nagashima, Appl. Opt. **47**, 1129 (2008).
11. S. Magnitskiy, N. Nagorskiy, A. Faenov, T. Pikuz, T. Tanaka, M. Ishino, M. Nishikino, M. Fukuda, M. Kando, T. Kawachi, and Y. Kato, Nature Commun. **4**, 2923 (2013).
12. A. Ya. Faenov, Y. Kato, M. Tanaka, T. A. Pikuz, M. Kishimoto, M. Ishino, M. Nishikino, Y. Fukuda, S. V. Bulanov, and T. Kawachi, Opt. Lett. **34**, 941 (2009).
13. T. Pikuz, A. Faenov, Y. Fukuda, M. Kando, P. Bolton, A. Mitrofanov, A. Vinogradov, M. Nagasono, H. Ohashi, M. Yabashi, K. Tono, Y. Senba, T. Togashi, and T. Ishikawa, Opt. Express **20**, 3424 (2012).
14. T. A. Pikuz, A. Ya. Faenov, Y. Fukuda, M. Kando, P. Bolton, A. Mitrofanov, A. V. Vinogradov, M. Nagasono, H. Ohashi, M. Yabashi, K. Tono, Y. Senba, T. Togashi, and T. Ishikawa, Appl. Opt. **52**, 509 (2013).

16

Stabilization of a high-order harmonic generation seeded extreme ultraviolet free electron laser by time-synchronization control with electro-optic sampling

H. Tomizawa[1], T. Sato[1], K. Ogawa[1], K. Togawa[1], T. Tanaka[1], T. Hara[1], M. Yabashi[1], H. Tanaka[1], T. Ishikawa[1], T. Togashi[2], S. Matsubara[2], Y. Okayasu[2], T. Watanabe[2], E.J. Takahashi[3], K. Midorikawa[3], M. Aoyama[4], K. Yamakawa[4], S. Owada[5], A. Iwasaki[5], and K. Yamanouchi[5]

[1] *RIKEN, SPring-8 Center, Kouto 1-1-1, Sayo, Hyogo 679-5148, Japan*

[2] *Japan Synchrotron Radiation Research Institute, Kouto 1-1-1, Sayo, Hyogo 679-5198, Japan*

[3] *RIKEN Center for Advanced Photonics, Hirosawa 2-1, Wako, Saitama 351-0198, Japan*

[4] *Japan Atomic Energy Agency, Umemidai 8-1-7, Kizugawa, Kyoto 619-0215, Japan*

[5] *The University of Tokyo, Hongo 7-3-1, Bunkyo-ku, Tokyo 113-0033, Japan*

Abstract

A fully coherent free electron laser (FEL) seeded with a higher-order harmonic (HH) pulse from high-order harmonic generation (HHG) is successfully operated for a sufficiently prolonged time in pilot user experiments by using a timing drift feedback. For HHG-seeded FELs, the seeding laser pulses have to be synchronized with electron bunches. Despite seeded FELs being non-chaotic light sources in principle, external laser-seeded FELs are often unstable in practice because of a timing jitter and a drift between the seeding laser pulses and the accelerated electron bunches. Accordingly, we constructed a relative arrival-timing monitor based on non-invasive electro-optic sampling (EOS). The EOS monitor made uninterrupted shot-to-shot monitoring possible even during the seeded FEL operation. The EOS system was then used for arrival-timing feedback with an adjustability of 100 fs for continual operation of the HHG-seeded FEL. Using the EOS-based beam drift controlling system, the HHG-seeded FEL was operated over half a day with an effective hit rate of 20%–30%. The output pulse energy was 20 μJ at the 61.2 nm wavelength. Towards seeded FELs in the water window region, we investigated our upgrade plan to seed high-power FELs with HH photon energy of 30–100 eV and lase at shorter wavelengths of up to 2 nm through high-gain harmonic generation (HGHG) at the energy-upgraded SPring-8 Compact SASE Source (SCSS) accelerator. We studied a benefit as well as the feasibility of the next HHG-seeded FEL machine with single-stage HGHG with tunability of a lasing wavelength.

Keywords: arrival-timing monitor; electro-optic sampling (EOS); high-gain harmonic generation (HGHG); high-order harmonic generation (HHG); self-amplified spontaneous emission (SASE); the soft x-ray water window region; wavelength-tunable seeded FEL; X-ray free electron laser (XFEL)

1. Introduction

A free electron laser (FEL) was proposed as one of the most promising coherent light sources with arbitrary wavelength in 1971[1]. Today, using self-amplified spontaneous emission (SASE) schemes, FELs are available in a wide region up to hard x-rays[2]. The SPring-8 Compact SASE Source (SCSS)[3], which is a prototype FEL machine at SPring-8, generates extreme ultraviolet (EUV) pulses at wavelengths from 50 to 60 nm. The SCSS was constructed for feasibility tests of new components to embody our x-ray FEL (XFEL) machine concept of the SPring-8 Angstrom Compact Free Electron Laser (SACLA)[4]. We are upgrading this EUV-FEL to a higher energy machine with the aim of lasing at even shorter wavelengths. It consists of a high-brightness thermionic electron gun with several bunching sections, a high-energy accelerator and in-vacuum undulators.

Correspondence to: H. Tomizawa, RIKEN, SPring-8 Center, Kouto 1-1-1, Sayo, Hyogo 679-5148, Japan. Email: hiro@spring8.or.jp

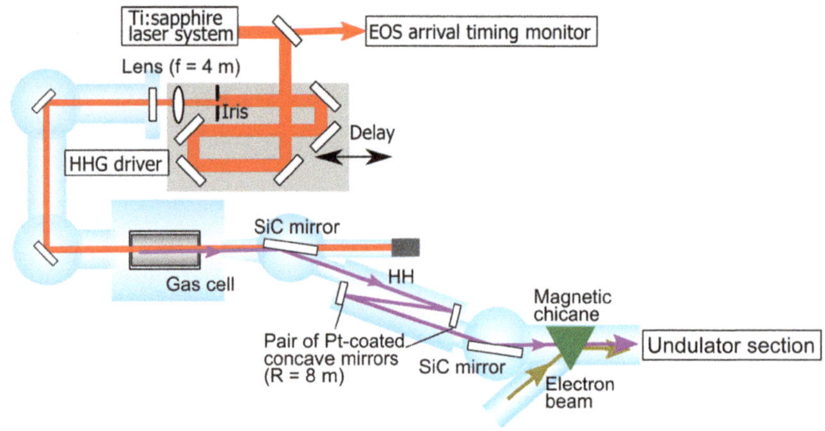

Figure 1. Schematic drawing of HHG and its beam transportation system for a seeded EUV-FEL. HH pulses are generated in a xenon gas cell with a lens (focal length: 4 m) and separated from the fundamental pulses of the Ti:S laser by the first SiC mirror. A pair of Pt-coated concave mirrors with an 8 m curvature radius is used for the loose focusing HH pulses. The seeding HH pulses are selectively reflected by the second SiC mirror, and fully overlap the electron bunch at the front-end of the first undulator (Undulator 1).

In the SASE scheme, spontaneous radiation is first generated at the front-end of the first undulator. Then, it is exponentially amplified up to the saturated intensity level in a single pass while travelling in the following undulators[5]. Due to this radiation mechanism, an SASE-FEL pulse is not temporally coherent. That is, the spectra have uncontrollable spikes in the frequency domain and fluctuate shot-to-shot. The fluctuation of SASE-FEL pulses makes interpretations of experimental results difficult. To provide a spectrum with a targeted single peak on demand for user experiments, a fully coherent seeding source is required without an optical cavity in shorter wavelengths below the EUV region.

To solve this problem, a number of seeding schemes have been proposed up to now. Seeding schemes are roughly categorized into two kinds of approach, those that use some external seed sources and those that use absolutely none. One is called self-seeding, in which an SASE pulse monochromatized just after the first undulator section is injected into the second undulator section as the seeding source[6, 7]. In the other scheme, an external optical laser pulse is independently provided and used as a fully coherent seeding source. Recently, the high-order harmonic generation (HHG) of a femtosecond laser pulse focused in a gaseous medium was expanded up to the water window region and became a practical fully coherent light source in this region[8]. It is possible to seed FELs in the EUV wavelength region. However, since the pulse energy is limited and is insufficient to overcome the shot noise at a shorter wavelength, the external seeding scheme becomes much more difficult. To extend FEL seeding to a shorter wavelength, the seeding scheme called high-gain harmonic generation (HGHG)[9] is often adopted. In the user facility at FERMI (Elettra), the third-order harmonic (TH) pulse of a Ti:sapphire (Ti:S) laser is used as an optical seeder and generates the seeded FEL at its 13th harmonic (20 nm) with single-stage HGHG[10] and furthermore at the fifth harmonic (4 nm) of the first HGHG pulse with a double cascade of HGHG[11]. In this system, the seeding TH pulse (~260 nm) is generated with a nonlinear crystal because of higher conversion efficiency. This allows the utilization of an optical seeder with a high pulse energy of the millijoule level, but a longer wavelength far from the final target wavelength of lasing.

In the hard x-ray region, self-seeding has been established as a robust and promising method. However, for the wavelength region longer than soft x-rays, self-seeding seems not to be so simple. In SPring-8, we made a decision to select the feasible self-seeding utilizing Bragg reflection from a thin diamond crystal[7] as the main seeding plan at SACLA for hard x-rays, and an HHG-based seeder for soft x-rays as an optional plan at SCSS. By directly seeding in the EUV region we developed an external seeding system with higher-order harmonic (HH) pulses from 50 to 62 nm (Figure 1). These correspond to the 13th (61.5 nm) and 15th (53.3 nm) harmonics of a Ti:S laser (central wavelength: 800 nm). In 2010, we achieved the first successful seeding of the SCSS EUV-FEL at the 13th harmonic of the Ti:S laser[12]. However, the low contrast ratio of the pulse energy of the seeded FEL pulses with respect to the SASE background noise as well as the low probability of successful seeding events were not favoured for user experiments. In addition, the best seeding conditions did not last more than ten minutes. The main reason for the small number of successful shots of the seeded FEL pulses was the poor temporal overlap between the 13th-order harmonic pulses and the electron bunches at the undulator. By applying a relative arrival-timing monitor based on electro-optic sampling (EOS) for the timing feedback, we improved the beam parameter setup and HHG seeding performance at SCSS. As a result, we observed a clear difference between the SASE-FEL and a seeded FEL with a high contrast.

In this paper, improvements in the HH-seeding quality at the SCSS EUV-FEL are described in detail. By applying a non-invasive EOS arrival-timing monitor with feedback, sufficient FEL pulses are continuously seeded in a narrow spectral band. In this first stage of our development, we choose direct HHG seeding to study the feasibility of an HGHG scheme started from the shortest possible HH wavelength. In the next stage, to extend the FEL photon energy to the water window region with wavelength tunability, we propose single-stage HGHG with continuum HHG by mixing multicycle two-colour laser fields. To reduce the noise degradation of seeding, the total harmonic number of HGHG should be as low as possible. Therefore, we do not choose a multistage HGHG scheme in order to avoid noise degradation due to the frequency multiplication process[13].

2. Setup of HHG direct seeding at SCSS

Our machine setup of the HHG-seeded FEL is shown in Figures 1 and 2. It consists of an accelerator equipped with a variety of beam monitors to optimize overlapping conditions, and an ultrafast optical laser source. HH pulses from a femtosecond Ti:S laser are generated in a xenon gas cell and transported into the undulator section. In the direct seeding scheme with an external HH laser pulse, maximization of the overlap between the seed laser pulse and the electron bunch is crucial in 6D phase space (x, x', y, y', t, E) at the front-end of the first undulator. Here, x and y are the horizontal and vertical centroid positions (sizes), x' and y' are their normalized momenta (divergences), and t and E are the time and the photon energy (central wavelength). The spatial overlap is checked with screen monitors at two points in the undulator section. Additionally, the EOS monitor checks the temporal overlap in real time. Utilizing the electro-optic (EO) probe pulse split from the HHG-driving laser pulse, the arrival time difference between the seeder pulse and the electron bunch is under control and is fixed at the optimal seeding condition with monitoring spectra and the pulse energy of the seeded FEL pulses.

2.1. SCSS prototype accelerator and operation parameters for seeding

The SCSS test accelerator consists of a thermionic electron gun, RF bunch compressors (lowest frequency of RF cavity: 238 MHz), a C-band (5712 MHz) main Linac and two in-vacuum undulators (each of 4.5 m length). Electron bunches are generated at the pulsed thermionic electron gun with a CeB_6 single crystal cathode[14] with a repetition rate of 30 Hz. The electron bunches are mainly compressed through a velocity bunching process in the RF bunch compressor cavities and accelerated by a C-band high-gradient Linac up to 250 MeV. At the end, the beam is injected into a series of in-vacuum undulators with 15 mm period lengths and 300 undulator periods. For the HHG seeding operation, a bunch charge is set to 300 pC and the transverse beam size is $\sim$200 μm (FWHM) in a diameter at the entrance of the first undulator. One notable difference in the accelerator setup between SASE and seeded FEL operations is the electron bunch length. For SASE, the bunch length is normally adjusted to be the shortest so that the FEL gain is maximized. For a seeded FEL, the highest gain may not be the best. Since the seed pulse length provided by the 13th harmonic is of the order of tens of femtoseconds, i.e., shorter than the nominal electron bunch length in our case, SASE is inevitably amplified where the seed pulse does not exist in the electron bunch. In order to generate single-spike, fully coherent FEL pulses, it is necessary to suppress SASE amplification as much as possible, yet a seeded FEL needs to be above a certain level. The first requirement for this is to prepare a seed laser pulse with a peak power higher than the shot noise in the lethargy regime. In our case, HH laser pulses are able to provide enough peak power. Second, the electron bunch length should be somehow optimized so that the contrast ratio between the SASE and the seeded FEL is maximized. According to numerical simulations in our case, a bunch length of 1 ps (FWHM) is too long with a lower peak current to obtain enough gain in a seeded FEL, while the nominal bunch length for SASE operations, 300 fs, is too short with a higher peak current to suppress SASE gain in our experimental conditions. As a result, it was a good compromise to fix the bunch length at 600 fs (FWHM). Note that the longer bunch length also helps to mitigate the shot-to-shot fluctuation of the output FEL power coming from the timing jitter between electron bunches and seed laser pulses.

The seeded FEL pulses are transported to the photon beamline of the SCSS EUV-FEL (Figure 2). During the seeding operations, we measure the spectra of the seeded FEL pulses with a single-shot EUV spectrometer, which has 10^3 resolving power $(\lambda/\Delta\lambda)$ around 60 nm, at the end of the photon diagnostics line. The pulse energy of the seeded FEL is detected by a gas monitor detector installed at the front-end of the beamline utilized by users. The gas monitor detector filled with argon gas has been calibrated with a calorie meter and it enables us to obtain the absolute pulse energy non-destructively[15]. These beamlines are switched by inserting a mirror at the entrance of the photon diagnostics line.

2.2. HHG driver and EO probe pulses generated from the common laser source

The external optical laser system for the driver of the HHG and the EOS probe pulse is based on a chirped pulsed amplification system of a femtosecond Ti:S laser (800 nm, 180 fs, 30 Hz). The system consists of a mode-locked oscillator (Spectra-Physics: Tsunami) that is synchronized to

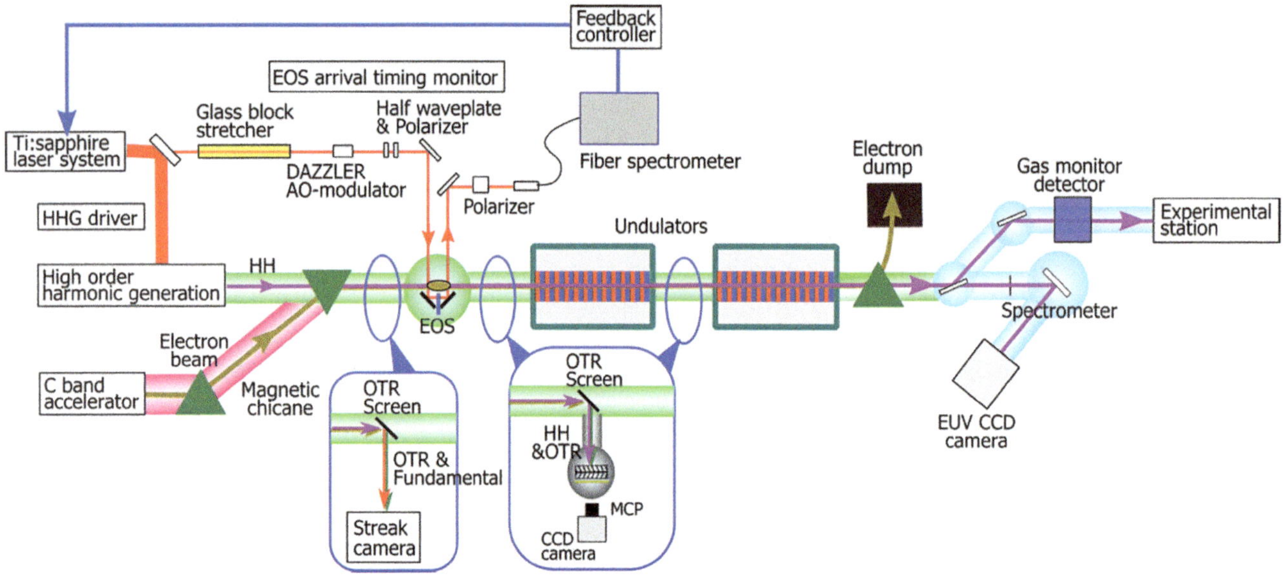

Figure 2. Schematic drawing of the HHG-seeded FEL system with a timing drift control. This seeded FEL system consists of an SCSS FEL machine (C-band accelerator, magnetic chicane and in-vacuum undulators), a Ti:S laser system (which is the common laser pulse source for an HHG seeding system) and an EOS-based arrival-timing monitor. The EOS arrival-timing monitor is installed before the first undulator. Utilizing the EO probe pulse optically split from the HHG-driving laser pulse, the arrival time difference of the seed laser pulse and the electron bunch is under control and is fixed at the optimal seeding condition. The spectra and pulse energy of the seeded FEL are measured with a single-shot spectrometer and a gas monitor detector at the end of the beamline, respectively. Insets: the spatial profile of the seeding pulse and the electron bunch on a phosphor screen are measured with microchannel plates at the entrance and end of the first undulator. Temporal overlap is roughly checked by a streak camera at sweep ranges from 1 ns down to 50 ps.

the 238 MHz master clock of the SCSS by feedback locking the cavity length, a regenerative amplifier (Spectra-Physics: Spitfire) and a home-built four-pass amplifier. The output pulse energy is up to 120 mJ at the laser source and 30% of this output is used for the seeding operation. More than 99% of the pulse energy is used for the HHG system, and the residual energy for the probe pulse of EOS. The EO probing pulse is stretched by passing through a bulk stretcher before arriving at the EO crystal. It gives enough dynamic range to measure the timing drift of the electron bunch with respect to the HH pulse. For FEL seeding with an external laser pulse, it is essential to establish the coincidence and to sustain the overlapping condition between the HH pulses and electron bunches.

On the other hand, the HHG-driving laser pulse with a pulse energy of 30 mJ is loosely focused by a plano-convex lens (focal length: 4 m) and delivered into the target chamber through a thin window. We set the focus around the entrance pinhole of an interaction cell that is filled with xenon gas. The target gas pressure is adjusted to balance the geometrical phase shift and the harmonic dipole phase. The HH pulse is selectively reflected with SiC mirrors set at the Brewster angle (69°) to allow the fundamental Ti:S laser through. By introducing a pair of Pt-coated, nearly normal-incidence mirrors with 8 m curvature radii, the HH pulses are loosely focused in the front-end of the first undulator.

After being reflected at the second SiC separator mirror, the HH pulse is combined with the electron beam at the magnetic chicane, and then both beam are introduced to the undulator section. Since the SiC mirrors reflect EUV light above 30 nm, a few orders of HH including the 13th are selected and sent to the undulator. The beam diameter of the 13th harmonic at the entrance of the first undulator is ~500 μm (FWHM). The pulse energy of the 13th harmonic is estimated to be 2 nJ, which is measured downstream and calibrated by the spectrometer as well as the gas monitor detector at the end station. The total optical throughput of the HH laser transportation is ~1% (the HH pulse is generated with ~200 nJ at the gas cell). The resulting peak power is estimated to be 40 kW in the seeding region, assuming a pulse duration of 50 fs (FWHM).

2.3. Measurement to keep overlapping between electron bunches and HH pulses

To optimize the seeding conditions, the HH pulses and electron bunches must overlap each other in the 6D phase space. Thus, beam diagnostics for achieving optimal 6D overlapping are key requirements for successful operation. To obtain a sufficient power gain of the FEL amplifier and overcome the shot noise under the restriction of seeding pulse energy, both the HH pulse and the electron bunch are compressed in the 6D phase space volume as much as possible. However, this makes it difficult to keep the overlapping condition continuously. Therefore, we measure the temporal and spatial sizes of both the electron bunches and HH pulses at the first undulator and relax the overlapping

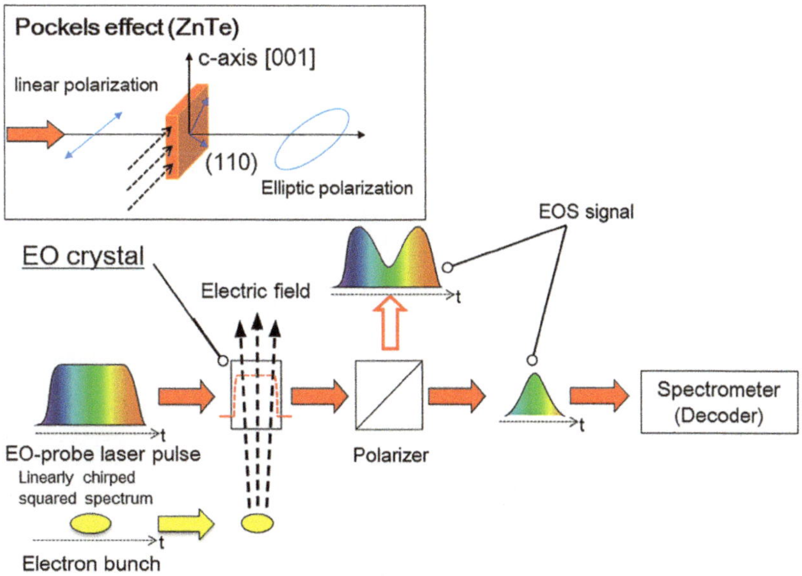

Figure 3. Principle of EOS measurement in the manner of spectral decoding. The probe pulse is linearly chirped and acts as a carrier wave for spectral decoding EOS. In addition, the use of a linear-chirped laser pulse with a flat-top spectrum to probe an ultrafast EO crystal (ZnTe) makes it possible to characterize the temporal bunch charge distribution precisely in real time. The adaptive AO modulator (Fastlite: Dazzler HR45-650-1100) is able to shape both the spectral phase and the intensity distributions of laser pulses with broadband spectra. The EO crystal is set near the electron beam. The linear polarization of the carrier wave changes into an elliptic polarization mainly due to the Pockels effect in the EO crystal under the electric field of the electron bunch. The information on the electron bunch charge distribution is encoded as the intensity modulation in the spectrum, and decoded bunch-by-bunch by a multichannel spectrometer.

conditions by enlarging both the sizes as long as the seed power keeps a sufficiently high contrast against the shot noise of the electron beam.

The spatial profiles and positions of the seeding pulse and the electron bunch are measured on the same area sensor cameras composed of microchannel plates (MCPs) with phosphor screens at the entrance and the end of the first undulator. The system allows simultaneous monitoring of the spatial profiles of the HH pulse and the optical transition radiation (OTR) from the electron beam to ensure spatial overlapping while travelling in the first undulator. The mismatch of the transverse centroid position and the angle of the direction between the HH pulses and the electron bunches are suppressed into ranges of less than 100 μm and 100 μrad by precisely steering the optical path of the HH pulse using the two Pt-coated mirrors.

On the other hand, temporal overlapping is also crucial for the seeding scheme. For this purpose, sequential tuning steps combining fine and coarse adjustments are applied. First, the timing difference between the OTR of the electron bunch and the HHG-driving Ti:S laser pulse is roughly measured with a streak camera (Hamamatsu Photonics K.K: FESCA-200). Then the timing difference is adjusted with an electrical delay unit (Candox Systems Inc.: 84DgR238B01) of the Ti:S laser, and consequently lowered to a certain extent (<1 ps). Finally, the peak positions of the EO signals are kept at the same wavelength as the optimal seeding condition which is defined with the monitoring spectra and pulse energy of the seeded FEL pulses downstream.

2.4. Arrival-timing measurement with EOS and feedback system

In the relative arrival-timing measurement system, the EOS technique is applied in the manner of spectral decoding (Figure 3). It consists of a probe laser, an EO crystal, a polarizer and a multichannel spectrometer for real-time measurements. We use a zinc telluride (ZnTe (110): 3 × 4 mm^2 area and 1 mm thick) as the EO crystal in this setup. The ZnTe crystal is set in the vicinity of the electron beam path, in which the probe laser pulse is passing and modulated. The probe pulse is linearly chirped and works as a carrier wave with a linear relation between frequency (wavelength) and time (relative arrival timing). When the electron bunch passes near the EO crystal, the orthogonal polarization components of the EO probe pulse are retarded, one against the other (birefringence effect), through the crystal under the high electric field of the electron bunch. Then the polarization modulation is converted to intensity modulation in the spectrum by the polarizer. As a result, the electron bunch timing with respect to the EO probe laser pulse is encoded as the intensity modulation in the spectrum, and is diagnosed by a multichannel spectrometer in real time. The relative timing drift is compensated with a feedback control.

In our EOS system, an EO probe laser pulse is optically split from the common laser source to drive the HHG in the accelerator tunnel. After splitting, the EO-probe pulse is primarily stretched through a bulk stretcher made of

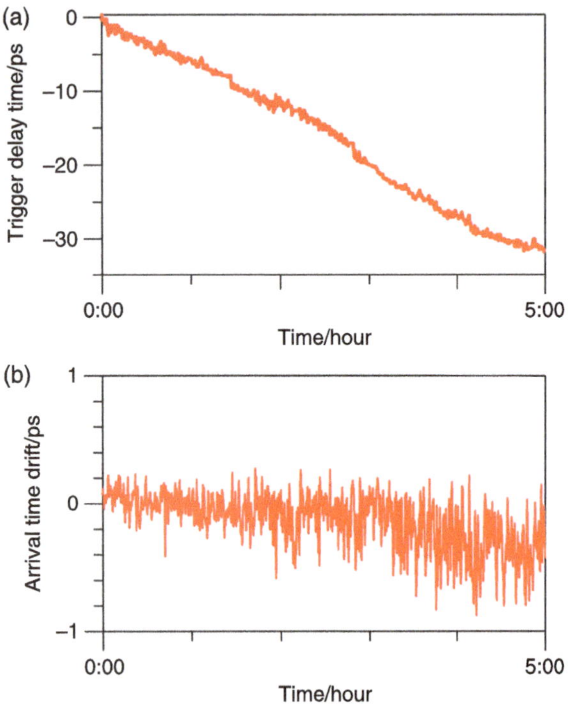

Figure 4. (a) The trigger delay time added to the Candox delay unit by using the home-built feedback system. (b) The relative arrival time of the electron bunch with respect to the optical laser pulse. Both (a) and (b) were measured for 5 h simultaneously.

high-index glass blocks (Ohara: S-NPH3). Next, to shape the spectrum flat-top and control the fine stretching factor with compensating non-linearity of the chirp, we install an adaptive AO modulator (Fastlite: Dazzler HR45-650-1100) as a fine adjustment tool. The arrival timing of the electron bunch with respect to the HHG driver laser pulse is decoded as the spectral peak of the EO signal. The EO signals are decoded by a multichannel spectrometer (Ocean Optics: QE65000). The relative timing between the HH pulse and the electron bunch is fixed at the optimal timing with a feedback system.

The delay time shown in Figure 4(a) is added to the Candox delay unit for five hours. As a consequence, the timing drift is suppressed, as shown in Figure 4(b). If the trigger delay had not been optimized by our feedback system, the relative arrival time drift that corresponds to the delay time shown in Figure 4(a) would have been observed.

3. Experimental results and evaluation

3.1. Experimental results for the seeded FEL pulses

Figure 5 compares typical spectra of the seeded FEL (red line) and the SASE (blue line) which was measured by the single-shot EUV spectrometer at the end of the beamline (shown in Figure 2). The spectrum of the seeded FEL has almost a single peak with a bandwidth of about 0.06 nm

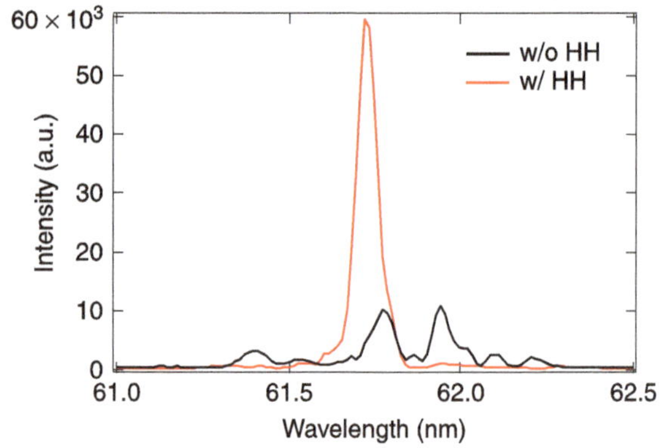

Figure 5. Comparison of the typical spectra of FEL pulses with (red line) and without (blue line) seeding HH pulses. The spectral bandwidth of the seeded FEL pulse was 0.06 nm (FWHM).

(FWHM). It follows that the bunch length of 600 fs was appropriate for providing users with single-spike, fully coherent FEL pulses. With the existing setup, however, not all the output FEL pulses came in such ideal FEL pulses with a single spike; some distributions with many spikes were also observed in some shots. Thus, we evaluated the success rate of the seeding by introducing the following criteria.

For a quantitative evaluation of the probability of a successful seeding effect, we shall define an 'effective' hit rate. When the spectral peak intensity of the output pulse exceeded the threshold indicated in the following formula, we defined the event as an 'effective' hit:

$$I_{\text{seed}} - \overline{I_{\text{SASE}}} > 4\sigma_{\text{SASE}},$$

where I_{seed} and $\overline{I_{\text{SASE}}}$ are the peak intensity with an HH pulse injected and the averaged peak intensity without the seeding source, respectively, and σ_{SASE} is the standard deviation of the statistical fluctuation without the seeding source. In the case of our former experiment in 2010[12], according to this definition, the effective hit rate was calculated as 0.3%.

Figure 6 shows the correlation data plot between the spectral peak intensity of the output pulse subtracted by the average SASE intensity and the wavelength of the peak intensity in the data for 10,000 shots with seeded FEL operation in 2012. In Figure 6, the red points represent the seeded pulses with effective hits, which exceeded 4 σ_{SASE}. The blue points, which are less than $4\sigma_{\text{SASE}}$, are defined as ineffective hits. Spectral peak intensities above the threshold are distributed from 61.5 to 62.0 nm. Conversely, the spectral peak intensities in the SASE-like region (blue points) are distributed over 2.5 nm. The standard deviation of the peak intensity wavelength is 0.08 nm for the seeded FEL pulse with effective hits. Consequently, this statistical data analysis shows that the standard deviation is comparable to the spectral bandwidth of the seeded FEL pulses mentioned above. Our definition of the effective hit rate is useful to

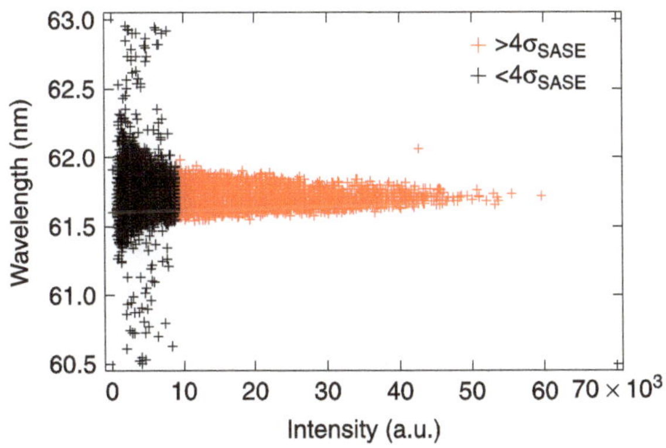

Figure 6. Correlation data plot between the normalized intensity and the spectral peak intensity with seeded operation. Here, σ_{SASE} is the standard deviation of the peak intensities without the HH pulses under the seeding condition of the FEL. Effective seeded FEL pulses are defined as being large as 4 σ_{SASE} (red) for our user experiments.

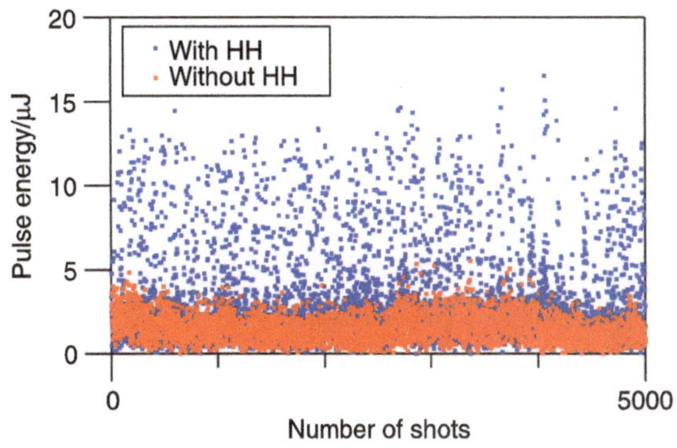

Figure 7. Trend graph of peak intensities of 5,000 FEL pulses in the seeded operations with (blue points) and without (red points) HH pulses. Experimental results of the experiment with feedback in 2012 are shown. The contrast ratio of the peak intensity was improved by a factor of ~3. In 2010, the seeded FEL pulse energy was 1.3 μJ[12]. During our experiments, we achieved a pulse energy of up to 20 μJ at maximum.

judge the seeding quality in our user experiments. For the 10,000 shot data in Figure 6, the effective hit rate was 30%.

3.2. Improvement of seeding condition with EOS feedback

The trend graph of the peak intensity is shown in Figure 7. The ability to not only increase the hit rate but also sustain a seeding effect and keep the output power as stable as possible is critical to practical user experiments. As presented in Figure 7, the average output power was kept almost constant for hours while the pilot user experiment was being implemented, although shot-to-shot fluctuation was obviously an issue, as already discussed. In principle, the EO feedback system is able to keep the seeding effect as long as it is needed. It is also worth noticing that the sustained seeding is beneficial not only to experiments, but also to accelerator operations. In our former experimental result in 2010[12], the output pulse energies and the seeded FEL gain were limited to 1.3 μJ and 650, respectively. After applying the EOS feedback system, the output energy was increased up to 15–20 μJ, and the seeded FEL gain reached as much as $\sim 10^4$. Such an improvement was mainly due to the capability of the reproducible seeding condition in the time domain while adjusting other dimensions. For seeded FEL tunings, it is important to keep a temporal overlapping while the spatial and angular overlapping is being adjusted. Thus, the feedback system contributed not only to increasing the hit rate, but also to improving the contrast ratio of output intensities between the seeded FEL and SASE.

4. Wavelength-tunable seeded FEL in the water window region of soft x-rays

Here, we discuss further extension of the wavelength of a seeded FEL scheme to shorter wavelength regions. Despite being one of the milestones of seeded FELs in the soft x-ray region, seeded FELs have not been realized with a sufficient level of pulse energy for user experiments in the water window region.

Figure 8 shows the numerical simulation results for the output pulse energy after the FEL amplification versus the photon energy (radiation wavelength) in the case of single-stage HGHG assuming that the HH pulse energy is between 30 and 100 eV. To reduce the noise degradation of the seeding effect, we selected the shortest HH wavelength (highest photon energy) possible in its plateau region and reduced the total harmonic number of HGHG to as low as possible.

In this simulation, the normalized emittance, energy spread and peak current of the electron beam were fixed to be 0.8 mm mrad (RMS), 8×10^{-5} (RMS) and 1.2 kA, respectively, while the beam energies were varied from 1.0 to 1.8 GeV. We assumed 5 cm period undulators with a fixed K-value of 3.2 for the modulator section, and 2.8 cm period undulators for the radiator section with the K-values being tuned from 1.06 to 2.99 according to the harmonic number. At the same time, the HH photon energy was assumed to be tuned from 30 to 100 eV, with a beam size of 100 μm (RMS), pulse length of 5 fs (FWHM) and maximum pulse energy of 15 nJ. Note that the input pulse energy was reduced under certain conditions to avoid oversaturation in the modulator section.

The results of this simulation show clearly that the present single-HGHG scheme can readily be applied to the wavelength region below 2 nm with a pulse energy level of a few tens of microjoules, where an HH pulse with sufficiently high intensity (>20 nJ/pulse) has already been achieved[16]. For seeding in the soft x-ray region, it is necessary to improve the total throughput of HH pulses by more than ten times. To improve the throughput of HH transportation, we designed the optics with Kirkpatrick–Baez mirrors (>80%) and Pt-coated mirrors with 1 degree of incident angle (>95%). This

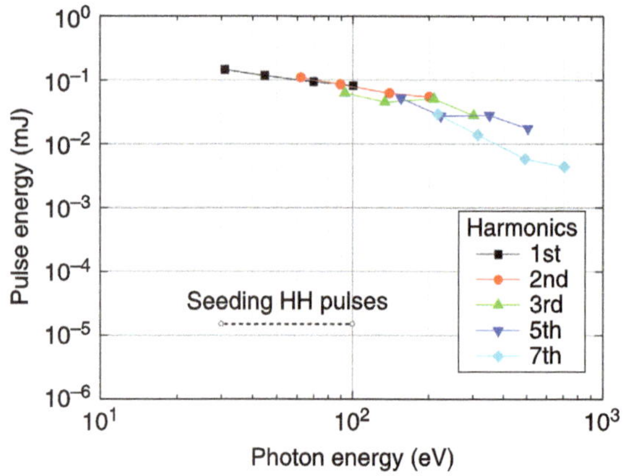

Figure 8. Expected pulse energies of the HHG-seeded HGHG FEL plotted as a function of the photon energy for different harmonic numbers. The harmonics of HGHG were calculated up to the seventh order. In this calculation, the photon energy of the seeding HH pulse was tuned from 30 to 100 eV, with a beam size of 100 μm (RMS), pulse length of 5 fs (FWHM) and maximum pulse energy of 15 nJ. The conditions of the seeding HH pulses are feasible even in continuum HHG to realize a wavelength-tunable seeded FEL in the soft x-ray water window region.

allowed us to obtain 15 nJ of a single harmonic pulse at the undulator.

We also considered that the seeded FEL in the water window region of soft x-rays with wavelength tunability is a useful light source for user experiments. Specifically, it is critically important to adjust the central wavelength of narrow-bandwidth seeded FELs for atomic, molecular and optical physics (AMO) experiments that aim at a specific wavelength for resonance absorptions. For wavelength-tunable operation, we proposed single-stage HGHG seeded with continuum HHG (high dense harmonics) generated by mixing multicycle two-colour laser fields with the fundamental (800 nm) and optical parametric amplifier (1300 nm) of a femtosecond Ti:S laser pulse[17]. By combining a two-colour field synthesis and an energy-scaling method of HHG, the isolated sub-femtosecond pulse reaches as high as 1.3 μJ at ~30 eV[18]. The bandwidths of individual HH pulses are wider than the bandwidth of SASE in the fundamental. According to our simulation, we can provide our users with a few tens of microjoules of seeded FEL at 500 eV. This seeding scheme is feasible with wavelength tunability and can be extended to be in the water window region.

5. Summary and future plan

By maintaining the best overlapping condition between the 50 fs HH pulse and the 600 fs electron bunch, the 13th harmonic of a Ti:S laser was significantly amplified with the SCSS test accelerator employed as an FEL amplifier. The temporal and spatial overlaps of the electron bunch and the HH pulse, as well as the adjustment of the wavelength of the undulator radiation to that of the seeding HH pulse, were precisely tuned to achieve the seeded FEL operation. Spectral narrowing was observed in our seeded conditions. Moreover, a relative arrival-timing monitor and a timing feedback system were introduced in this experiment. With the system, the seeded FEL performance was significantly improved over the long term. Compared with our former seeding condition in 2010, the pulse energy and the FEL gain of the seeded FEL were improved from 1.3 to 20 μJ and from 650 to 10^4 (15 times improvement), respectively. The effective hit rate was improved from 0.3% to 20%–30% (improvement of two orders). Consequently, operation for pilot user experiments with the seeded FEL was attained at SCSS.

For an even higher effective hit rate rising towards to 100% in long-term operation, it is necessary to improve the homogeneity of the electron bunch charge density, the spectral flatness of the EOS probe laser and the relative pointing stability in the transverse overlapping. In order to maintain the best overlapping, the HH pulses and electron bunches must overlap each other in the 6D phase space. We adopted a strategy to overcome the uncontrollable timing and pointing jitter by covering it with large beam sizes. However, this strategy is guaranteed only in the case of a homogeneous electron bunch distribution and uniform current distribution within the overlapping region. In our seeding experiment, the electron bunch was de-bunched by up to 600 fs, but it seems that the effective longitudinal overlapping region was not longer than 400 fs according to scanning of the relative timing delay. The fine distribution of electron bunches needs to be further investigated to effectively suppress the jitter effects. To monitor the relative pointing to keep spatial overlap including their divergences, we are developing a single-shot 3D-BCD (three-dimensional bunch charge distribution) monitor at SPring-8[19]. This extended EOS monitor and a sophisticated feedback system may suppress the long-term drift components of relative pointing stability. In this development, the EO probe laser has been developed as a lineally chirped broadband probe laser with a square spectrum with a large spectral flat-top range. By characterizing the nonlinear chirp of this EO probing laser pulse with a Dazzler-based chirp scanning measurement, we can finely adjust the pulse duration of an octave-broadband probe laser with a linear chirp rate of 1 fs nm^{-1} for probing EO crystals[19]. The octave-broadband flat-top spectrum of the EO probe laser gives not only the upper limit of temporal resolution, but also a large dynamic range for arrival timing (trends in timing drift and pointing data over time). It will be a great contribution to improving the seeded FEL operation with the sophisticated feedback system of EOS measurements.

In addition, further compression of the HH pulse under the restriction of the seeding pulse energy and the electron bunch is required for a single-stage HGHG aiming at the water window region. To achieve significant difference from the SASE, the seeding pulses must have the same wavelength as the resonance wavelength of the FEL with a power level well above the shot noise power. Accordingly, depending on the magnitude of the timing jitter, it becomes necessary to find an appropriate electron bunch duration with which the effective hit rate, the contrast ratio and the output power should be all in the acceptable ranges for the user experiments. With progress to even shorter wavelengths, the temporal resolution of EOS and the temporal response of EO crystal have to be improved to tens of femtoseconds (FWHM). For this purpose, we are developing a novel EOS system with the organic EO crystal DAST (4-N, N-dimethylamino-4′-N'-methyl stilbazolium tosylate) with a temporal response of 30 fs (FWHM)[20, 21].

We seriously consider the generation of spatially and temporally coherent soft x-rays by the proposed scheme as one of the candidates for our FEL upgrades. In order to enable a wavelength-tunable seeded FEL in the water window region of soft x-rays to work effectively and constantly, it is necessary to develop a high throughput of optics to transport HH pulses and to achieve a high conversion efficiency of continuum HHG due to a two-colour field synthesis.

References

1. J. M. J. Madey, J. Appl. Phys. **42**, 1906 (1971).
2. P. Emma, R. Akre, J. Arthur, R. Bionta, C. Bostedt, J. Bozek, A. Brachmann, P. Bucksbaum, R. Coffee, F.-J. Decker, Y. Ding, D. Dowell, S. Edstrom, A. Fisher, J. Frisch, S. Gilevich, J. Hastings, G. Hays, Ph. Hering, Z. Huang, R. Iverson, H. Loos, M. Messerschmidt, A. Miahnahri, S. Moeller, H.-D. Nuhn, G. Pile, D. Ratner, J. Rzepiela, D. Schultz, T. Smith, P. Stefan, H. Tompkins, J. Turner, J. Welch, W. White, J. Wu, G. Yocky, and J. Galayda, Nat. Photon. **4**, 641 (2010).
3. T. Shintake, H. Tanaka, T. Hara, T. Tanaka, K. Togawa, M. Yabashi, Y. Otake, Y. Asano, T. Bizen, T. Fukui, S. Goto, A. Higashiya, T. Hirono, N. Hosoda, T. Inagaki, S. Inoue, M. Ishii, Y. Kim, H. Kimura, M. Kitamura, T. Kobayashi, H. Maesaka, T. Masuda, S. Matsui, T. Matsushita, X. Maréchal, M. Nagasono, H. Ohashi, T. Ohata, T. Ohshima, K. Onoe, K. Shirasawa, T. Takagi, S. Takahashi, M. Takeuchi, K. Tamasaku, R. Tanaka, Y. Tanaka, T. Tanikawa, T. Togashi, S. Wu, A. Yamashita, K. Yanagida, C. Zhang, H. Kitamura, and T. Ishikawa, Nat. Photon. **2**, 555 (2008).
4. T. Ishikawa, H. Aoyagi, T. Asaka, Y. Asano, N. Azumi, T. Bizen, H. Ego, K. Fukami, T. Fukui, Y. Furukawa, S. Goto, H. Hanaki, T. Hara, T. Hasegawa, T. Hatsui, A. Higashiya, T. Hirono, N. Hosoda, M. Ishii, T. Inagaki, Y. Inubushi, T. Itoga, Y. Joti, M. Kago, T. Kameshima, H. Kimura, Y. Kirihara, A. Kiyomichi, T. Kobayashi, C. Kondo, T. Kudo, H. Maesaka, X. Maréchal, T. Masuda, S. Matsubara, T. Matsumoto, T. Matsushita, S. Matsui, M. Nagasono, N. Nariyama, H. Ohashi, T. Ohata, T. Ohshima, S. Ono, Y. Otake, C. Saji, T. Sakurai, T. Sato, K. Sawada, T. Seike, K. Shirasawa, T. Sugimoto, S. Suzuki, S. Takahashi, H. Takebe, K. Takeshita, K. Tamasaku, H. Tanaka, R. Tanaka, T. Tanaka, T. Togashi, K. Togawa, A. Tokuhisa, H. Tomizawa, K. Tono, S. Wu, M. Yabashi, M. Yamaga, A. Yamashita, K. Yanagida, C. Zhang, T. Shintake, H. Kitamura, and N. Kumagai, Nat. Photon. **6**, 540 (2012).
5. W. Ackermann, G. Asova, V. Ayvazyan, A. Azima, N. Baboi, J. Bähr, V. Balandin, B. Beutner, A. Brandt, A. Bolzmann, R. Brinkmann, O. I. Brovko, M. Castellano, P. Castro, L. Catani, E. Chiadroni, S. Choroba, A. Cianchi, J. T. Costello, D. Cubaynes, J. Dardis, W. Decking, H. Delsim-Hashemi, A. Delserieys, G. Di Pirro, M. Dohlus, S. Düsterer, A. Eckhardt, H. T. Edwards, B. Faatz, J. Feldhaus, K. Flöttmann, J. Frisch, L. Fröhlich, T. Garvey, U. Gensch, Ch. Gerth, M. Görler, N. Golubeva, H.-J. Grabosch, M. Grecki, O. Grimm, K. Hacker, U. Hahn, J. H. Han, K. Honkavaara, T. Hott, M. Hüning, Y. Ivanisenko, E. Jaeschke, W. Jalmuzna, T. Jezynski, R. Kammering, V. Katalev, K. Kavanagh, E. T. Kennedy, S. Khodyachykh, K. Klose, V. Kocharyan, M. Körfer, M. Kollewe, W. Koprek, S. Korepanov, D. Kostin, M. Krassilnikov, G. Kube, M. Kuhlmann, C. L. S. Lewis, L. Lilje, T. Limberg, D. Lipka, F. Löhl, H. Luna, M. Luong, M. Martins, M. Meyer, P. Michelato, V. Miltchev, W. D. Möller, L. Monaco, W. F. O. Müller, O. Napieralski, O. Napoly, P. Nicolosi, D. Nölle, T. Nuñez, A. Oppelt, C. Pagani, R. Paparella, N. Pchalek, J. Pedregosa-Gutierrez, B. Petersen, B. Petrosyan, G. Petrosyan, L. Petrosyan, J. Pflüger, E. Plönjes, L. Poletto, K. Pozniak, E. Prat, D. Proch, P. Pucyk, P. Radcliffe, H. Redlin, K. Rehlich, M. Richter, M. Roehrs, J. Roensch, R. Romaniuk, M. Ross, J. Rossbach, V. Rybnikov, M. Sachwitz, E. L. Saldin, W. Sandner, H. Schlarb, B. Schmidt, M. Schmitz, P. Schmüser, J. R. Schneider, E. A. Schneidmiller, S. Schnepp, S. Schreiber, M. Seidel, D. Sertore, A. V. Shabunov, C. Simon, S. Simrock, E. Sombrowski, A. A. Sorokin, P. Spanknebel, R. Spesyvtsev, L. Staykov, B. Steffen, F. Stephan, F. Stulle, H. Thom, K. Tiedtke, M. Tischer, S. Toleikis, R. Treusch, D. Trines, I. Tsakov, E. Vogel, T. Weiland, H. Weise, M. Wellhöfer, M. Wendt, I. Will, A. Winter, K. Wittenburg, W. Wurth, P. Yeates, M. V. Yurkov, I. Zagorodnov, and K. Zapfe, Nat. Photon. **1**, 336 (2007).
6. J. Feldhaus, E. L. Saldin, J. R. Schneider, E. A. Schneidmiller, and M. V. Yurkov, Opt. Commun. **140**, 341 (1997).
7. G. Geloni, V. Kocharyan, and E. Saldin, *DESY10-008* (2010).
8. E. J. Takahashi, T. Kanai, K. L. Ishikawa, Y. Nabakawa, and K. Midorikawa, Phys. Rev. Lett. **101**, 253901 (2008).
9. L.-H. Yu, M. Babzien, I. Ben-Zvi, L. F. DiMauro, A. Doyuran, W. Graves, E. Johnson, S. Krinsky, R. Malone, I. Pogorelsky, J. Skaritka, G. Rakowsky, L. Solomon, X. J. Wang, M. Woodle, V. Yakimenko, S. G. Biedron, J. N. Galayda, E. Gluskin, J. Jagger, V. Sajaev, and I. Vasserman, Science **289**, 932 (2000).
10. E. Allaria, R. Appio, L. Badano, W. A. Barletta, S. Bassanese, S. G. Biedron, A. Borga, E. Busetto, D. Castronovo, P. Cinquegrana, S. Cleva, D. Cocco, M. Cornacchia, P. Craievich, I. Cudin, G. D'Auria, M. Dal Forno, M. B. Danailov, R. De Monte, G. De Ninno, P. Delgiusto, A. Demidovich, S. Di Mitri, B. Diviacco, A. Fabris, R. Fabris, W. Fawley, M. Ferianis, E. Ferrari, S. Ferry, L. Froehlich, P. Furlan, G. Gaio, F. Gelmetti, L. Giannessi, M. Giannini, R. Gobessi, R. Ivanov, E. Karantzoulis, M. Lonza, A. Lutman, B. Mahieu, M. Milloch, S. V. Milton, M. Musardo, I. Nikolov, S. Noe, F. Parmigiani, G. Penco, M. Petronio, L. Pivetta, M. Predonzani, F. Rossi, L. Rumiz, A. Salom, C. Scafuri, C. Serpico, P. Sigalotti, S. Spampinati, C. Spezzani, M. Svandrlik, C. Svetina, S. Tazzari, M. Trovo, R. Umer, A. Vascotto, M. Veronese, R. Visintini, M. Zaccaria, D. Zangrando, and M. Zangrando, Nat. Photon. **6**, 699 (2012).
11. E. Allaria, D. Castronovo, P. Cinquegrana, P. Craievich, M. Dal Forno, M. B. Danailov, G. D'Auria, A. Demidovich, G. De Ninno, S. Di Mitri, B. Diviacco, W. M. Fawley, M. Ferianis, E. Ferrari, L. Froehlich, G. Gaio, D. Gauthier, L. Giannessi, R. Ivanov, B. Mahieu, N. Mahne, I. Nikolov, F. Parmigiani, G. Penco, L. Raimondi, C. Scafuri, C. Serpico, P. Sigalotti, S. Spampinati, C. Spezzani, M. Svandrlik, C. Svetina, M. Trovo, M. Veronese, D. Zangrando, and M. Zangrando, Nat. Photon. **7**, 913 (2013).

12. T. Togashi, E. Takahashi, K. Midorikawa, M. Aoyama, K. Yamakawa, T. Sato, A. Iwasaki, S. Owada, T. Okino, K. Yamanouchi, F. Kannari, A. Yagishita, H. Nakano, M. E. Couprie, K. Fukami, T. Hatsui, T. Hara, T. Kameshima, H. Kitamura, N. Kumagai, S. Matsubara, M. Nagasono, H. Ohashi, T. Ohshima, Y. Otake, T. Shintake, K. Tamasaku, H Tanaka, T. Tanaka, K. Togawa, H. Tomizawa, T. Watanabe, M. Yabashi, and T. Ishikawa, Opt. Express **19**, 317 (2011).
13. E. L. Saldin, E. A. Schneidmiller, and M. V. Yurkov, Opt. Commun. **202**, 169 (2002).
14. K. Togawa, T. Shintake, T. Inagaki, K. Onoe, T. Tanaka, H. Baba, and H. Matsumoto, Phys. Rev. ST Accel. Beams **10**, 020703 (2007).
15. M. Kato, N. Saitoa, T. Tanaka, Y. Morishita, H. Kimura, H. Ohashi, M. Nagasono, M. Yabashi, K. Tono, T. Togashi, A. Higashiya, and T. Ishikawa, Nucl. Instrum. Methods Phys. Res. A **612**, 209211 (2009).
16. E. J. Takahashi, Y. Nabekawa, and K. Midorikawa, Appl. Phys. Lett. **84**, 4 (2004).
17. E. J. Takahashi, P. Lan, O. D. Mücke, Y. Nabekawa, and K. Midorikawa, Phys. Rev. Lett **104**, 233901 (2010).
18. E. J. Takahashi, P. Lan, O. D. Mücke, Y. Nabekawa, and K. Midorikawa, Nat. Commun. **3691**, 1 (2013).
19. H. Tomizawa, T. Togashi, K. Ogawa, H. Tanaka, M. Yabashi, S. Matsubara, Y. Okayasu, and T. Ishikawa, in *Proceedings of the 2nd International Beam Instrumentation Conference (IBIC2013)* MOAL2 (2013).
20. H. Tomizawa, H. Hanaki, and T. Ishikawa, in *Proceedings of the 29th International Free Electron Laser Conference* 472 (2007).
21. Y. Okayasu, H. Tomizawa, S. Matsubara, N. Kumagai, A. Maekawa, M. Uesaka, and T. Ishikawa, Phys. Rev. ST Accel. Beams **16**, 052801 (2013).

17

Electromagnetic beam propagation in nonlinear media

V.V. Semak[1] and M.N. Shneider[2]

[1]*Signature Science, LLC, NJ 08234, USA*

[2]*Department of Mechanical and Aerospace Engineering, Princeton University, NJ 08544, USA*

Abstract

We deduce a complete wave propagation equation that includes inhomogeneity of the dielectric constant and present this propagation equation in compact vector form. Although similar equations are known in narrow fields such as radio wave propagation in the ionosphere and electromagnetic and acoustic wave propagation in stratified media, we develop here a novel approach of using such equations in the modeling of laser beam propagation in nonlinear media. Our approach satisfies the correspondence principle since in the limit of zero-length wavelength it reduces from physical to geometrical optics.

Keywords: light propagation; self-focusing; wavefront correction

1. Introduction

The science of nonlinear optics is one of the rapidly growing fields driven by multiple important technological applications. Since the invention of lasers this field has experienced revolutionary progress fueled by splendid experimental and theoretical results that provide deep understanding of the nonlinear response of matter to high intensity electromagnetic waves. However, as we will demonstrate here, this achievement was accompanied by the fundamental failure of one particular subdiscipline – the propagation of a laser beam in nonlinear media.

Many theoretical works describing laser propagation in nonlinear media have been published since the concept of laser beam self-focusing and self-trapping was proposed[1]. After more than 50 years of intense research the theoretical concepts and models of self-focusing, beam self-trapping, filamentation and filament plasma defocusing and the corresponding mathematical models were formulated. The books and extensive reviews (for example, Refs. [2–8]) written within the past two decades devote chapters to the detailed description of the peculiarities of this physical phenomenon. Current frontier research of laser beam propagation in nonlinear media deals with ingenious formulations and creative solutions of the nonlinear Schrodinger equation that describes laser beam collapse, self-trapping, dispersion, filamentation, modulation instability, pulse splitting and other extraordinary particularities of nonlinear beam propagation[6, 9, 10]. Therefore, understandably, we did not expect to discover that the results obtained with our recent straightforward theoretical model and numerical simulation of ultrahigh intensity laser pulse propagation in gases[11, 12] contradict the established models of self-focusing, beam self-trapping, filamentation and continuum generation.

Being unable to find either errors or invalidating assumptions in our forthright approach we journeyed back to the source (Maxwell's equations) in order to review the foundational scientific principles. This examination exposed the assumptions in the original physical concept that, in our opinion, are inconsistent and self-contradictory. Also, this examination provided theoretical justification and supports the validity of our previously published approach[11, 12]. Below is the account of the results of this journey.

Correspondence to: M.N. Shneider, Department of Mechanical and Aerospace Engineering, Princeton University, NJ 08544, USA. Email: m.n.shneider@gmail.com

2. Formulation of general equation for beam of electromagnetic wave propagation in nonlinear media

As is well known, the electric and magnetic fields in dielectric media are described by the macroscopic Maxwell's equations as follows:

$$\nabla \cdot \vec{D} = 0, \tag{1}$$

$$\nabla \cdot \vec{B} = 0, \tag{2}$$

$$\nabla \times \vec{E} = -\frac{\partial \vec{B}}{\partial t}, \tag{3}$$

$$\nabla \times \vec{H} = \frac{\partial \vec{D}}{\partial t}, \tag{4}$$

where $\vec{E}$ and $\vec{B}$ are the electric and the magnetic fields, correspondingly, while $\vec{D}$ and $\vec{H}$ are, respectively, the displacement and magnetization fields. The latter fields, sometimes called 'macroscopic' fields, reflect the effect from matter and are defined using phenomenological constituent equations that relate them to the 'microscopic' electric field, $\vec{E}$, and the magnetic field, $\vec{B}$:

$$\vec{D} = \varepsilon_0 \varepsilon \vec{E}, \tag{5}$$
$$\vec{B} = \mu_0 \mu \vec{H}, \tag{6}$$

where ε_0 and μ_0 are the permittivity and the permeability of free space and ε and μ are the permittivity and the permeability of material.

Following the known procedure we take the *curl* vector operator of both sides of Equation (3) and the time derivative of both sides of Equation (4) and assume that the magnetic effect of the media is negligible, i.e., $\mu = 1$. Then, using Equations (5) and (6), we can eliminate the magnetic and magnetization fields, obtaining the equation for the electric field

$$\nabla \times (\nabla \times \vec{E}) + \varepsilon_0 \mu_0 \frac{\partial^2 (\varepsilon \vec{E})}{\partial t^2} = 0. \tag{7}$$

Assuming that the material effect on the electric field is slow compared to the period of optical oscillation or the laser pulse duration, i.e., assuming time independence of the permittivity of material, assuming that the material is weakly absorbing, and recalling that the speed of the electromagnetic wave in a vacuum is $c = 1/\sqrt{\varepsilon_0 \mu_0}$ and the index of refraction of the material is $n = \sqrt{\varepsilon}$, we rewrite Equation (7) as

$$\nabla \times (\nabla \times \vec{E}) + \frac{n^2}{c^2} \frac{\partial^2 \vec{E}}{\partial t^2} = 0. \tag{8}$$

Now, using the identity $\nabla \times (\nabla \times \vec{A}) = \nabla(\nabla \cdot \vec{A}) - \Delta \vec{A}$ we rewrite Equation (8) in the following form:

$$\Delta \vec{E} - \frac{n^2}{c^2} \frac{\partial^2 \vec{E}}{\partial t^2} - \nabla(\nabla \cdot \vec{E}) = 0. \tag{9}$$

From this point the derivations will significantly deviate from the procedure commonly performed in all of the books and journal publications on nonlinear optics since we will be considering the permittivity of the material and, consequently, the index of refraction, as a coordinate dependent function. In contrast to our approach, customary considerations treat permittivity as a constant, zeroing the third term in the left-hand side of Equation (9) and reducing this equation to the commonly known form that contains only two first terms and is called the wave equation. As we show below, neglecting the third term is a significant mistake that leads to an inadequate description of wave propagation in nonlinear media.

For the conditions of inhomogeneity of the properties of electrically neutral media, such as in the case of propagation of short wavelength waves in the ionosphere[13], Equation (9) can be rewritten in the following form:

$$\Delta \vec{E} - \frac{n^2}{c^2} \frac{\partial^2 \vec{E}}{\partial t^2} + \nabla \left(\frac{\nabla \varepsilon \cdot \vec{E}}{\varepsilon} \right) = 0, \tag{10}$$

where we use Equations (1) and (5) from which it follows that $\nabla \varepsilon \cdot \vec{E} + \varepsilon \nabla \cdot \vec{E} = 0$ and, therefore, $\nabla \cdot \vec{E} = -(\nabla \varepsilon \cdot \vec{E}/\varepsilon)$. Finally, recalling that $\varepsilon = n^2$ and using the convenient expression for the displacement field $\vec{D} = \varepsilon_0 \varepsilon \vec{E} = \varepsilon_0 (1 + \chi) \vec{E} = \varepsilon_0 \vec{E} + \vec{P}$, where χ is the electric susceptibility and $\vec{P}$ is the polarization density, we rewrite Equation (7) in the following form:

$$\Delta \vec{E} - \frac{1}{c^2} \frac{\partial^2 \vec{E}}{\partial t^2} + 2\nabla \left(\frac{\nabla n \cdot \vec{E}}{n} \right) = \mu_0 \frac{\partial^2 \vec{P}_L}{\partial t^2} + \mu_0 \frac{\partial^2 \vec{P}_{NL}}{\partial t^2}, \tag{11}$$

where the polarization density vector is represented by the sum of linear and nonlinear components denoted using the subscripts 'L' and 'NL', respectively.

The solution of the propagation equation expressed in the form of either Equation (10) or (11) can be expressed in terms of the slowly varying amplitude function

$$\vec{E}(\vec{r}, t) = \vec{A}(\vec{r}) e^{i(\vec{k}(\vec{r}) \cdot \vec{r} - \omega t)} + \text{c.c.} \tag{12}$$

In this solution the vector amplitudes of the electric field and the wavevector are coordinate dependent, i.e., expression (12) represents the electric field of a nonplanar wave.

Below we will demonstrate that, within paraxial approximation and considering the propagation range in which variation of the spatial profile of laser beam irradiance due to the effect of nonlinear induced refraction is small and, thus, can be considered as perturbation, the propagation Equations (10) or (11) can be straightforwardly modified into an equation that, as proposed in Refs. [11, 12], has a solution represented by the blending of the solution of the Helmholtz equation for propagation of a laser beam in a medium with a uniform and irradiance independent refractive index similar to the one obtained by Kogelnik and Li[14] and a correction term that represents nonlinear field perturbation expressed in terms of paraxial ray optics (the eikonal equation)[15].

3. Examination of current theory for laser beam propagation in nonlinear media

Now we will demonstrate that the current formulations of the propagation equation in nonlinear (self-induced inhomogeneity) media are based on two inconsistent and self-contradictory assumptions. First, in all theoretical works known to us it is assumed that the laser beam has a plane wavefront. Second, as mentioned above, the term responsible for refraction due to media inhomogeneity is disregarded, i.e., the media are assumed as homogeneous.

Under the first assumption, i.e., assumption of a plane wavefront, the solution of the propagation equation is expressed in the form of the slowly varying amplitude function

$$\vec{E}(x, y, z, t) = A_x(x, y, z)e^{\mathrm{i}(k_z z-\omega t)}\hat{x} + A_y(x, y, z)e^{\mathrm{i}(k_z z-\omega t)}\hat{y} + \text{c.c.} \quad (13)$$

Note that here, in accordance with the plane wave assumption, the scalar product of vectors $\vec{k}\cdot\vec{r}$ from Equation (12) is substituted with the product of scalars – $k_z z$, where k_z is the component of the wavevector along the z-axis. The solution form of Equation (13) assumes that the x- and y-components of the wavevector and z-component of the electric field are zero, i.e., the beam propagates exactly along the z-axis.

According to the second assumption, the term describing refraction on the gradient of the refractive index is neglected. Then, assuming linear polarization and, since the requirement of slowly varying amplitude implies that

$$\left|\frac{\partial^2 A}{\partial z^2}\right| \ll \left|k\frac{\partial A}{\partial z}\right|, \quad (14)$$

Equation (10) can be reformulated in a form retaining only the amplitude $A(x, y, z)$ of the electric vector aligned along either the x-axis or the y-axis,

$$\Delta_T A + \left(\frac{\omega^2 n^2}{c^2} - k_z^2\right) A + 2\mathrm{i}k_z\frac{\partial A}{\partial z} = 0, \quad (15)$$

where Δ_T is the transverse Laplace operator.

By definition, the wavevector $k = \omega n/c$, and since a plane wave is assumed, the second term in Equation (15) must be zero. However, all textbooks and scientific articles at this stage of consideration submit that the wavefront deviates from a plane. This, of course, contradicts the initial assumption of a strictly plane front; however, is necessary, as otherwise the self-focusing and self-trapping would not follow from Equation (15). Thus, the second term of propagation Equation (15) in all current models is assumed to be nonzero. We will show below that this manipulation has a devastating consequence.

Another inconsistent assumption of the current theory of electromagnetic wave propagation in nonlinear media is that the third term in complete propagation Equation (11) is negligible and can be omitted, leading to the commonly used Equation (15). As far as we know a justification for such an omission was never provided. It appears that the origin of this assumption can be traced to the original work[16]. All of the subsequent works, except for our recent publications[11, 12], followed the path laid[16] and labored on various mathematical treatments of equations that can be traced to the propagation equation deduced for homogeneous media. Thus, the 'foundational' propagation equation in nonlinear optics was deduced while disregarding media inhomogeneity (inherent, induced or self-induced), and has the form of Equation (15) missing the 'refraction' term (see for example, Equation 7.2.9 in Ref. [3]).

At this point one should wonder how an equation with a term missing the refraction due to media inhomogeneity can describe self-focusing. The answer is hidden in the second term of Equation (15). The 'nonzeroing' of the second term in Equation (15) is crucial for constructing all nonlinear propagation effects out of this oversimplified equation. Indeed, introducing nonlinearity of the refractive index, $n = n_0 + n_2\langle|E|^2\rangle_t$, Equation (15) can be modified into the following form[17, 18]:

$$\frac{\partial^2 E}{\partial x^2} + 2\mathrm{i}k\frac{\partial E}{\partial z} = -k^2\frac{n_2}{n_0}|E|^2 E, \quad (16)$$

that is equivalent to the infamous (in the realm of nonlinear wave propagation) nonlinear Schrodinger's equation[19]

$$\frac{\partial^2 \psi}{\partial x^2} + \mathrm{i}\frac{\partial \psi}{\partial z} = -\kappa|\psi|^2\psi. \quad (17)$$

4. Revised propagation equation for slowly varying amplitude

Let us revise the foundation of the current theory for the propagation of a beam of an electromagnetic wave in a nonlinear medium. In this consideration we will use the general form of 3D propagation Equation (7) that follows directly from Maxwell's equations. Following a traditional approach, we will look for the solution of this equation in the following form of wave with varying amplitude:

$$\begin{aligned}\vec{E} &= E_x\hat{x} + E_y\hat{y} + E_z\hat{z} = \vec{A}(x, y, z)p(t)e^{\mathrm{i}(k_x\hat{x}+k_y\hat{y}+k_z\hat{z}-\omega t)} \\ &= (A_x(x, y, z)\hat{x} + A_y(x, y, z)\hat{y} + A_z(x, y, z)\hat{z})p(t) \\ &\quad\times e^{\mathrm{i}(k_x\hat{x}+k_y\hat{y}+k_z\hat{z}-\omega t)}, \end{aligned} \quad (18)$$

where function $p(t)$ is the dimensionless pulse shape such that the time integral of this function from minus to plus infinity equals unity.

Let us now modify propagation Equation (9) assuming that the pulse shape is a slowly varying function compared to the period of wave oscillation and that the dielectric constant, ε, and therefore the refractive index, n, are time independent. Then, recalling that $\varepsilon = n^2$, and substituting solution of Eqaution (18) into Equation (9), neglecting the time derivative of pulse shape $p(t)$ as it is a slow function, and eliminating the exponent of phase, Equation (9) can be transformed into the equation for the amplitude,

$$\begin{aligned}&\left(k_x^2+k_y^2+k_z^2-\frac{\omega^2n^2}{c^2}\right)A_x\hat{x}\\&+\left(k_x^2+k_y^2+k_z^2-\frac{\omega^2n^2}{c^2}\right)A_y\hat{y}\\&+\left(k_x^2+k_y^2+k_z^2-\frac{\omega^2n^2}{c^2}\right)A_z\hat{z}\\&-2\mathrm{i}\left(k_x\frac{\partial A_x}{\partial x}+k_y\frac{\partial A_x}{\partial y}+k_z\frac{\partial A_x}{\partial z}\right)\hat{x}\\&-\left(\frac{\partial^2 A_x}{\partial x^2}+\frac{\partial^2 A_x}{\partial y^2}+\frac{\partial^2 A_x}{\partial z^2}\right)\hat{x}\\&-2\mathrm{i}\left(k_x\frac{\partial A_y}{\partial x}+k_y\frac{\partial A_y}{\partial y}+k_z\frac{\partial A_y}{\partial z}\right)\hat{y}\\&-\left(\frac{\partial^2 A_y}{\partial x^2}+\frac{\partial^2 A_y}{\partial y^2}+\frac{\partial^2 A_y}{\partial z^2}\right)\hat{y}\\&-2\mathrm{i}\left(k_x\frac{\partial A_z}{\partial x}+k_y\frac{\partial A_z}{\partial y}+k_z\frac{\partial A_z}{\partial z}\right)\hat{z}\\&-\left(\frac{\partial^2 A_z}{\partial x^2}+\frac{\partial^2 A_z}{\partial y^2}+\frac{\partial^2 A_z}{\partial z^2}\right)\hat{z}\\&-2\left[\nabla\left(\frac{\nabla n}{n}\cdot(\vec{A}e^{\mathrm{i}(k_xx+k_yy+k_zz)})\right)\right]e^{-\mathrm{i}(k_xx+k_yy+k_zz)}\\&=0.\end{aligned}\tag{19}$$

Note that the first three terms in the left-hand side of Equation (19) are null, since by definition $k_x^2+k_y^2+k_z^2-(\omega^2n^2/c^2)=0$, and, as in the previous deduction of Equation (10), we used equation $\nabla\cdot\vec{E}=-(\vec{E}\nabla\varepsilon/\varepsilon)$ in order to formulate the right-hand side of Equation (19). Thus, the propagation equation in the nonlinear media has the following form:

$$\begin{aligned}&\Delta\vec{A}+2\mathrm{i}\left(k_x\frac{\partial\vec{A}}{\partial x}+k_y\frac{\partial\vec{A}}{\partial y}+k_z\frac{\partial\vec{A}}{\partial z}\right)\\&+2\left[\nabla\left(\frac{\nabla n}{n}\cdot(\vec{A}e^{\mathrm{i}(k_xx+k_yy+k_zz)})\right)\right]e^{-\mathrm{i}(k_xx+k_yy+k_zz)}\\&=0.\end{aligned}\tag{20}$$

This equation is cardinally different from the currently used Equations (15)–(17) in both aspects of physics and of mathematics. From the point of view of a physicist, Equation (20) straightforwardly shows that beam refraction is produced by self-induced inhomogeneity of the refractive index (that can be a result of the Kerr effect, material ionization etc.). Also, a physicist will find appealing that the presented theory expressed by Equation (20) satisfies the correspondence principle since, as we will demonstrate below, it contains geometrical optics and Equation (20) transforms into the ray-optics equation under the assumption of infinitely small wavelength. In contrast, Equations (15)–(17) do not lead to the geometrical optics in the limiting case of infinitely small wavelength and thus do not satisfy the correspondence principle.

From the point of view of a mathematician new equation of propagation Equation (20) dramatically differs from the current propagation equation in one very peculiar aspect – it has no self-similar solution. In contrast, Equation (16) rewritten as nonlinear Schrodinger's Equation (17) has a self-similar solution, or so called soliton solution, that serves as the foundation for the prediction of laser beam self-trapping and all the current 'filament' theories that predict mind boggling lengths of self-trapped laser 'filament'.

5. Modification of the revised propagation equation for cases of slowly variable amplitude and paraxial laser beam propagation at distances shorter than or comparable to the Rayleigh length

Let us explore the complete 3D propagation Equation (20) within paraxial beam approximation while considering a propagation range in which perturbation of the spatial profile of laser beam irradiance by the nonlinear induced refraction is negligible. The latter condition is realized within the range of several Rayleigh lengths for a focused laser beam with pulse energy that is below a certain value (see detailed discussion in Refs. [11, 12]).

Here we will demonstrate that the propagation Equation (20) leads, as proposed in Refs. [11, 12], to the solution represented by the blending of the solution of the Helmholtz equation for propagation of a laser beam in a medium with uniform and irradiance independent refractive index similar to the one obtained by Kogelnik and Li[14] and a correction term that represents nonlinear field perturbation expressed in terms of a paraxial ray-optics (eikonal) equation[15].

Assuming paraxial beam propagation it is easy to see that one can neglect in propagation Equation (20) the terms containing x- and y-components of the wavevector and the terms containing the z-component of the electric field amplitude as well as their derivatives (see the schematic of the paraxial beam propagation in Figure 1).

Assuming linear polarization in the x–z plane and neglecting smaller terms as described above allows the following simplification of propagation Equation (20):

$$\begin{aligned}\Bigg[&\left(\frac{\partial^2 A_x}{\partial x^2}+\frac{\partial^2 A_x}{\partial y^2}\right)+2\mathrm{i}\left(k_x\frac{\partial A_x}{\partial x}+k_z\frac{\partial A_x}{\partial z}\right)\\&-\mathrm{i}\left(\frac{\partial k_x}{\partial x}+\frac{\partial k_z}{\partial z}\right)A_x+2\mathrm{i}\left(x\frac{\partial k_x}{\partial x}\frac{\partial A_x}{\partial x}+z\frac{\partial k_z}{\partial z}\frac{\partial A_x}{\partial z}\right)\\&-\mathrm{i}\left(x\frac{\partial^2 k_x}{\partial x^2}+z\frac{\partial^2 k_z}{\partial z^2}\right)A_x+2\left(xk_x\frac{\partial k_x}{\partial x}+zk_z\frac{\partial k_z}{\partial z}\right)A_x\end{aligned}$$

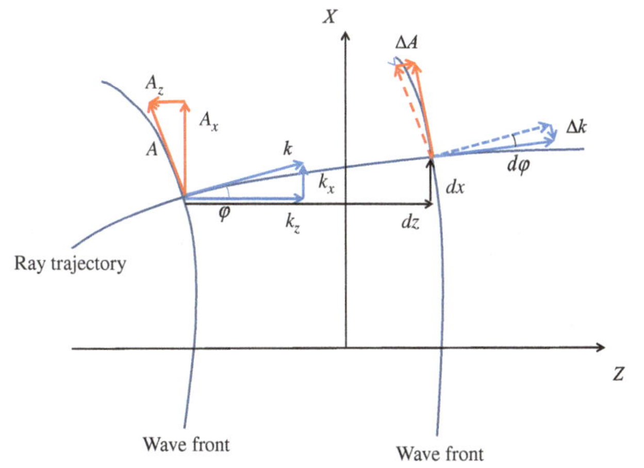

Figure 1. Schematic of evolution of electric field amplitude and wavevector during laser beam propagation.

$$\begin{aligned}
&+\left(x^2\left(\frac{\partial k_x}{\partial x}\right)^2+z^2\left(\frac{\partial k_z}{\partial z}\right)^2\right)A_x\Bigg]\hat{x}\\
&+\Bigg[\left(\frac{\partial^2 A_z}{\partial x^2}+\frac{\partial^2 A_z}{\partial y^2}\right)+2\mathrm{i}\left(k_x\frac{\partial A_z}{\partial x}+k_z\frac{\partial A_z}{\partial z}\right)\\
&-\mathrm{i}\left(\frac{\partial k_x}{\partial x}+\frac{\partial k_z}{\partial z}\right)A_z+2\mathrm{i}\left(x\frac{\partial k_x}{\partial x}\frac{\partial A_z}{\partial x}+z\frac{\partial k_z}{\partial z}\frac{\partial A_z}{\partial z}\right)\\
&-\mathrm{i}\left(x\frac{\partial^2 k_x}{\partial x^2}+z\frac{\partial^2 k_z}{\partial z^2}\right)A_z+2\left(xk_x\frac{\partial k_x}{\partial x}+zk_z\frac{\partial k_z}{\partial z}\right)A_z\\
&+\left(x^2\left(\frac{\partial k_x}{\partial x}\right)^2+z^2\left(\frac{\partial k_z}{\partial z}\right)^2\right)A_z\Bigg]\hat{z}\\
&-2\left[\nabla\left(\frac{\nabla n}{n}\cdot(\vec{A}e^{\mathrm{i}(k_x x+k_y y+k_z z)})\right)\right]e^{-\mathrm{i}(k_x x+k_y y+k_z z)}\\
&\approx\left[\left(\frac{\partial^2 A_x}{\partial x^2}+\frac{\partial^2 A_x}{\partial y^2}\right)+2\mathrm{i}k_z\frac{\partial A_x}{\partial z}\right]\hat{x}+\left[2\mathrm{i}k_z\frac{\partial A_z}{\partial z}\right]\hat{z}\\
&-2\left[\nabla\left(\frac{\nabla n}{n}\cdot(\vec{A}e^{\mathrm{i}(k_x x+k_y y+k_z z)})\right)\right]e^{-\mathrm{i}(k_x x+k_y y+k_z z)}\\
&=0. \qquad (21)
\end{aligned}$$

Then, from the simplified propagation Equation (21) we can extract two equations: one for the x-coordinate

$$\begin{aligned}
&\left(\frac{\partial^2 A_x}{\partial x^2}+\frac{\partial^2 A_x}{\partial y^2}\right)+2\mathrm{i}k_z\frac{\partial A_x}{\partial z}\\
&+2\left[\nabla_x\left(\frac{\nabla n}{n}\cdot\vec{A}e^{\mathrm{i}(k_x x+k_y y+k_z z)}\right)\right]e^{-\mathrm{i}(k_x x+k_y y+k_z z)}\\
&=0, \qquad (22)
\end{aligned}$$

and the second for the z-coordinate

$$\begin{aligned}
&\mathrm{i}k_z\frac{\partial A_z}{\partial z}+\left[\nabla_z\left(\frac{\nabla n}{n}\cdot(\vec{A}e^{\mathrm{i}(k_x x+k_y y+k_z z)})\right)\right]e^{-\mathrm{i}(k_x x+k_y y+k_z z)}\\
&=0. \qquad (23)
\end{aligned}$$

The third term in the left-hand side of Equation (22) for the x- (transverse) component of the amplitude of the electric field is small compared to the first two terms, and for the short propagation distances it can be neglected. Then, Equation (22) acquires a form similar to the equation obtained for diffraction dominated laser beam propagation in empty resonators[14]. The solutions of Equation (22) while neglecting the third term can be found in the classical article of Kogelnik and Li[14], and according to this work, the fundamental mode of the solution is represented by the field that has Gaussian distribution of amplitude in the radial direction with a beam width that changes along the z-axis and has a spherical shape of the wavefront with a radius that is also a function of z. Of course, for far field propagation the third term in the left-hand side of Equation (22) must be accounted for, since the relatively small deviations of the amplitude of the electric field and the shape of the wavefront should accumulate while propagating at long distances, resulting in significant modification of both the beam intensity distribution and the wavefront shape.

Both terms in the left-hand side of Equation (23) have similar magnitude. Now, let's demonstrate that the solution of Equation (23) describes perturbation of the wavefront of the 'carrier' field given by the approximate solution of Equation (22) described above.

$$\begin{aligned}
\mathrm{i}k_z\frac{\partial A_z}{\partial z}\hat{z}&=-\left[\nabla_z\left(\frac{\nabla n}{n}\cdot(\vec{A}e^{\mathrm{i}(k_x x+k_y y+k_z z)})\right)\right]\\
&\quad\times e^{-\mathrm{i}(k_x x+k_y y+k_z z)}\\
&=-e^{-\mathrm{i}(k_x x+k_y y+k_z z)}\frac{\partial}{\partial z}\left(\frac{1}{n}\frac{\partial n}{\partial x}A_x e^{\mathrm{i}(k_x x+k_y y+k_z z)}\right)\hat{z}\\
&=-\Bigg[\left(\frac{1}{n}\frac{\partial^2}{\partial z\partial x}-\frac{(\partial n/\partial z)(\partial n/\partial x)}{n^2}\right)A_x\\
&\quad+\frac{1}{n}\frac{\partial n}{\partial x}\left(\frac{\partial A_x}{\partial z}+\mathrm{i}k_z A_x\right)\Bigg]\hat{z}\\
&\approx-\mathrm{i}k_z\frac{1}{n}\frac{\partial n}{\partial x}A_x\hat{z} \qquad (24)
\end{aligned}$$

or

$$\frac{\partial A_z}{\partial z}\approx-\frac{1}{n}\frac{\partial n}{\partial x}A_x. \qquad (25)$$

Finally, recalling that in paraxial approximation change of the angle between the wavevector and z-axis, $d\varphi$, equals the ratio of the x and z components of the wavevector or the ratio of the z- and x-components of the electric field amplitude, Equation (25) can be rewritten as

$$d\varphi=\frac{-\delta A_z}{A_x}\approx\frac{1}{n}\frac{\partial n}{\partial x}dz, \qquad (26)$$

which is the eikonal equation[15]. We use integration of this equation (see, for example, Equation (7) in Ref. [12]) in our previous works[11, 12] in order to compute radial distribution of the angle of the wavevector after the propagation of a focused laser beam through the caustic under conditions when focusing due to the Kerr effect and defocusing due to ionization take place (see, for example, Equation (8) in Ref. [12]).

At the same time, the local projections of the wavevector are uniquely determined from the system of equations

$$k_x^2 + k_z^2 = \frac{\omega^2 n^2}{c^2}$$
$$\frac{k_x}{k_z} = \text{tg}\varphi, \tag{27}$$

with the solution

$$k_z(x,z) = \frac{\omega n(x,z)}{c(1+\text{tg}\varphi(x,z))^{1/2}},$$
$$k_x(x,z) = \frac{\omega n(x,z)\text{tg}\varphi(x,z)}{c(1+\text{tg}\varphi(x,z))^{1/2}}. \tag{28}$$

Finally, complete Equation (20) for electromagnetic wave propagation written, without affecting its generality, for linear polarization in the x–z plane (or instead its simplified equivalent Equation (21) deduced for paraxial and near field propagation) in combination with Equations (26)–(28) and with the addition of the equation describing inhomogeneity of the index of refraction (either self-induced due to the Kerr effect and ionization or externally induced due to thermal effect, large scale turbulence, etc. or inherent due to spatially variable material properties) represents a closed system of equations that provides a unique solution describing electromagnetic wave propagation in an inhomogeneous medium.

6. Example of computational simulation of paraxial propagation of a focused laser beam in nonlinear media

The results of the simulation of laser beam propagation through air (normal conditions) assuming Gaussian temporal shape of the laser pulse with duration $\tau = 100$ fs and a Gaussian spatial beam profile with the beam radius on $1/e^2$ level in the waist, $w_0 = 50$ μm, are shown in Figure 2, that presents the local angle of the wavevector, φ_{z_s}, computed for propagating from $z_0 = -2z_R$ to $z_s = 2z_R$ (here Rayleigh length is $z_R = 19.62$ mm) as a function of dimensionless laser radius, $r/w(z_s)$ (ratio of radial coordinate to the Gaussian beam radius $w(z_s)$ at the end of the computational range, z_s), for different moments of dimensionless time, η/τ (negative – before the pulse), and for two pulse energies of 0.2 and 0.4 mJ.

The simulation results show that the contribution to the refractive index due to the Kerr effect (Figure 2a) results in focusing of the central part of the beam. The diffraction divergence of the outer area of the beam is partially compensated and in the locations with dimensionless laser

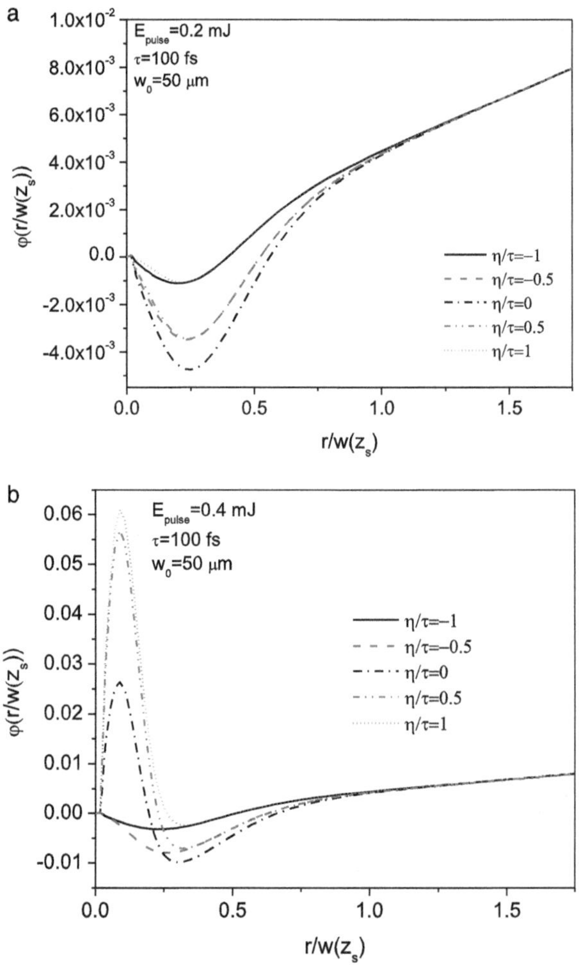

Figure 2. The angle of individual rays at the exit of the zone with high beam intensity where induced refraction is large computed for laser beam propagation through air (normal conditions) assuming Gaussian temporal shape of the laser pulse with duration $\tau = 100$ fs and a Gaussian spatial beam profile with the beam radius on $1/e^2$ level in the waist, $w_0 = 50$ μm, shown as a function of dimensionless laser radius, $r/w(z_s)$, for different moments of dimensionless time, η/τ, (negative – before the pulse): (a) total radiated energy per pulse $E_{\text{pulse}} = 0.2$ mJ (maximum irradiance $I_0 = 2.87 \times 10^{17}$ W m^{-2}); (b) total radiated energy per pulse $E_{\text{pulse}} = 0.4$ mJ (maximum irradiance $I_0 = 5.75 \times 10^{17}$ W m^{-2}). All computational results correspond to the laser wavelength $\lambda = 800$ nm.

radius $r/w(z_s)$ larger than approximately unity, the self-induced refraction is negligible and the beam divergence is governed predominantly by diffraction. For dimensionless laser radius $r/w(z_s) < 1$ the nonlinear focusing is dominant. In this region the optical power of the Kerr self-focusing 'lens' is increasing with time, reaching its maximum when laser power is at the maximum value ($\eta = 0$) and then symmetrically (in time) decreasing. Simultaneously with refraction the nonlinear Kerr produces a shift of the laser frequency numerically described in Ref. [12].

In nonlinear media that are ionized by the laser radiation the contribution of the Kerr effect that produces focusing is combined with the defocusing effect due to nonuniform ionization (Figure 2b). The ionization is considered as in

Ref. [12]. We assume that the main ion produced in air is O_2^+. For the laser wavelength $\lambda = 800$ nm and typical values of the radiation intensity at which self-focusing occurs, the Keldysh parameter $\gamma > 1$ and the limit of multiphoton ionization are valid. The electron number density is given by the equation

$$N_e(r, z, t) = \int_{-\infty}^{t} \sigma_k [N_0 - N_e(r, z, \xi)][I(r, z, \xi)]^k d\xi \quad (29)$$

where N_0 is the number density of neutrals, and $I(r, z, t)$ is the laser pulse intensity; σ_k is the ionization rate due to absorption of k photons such that $k = \text{int}[I_i/\hbar\omega] + 1$, where I_i is the potential ionization of gas and ω is the laser angular frequency.

The effect of media ionization on the beam propagation results in strong divergence of the near-axis part of the beam (Figure 2b). The maximum angle of divergence is increasing in time as the degree of ionization grows, reaching its maximum at the end of the laser pulse. Simultaneously, the outer part of the beam behaves similarly to the above described dynamics, dominated by the focusing due to the Kerr effect.

It was shown in our previous work[11] that the critical laser power for self-focusing depends on the laser intensity, which is determined by the effective radius of a Gaussian beam. This power is substantially less than independent of laser intensity critical power computed in accordance with the current theory[16]. For example, our theory predicts that for the interaction conditions as in Figure 2 the critical power is $P_{cr} \approx 0.135$ GW. Whereas, for the same conditions, the current theoretical model predicts the almost order of magnitude larger value $P_{cr} \approx 1.7$ GW.

7. Concluding remarks

For obvious reasons, it is usual practice in the majority of nonspecialized educational courses and textbooks to ignore the contribution of the term containing $\nabla\varepsilon$ and describe electromagnetic wave propagation using the equation deduced for uniform media. Unfortunately, without much consideration, this propagation equation was adopted in nonlinear optics and after modifications it took the form of Equations (15) and (16). Then, a relatively recent trend mandated further modification of this simplified equation in order to acquire an appearance similar to the nonlinear Schrodinger's equation (i.e., Equation (17)). Here we would like to reinforce that omitting the third term in the left-hand side of the above derived complete propagation Equation (9) leads to inadequate description of the physics involved. Indeed, it is self-contradictory to disregard the inhomogeneity of optical properties of media (both induced and inherent) at the stage of deduction of the propagation equation and then to reintroduce into the obtained simplified propagation equation the nonlinear dependence of the index of refraction, n, on the laser irradiance. Trivial estimates show a non-negligible contribution of the term containing $\nabla\varepsilon$ in complete Equation (10) for laser beam propagation in nonlinear media. Therefore, ignoring this term leads to false predictions. One of the examples of such false prediction is waveguide like propagation of a laser beam in nonlinear media[16] that leads to the development of fascinating concepts of optical soliton and laser beam filamentation that recently produced a flurry of extensive theoretical research. It is easy to see that for a laser beam with intensity distribution that is near-Gaussian a possible solution of an incorrectly abbreviated propagation equation, such as Equations (15)–(17), indeed has a self-similar form. The currently accepted interpretation of this self-similarity is that the diffraction divergence and divergence produced due to media ionization is compensated by self-focusing[2–4]. One of the results, following on from the self-similarity of the solution of this inadequate propagation Equations (15)–(17), is a captivating (however, contradictory to experimental observations) prediction that kilometers long transmission of the laser beam can be achieved in an atmosphere without the beam diverging[6].

Our work demonstrates that the solution of complete Equation (10) that adequately describes laser beam propagation in nonlinear media does not have self-similar form. As a demonstration we solved a complete propagation equation for the conditions when input from the nonlinear refraction can be treated as perturbation of the solution of the linear Helmholtz equation describing propagation of a focused laser beam[11, 12]. This solution demonstrated that laser beam divergence is affected by Kerr self-focusing and plasma defocusing differently in different radial locations of the laser beam and in different times during the laser pulse, i.e., self-similar beam propagation does not occur.

Another inconsistency of the customary approach in which Equation (17) or its derivatives are utilized for modeling of diverging or converging laser beam propagation in nonlinear media is that Equation (17) is derived under the assumption of a plane wavefront. Consequently, it is inapplicable for the description of nonplanar wavefronts. However, in a self-contradictory manner similar to the above illustrated reinsertion of the nonlinear effect into Equation (17), the complex nonplanar wave propagation is described in all current theoretical models using plane wave propagation Equation (17) or its derivatives. At this point it is worth mentioning that the Poynting vector maintains its direction in the approximation of a plane wavefront. In contrast, for a converging or diverging laser beam the local direction of the Poynting vector varies as a function of the distance from the axis. Our previous eikonal-paraxial model[11, 12] correctly reflects this behavior of the Poynting vector.

The effect of inhomogeneity of the dielectric constant on electromagnetic wave propagation is known in several relatively narrowly specialized fields. So far, the practical application of this concept was limited to the theory of radio wave propagation in the ionosphere (see Ref. [13]) and electromagnetic and acoustic wave propagation in stratified media[20], such as radar propagation in atmospheric boundary layers[21].

In conclusion we summarize the contribution of this work to the field of nonlinear optics as follows. (1) Realization that the gradient of dielectric constant always provides a non-negligible contribution in the propagation equation of a laser beam with realistic beam profile because the characteristic length of change of irradiance is comparable to the 'beam size' (for any reasonable definition of this physical property). (2) Development of the method described in Refs. [11, 12] in which we integrated diffractive and geometrical optics by blending the solution of the linear Maxwell's equation and a correction term that represents nonlinear field perturbation expressed as solution of a paraxial ray-optics (eikonal) equation that opens an elegant means of numerical computation of the ray trajectories (avoiding singularities) as the focused laser beam propagates in a nonlinear and ionized medium through its caustic (the area near the focal plane that extends several Rayleigh lengths).

The realm of nonlinear optics that deals with laser beam propagation has benefited from a multitude of experimental works and significant experience has been accumulated in solving complex mathematical problems. However, it seems that substantial improvement of the theoretical part of nonlinear optics is needed and revision of the fundamentals of the theoretical model provided in our work can revitalize and substantially advance this field.

Acknowledgements

One of the authors (V.V.S.) would like to thank DARPA and, in particular, Dr R. B. for support of a project that led to the realization of inconsistencies of the currently accepted theoretical model of laser filamentation. M.N.S. acknowledge partial support from the NSF/DOE Partnership in Basic Plasma Science and Engineering (NSF award PHY-1418845 and DOE award DE-SC0012454). Also, the authors would like to thank K. Semak for her contribution in formulating the mathematical model and encouragement during our work on this manuscript.

References

1. G. A. Askaryan, Sov. Phys. JETP **15**, 1088 (1962).
2. G. Agrawal, *Nonlinear Fiber Optics*, 5th edn (Academic Press, 2013).
3. R. W. Boyd, *Nonlinear Optics*, 2nd edn (Academic Press, 2003).
4. S. L. Chin, *Femtosecond Laser Filamentation* (Springer, 2010).
5. R. Y. Chiao, T. K. Gustafson, and P. L. Kelley, in *Topics in Applied Physics, Vol. 114*, R. W. Boyd, S. G. Lukishova and Y. R. Shen (eds) (Springer, 2009).
6. A. Couairon and A. Mysyrowicz, Phys. Rep. **441**, 47 (2007).
7. S. L. Chin, S. A. Hosseini, W. Liu, Q. Luo, F. Théberge, N. Aközbek, A. Becker, V. P. Kandidov, O. G. Kosareva, and H. Schroeder, Canad. J. Phys. **83**, 863 (2005).
8. P. Sprangle, J. R. Penano, and B. Hafizi, Phys. Rev. E **66**, 046418 (2002).
9. B. Shim, S. E. Schrauth, and A. L. Gaeta, Opt. Express **19**, 9118 (2001).
10. Y. Sivan, G. Fibich, B. Ilan, and M. I. Weinstein, Phys. Rev. E **78**, 046602 (2008).
11. V. V. Semak and M. N. Shneider, J. Phys. D **46**, 185502 (2013).
12. V. V. Semak and M. N. Shneider, J. Phys. D **47**, 045503 (2014).
13. K. G. Budden, *The Propagation of Radio Wave: The Theory of Radio Waves of Low Power in the Ionosphere and Magnetosphere* (Cambridge University Press, 1985).
14. H. Kogelnik and T. Li, Appl. Opt. **5**, 1550 (1966).
15. M. Born and E. Wolf, *Principles of Optics*, 7th expanded edn (Cambridge University Press, 2003).
16. R. Chiao, E. Garmire, and C. H. Townes, Phys. Rev. Lett. **13**, 479 (1964).
17. V. I. Talanov, JETP Lett. **2**, 138 (1965).
18. P. L. Kelley, Phys. Rev. Lett. **15**, 1005 (1965).
19. V. E. Zakharov and A. B. Shabat, Sov. Phys. JETP **34**, 62 (1972).
20. L. M. Brekhovskikh, *Waves in Layered Media* (Academic Press, 1965).
21. G. D. Dockery, IEEE Trans. Antennas Propag. **36**, 1464 (1988).

18

Target fabrication for the POLAR experiment on the Orion laser facility

C. Spindloe[1], D. Wyatt[1], D. Haddock[1], I. East[1], J.E. Cross[2], C.N. Danson[2,3], E. Falize[4], J.M. Foster[2,3], M. Koenig[5], and G. Gregori[2]

[1]*Target Fabrication Group, Central Laser Facility, Rutherford Appleton Laboratory, Harwell Science and Innovation Campus, Didcot OX11 0QX, UK*

[2]*Atomic and Laser Physics, Clarendon Laboratory, University of Oxford, OX1 3PU, UK*

[3]*AWE, Aldermaston, Reading RG7 4PR, UK*

[4]*CEA-DAM-DIF, F-91297 Arpajon, France*

[5]*Laboratoire pour l'Utilisation Des Laser Intense, UMR 7605 CNRS-CEA-Ecole Polytechnique-Université Paris VI, Palaiseau, France*

Abstract

This article describes the fabrication of a suite of laser targets by the Target Fabrication group in the Central Laser Facility (CLF), STFC Rutherford Appleton Laboratory for the first academic-access experiment on the Orion laser facility (Hopps *et al.*, Appl. Opt. **52**, 3597–3601 (2013)) at Atomic Weapons Establishment (AWE). This experiment, part of the POLAR project (Falize *et al.*, Astrophys. Space Sci. **336**, 81–85 (2011); Busschaert *et al.*, New J. Phys. **15**, 035020 (2013)), studied conditions relevant to the radiation-hydrodynamic processes occurring in a remarkable class of astrophysical star systems known as magnetic cataclysmic variables. A large number of complex fabrication technologies and research and development activities were required to field a total of 80 high-specification targets. Target design and fabrication procedures are described and initial alignment and characterization data are discussed.

Keywords: astrophysics; microtargets

1. Introduction

The target-fabrication group within the Central Laser Facility (CLF) has extensive expertise in the areas of micro-assembly, thin-film coating, target characterization and micromachining. The group was established to support the academic user community having access to the lasers within the CLF. The Atomic Weapons Establishment (AWE) Orion laser–plasma interaction facility[1] became operational in 2013 and as part of its remit makes up to 15% of its operational time available for experiments by the UK academic community and their international collaborators. The CLF supports this access by providing laser targets for these experiments.

The POLAR project[2, 3] aims to replicate in the laboratory some of the hydrodynamic processes occurring in X-ray binary star systems known as magnetic cataclysmic variables[4, 5]. In these binary stars, matter from a cold star is gravitationally accelerated towards a dense, white-dwarf companion star. Collimation of this flow by the white-dwarf's intense magnetic field results in a so-called accretion column of material at the poles of the star, instead of the usual accretion disc. The flow of material towards the star is impeded by the white-dwarf atmosphere and a reverse shock wave that is established in this column, near the surface of the white dwarf. The exceptionally high temperatures generated by this shock wave result in X-ray emission, and so give rise to the classification X-ray binary star. The purpose of the Orion experiment was to build a laboratory analogue of this process, in which the hydrodynamic processes could be studied in a controlled manner. The experiment uses high-power laser pulses to generate high-velocity plasma flow (collimated not by a magnetic field as in the star, but by a miniature surrounding tube) which impacts on the surface of an obstacle that mimics the white dwarf, resulting in the reverse shock wave that the experiment aims to study. The experimental layout is shown in Figure 1.

The main interaction target was the POLAR target, designed to produce the high-temperature plasma conditions

Correspondence to: C. Spindloe, Science and Technology Facilities Council, Rutherford Appleton Laboratory, Harwell Oxford, Chilton, Didcot, Oxon OX11 0QX, UK. Email: chris.spindloe@scitechprecision.com

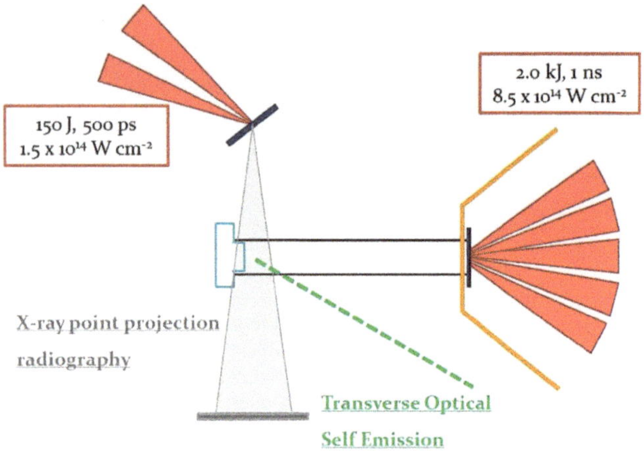

Figure 1. Experimental layout for the Orion POLAR campaign.

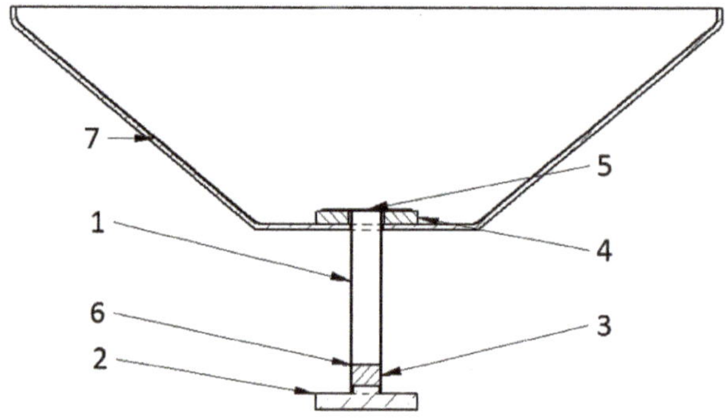

Figure 2. A CAD representation of the target sliced through the centreline (annotations are described in the text above).

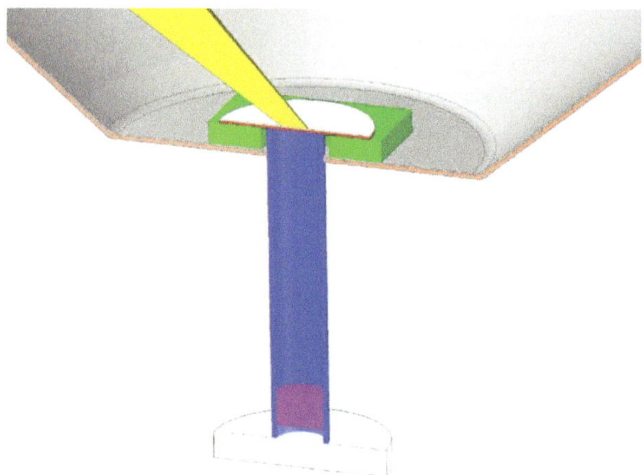

Figure 3. Shows a 3D CAD cut away of the target and indicates the positions of the pusher foil position and also shows the foam insert at the bottom of the tube.

that were of interest to the experimental group. This is shown in Figure 1 as the horizontal tube which is 3.5 mm long and 550 μm internal diameter (650 μm outside diameter). Five nanosecond beams of the Orion laser are directed on to an ablator-pusher laser target, consisting of a multi-layered gold–plastic, or brominated plastic, foil with CH ablator, producing the plasma flow down the tube. The grey block at the end of the tube is the obstacle, made of quartz or steel. This target was mounted onto the Orion target positioner.

Secondary to this, a backlighter target was required to provide an X-ray image of the target during the interaction[6, 7]. This target is shown at the top of Figure 1, where two of the Orion beamlines are used to generate plasma whose X-ray emission was the backlighting source. The resulting X-ray image was recorded on the other side of the POLAR target using X-ray imaging plates. This backlighting target was positioned on a separate target mount designed to be inserted from a Ten Inch Manipulator (TIM), a diagnostic-insertion device mounted in the Orion target chamber.

The targets utilized the full scope of the target-fabrication group's capabilities and also required research and development into new, cost-effective materials.

2. POLAR interaction target

2.1. Specification

The POLAR interaction target was a combination of:

(1) a polyimide (PI) tube inside which the material could flow, and which was used to mimic the magnetic collimation that occurs in the binary star system;

(2) a quartz or steel obstacle, against which the material flow stagnates to generate the required reverse shock wave;

(3) a low-density foam sample;

(4) a washer for holding the target material;

(5) a pusher material that was either a brominated plastic, a gold/plastic multi-layer assembly, or a chlorinated plastic;

(6) a slit in the wall of the tube to allow a diagnostic view of the flow and of the shock wave whose X-ray emission was recorded by the AWE 'Dante'[8] diagnostic, and

(7) a copper cone to shield the diagnostics from scattered light from the incident laser beams.

These numbers correspond to the features in Figure 2.

The detailed design of this target underwent a number of modifications during the campaign. A number of alignment features and coatings were added to protect the target and to allow for the positioning of the target in the Orion chamber. These included small-diameter copper wires placed transversely across the tube as fiducial reference points to measure the propagation of the shock, and fibres on the target stalk to provide reference points for very precise alignment in the Orion target chamber. A 3D representation is shown

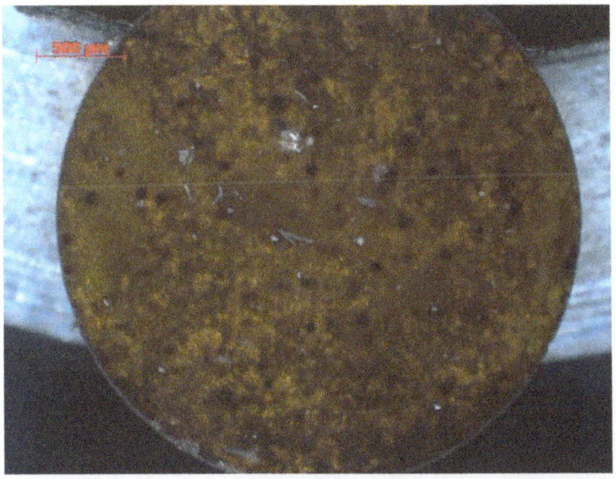

Figure 4. A fully processed Brominated plastic disk showing internal structure (dark areas are the un-melted material).

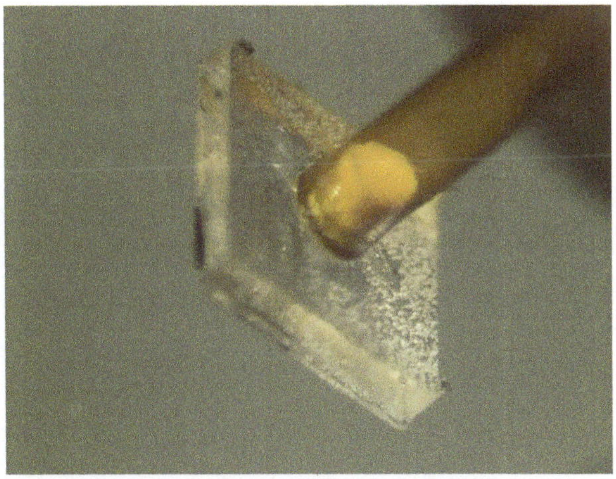

Figure 5. A low-density sample placed at the end of the PI tube next to the quartz obstacle.

in Figure 3, depicting just one of the five laser beams, shown in yellow, incident onto the pusher foils.

2.2. Brominated foil

Brominated plastic for the pusher material was not commercially available in the thickness required and purchasing a bespoke made foil was not possible within the budget. A research programme was initiated to press plastic films of brominated plastic from a commercially available powder. A heated press was used that had previously been used to manufacture deuterated plastic films. A number of tests were carried out to heat and mould bromopolystryene (C_8H_7Br) that had a bromine content of 43% by weight.

Initial results of pressing CHBr had mixed success, and temperature, pressure and mass were all important variables in producing a good quality film. At low temperatures (<200 °C) the powder did not show significant melt and was just fused together to form a brittle pressed pellet that was too thick. When the temperature was increased to ~ 280 °C the sample showed signs of melt and the use of the correct amount of material produced a film ~ 25 μm thick. However it can be seen in Figure 4 that the sample has not undergone a full melt as there are features of the order of 50 μm that are visible as dark areas in the disk and these correspond to the initial particle size of the powder used to manufacture the films.

To produce more uniform films a temperature of >300 °C would be required. Initial scanning electron microscope (SEM) analysis using energy-dispersive X-ray spectroscopy (EDX) confirms that there is bromine content still present after processing although there is some structure that is present in the pressed disk that is due to the target foil not going through a full melt when forming as shown in Figure 4.

Further characterization work using oxygen flask combustion, followed by ion chromatography has shown that the bromine levels of the samples analysed by mass percent was 31.10 (m/m%)[9] in correlation with the initial brominated powder that was 43% by weight. There is some loss of Bromine that is seen that requires further investigation.

2.3. Foam insert

Low-density foam of density 300 mg/cm^3 was to be placed at the end of the PI tube. The foam was required to be positioned next to the obstacle and was required to be 400 μm in length. The foam was manufactured using a critical point drying (CPD) technique[10] and a number of techniques were trialled to form the foam inside the tube. Firstly to make the foam *in situ* the PI tube was filled with the production liquid and UV cured. However filling of the liquid in the tube was not reproducible and therefore the foam did not cure to the correct length. The foam was then manufactured in a PMMA tube which transmits UV; however when running the CPD process the tube is incompatible with the chemicals and therefore lost shape and the foam produced was contaminated and distorted. The solution was to manufacture the foam in a tube that would survive the process (PI) and was UV transmissive. The foam was then removed from the tube, cut to length and re-inserted inside the final assembly tube. An example of the foam *in situ* can be seen in Figure 5.

2.4. Final target assembly

The remaining target components were produced using precision micromachining (washer and cone shield) and thin-film coating techniques (gold–plastic multi-layered pusher), and were integrated into the final target assembly as shown in Figure 6. The target is mounted on an AWE Orion mount with a cross brace to stabilize the cone. There

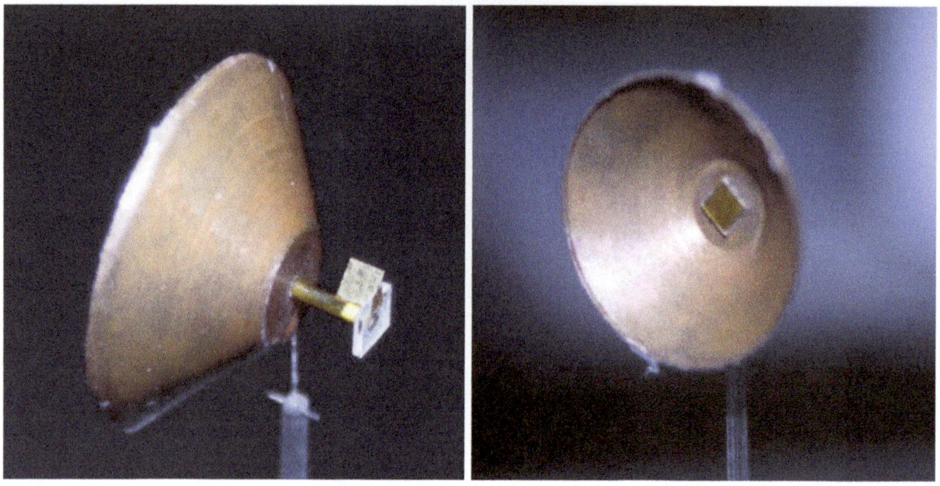

Figure 6. Images of the POLAR target from the obstacle side (left) the Au/CH pusher (right).

Figure 7. A side view of the foam inside the tube with the grid above.

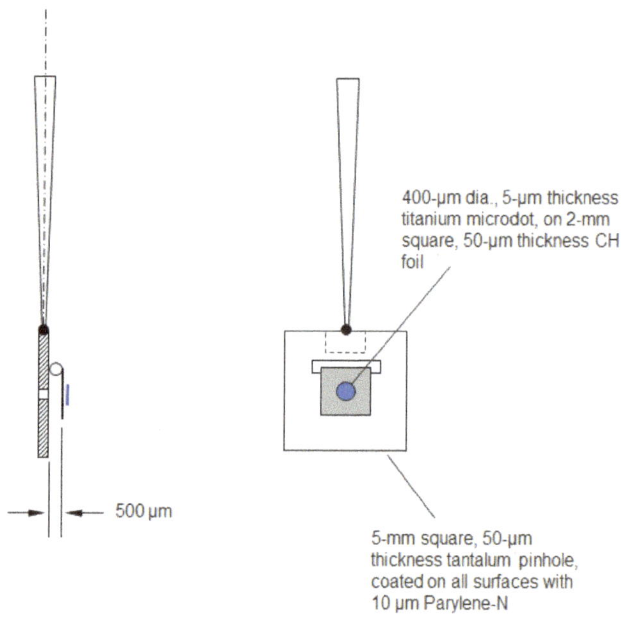

Figure 8. The backlighter target design.

is a 400 lines/inch (63 μm spacing) gold grid for spatial calibration of the X-ray backlighting images, as shown in Figure 7. The Dante diagnostic slit, although not visible, has been laser machined into the wall of the tube.

3. Backlighter target

The backlighter target assembly design is shown in Figure 8. The X-ray backlighting source was the helium-like resonance line radiation from laser targets of chlorine (Parylene D), scandium, and titanium (respectively 2.79, 4.32 and 4.75 keV X-ray energy), chosen to highlight different features of the hydrodynamic flow in the X-ray absorption images. These different backlighter materials were cut into 400 μm diameter dots using laser machining and glued onto a CH support that was purchased commercially. This was then placed 500 μm from a tantalum pinhole that had been coated with 10 μm of Parylene-N (to prevent hydrodynamic closure of the tantalum pinhole, following X-ray ablation by the nearby backlighting plasma). This assembly was then attached to an Orion target mount, as shown in Figure 9, for positioning in the Orion target chamber.

4. X-ray target imaging

Sample backlighter images are shown in Figure 10. The Orion laser parameters together with the target design used to produce these images are detailed in Table 1. These images are produced from targets that are laser driven, but with a very short delay on the drive beams of only 4.75 ns. As such, no plasma has yet reached the end of the tube and they can

Table 1. Details of the Orion laser and POLAR target parameters for the images in Figure 10.

Image	Main target type	Drive beams total energy (kJ)	BL beams total energy (J)	BL beams pulse length (ns)	Delay between cones (ns)	Filters on SPCA	Filters on image plate	Step wedge
10(a)	CHBr Dante slit and grid	2.1	710	1	4.75	10 μm Ti + 2 ×50 μm plastic	8 μm aluminized mylar	50 μm, 100 μm, 150 μm C_2H_6
10(b)	CHBr foam and Al flash on quartz	2.4	780	1	4.75	10 μm Ti + 2 ×50 μm plastic	8 μm aluminized mylar	N/A

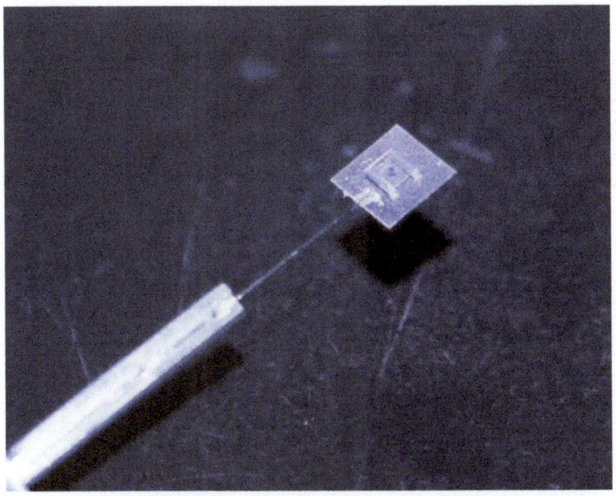

Figure 9. The fully assembled backlighter target.

therefore be used to distinguish the target features described earlier. The two images used in Figure 10 were produced on two independent shots with two different targets. In Figure 10(a) the target features the gold calibration grid (shown optically in Figure 7) and Dante slits in the PI tube; whilst in Figure 10(b) the target is seen with the foam in side the PI tube. The clearly defined grid structure evident in Figure 10(a) gives a resolution of the backlighting image of ~10 μm.

5. Conclusions

A total of 80 targets (34 POLAR and 46 backlighter) were delivered for two periods of two weeks of academic access on Orion during November 2013 and February 2014. The development of a number of new target-fabrication techniques at the CLF, Rutherford Appleton Laboratory has enhanced the capabilities that are available and has contributed to a successful academic-access campaign. Further work is to be carried out to enhance the quality of the brominated plastic pushers to remove defect points in the target that may lead to instabilities in the plasma as it propagates down the tube and also to produce more uniform foam next to the obstacle.

References

1. N. Hopps, C. Danson, S. Duffield, D. Egan, S. Elsmere, M. Girling, E. Harvey, D. Hillier, M. Norman, S. Parker, P. Treadwell, D. Winter, and T. Bett, Appl. Opt. **52**, 3597 (2013).
2. E. Falize, B. Loupias, A. Ravasio, C. D. Gregory, A. Diziere, M. Koenig, C. Michaut, C. Cavet, P. Barroso, J.-P. Leidinger, X. Ribeyre, J. Breil, H. Takabe, Y. Sakawa, Y. Kuramitsu, T. Morita, N. C. Woolsey, W. Nazarov, and S. Pikuz, Astrophys. Space Sci. **336**, 81 (2011).
3. C. Busschaert, E. Falize, B. Loupias, C. Michaut, A. Ravasio, A. Pelka, R. Yurchak, and M. Koenig, New J. Phys. **15**, 035020 (2013).

(a) (b)

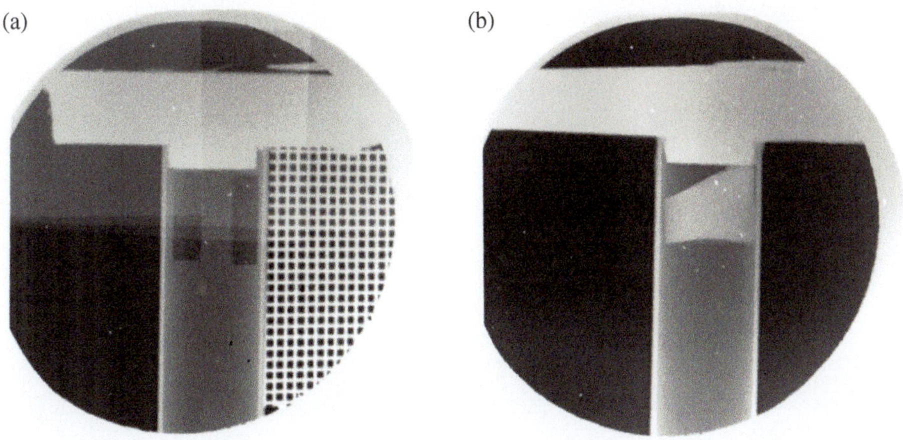

Figure 10. Sample backlighter images showing: (a) the PI tube with the quartz/steel obstacle at the top, the diagnostic slits cut in to the tube and the gold resolution grid attached to the target and (b) the PI tube with the quartz/steel obstacle at the top and a low-density foam insert below the stop.

4. D. Ryutov, R. P. Drake, J. Kane, E. Liang, B. A. Remington, and W. M. Wood-Vasey, Astrophys. J. **518**, 821 (1999).
5. E. Falize, C. Michaut, and S. Bouquet, Astrophys. J. **730**, 96 (2011).
6. R. R. Whitlock, S. P. Obenschain, and J. Grun, Appl. Phys. Lett. **41**, 429 (1982).
7. C. C. Kurantz, B. E. Blue, R. P. Drake, H. F. Robey, J. F. Hansen, J. P. Knauer, M. J. Grosskopf, C. Krauland, and D. C. Marion, Rev. Sci. Instrum. **77**, 10E327 (2006).
8. C. D. Bently, R. D. Edwards, J. E. Andrew, S. F. James, M. D. Gardner, A. J. Comley, K. Vaughan, C. J. Horsfield, M. S. Rubery, S. D. Rothman, S. Daykin, S. J. Masoero, J. B. Palmer, A. L. Meadowcroft, B. M. Williams, E. T. Gumbrell, J. D. Fyrth, C. R. D. Brown, M. P. Hill, K. Oades, M. J. Wright, B. A. Hood, and P. Kemshall, Rev. Sci. Instrum. **83**, 10D732 (2012).
9. Analysis report from Loughborough Surface Analysis Ltd (Oct 2014).
10. J. W. Falconer, W. Nazarov, and C. J. Horsfield, J. Vac. Sci. Technol. A **13**, 1941 (1995).

Design and current progress of the Apollon 10 PW project

J.P. Zou[1], C. Le Blanc[1], D.N. Papadopoulos[1], G. Chériaux[2], P. Georges[3], G. Mennerat[4], F. Druon[3], L. Lecherbourg[1], A. Pellegrina[1,3], P. Ramirez[1,3], F. Giambruno[1,2], A. Fréneaux[1,2], F. Leconte[1,2], D. Badarau[1], J.M. Boudenne[1], D. Fournet[1], T. Valloton[1], J.L. Paillard[1], J.L. Veray[1], M. Pina[1], P. Monot[4], J.P. Chambaret[4], P. Martin[4], F. Mathieu[1], P. Audebert[1], and F. Amiranoff[1]

[1]*Laboratoire pour l'Utilisation des Lasers Intenses, CNRS, Ecole Polytechnique, Palaiseau, France*

[2]*Laboratoire d'Optique Appliquée, ENSTA ParisTech, CNRS, Palaiseau, France*

[3]*Laboratoire Charles Fabry, UMR 8501, Institut d'Optique, CNRS, Université Paris Sud, Palaiseau, France*

[4]*CEA, Iramis, SPAM, Saclay, France*

Abstract
The objective of the Apollon project is the generation of 10 PW peak power pulses of 15 fs at 1 shot/minute. In this paper the Apollon facility design, the technological challenges and the current progress of the project will be presented.

Keywords: titanium lasers; PW class facilities

1. Introduction

The main goal of the Apollon project is to provide to users 10 PW peak power pulses at a repetition rate of 1 shot/minute. To reach this extreme peak power level, Apollon will produce extremely short pulses of 15 fs and a corresponding moderate energy of 150 J after compression. This is in fact the principal difference of Apollon when compared with other multi-PW projects internationally (ELI-NP, Vulcan 10 PW, SIOM 10 PW project etc.) where pulse duration in the range of 25–40 fs is targeted.

The main 10 PW beam, combined with three secondary beams (a 1 PW beam, a 10 TW probe and the uncompressed high energy beam), will generate intensity on target surpassing the barrier of 2×10^{22} W cm^{-2}, capable of producing ultra-intense and ultra-short sources of particles (electrons, protons...), and coherent and highly energetic X rays. The design and the construction of the Apollon sub-laser systems have been performed in the framework of a French multi-laboratory research consortium. Apollon will be integrated as an upcoming user facility at the end of 2014 at l'Orme des Merisiers, Saclay, France.

Correspondence to: J.P. Zou, Laboratoire pour l'Utilisation des Lasers Intenses, CNRS, Ecole Polytechnique, Palaiseau, France.
Email: ji-ping.zou@polytechnique.fr

2. Apollon architecture strategy

The Apollon architecture has been defined taking into account the current advances in research and development of the laser technology, laser components and laser materials for ultra-short, ultra-intense laser systems. To produce 10 PW pulses, it is important to determine the output energy level and the ultra-short pulse duration.

Apollon will produce a relatively moderate output energy of 150 J, limited by the size and the damage threshold of the broadband gratings used for pulse compression. In addition, for higher output energy, even larger gain media and more significant pump energy would be required. As a consequence, the transverse lasing risk in such media would be more accentuated and the shot repetition rate compromised.

Apollon will produce high-contrast ultra-short pulses of 15 fs, based on excellent spectral bandwidth management and spectral phase minimization all along the chain. In fact, a large spectral bandwidth of about 200 nm and active spectral phase compensation are needed to support high energy 15 fs pulse generation. However, the main aspects responsible for the pulse duration enlargement and contrast degradation are as follows: the spectral gain narrowing and deformation during energy amplification; the time delay

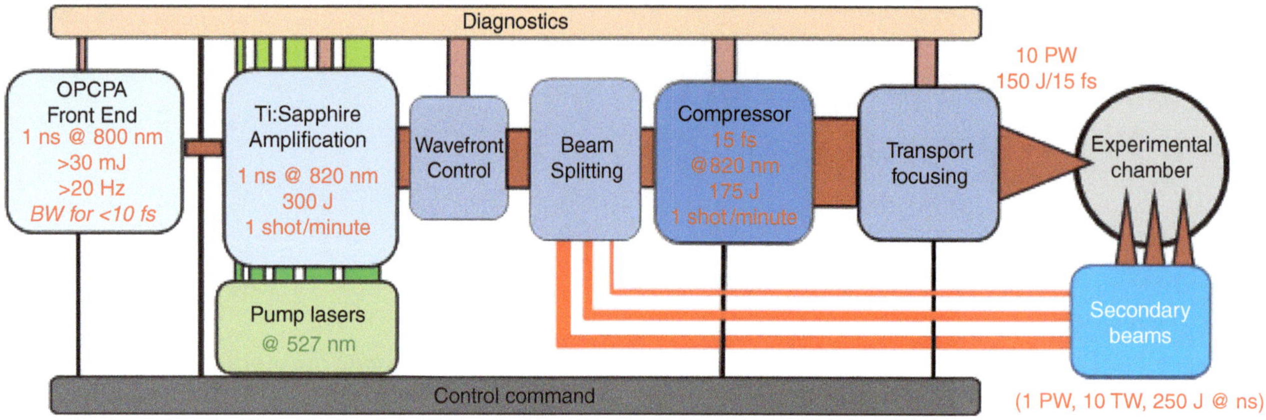

Figure 1. Global schematic of the Apollon 10 PW laser installation.

induced by the longitudinal chromatism generated in spatial filters; the spectral phase induced by material dispersion; and the broadband reflectance limitation and the group delay dispersion (GDD) of the beam-steering mirrors in the chain. For such a laser system, relying on a very broad spectrum, the spatio-temporal coupled effects are of particular importance to study. In the Apollon sub-system design, all of the factors explained in this section will be taken into consideration.

3. Sub-system design

To meet the complete set of requirements, the Apollon design is based on a hybrid OPCPA-Ti:sapphire architecture. A simplified schematic of the Apollon laser is shown in Figure 1. The design of the Apollon sub-systems including the front-end, the power amplification section (PAS) and the 10 PW compression will be described in the three following sections.

3.1. Apollon front-end

The OPCPA-based Apollon front-end employs state of the art technology to generate high quality and high temporal contrast pulses with a spectrum supporting sub-10 fs pulse duration. A non-collinear OPCPA architecture using BBO crystals has been chosen to allow broad-bandwidth amplification. A particular feature of our design is that the 2-grating aberration-free Offner stretcher is integrated in the front-end between two OPCPA stages. The first OPCPA stage operating in the picosecond regime provides high gain (1000×) and temporal contrast enhancement, while the second OPCPA stage operates in a low gain (50–100×) but energetic nanosecond regime.

Both the injection pulses at 800 nm and the pump pulses (SHG of 1030 nm) are provided by a broadband Ti:sapphire oscillator (Rainbow, Femtolasers) for optimal temporal synchronization.

The mJ, 25 fs injection pulses (femtopower) are first temporally cleaned and spectrally broadened in a novel double crossed polarized wave (XPW) configuration[1], and then stretched in a bulk stretcher (BK7). The spectral phase is actively controlled by a double passed Dazzler (Fastlite). After a final spatial filtering stage, about 2 μJ, 6 ps stretched pulses with excellent spatial, temporal and spectral quality and stability are injected into the ps-OPCPA stage.

For the OPCPA pump lasers, diode pumped Yb-doped crystal technology is used. The pJ level pulses at 1030 nm from the Ti:sapphire oscillator are first temporally stretched and pre-amplified to the mJ level in the nanosecond regime. Amplification then takes place in two parallel chains: (a) the picosecond line based on an Yb:CaF_2 multipass amplifier providing >15 mJ of pump energy at 515 nm after compression and SHG, and (b) the nanosecond line based on an Yb:YAG regenerative amplifier[2] followed by an Yb:CaF_2 booster to reach >300 mJ at 515 nm at 20–100 Hz repetition rate[3].

So far we have performed OPCPA experiments in the picosecond regime resulting in more than 7 mJ pulses at 100 Hz. Figure 2 shows an OPCPA output spectral profile and phase (left) and corresponding compressed pulse of 9.5 fs (right) with a contrast better than 10^{-12}.

3.2. Apollon power amplification

The PAS is based on 5 multipass Ti:sapphire amplifiers to obtain 300 J before compression. Four main points are investigated.

(1) The spectral evolution management, aiming to optimize spectral profile and width compatible with high-contrast 15 fs pulses: spectral filters are designed to fight against gain narrowing and spectral shifting due to the saturation[4]. Two sets of these filters will be set up at the entrance of the PAS

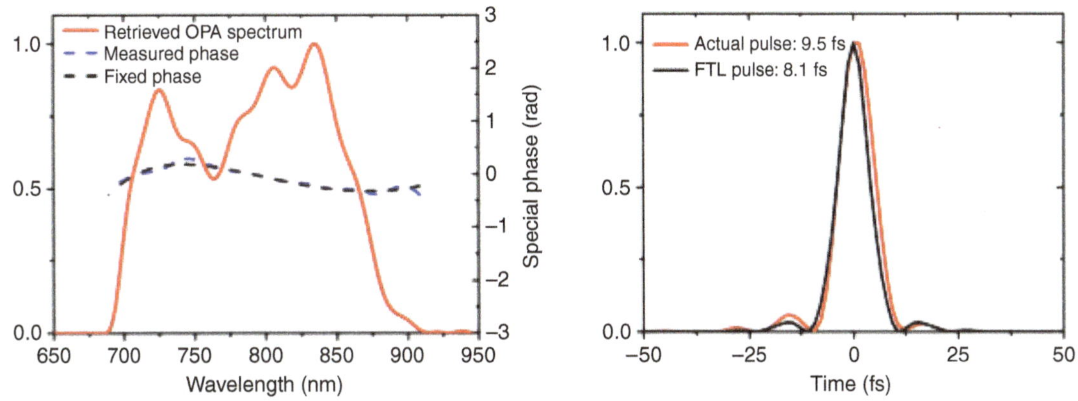

Figure 2. Measurement of the compressed ps-OPCPA pulses: spectral profile and phase (left) and pulse intensity (right).

and before the 3rd amplifier to allow optimal energy amplification while preserving ~70 nm bandwidth (FWHM) centered at 820 nm. Using another front-end with 50 nm spectral bandwidth, we have demonstrated spectral management with such filters at the 3 J level and compressibility down to 23.5 fs of the amplified pulses.

(2) The beam quality control including beam energy distribution optimization and beam wavefront correction. This will be achieved by performing image relay through the whole chain using the well studied off-axis parabolic telescopes and an optical adaptive closed-loop. An optimized spatial filtering will also be implemented to minimize the intensity modulations at high spatial frequencies. In this way, the focal spot quality characterized by a Strehl ratio >0.5 should be effectively obtained.

(3) Transverse lasing suppression: Apollon employs the largest ever grown crystals, reaching 175 mm in diameter (GT Crystal Systems) and pumped areas as large as 150 mm for the last stage. Diameter-to-thickness ratios even greater than 3.5 and high pump energy densities of more than 1 J cm^{-2} make the management of parasitic transverse effects a crucial and challenging task. For the suppression of transverse lasing, novel techniques have been explored based on the use of index-matched liquid. The originality of our approach is related to the online, active control of the minimization of the refractive index difference between the crystal and the liquid. Preliminary experimental tests, in excellent accordance with our simulations, have recently provided solid proof of the efficiency of our method.

(4) High energy pump lasers: a total energy amount of 800 J in the green is required for pumping the last three stages of the PAS. A commercial pump system developed by Continuum & National Energetics will be used in Apollon. This laser uses flashlamp pumped large Nd-glass multi-slab amplifiers and liquid cooling allowing an operation rate of at least 1 shot/minute. The first module of the pump system, providing 400 J, will be delivered at the end of 2014.

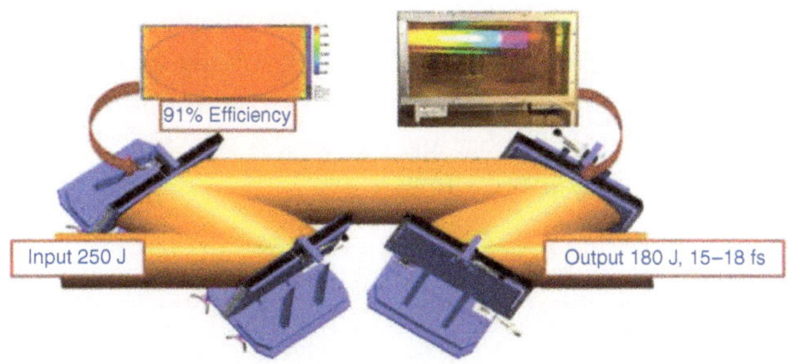

Figure 3. Compressor design with four monolithic gratings. Typical efficiency map for a grating.

3.3. Apollon pulse compression

The main beam line compressor of Apollon is based on a typical four grating unfolded configuration (Figure 3). It has been designed on the basis of three principal considerations. The targeted ultra-short pulse duration after compression corresponds to a very large spectral bandwidth (200 nm full bandwidth at 820 nm) and needs accurate spectral phase management. The damage threshold of the gratings is also an important issue. In fact, to deliver 10 PW pulses to the experiments, energy of 180 J is required at the output of the compressor, corresponding to about 250 J at the input for a compression assuming an efficiency of >70%. Existing grating technology, however, limits our choice to gold-coated gratings as the only compatible solution with the required bandwidth of 720–920 nm and a damage threshold in the range of 200–300 mJ cm^{-2}. In our design, the peak fluence on the first grating of <110 mJ cm^{-2} imposes the use of a 400 mm beam size and therefore meter-long gratings.

The Apollon compressor employs gratings of 910 × 455 mm^2 manufactured by Laurence Livermore National Laboratory (LLNL). Since January 2012, six of these gratings have been received and fully characterized. The gratings have 1480 l mm^{-1} and their diffraction efficiency has been measured to be better than 92% on three wavelengths (780, 800, and 840 nm). The wavefront error of the diffracted beam is lower than $\lambda/3$ PtV over the whole effective surface. Taking into account the geometrical limitations due to the large beam section, the incident angle has been set at 56°. The spectral phase induced by the material dispersion in the laser chain, as well as the GDD of the mirror coating, has then been minimized based on the design of an unmatched stretcher in the front-end which uses two 1450 l mm^{-1} gratings at the incident angle of 52.4°. The compensation of the residual higher order spectral phase by Dazzler will enable us to reach a final duration below 15 fs after pulse compression.

The compressor will be implemented in a large stainless steel vacuum chamber ($6.2 \times 3 \times 3.1$ m^3) specified to provide a vacuum level of 10^{-7} mbar and cleanliness of ISO6. The large size of the chamber allows entry to the compressor from a single entrance and handling and alignment of the gratings inside the chamber. Delivery of the chamber is scheduled for September 2014.

Acknowledgements

The authors gratefully acknowledge financial support from the ILE-APOLLON 07-CPER 017-01 contract.

References

1. L. P. Ramirez, D. N. Papadopoulos, A. Pellegrina, P. Georges, F. Druon, P. Monot, A. Ricci, A. Jullien, X. Chen, J. P. Rousseau, and R. Lopez-Martens, Opt. Express **19**, 93 (2011).
2. J. Tümmler, R. Jung, H. Stiel, P. V. Nickles, and W. Sandner, Opt. Lett. **34**, 1378 (2009).
3. D. N. Papadopoulos, F. Friebel, A. Pellegrina, M. Hanna, P. Camy, J.-L. Doualan, R. Moncorge, P. Georges, and F. Druon, IEEE JSTQE **21**, 1 (2015).
4. F. Giambruno, C. Radier, G. Rey, and G. Chériaux, Appl. Opt. **50**, 2617 (2011).

20

Influence of thermal reduced depolarization on a repetition-frequency laser amplifier and compensation

Xin-ying Jiang, Xiong-wei Yan, Zhen-guo Wang, Jian-gang Zheng, Ming-zhong Li, and Jing-qin Su
Research Center of Laser Fusion, CAEP, P.O. Box 919-988, Mianyang 621900, China

Abstract
Thermal stress can induce birefringence in a laser medium, which can cause depolarization of the laser. The depolarization effect will be very severe in a high-average-power laser. Because the depolarization will make the frequency doubling efficiency decline, it should be compensated. In this paper, the thermal characteristics of two kinds of materials are analyzed in respect of temperature, thermal deformation and thermal stress. The depolarization result from thermal stress was simulated. Depolarization on non-uniform pumping was also simulated, and the compensation method is discussed.

Keywords: depolarization; depolarization compensation; thermal birefringence

1. Introduction

Inertia fusion energy (IFE) is recognized as a clean energy. The solid-state laser is one of the most competitive technical approaches for fusion energy drivers. IFE puts forward new and higher technology requirements for solid-state laser drivers, which are different from the inertial confinement fusion (IFC) laser driver. IFE includes the following main characteristics: high power, high repeat frequency/high efficiency and high reliability. These characteristics constitute the key advantages of the new generation of solid-state laser drivers. The requirement of repetition frequency is the prominent difference between ICF and IFE laser drivers. The thermal effect is a key technical problem which must be solved[1–5]. The current international development of research on fusion energy drivers demonstrates that each has its merits. The choices of materials and thermal management are also different. Yb:YAG and Nd:glass are two kinds of suitable materials. Yan has researched the energy storage characteristics of the two kinds of materials[6]. In this paper, the thermal characteristics of these two kinds of materials are analyzed, in respect of temperature, thermal deformation and thermal stress. The depolarization result from thermal stress was simulated. Because the depolarization will make frequency doubling efficiency decline, so it should be compensated. In classical literature, the method of depolarization compensation was always the placing of a polarization rotator between two identical amplifiers[7, 8]. A rotator rotates the polarization by 90°. For full compensation, the two amplifiers should accord and the laser should pass through them in a symmetrical position. However, for a practical laser, realistically non-uniform pumping will exist, and it will be difficult to make two amplifiers completely the same. In this paper, we consider another method to compensate the depolarization.

Correspondence to: J.-G. Zheng, Mianshan Road No.64, Mianyang, Sichuan province, China, ZIP code: 621900. Email: zjg8861@163.com

2. Theory and numerical analysis

The temperature distribution of laser disks was determined by heat production rate, heat transfer coefficient and thermal performance of the laser medium. The expressions are

$$\begin{cases} \rho C \dfrac{\partial T(x,y,z)}{\partial t} = k \left(\dfrac{\partial^2 T}{\partial x^2} + \dfrac{\partial^2 T}{\partial y^2} + \dfrac{\partial^2 T}{\partial z^2} \right) + q_v \\ T(x,y,z;0) = T_0 \\ -\kappa \dfrac{\partial T}{\partial \vec{n}} \Big|_{\sum i} = h_{c_i}(T - T_{\text{gas}}), \end{cases} \tag{1}$$

where $\sum$ is the surface of the laser slab, κ, ρ, c are, respectively, thermal conductivity coefficient, mass density and heat capacity of the laser slab, h_c is the coefficient of heat transfer and q_v is the heat production rate.

With the increase of temperature, thc laser disk should expand, but it will always be restricted, which will generate thermal stress. In addition the generated stresses induce

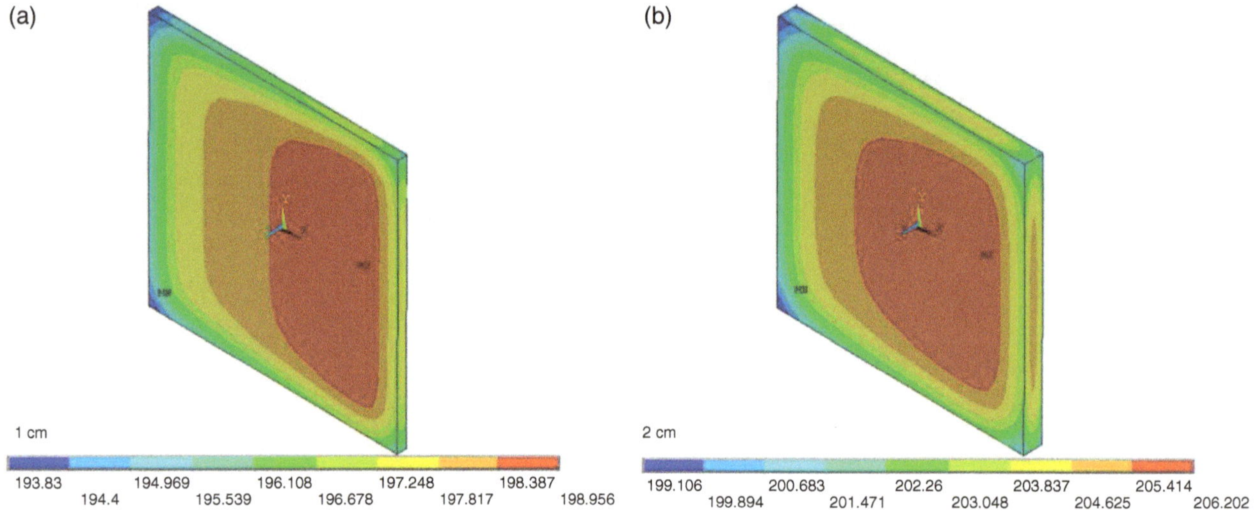

Figure 1. Temperature distribution of Yb:YAG: (a) temperature of 1 cm Yb:YAG; (b) temperature of 2 cm Yb:YAG.

birefringence, and then the laser will be depolarized and the wavefront will be aberrant.

Firstly, the temperature of the laser medium was calculated by a finite-element program; secondly, the thermal stress and displacement were calculated through a coupled program; thirdly, the raise of temperature, displacement, and thermal stress were exported to an optical program with a MATLAB code which can calculate the wavefront and depolarization.

To calculate the temperature, the pumping power and distribution should be given. Two kinds of pumping distribution were calculated, one is a uniform distribution, the other is a non-uniform distribution which was measured by CCD in a practical laser.

3. Thermal analysis of Yb:YAG and Nd:glass with uniform pumping

The advantages of Yb:YAG are high thermal conductivity, high thermal shock resistance and low heat generation rate. The disadvantage of Yb:YAG, however, is the low absorption cross section and emission cross section at room temperature, which reduces the requirement of high pumping and high laser fluence. Yb:YAG has a high pumping saturation intensity (28 kW cm^{-2}). Therefore high pumping intensity is needed to improve optical efficiency and this results in high thermal power density, particularly in high-repetition-rate operation. The Boltzmann thermal population of the lower laser level is affected significantly by temperature. Cryogenically cooled Yb:YAG has significant potential for high-average-power solid-state lasers with good spatial beam quality because of the material's excellent thermo-optic and spectroscopic properties at low temperatures. For example, YAG has a significantly higher thermal conductivity, lower thermo-optic coefficient dn/dT, and lower coefficient of thermal expansion than at room temperature. Each of these properties results in a reduction of thermo-optic effects at low temperature. For example, at 100 K Yb:YAG energy levels become four level, the saturation intensity decreases, the quantum defect is only 9%, and an absorption band at 940 nm remains broad enough for diode pumping. According to the energy fluence design[6], the optimized working temperature is from 190 to 200 K, the pumping power density on a single surface is 22.5 kW cm^{-2}, the pumping pulse duration is 300 μs, the beam aperture is 26 cm × 26 cm, and the total thickness is 20 cm. As a result of the total thickness being too large, it is necessary to cut the medium into several pieces to improve cooling capability.

Different thicknesses of Yb:YAG were calculated which included 1 and 2 cm. The temperature distribution of Yb:YAG is shown in Figure 1. Because the temperature of 2 cm Yb:YAG was too high, we adopted a 1 cm medium as the design point. The thermal stress and depolarization of 1 cm Yb:YAG were calculated and the results are shown in Figure 2.

For 1 cm thickness Yb:YAG, the maximum temperature rise was 9 K, the maximum xy shear stress was 1.29×10^6 Pa, and the maximum depolarization of electric field through one disk was less than 0.1.

The advantages of Nd:glass include the ease of obtaining large sizes and the fact that the temperature to translation effect can be ignored. The disadvantages of Nd:glass include low thermal conductivity and low quantum efficiency. The thermal effect of Nd:glass was calculated, with a thermal power density of 2.27 W cm^{-3} and a surface heat transfer coefficient of 500 W m^{-2} K^{-1}, for a thickness of Nd:glass

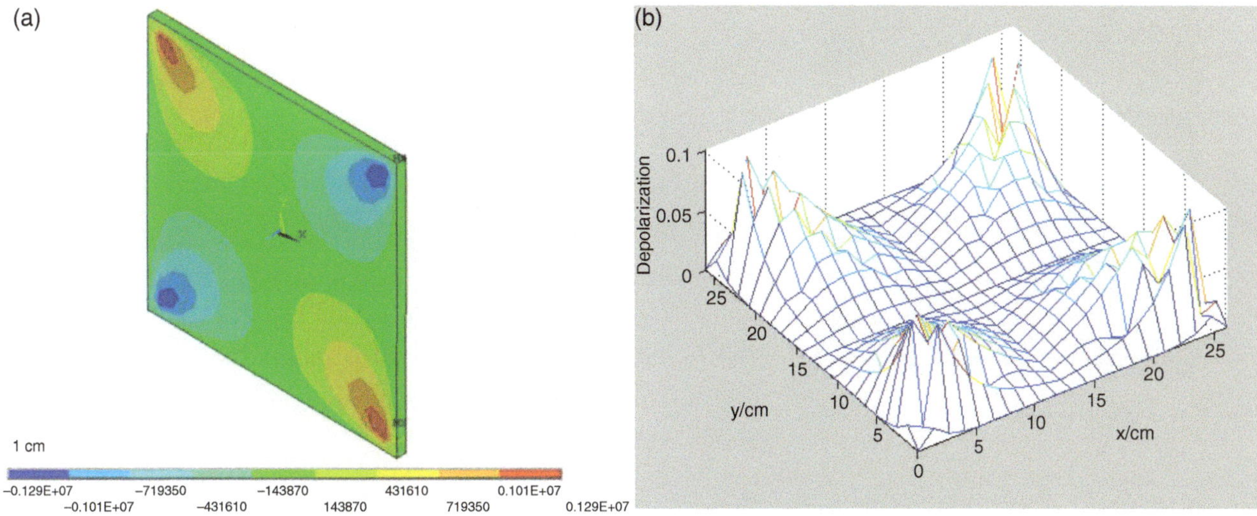

Figure 2. (a) xy shear stress of Yb:YAG; (b) depolarization of one disk.

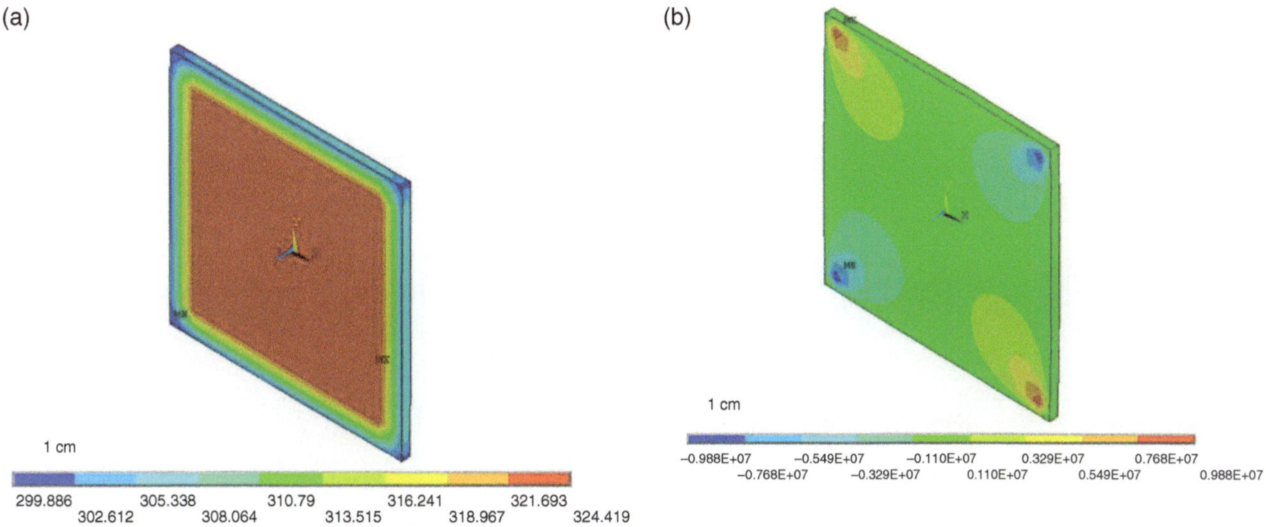

Figure 3. (a) Temperature distribution and (b) xy shear stress of Nd:glass.

of 1 cm. The temperature and shear stress are shown in Figure 3. The depolarization is shown in Figure 4.

For a 1 cm Nd:glass disk, the maximum temperature rise was 59 K and the maximum xy shear stress was 9.88×10^6 Pa; the maximum depolarization of electric field through one disk was 0.9.

From the calculated results, the thermal effect of Yb:YAG was less than Nd:glass. This is because Yb:YAG has higher thermal conductivity. In addition, in the high-average-power mode, the temperature gradient and the depolarization inside the medium cannot be avoided. The depolarization loss makes Brewster's angle slab amplifiers unsuitable for IFE lasers, because they can cause near field beam modulation and energy loss.

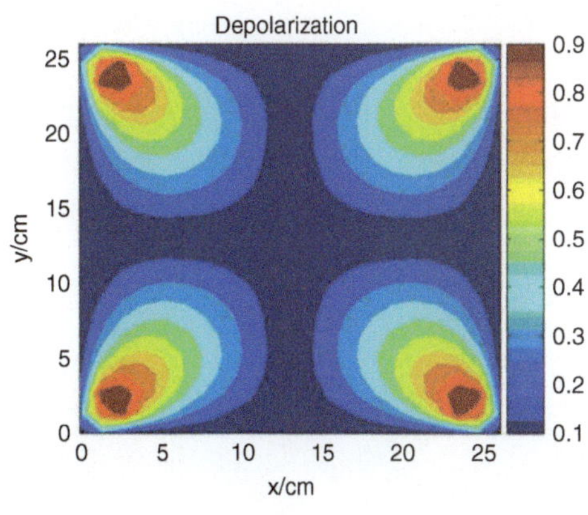

Figure 4. Depolarization of one Md:glass disk.

Figure 5. Non-uniform pumping distribution measured by CCD.

4. Thermal analysis of Yb:YAG with non-uniform pumping

In a practical laser, because it is difficult to make the pumping distribution very uniform, the pumping distribution will always be modulated. The non-uniform pumping will make the temperature of the laser medium more asymmetric and then the stress and depolarization will be different from a medium pumped uniformly.

An experimental pumping distribution was used to calculate the thermal effect of Yb:YAG. The pumping distribution is shown in Figure 5. The temperature and xy shear stress are shown in Figure 6. As you can see, the distribution of temperature and xy shear stress were not uniform and were asymmetric. The depolarization of the laser passing through the Yb:YAG was calculated. The depolarization of the laser passing through one disk and through twenty disks is shown in Figure 7. As you can see, the depolarization in Figure 7 is not inerratic or symmetrical. The undulation of depolarization in space is more acute than it is in Figure 2 which was pumped with a uniform distribution. We can deduce that if the medium was pumped with another non-uniform pumping distribution, the distribution of depolarization would be different from Figure 7.

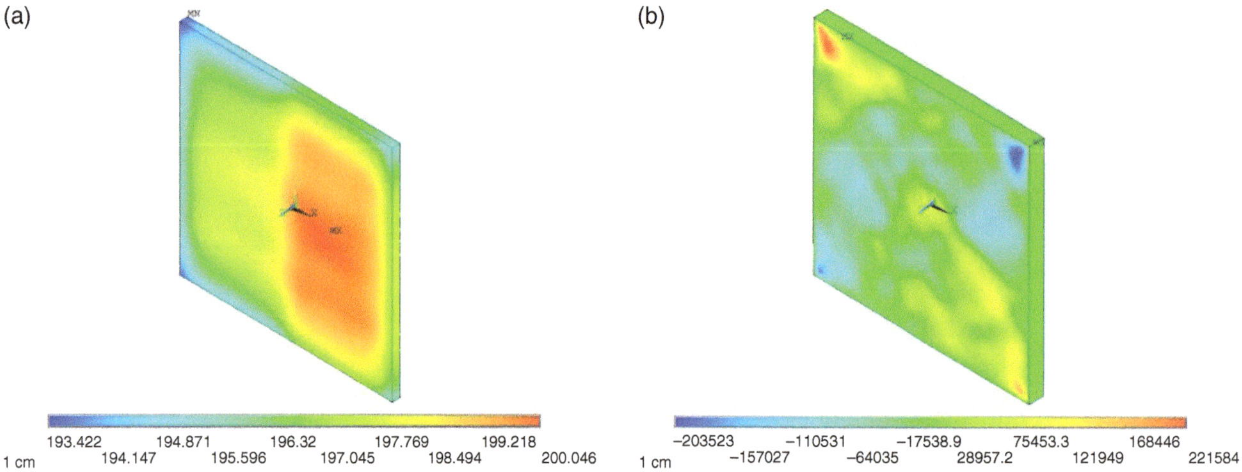

Figure 6. (a) Temperature distribution and (b) xy shear stress of Yb:YAG with non-uniform pumping.

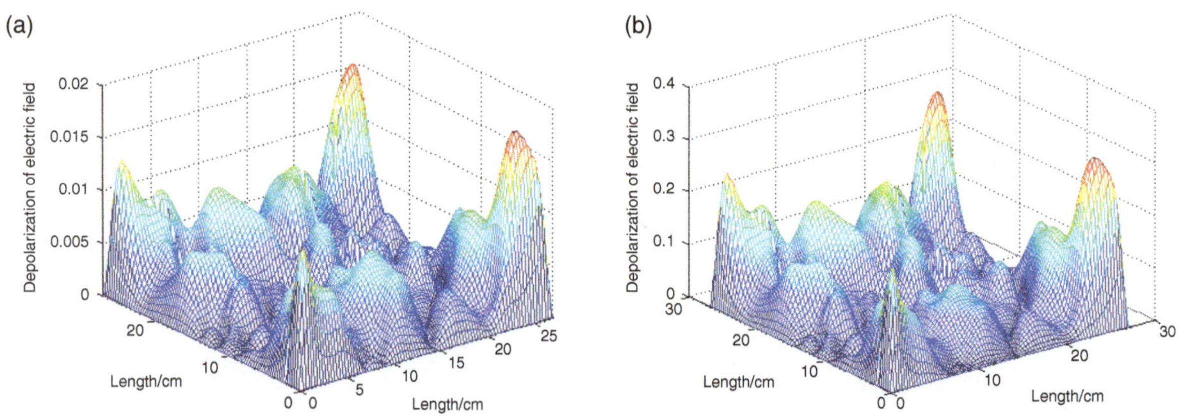

Figure 7. The depolarization of the laser passing through (a) one disk and (b) twenty disks.

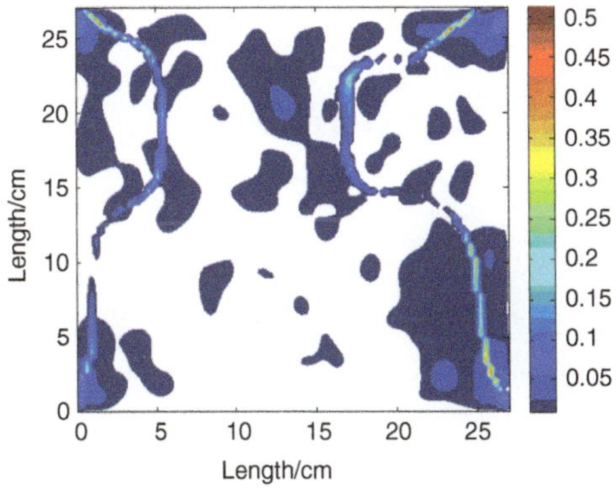

Figure 8. The residual depolarization caused by light excursion.

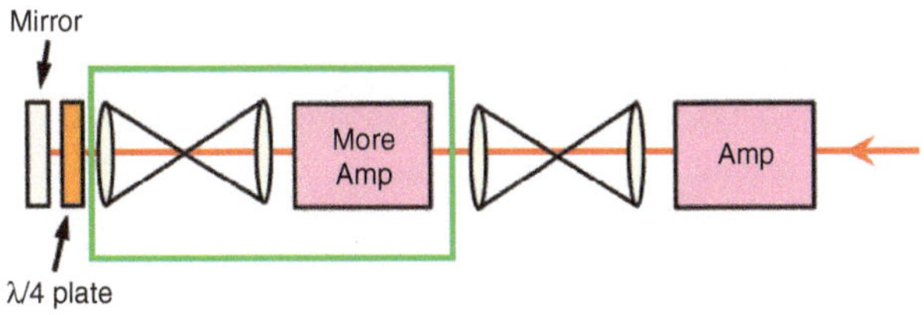

Figure 9. Schematic of depolarization compensation.

5. Depolarization compensation

In classical literature, the method of depolarization compensation was always the placing a polarization rotator between two identical amplifiers. A rotator rotates the polarization by 90°. For full compensation, the two amplifiers should accord and the laser should pass through them in a symmetrical position. However, for a practical laser, realistically non-uniform pumping will exist, and it will be difficult to make the two amplifiers completely the same.

In this condition, we consider compensating the depolarization of one amplifier by itself. Using a Quartz rotator after the amplifier and putting a high reflection mirror after the rotator, the laser will pass through the amplifier twice with the polarization rotated 90°. For a parallel beam, the depolarization will be compensated completely. But if the beam has a large aberrance, the effect of compensation will be reduced. The deterioration of compensation was affected by the gradient of the depolarization in space and the aberration of the beam. An elementary simulation of the deterioration of compensation caused by light excursion was performed. Figure 8 shows the residual depolarization caused by light excursion which was 0.014 rad (this is a large value determined by the gridding of my calculation). As you can see, the deterioration of compensation is not uniform in space, and is severe in the area of the large depolarization gradient.

A good wavefront and image relay will be necessary. Figure 9 shows the compensation schematic. In this way, we can use more amplifiers with different parameters.

6. Conclusion

The thermal effects of two kinds of laser material were simulated. The temperature gradient and the depolarization inside the medium cannot be avoided. The depolarization loss makes Brewster's angle slab amplifiers unsuitable for IFE lasers, because they can cause near field beam modulation and energy loss. In a practical laser, the pumping distribution is not very uniform and one cannot make all amplifiers identical to each other, so we consider a compensation method which compensates the depolarization by itself.

Acknowledgement

This work was supported by the Fusion Energy Center of CAEP (R-2014-0202-02).

References

1. J.-C. Chanteloup, D. Albach, M. Arzakantsyan, T. Novo, and B. Vincent, in *7th International HEC-DPSSL Workshop* (2012).
2. J. Kawanaka, in *5th High Energy Class Diode Pumped Solid State Laser Workshop* (2009).
3. J. Collier, in *5th High Energy Class Diode Pumped Solid State Laser Workshop* (2009).
4. J. Zheng, Z. Wang, X. Yan, X. Jiang, D. Wu, X. Tian, X. Zhang, M. Li, Q. Zhu, J. Su, K. Zheng, D. Hu, F. Jing, and W. Zheng, in *8th High Energy Class Diode Pumped Solid State Laser Workshop* (2014).
5. J. Zheng, X. Jiang, X. Yan, J. Zhang, Z. Wang, D. Wu, X. Tian, X. Zhang, M. Li, Q. Zhu, J. Su, F. Jing, and W. Zheng, High Power Laser Sci. Eng. **2**, e27 (2014).
6. X. Yan, J. Zheng, X. Jiang, Z. Wang, Y. Zhang, and M. Li, Chin. J. Lasers **29**, 0702008 (2012).
7. M. Frede, R. Wilhelm, M. Brendel, C. Fallnich, F. Seifert, B. Willke, and K. Danzmann, Opt. Express **12**, 3581 (2004).
8. M. Frede, R. Wilhelm, R. Gau, M. Brendel, I. Zawischa, C. Fallnich, F. Seifert, and B. Willke, Classical Quantum Grav. **21**, 895 (2004).

An important effect of filamentation instability on laser fusion physical processes

Zunqi Lin, Anle Lei, Wei Fan, Shenlei Zhou, and Li Wang

National Laboratory on High Power Laser and Physics, Chinese Academy of Sciences, Shanghai 201800, China

Shanghai Institute of Optics and Fine Mechanics, Chinese Academy of Sciences, Shanghai 201800, China

Abstract

The process of high power laser interaction with the large scale length corona plasma produced by the leading edge of the laser pulse has been investigated. Early experimental results are re-analyzed and conclusions drawn. In particular, studies of the close connection of unstable filamentation instability with – mainly – two-plasmon decay and – partly – stimulated Raman scattering, stimulated Brillouin scattering, and resonance absorption are carried out in this paper. The positive and negative effects of filamentation instability are also discussed.

1. Introduction

Recent laser fusion research, represented by the remarkable exploitation of the NIF (National Ignition Facility) Program and NIC (National Ignition Campaign) Program at LLNL (Lawrence Livermore National Laboratory) in America, has made excellent progress towards attaining the conditions required for ignition. Achieving ignition, however, is still a challenge. Some scientific cliff difficulties[1] appeared in NIC ignition experiments and need to be further explored. These difficulties cover a number of physical areas. Firstly, the value of the efficiency of the coupling of the input $3\omega_0$ laser to the ignition fuel target was lower than the expected simulation values; for example, the implosion velocity and fuel target stagnation pressure were 13% and ~50% lower. Secondly, the radiation driving target ablation ratio was ~20%–60% higher than the simulation value. Thirdly, the target deuterium-tritium (DT) fuel met an earlier "mix cliff" limitation at the time when the residual ablator's mass is 0.4 mg instead of the designed 0.28 mg. Fourthly, the laser–plasma interaction (LPI) process occurs during the $3\omega_0$ energy deposition on the corona of the hohlraum inner surface. In addition, it should be noted that LPI might have a significant effect on the other three difficulties, though many efforts have been made to increase the ignition threshold factor $\exp(\mathrm{ITF}_x)$ and the experimental results have been getting better in the past two years.

Until now, the under-used classical theory of LPI (CTLPI)[2–6] has been based on the boundary condition hypothesis of uniform incident light and bulk homogeneous plasmas. However, it is known that the far-field profile of the incident laser beam has a strong speckle irradiance contrast with severe self-phase modulation on the wavefront. Meanwhile, laser produced plasma jetting always appears and develops, which definitively spoils the corona plasma homogeneity. Therefore, it is necessary to investigate and analyze precisely in detail the features of the production and development of small scale filamentation instability, which may spatially and temporally modify the plasma homogeneity and can be easily induced by the leading laser interaction with the plasmas.

Self-focusing filamentation instability was studied separately and individually together with the other important nonlinear three-wave interaction processes in the 1970s and 1980s[2–6]. In the years of 1976 and 1977, Eidman *et al.*[7] and Hass *et al.*[8] found in the X-ray pinhole camera images from laser-irradiated targets that the whole X-ray spot size was smaller than the incident laser hot spot size, which indicated that the laser hot spots had self-focused due to the thermal effect to a smaller size, resulting in a stronger X-ray emitting spot. Baldis *et al.*[9] measured the filaments in 1980 using Thomson scattering and optical probe beam reflection. Following the early investigations, Lin *et al.*[10, 12–16], Willi *et al.*[11], Giulietti *et al.*[17], and Jones *et al.*[6] investigated the small scale filamentation instability further in the 1980s and 1990s.

Recently the effect of phase front deformation on the growth of the filamentation instability was studied[18]. Filamentation instability and its connection to two-plasmon decay (TPD) instability were also investigated by using energetic proton radiography[19, 20], a diagnostic technique appropriate for the observation of electromagnetic fields[19–21].

Correspondence to: Zunqi Lin, National Laboratory on High Power Laser and Physics, Chinese Academy of Sciences, Shanghai 201800, China. Email: zqlin@mail.shcnc.ac.cn

2. Collection and re-analysis for some of the earlier experimental results on small scale unstable filamentation

The early experimental investigation of the connection of TPD instability with filamentation instability is re-analyzed in a new way. Differing from the old approach, which considered more the details of the $3\omega_0/2$ emission features and separately the link of some contribution from the filamentation process, our approach provides in this paper more qualitative filamentation features, from the original experimental graphs, and more detailed analysis for the experiments. Several important pieces of evidence are given now to show that the unstable small scale self-focusing filamentation instability (SSSFI) should be the first important factor for stimulating the TPD instability and, as a deduction, the other three instabilities (i.e., stimulated Raman scattering (SRS), stimulated Brillouin scattering (SBS), and resonance absorption (RA)) in laser plasma corona, though the evidence was not completely recognized in the previous articles. Following the new recognition of LPI physics, the contribution of LPI to some of the branches of scientific difficulties of laser fusion ignition physics are discussed.

2.1. The observation of the connection of filamentation instability to the TPD instability

Lin *et al.*[10] obtained the time integrated $3\omega_0/2$ bunched filament structures from a cubic six-laser-beam-irradiated spherical glass micro-balloon target corona in 1981. Figure 1 shows the 2D space resolved experimental images of time integrated $3\omega_0/2$ filament emission bunches, each defined as the integration or overlapping of many individual filaments in space and/or time dimensions. The earlier work[10]

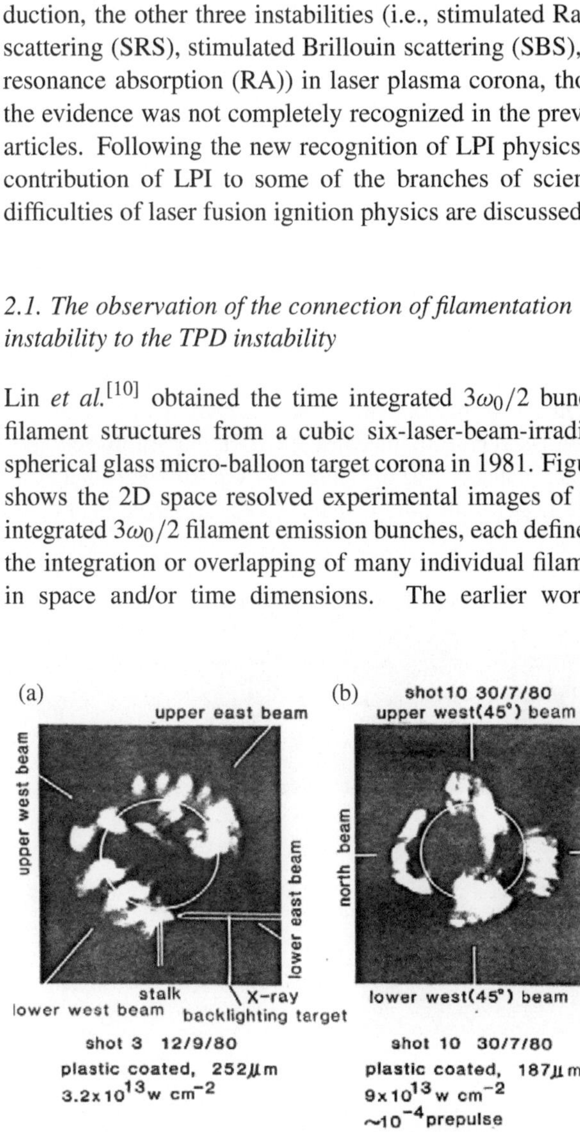

Figure 1. Photographs of the $3\omega_0/2$ harmonic emission from six-beam-irradiated spherical targets taken from two perpendicular directions. (a) 1 μm polymer coated target, initial diameter 252 μm, irradiance 3×10^{13} W/cm^2. (b) 1 μm polymer coated target, initial diameter 187 μm, irradiance 9×10^{13} W/cm^2[10].

concluded that "The $3\omega_0/2$ emission structures imply that the plasma corona is no longer homogeneous in which an $n_c/4$ layer can exist within a background plasma of much higher density. As the $3\omega_0/2$ emission structures are aligned along the laser beam directions, the results reported here are clear evidence for self-focusing and filamentation in the underdense corona plasma surrounding the target." However, many succeeding investigations showed that the fine $3\omega_0/2$ emission structures with rather high time, space, and spectrum resolutions opened some new phases. Some questions should be proposed and answered, such as:

Is the small scale self-focusing filamentation instability (SSSFI) stable or unstable?

What are the effective physical boundary conditions for SSSFI? Classical bulk plasmas interacting with classically uniform incoherent light? Or the leading section of the laser pulse, or the added prepulse, produced plasmas interacting with the real incident laser light profile with severe phase modulation and strong amplitude modulation?

What is the connection between SSSFI and TPD, SRS, SBS, and RA?

What role does the SSSFI play in LPI? Is it very important or negligible?

The filaments in figure 1 could be composed of many temporarily existing $n_c/4$ layers, which indicate that the corona plasma is already very inhomogeneous. These temporarily existing $n_c/4$ layers are produced transiently by the self-phase modulation (SPM) of the incident laser light driving forward the filament ends and increasing the electron density to higher than $n_c/4$, with much higher laser intensity inside the filaments than in background plasmas. That perhaps was the first experiment to study the connection of filamentation instability with the other kinds of instabilities in the laser plasma corona. On the foundation of CTLPI, another unexpected result was that there was no uniform $3\omega_0/2$ emission around the bulk spherical shell plasma $n_c/4$ density layer during the whole laser pulse period, except the strong $3\omega_0/2$ filament bunches which are just along the direction of the laser light normally or tangentially incident on the glass micro-balloon target. One explanation is that the laser average power intensities outside the $n_c/4$ layer of the individual and the integrated filament bunches were still lower than the predicted threshold of TPD instability, which was considered as 5×10^{14} W/cm^2 according to the earliest CTLPI. This implies that the laser intensity in the laser produced filament ends should be several tens of times higher than that outside. In addition, since the irradiance at the ends ("bottoms" was adopted on the original paper[10]) of the filament channels was considerably larger than that in the surrounding areas, one would expect the emission to have been caused by the modulated moving $n_c/4$ trajectories of the filament self-focusing ends.

2.2. The experimental evidence for the unstable SSSFI phenomenon and the features of spatially and temporally resolved $3\omega_0/2$ emission

The spatially and temporally resolved $3\omega_0/2$ emission was investigated in 1986[15] and 1992[16]. The optical streak camera slit was horizontally aligned parallel to the driving laser central axis for observing the laser focal center produced filaments along the filament channels by imaging optics. The experimental arrangement is shown in figure 2(a)[15].

It can be found in the spatially and temporally resolved $3\omega_0/2$ emission photograph in figure 2(d) that two individual filaments were caught with filament lengths of 120 μm and 43 μm, respectively. The longest and the shortest filament lengths in figures 2(b) and (c) were 130 μm and 15 μm, respectively. The temporal existence duration at all of the spatial filament positions was ~5 ps, and along with the lifetime of each filament from the beginning of the stimulated $3\omega_0/2$ emission to the ending of the emission, came to around ~15 ps–20 ps. One should pay attention here to the difference of the real inner aperture of a filament from the outer aperture of the $3\omega_0/2$ emission filaments. The former could be reduced to the minimum diameter scale of the incident laser wavelength at the ends of the second section of the filament duration of the lifetime, while the latter $3\omega_0/2$ emission aperture and the existence time mainly or only depend on the interaction region and the period in terms of the TPD produced $\omega_0/2$ plasmon interaction with the incident laser photons at the transient ends or nodes of filaments.

We may define, among the filaments with different lengths, the long length filaments as the strong filaments, which have higher laser intensities on the filament ends with stronger laser self-phase modulation drive (SPMD) or pondermotive force drive (PFD) ability to drive the filament ends to penetrate deeper into the plasma corona, even to the region of a quarter of the critical layer. We may, in the same way, define the moderate and the weak filaments. It should be noted here that the above defined $3\omega_0/2$ emission filament length could be the temporally first half-section of the whole filament temporal development process, in which the self-phase modulation will be enhanced by further PFD self-focusing such that the axial plasmas are pushed forward to build up the filament ends or nodes with higher plasma density to a level higher than $n_c/4$, or even n_c, to construct the limited laser imprinting.

From figures 2(b), (c), and (d) it is found that: (i) The individually weak filaments appeared at the temporal pulse profile valleys of the amplitude modulated laser pulse. At the pulse profile peak region, the total filament sampling numbers integrated along the thickness direction, shown in figure 2(a), increased quickly in an overlapped short transient period, even bringing about a signal saturation. However, the distinct features of unstable filamentation effects, such as the temporal existence period and the long spatial length, and the position of the low density corona area, are displayed and easy to examine. (ii) From the analysis of above experimental results, there is still no evidence for CTLPI produced $3\omega_0/2$ emission from bulk homogeneous plasma corona. (iii) It seems that the $3\omega_0/2$ emission intensity arising from laser–plasma interaction of TPD instability for high Z target material (Ta) in figure 2(b) was stronger than that for the target with intermediate Z (Al) in figure 2(c).

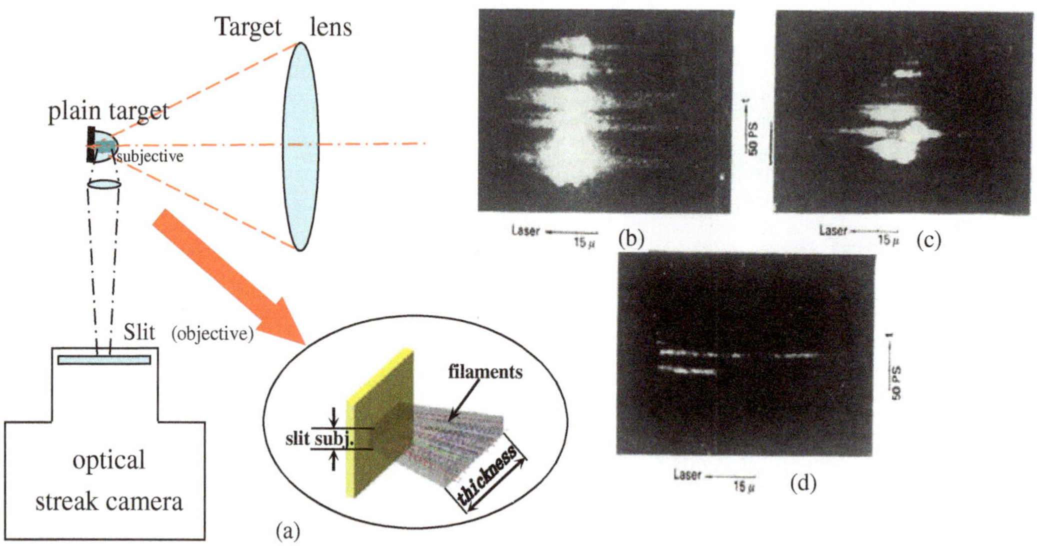

Figure 2. $(3/2)\omega_0$ time and space resolved photographs, along the filament channels. (a) The experimental setup. The magnification of the imaging optics was 19, with the temporal resolution of 5–10 ps and the spatial resolution of 5 μm. (b) Shot No. 107; narrow band laser irradiation on a Ta plain target with laser incidence angle ~ 10°, energy 8.3 J and pulse width 250 ps. (c) Shot No. 124; narrow band laser irradiation on an Al plain target with laser incidence angle ~ 20°, energy 8.1 J and pulse width 250 ps. (d) Shot No. 113; broad band laser irradiation on a Ta target with laser incidence angle ~ 10°, energy 3.6 J and pulse width 250 ps [15].

(iv) Figure 2(d) illustrates that the power of each single $3\omega_0/2$ filament depends on the total fraction of the incident laser power assigned to the related first half-section of the filaments. And the fraction of assignment depends on the random distribution of the very inhomogeneous bulk plasmas.

The work in the article[15] adopted a group of ~3 nm bandwidth incident laser interaction experiments to examine the $3\omega_0/2$ emission feature. It is described in the paper that the wide band laser focal spot size is much larger and more uniform than the narrow one[15], resulting in quite a large percentage laser irradiance reduction, leading to the $3\omega_0/2$ filamentation being seldom produced[15]. For many shots of the broad band laser, there were no appearances of $3\omega_0/2$ filaments. When the broad band laser irradiance increased, the $3\omega_0/2$ filaments seldom appeared and were seldom caught.

Comparing the observed $3\omega_0/2$ filament images in figures 2(b), (c) and (d), one should find that all of the $3\omega_0/2$ signal was produced and ~90° side scattered from the individual or bunched filament channels, with intense node modulations along their own filament channel axes, though the intensity modulation appeared a bit weaker than that observed from the backscattered spectrum node modulations of filaments, which will be described and re-analyzed below. The weakening mechanism could be caused by the transformation astigmatic distortion of the $3\omega_0/2$ node signals during the imagery process while the light emitted from one node subject passed through a large gradient plasma density area, in the left side in figures 2(b) and (c), along different directions transversely, at about 80° orientation relative to the incident laser major axis.

The above time and space resolved $3\omega_0/2$ emission experiments have provided further clear evidence for the origin, existence, and features of the unstable SSSFI, qualitatively, in detail. All the results directly further support the assertion that the unstable filament instability provides a decisive contribution to the TPD instability process in laser produced plasmas. It is unlikely that the filamentation instability behaves as a transient phenomenon "flicker" effect, predicted by the simulation.

It should be recognized that in the above analyses of Sections 2.1 and 2.2 the $3\omega_0/2$ filament channels appeared mainly in the bulk plasma corona area with average electron density from $n_c/4$ to $n_c/10$–$n_c/20$. This implies that, firstly, the TPD process is induced in the rather low electron density area instead of the macroscale $n_c/4$ density region which is predicted by early classical theory of TPD, and, secondly, the transient $3\omega_0/2$ emission and decay process, including all the TPD stimulation, the three-wave interaction of $\omega_0/2$ plasmons with the incident laser photons, the $\omega_0/2$ plasma electron wave fast decay by Landau damping and the accompanying fast electron and hard X-ray emission from the ends, nodes, or channels, occurred in the low density region of laser produced plasmas in a 5–10 ps period.

2.3. The experimental study of the dynamics of a small scale unstable filament growth and damping process

The features of space and spectrum resolved $3\omega_0/2$ emission

Some earlier work[12, 13] investigated spatially resolved spectra of $3\omega_0/2$ emission, induced by the TPD instability process, in 1981, and obtained a finest $3\omega_0/2$ emission photograph with the spatial and spectral resolutions of 5 μm and 0.3 nm, respectively. The experimental setup is shown in figure 3 and the spectrum was obtained at the imagery output plane position of the 0.5 m spectrometer with the spectrum entrance slit at around 100 μm. The backscattering spectrum structures for single and bunched filaments (figure 4) were observed with a single hump and/or double humps corresponding to the differently sampled spatial filament positions. In the experiment, the 153 μm diameter glass micro-ball was irradiated with a laser of intensity 2.5×10^{13} W/cm^2. The moderately low laser irradiance led to relatively few sampled filament numbers. Therefore the spatial position of the $3\omega_0/2$ filament spectral signals could be easily distinguished on the target corona area.

The features of the filament spectral structures from figure 4 can be derived as:

(i) All of the $3\omega_0/2$ spectrum signals obtained from the experiment originated from individual single filaments or a number of filament bunches. No sign of a homogeneous bulk plasma produced signal was found at the $n_c/4$ density layer.

(ii) Spectrum filaments No. 1 and No. 2 are double-humped strong filaments with a large gap width between the red spectrum hump peak and the blue one, which are at 3.1 nm and 5.6 nm, respectively. They can be denominated as strong spectrum filaments. The weak spectrum filaments No. 3 and No. 4 with the single red humps and the almost disappearing blue humps have the narrow gap widths of 3.4 nm and 3.1 nm, respectively.

(iii) The spatial node size from all individual filament emissions in the time and spectrum resolved photograph was measured as 4–5 μm in diameter, which coincides with the Willi[11] X-ray pinhole camera experimental results described in Section 2.4 below, with the 1.5 keV band X-ray wavelength and with each of the spot sizes in the 4–5 μm diameter range.

The space and spectrum resolved filamentation instability investigation is an important diagnostic means for not only $3\omega_0/2$ (TPD), but also SRS, ω_0 (SBS), and even $2\omega_0$ (RA). In particular, when the spatial slit width of the spectrometer can be adjusted to as narrow as possible, the total number of samples of filaments is decreased to the smallest, and the laser irradiance on the target surface is moderate.

The time and spectrum resolved $3\omega_0/2$ emission[12, 14, 15]

The experimental setup is given in figure 3 and the spectrum photograph is in figure 5. The temporally and spectrally

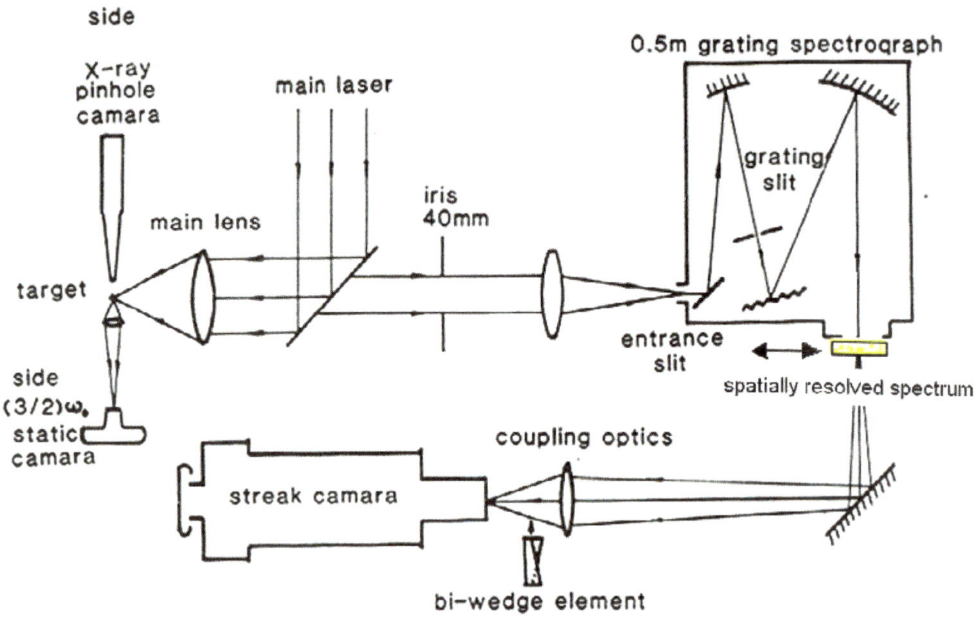

Figure 3. Experimental arrangement for space and time resolved spectroscopy of the $(3\omega_0/2)$ emission[12].

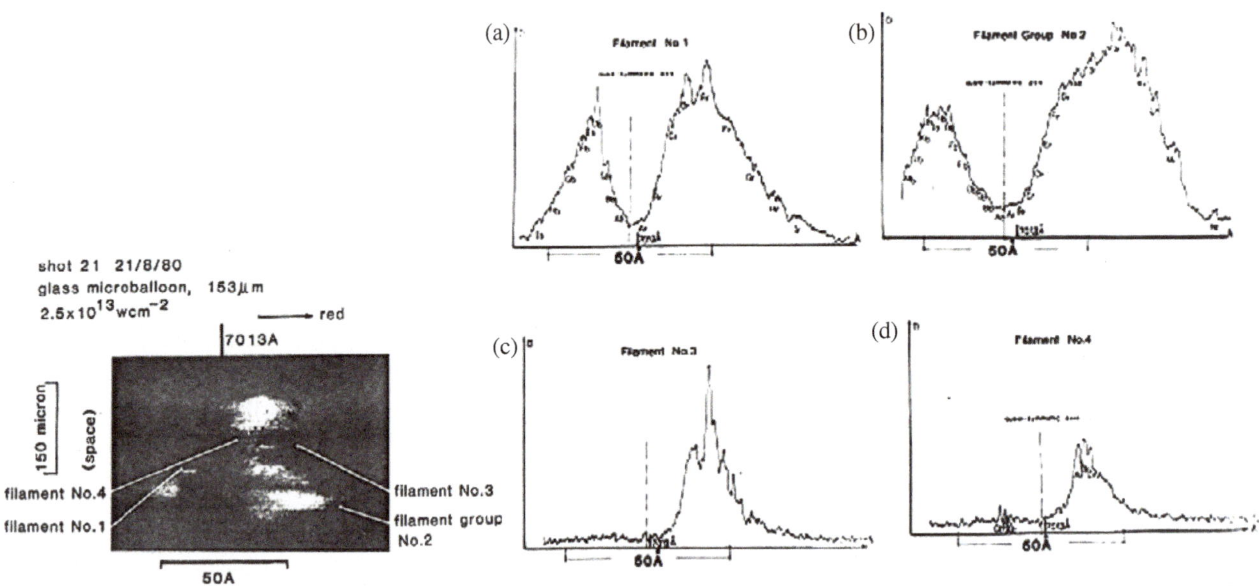

Figure 4. The $3\omega_0/2$ spatially resolved spectrum[13].

resolved $3\omega_0/2$ emission appeared a bit ambiguous in the photographs, making it difficult to distinguish a single-filament feature due to the much higher laser irradiance (2.8–14×10^{14} W/cm^2) on the micro-balloon target surface, which led to the emissions of too many filaments being squeezed together in a narrow picosecond range.

The intensity scanning for the two distinguishable spectrum filaments is shown in figure 5(b). The temporal widths of the spectrum trajectories from the two spectrum filaments are both around 4–5 ps. The spectrum scanning results support a spectrally quasi-symmetrical feature for the $3\omega_0/2$ backscattering spectrum, lending support to the Doppler shift model, which described the dynamics of the filament production and decay. Meanwhile, the theoretical framework and the complicated simulation of the generation of a series of strong $3\omega_0/2$ nodes along the filament channels are still awaiting construction. However, the $3\omega_0/2$ emission inside the unstable filament is always accompanied with filament node or end initiation and very fast collapse due to the low density plasma environment. Those processes will produce a large amount of intense X-ray node bursts in a very underdense surrounding corona plasma area. This picture

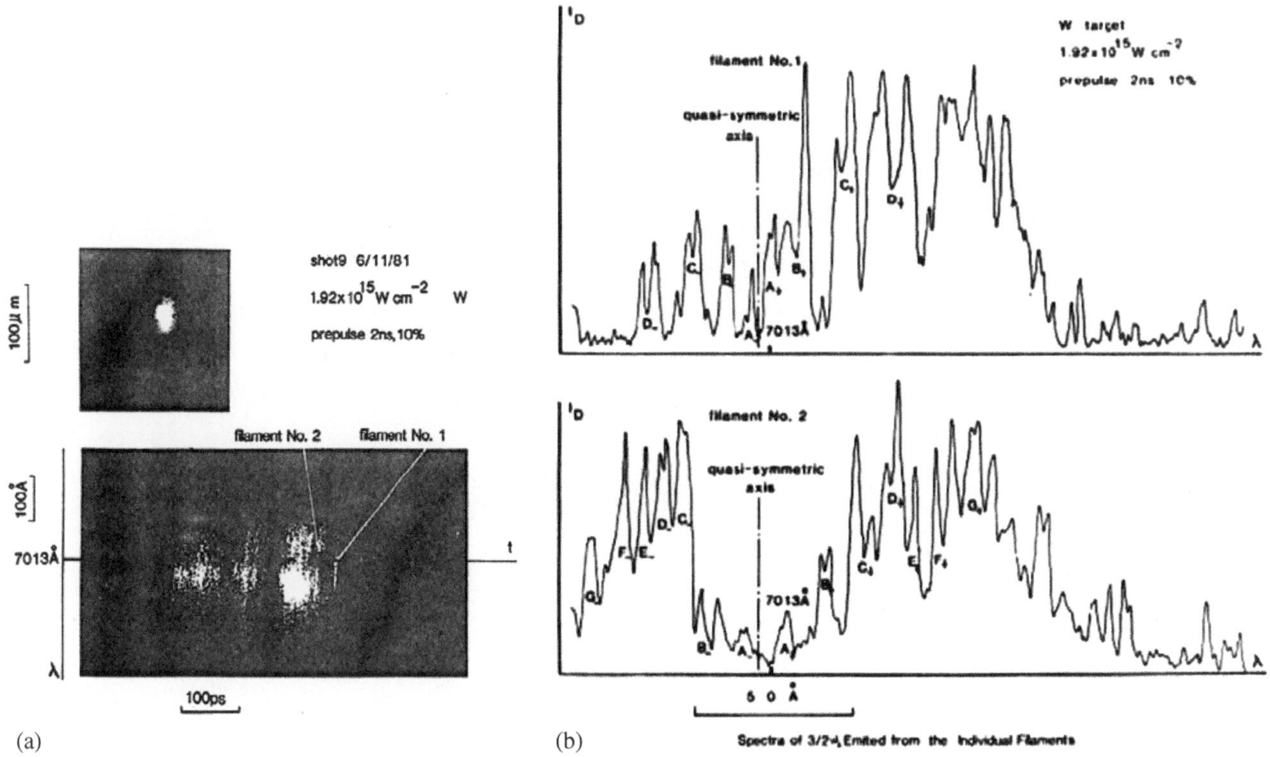

Figure 5. The temporally resolved spectrum[12].

is at variance with the previous common scenario wherein the intense X-rays, including X-ray spots or nodes, should be produced in the critical or ablation layer plasma area[11]. In addition, the picture may interpret the above mentioned 1.5 keV X-ray spotty spots, which are scattered and are larger than those of the laser-irradiated far-field intensity profile.

From figure 5, the small temporal difference of ~10–20 ps could still be seen clearly, from the very beginning of the red shift spectrum to the end of the blue one. The single and double humps with quasi-symmetrical distribution were also analyzed in earlier work[12, 14].

2.4. Evidence of the filament end or node dynamic collapse process induced large amount of 1.5 keV hot X-ray spot bursts

The large amount of spotty 1.5 keV X-ray hot spots with each spot size about 4–5 μm is shown in figure 6(c)[11]. The target was a smooth glass plate with the laser irradiance 3.3×10^{13} W/cm^2 and the laser spot size 250 μm in diameter. On the basis of the unstable SSSFI process, this spotty X-ray photograph can be attributed to the laser hot speckles, shown in figure 6(b), having split or been trapped randomly into the huge amount of self-focusing small scale filaments and the subsequent TPD decay induced X-ray emission with each spotty size in the range 4–5 μm, contrasting with the early description[11] which stated "the spotty X-ray spots originate from dense regions of the expanding plasma with similar temperature as the background plasma and are caused by higher ablation rates in the more intense spots of the laser beam". These SPMD and/or finally PFD small scale filaments will be stimulated and collapsed, resulting in the generation of the ~1.5 keV X-ray band bursts, and even the superhard X-ray bursts at the filament ends or nodes, mostly in underdense corona plasmas. The deduction shows that most of the 1.5 keV X-ray hot spots were generated in quite a large area of the very low density plasma region, which illustrates why the X-ray hot spot distribution area from figure 6(c) is larger than the incident laser speckle distribution area from figure 6(b) and why the plasma jetting shadowgraph areas from figures 6(a) and (d) are also obviously larger than those from figure 6(b).

Another interesting phenomenon is displayed in figure 6(a), where the shadowgraph with 50 ps temporal gate duration was just at the pulse trailing edge position of the incident heating laser, shown in figure 6(e), with a sixth of the laser power intensity, 5.5×10^{12} W/cm^2, on the target surface. In this case the LPI produced filamentation process still existed, bringing about the display of a clear filament, and the filament node or end modulation structures for the heating laser reached the rather underdense plasma region.

Inferring from the shadowgraphs in figure 6(a), the experimental result shown in figure 6(d) showed no filament node or end modulation structures at all due to the probe laser gate position being 1.4 ns delayed from the 1.1 ns full width main laser starting time. At that time, there was no main

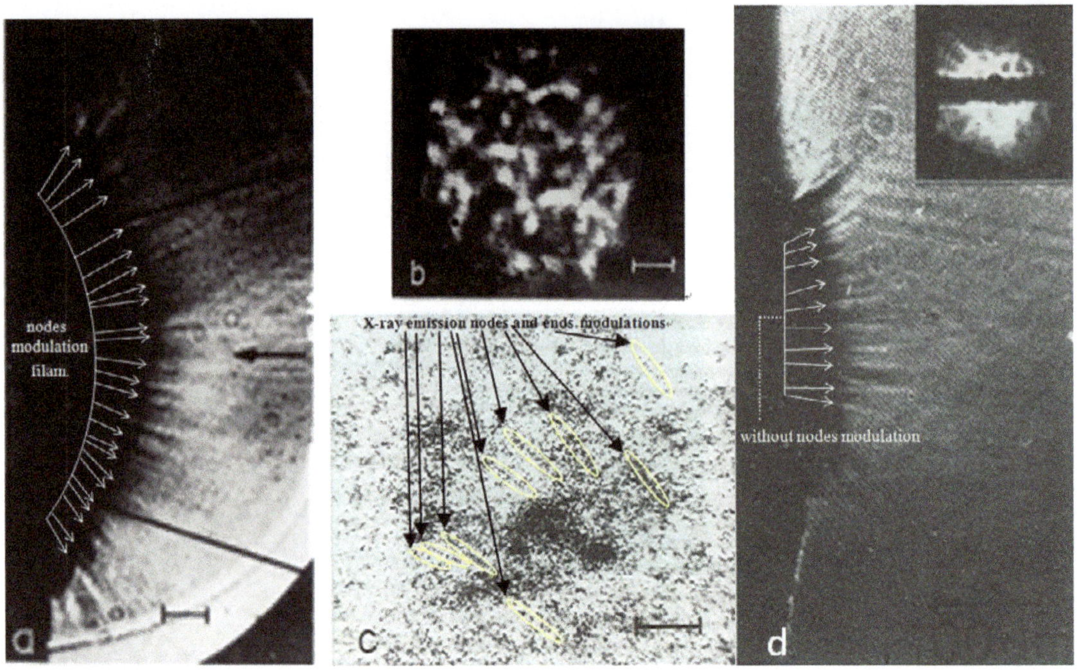

1.0 ns delayed shadowgraph X-ray hot spots distributions, laser from up-left-front 1.4 ns delayed shadowgraph

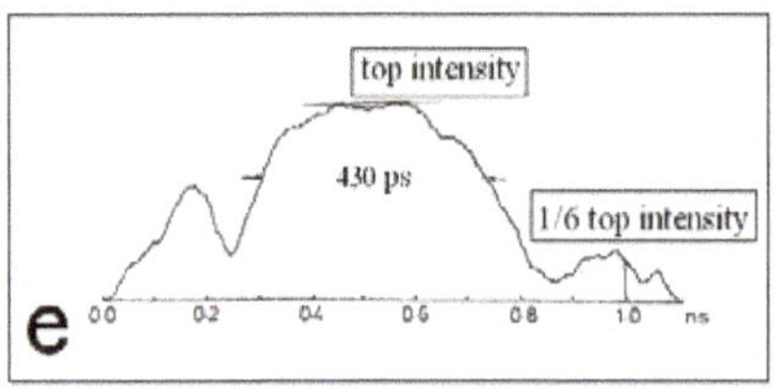

temporal profile of the green heating laser pulse

Figure 6. (a) Shadowgraph recorded on a glass target 1 ns after the beginning of the laser pulse, showing pronounced plasma jets with the clear filament node modulation. (b) Laser beam profile taken in the equivalent target plane. (c) X-ray pinhole camera image in the 1.5 keV energy band (pinhole filtered with 40 μm of beryllium). The scale for all the images is 50 μm, for (a), (b), and (c). (d) Shadowgraph 1.4 ns delayed from the beginning of the laser pulse, showing pronounced plasma jets without filament node modulation. (e) Temporal profile of the green heating laser pulse[11].

heating laser energy irradiated on the target. This was further important evidence for the massive filaments displayed with the end or node modulation on the very underdense plasma corona.

A sign of X-ray emissions from the modulated nodes of filaments along with the incident light could also be distinguished in figure 6(c) if the heating laser light came from the upper left front corner direction of the picture's central region. One may carefully resolve a number of spot linked strings along with the above mentioned incident laser light lines.

Therefore, there is actually, for the first time, simultaneously obtained new evidence of filament node modulation from figure 6(a), obtained by optical shadowgraph technology, and an extremely feasible proof of hot X-rays emitting from the modulated filament nodes and ends in an extremely low density plasma corona, without any other reasonable options for consideration, as shown in figure 6(c), from the 1.5 keV band X-ray spot pinhole imagery, though this phenomenon was not recognized in the originally published article[11].

2.5. The time and spectrum resolved SRS and SBS emission

Giulietti *et al.*[17] investigated the SRS and SBS emission features, in 1988. The temporally and spectrally resolved SRS and SBS emissions were observed using a $2\omega_0$ green heating laser as the incident laser in the experiments. The experimental results are shown in figures 7(a), (b), (c), and (d). The experimental parameters are labeled in the picture.

Figure 7(a) shows the low temporally resolved SBS spectrum signal. The spectrum was divided to both red and blue shift macroscale humps. The spectrum double-hump separation increases and decreases following the heating laser power intensity changes. On the basis of the unstable SSSFI model, integration of the total transient individual filament intensities will yield the SBS spectrum. However, the temporal resolution of the diagnostics was rather low,

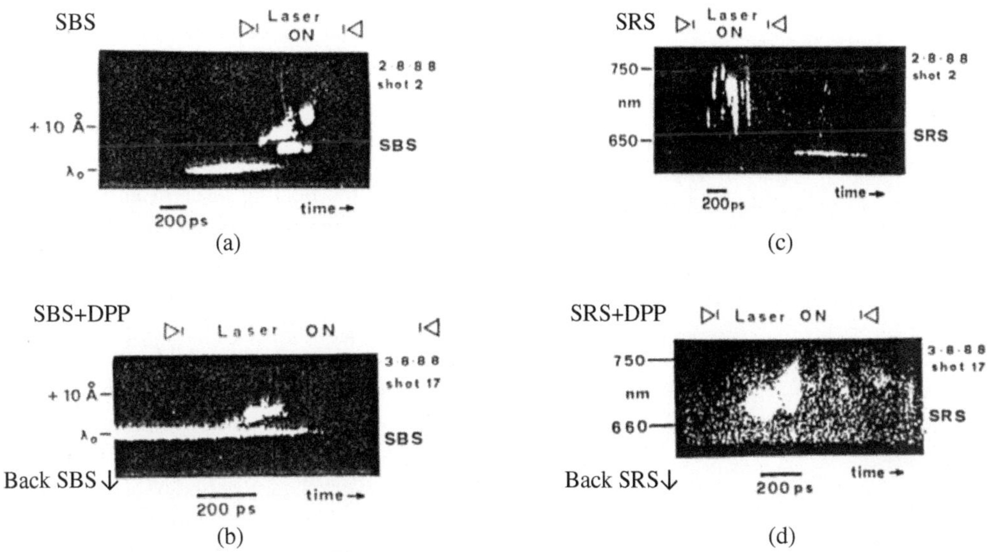

Figure 7. (a) SBS time resolved spectrum with an unsmoothed beam. (b) SBS time resolved spectrum with an RPP smoothed beam. (c) SRS time resolved spectrum with an unsmoothed beam. (d) SRS time resolved spectrum with an RPP smoothed beam[17].

and the laser irradiance on the target with intermediate Z was too high, so the experiment could not show the detailed filamentation effect in that experiment.

Figure 7(b) shows the experimental result on the random phase plate (RPP) application for the field beam smoothing. The random phase plate breaks the laser field speckles into a quite large number of fine speckles with much less energy and intensity for each speckle, resulting in a weakened filamentation process, if it should be stimulated, leading to the spectrally blue hump nearly vanishing and the red one being weakened.

Figure 7(c) displays the temporally resolved SRS spectrum emission with a serious temporal modulation, as a sign of the appearance of filamentation instability. However, it might be difficult to distinguish the filament temporal width and the other features, also due to the temporally low resolving power.

Figure 7(d) displays the spectrum feature weakened due to the application of the RPP.

Although the filamentations are weakened with the distributed phase plate (DPP) or RPP, this was not because of the action of the DPP or RPP itself, but because of the much larger heating region with the completely decreased incident laser average power intensity on the target surface.

2.6. The transfer from the SPMD unstable SSSFI to the PFD unstable SSSFI

The very serious node modulated damage produced inside the laser glasses and the final optics of the high power laser driver have, at a very early time, been clearly explained, while the laser output average power intensities, in the damage region, reach a level of about 10^{10} W/cm^2 or less[22, 23]. Figure 8(a) shows a typical example of damage to a few individual and/or several bunches of small scale self-focusing filamentation (SSSFF) produced inside the material output position of the Nd doped glass rod amplifier at the laser power intensity of 10^{10} W/cm^2 on the damage region.

The enhancement of the nonlinear refractive index can be expressed as

$$n = n_0 + n_2 \gamma I = n_0 + \Delta n. \tag{1}$$

Here n_0 represents the linear refractive index of solid material, and n_2 is the nonlinear index. As $n_0 \sim 1.5$, $n_2 = 1.13 \times 10^{-13}$ esu (Shanghai, Nd:glass), $\gamma = 4.19 \times 10^{-3}$ cm^2/W[24] is the unit transfer factor. Assuming $I = 10^{10}$ W/cm^2, which is strong enough to decisively produce, for the filaments of SSSFF, serious damage in the laser glass and the final optic quartz solid materials, we estimate the index enhancement $\Delta n = 4.7 \times 10^{-6}$.

In the case of laser produced plasmas, the refractive index n_{pe} can be described as

$$n_{pe} = (1 - n_e/n_c)^{1/2}. \tag{2}$$

Here n_e is the locally sampled electron density in corona plasma, and n_c is the critical density, where the frequency of the plasma electron wave is just the heating laser frequency. Assuming n_e to be equal to $n_c/30$, then n_{pe} will be approximately equal to 0.96667 and the refractive index enhancement Δn_{pe} will be $\sim 3.45 \times 10^{-2}$. The n_{pe} enhancement amplitude is four orders of magnitude larger than the SSSFI modulation induced index enhancement Δn in solid-state material. The above estimation implies that the generation of SSSFI with node modulation in the laser plasma corona should be much easier and the result more

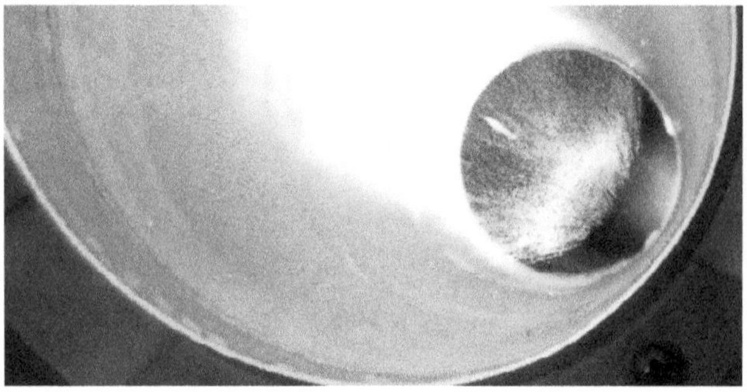

Figure 8. Small scale self-focusing filaments in the rod amplifier[25].

massive than in the glass or final optics of the heating laser driver.

Figure 8 displays the SSSFF process induced strong or weak, individual or bunched, filaments with node modulations inside the output area of a laser Nd:glass rod amplifier[25]. It should be noted that the uniformity and the other qualities of the laser glass and the final optics are extremely good, with the material homogeneity of 1–5 × 10^{-6}/cm and with no bobbles and defects affecting the SSSFI process.

To keep the laser facility safe and to maintain the output capability, laser scientists usually employ all of the methods that will prevent the laser facility from sustaining self-focusing filament damage. However, SPM damage effects are still seldom presented seriously. This is because there are deep phase and amplitude modulations at the laser near-field and far-field output, leading to the focal spot being full of highly intense spikes, which are quite evident from the reported results from NIF[25].

Considering the physical boundary conditions for LPI processes, it is well known that when a typical $3\omega_0$ heating laser irradiates a target surface, the irradiance distribution profile can be selected as shown in figure 9. These simulations and experimental results display quite heavy amplitude modulations with seriously non-erasable self-phase modulations. Moreover, the absolutely heaviest burden of intensity for solid-state material has already been increased to 3–5 orders of magnitude for the laser produced plasmas, by the laser focusing optics, from the intensity 10^{10} W/cm^2 to 10^{13-15} W/cm^2. As the central ignition mechanism requires four steps of incident driving laser pulse profile, which cover the laser power intensity range from 10^{12} W/cm^2 to 10^{15} W/cm^2, the study of LPI physical processes in these ranges is essential. This is re-analyzed above Section 2.4 of this paper for the filament generation feature, shown in figure 6(a), for the power intensity of 5.5 × 10^{12} W/cm^2. In the picture of figure 6(a), the gaps between the neighboring modulated nodes are estimated as around 20 μm and the scales of individual filament apertures for very low density plasma are around 10 μm. Therefore it is easy to estimate the light focusing angle inside each of the filament channels as approximately 20°–25° with respect to the filament channel axes, implying that ~25% of the total energy of laser light inside each filament could be used to generate filament walls and ends or nodes with plasma density higher than the surrounding plasma, bringing about the generation of the new transient denser layer pieces including $n_c/4$ density plasmas.

If we regard the heating laser in an earlier article[11] as the preheating laser, the shadowgraph of figure 6(d) in fact provides the physical boundary condition for the under-irradiated plasma corona, at 300 ps after the complete ending of the preheating laser pulse. The photograph illustrates firstly that the corona plasma could never be homogeneous, and is seriously jetting modulated. Secondly, the distribution of the energy assignment of each filament will to a considerable degree be decided by the jet structures, at least in the transient shadowgraph gate period of 50 ps, while the following main heating laser could be arranged to be just arrived.

It could be concluded that the SPMD induced SSSFI process has settled a foundation and framework for transferring to the next ponderomotive force driving SSSFI process, while the average laser irradiances reach a level larger than approximately 5 × 10^{14} W/cm^2. At such levels, the self-focusing filaments have the ability to focus further to the minimum level of the incident laser wavelength scale under the required ponderomotive force driving, resulting in the extremely high irradiance of 10^{16-17} W/cm^2 at the filament nodes or ends, bringing about a huge amount of plasmon generation and decay bursts, accompanying the relativistic electrons and superhard X-ray bursts. All of these effects could stir the plasma into an extremely inhomogeneous form, inducing a strong scattering of the heating laser light, and resulting in a serious reduction of the inverse bremsstrahlung absorption for the designed black cavity (Hohlraum) inner surface.

3. Discussion

3.1. Outcomes and judgments

To summarize the above early precise experimental results and the re-analyzed data, several outcomes and judgments can be extracted as follows.

(1) All of the $3\omega_0/2$ emission features were carefully re-studied, examined, and re-analyzed for the above early experiments, including 2D space, space spectrum, time spectrum, and space–time resolved experimental data. The $3\omega_0/2$ filament diameter was about 4–5 μm. The length was 15 μm–130 μm. And the $3\omega_0/2$ emission existence time at any local position along the filament channel from each individual filament is about 5 ps. Each filament's stimulated to collapsed period is about 10–20 ps.

(2) There were no signs of the $3\omega_0/2$ signal in terms of the CTLPI predicted features that have appeared at the average bulk plasma density $n_c/4$ shell layer until now. These results

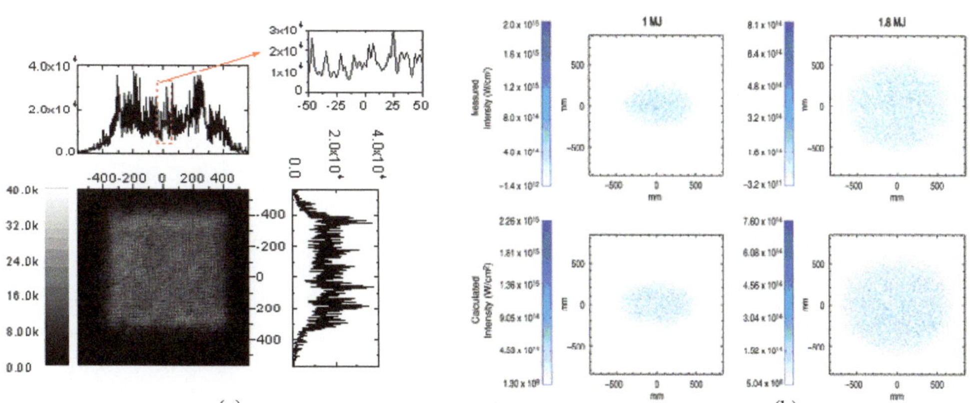

(a) (b)

Figure 9. Two samples of focal spot intensity distribution are given with different spatial smoothing mechanisms. (a) The measured focal spot and profiles along two dimensions of SG-II laser facility from a CCD (charge coupled device) of 1024 × 1024 elements with each pixel size being 13 μm at ~12× magnification when a lens array is applied to the laser beam, and the measured speckle contrast is 0.46 rms. (b) The measured (top) and calculated (below) far-field speckle profile on NIF when CPP is applied to the laser beam, and the measured contrast is 0.79 rms[26, 27].

indicate that the TPD process was not generated, or did not exist, in "the bulk homogeneous plasma environment", and was only stimulated inside the transient small scale self-focused filaments. In this sense, the filamentation instability could be extremely unlikely to show a "flicker" effect.

Moreover, because of the self-focusing effects, a very large part of the most intense heating laser light energy with higher light irradiance has been almost all trapped inside the filament channels transiently, and this could have no opportunity to go over the filament channels horizontally outside to realize their production, growth and saturation process for bulk homogeneous plasma, although, via spatial and temporal averaging, the corona laser plasma seems to be smoothing. In this sense, the unstable small scale filamentation instability should be regarded as the instability of first priority in the LPI corona area.

(3) Each filament can be temporally divided into two sections. For the leading first section, the focusing filament is constructed in terms of a serious self-phase modulation and also serious amplitude fluctuations from the heating laser beam profile with a bit lower power intensity with a larger inner aperture at the filament ends. As soon as the end power intensity increased to the threshold of the ponderomotive force production in the second section, the self-focus process will be accelerated to push forward the leading end electron density to much higher density, to the level of $n_c/4$ or even more, resulting in a transient $3\omega_0/2$ emission at the filament surrounding plasma density of approximately $n_c/20$ to $n_c/30$, shown in figures 2(b), (c), and (d). The appearance of a transient $3\omega_0/2$ emission spatially extending filament line at such a low surrounding plasma density, as one of the most important features of the unstable SSSFI, is among the solid evidence of the massive generation of mainly SPMD and PFD filaments in a low average laser irradiance of the 10^{13} W/cm^2 scale. In addition, it should be understandable that the CTLPI filamentation model would not be able to stimulate the TPD process and the $3\omega_0/2$ emission in the region of very low surrounding plasma density, about $n_c/20$ to $n_c/30$, for the corona area.

(4) After the stimulation of TPD and $3\omega_0/2$ emission in a filament, the rapid decay of the stimulated plasma electron wave by Landau damping or wave breaking occurs, and fast electrons and hot X-rays are emitted, leading to stirring of the corona plasma by the convection between the hot and/or superhot electrons and the surrounding cold electrons, producing more turbulent plasma jets, which will cause more seriously inhomogeneous distribution of plasma, and the acceleration of the next process of the succeeding filamentation process.

(5) The hot X-ray emission phenomena were studied with very high spatial resolution of an X-ray pinhole camera by Willi *et al.*[11] in 1982. A large amount of very spotty small 1.5 keV hot X-ray spots with different distribution to the large scale laser speckles was obtained. The target was a smooth glass plate with the laser irradiance of 3.3×10^{13} W/cm^2. The heating laser far-field irradiation size was 250 μm in diameter and each of the 1.5 keV hot X-ray spot sizes were 4–5 μm. The X-ray spot intensities were varied from weak to strong randomly, which just coincided very well with the SSSFI induced filament features for both sizes and intensities. These phenomena indicated that the heating laser speckles had further split into a huge amount of self-focused small scale filaments with the outer diameter of 4–5 μm. These small scale filaments will be stimulated and collapsed, resulting in the production of the ~1.5 keV X-ray bursts at the ends or nodes.

The further re-analysis of previous work by Willi *et al.*[11] proves that the precise experiments have simultaneously obtained both new evidence of filament node modulation, from figure 6(a), by temporally gated optical shadowgraph technology, and also a feasibly reasonable deduction of the extremely important proof of the hot X-ray emission from the modulated filament nodes or ends in an underdense plasma corona, shown in figure 6(c), from the 1.5 keV band X-ray pinhole imagery.

We note that the recent work by Séguin *et al.* discussed the relation between the filaments and the TPD instability[19], and they concluded that the filaments were not caused by TPD. Firstly, we have not declared that the filaments were caused by TPD. Actually, from the above analysis, we have recognized that the plasma density distribution is extremely inhomogeneous. Even if the laser spot intensity profile is uniform, it is easy to produce small scale self-focusing filaments (see figure 6(d)). Moreover, the laser phase front at the focus plane has already been severely modulated by self-phase modulation. Therefore, in complete LPI processes the first priority should be the filaments driven by self-phase modulation (small scale). TPD will be generated inside some strong filaments where the laser intensity reaches the TPD threshold (usually considered to be about 5×10^{14} W/cm^2). In this case, we regard the serious self-focusing of the filaments as the origin of the TPD process. And TPD will then stir the plasma, further leading to more serious plasma density inhomogeneity. As a result, the laser intensities at the filament channels are able to reach levels of even up to 1–5 $\times 10^{17}$ W/cm^2, resulting in ponderomotive force driven strong filament bunches. For figure 2(d), we consider the filaments in terms of $3\omega_0/2$ light emission. Only when the drive laser is on and interacts with the plasma is there (filamented) $3\omega_0/2$ light emission (via TPD). The temporal existence duration of filaments is ~10 ps, obtained by an optical streak camera with the temporal resolution of around 3 ps (see figures 2(b), 2(c), 2(d), and figure 5(a)). Long time strong small scale filaments were not observed. The filament (Fig. 3 in Ref. 19) observed by Séguin *et al.* is considered in terms of the electromagnetic field. This electromagnetic field is associated with the (filamented) plasma jet. The plasma jet may exist for a long time during the drive laser interaction with the plasma and even long after the end of the drive laser part. The existence of a large number of filamented plasma jets and their long lifetime indicate that the plasma density is very inhomogeneous during and after the drive laser part. Though the average laser intensity outside the filaments is lower than the predicted threshold of TPD instability, the laser intensity in the laser produced filament channels should be several tens of times higher than that outside due to laser self-focusing, and reach the TPD threshold. While Séguin *et al.* consider that the filaments are not caused by the TPD instability due to the average laser intensity being below the TPD threshold, they have not considered laser self-focusing.

3.2. Discussion and comments

The negative and positive effects on the laser fusion physical processes

The unstable SSSFI induced transient TPD and SRS processes play some essentially negative roles on the LPI at corona plasma with respect to the roles played by SBS and RA due to the plasma electron wave collapse processes generating hot and superhot electrons which are accompanied with hard and superhard X-ray emission. The investigation of the connection of the SRS to the unstable SSSFI has not been so precise and deep as that of TPD processes due to the lower space and time resolution of the SRS experiments. However, the temporally resolved SRS spectrum from figure 7(c) and (d) still clearly displays an SRS spectrum with seriously short temporal period filament modulation. These results imply that all of the conclusions from the connection of the SSSFI to the TPD instability could be transformed to qualitative conclusions regarding SRS instability.

For the indirect drive experiment, the heating laser will unavoidably, from the very beginning, construct a huge number of filaments with extremely long lengths, by SPMD induced SSSFI, for the extremely low density plasma inside the cylindrical black cavity. However, each individual filament will hold a much larger fraction of the total of the incident heating laser energy caused by the high Z gold plasma induced stronger inhomogeneous plasma jetting with a relatively long jetting scale horizontally. The succeeding process will accelerate the SRS and TPD instability being transiently stimulated and collapsed at the filament ends, nodes or channel walls, all around the fusion micro-balloon target. Finally, the huge amount of ends or nodes of plasma electron wave collapse, this being accompanied with hot electrons and/or hard X-ray emission, resulting in an increased ablation rate of the fusion target ablator arising from the burst hot electrons and the entropy enhancement of the iced target fuel for those several keV hard X-rays, leading possibly to the scaling of some of the scientific cliffs.

It should be mentioned that it may be the case that the backscattered SRS signals can represent the accurate SRS fraction (percentage) of the total heating laser energy. In this case, is the optical SRS production mechanism completely confirmed for the percentage?

It is well known that a single filament, produced in a very low Z target plasma corona, is able to occupy a rather small fraction of the incident laser energy. This is because the ions with light weight and low charge in laser produced plasmas possess a better balance ability for the plasma electrical field. This is very important for constructing a more homogeneous plasma to avoid the strong filament production for direct drive fusion. In the direct drive case, the low Z ablator material of the fusion target micro-balloon will generate much more electrically balanced corona plasma, leading also to the production of unavoidable, but weak, filaments via the heating laser far-field speckle smoothing technology, resulting in a smaller influence on the implosion target physics process, including weaker fuel preheating and a lower ablation rate for the ablator. Therefore, the smallest energy filaments produced in the lowest Z laser plasma may have the most positive effect for direct drive ignition.

The heating laser wavelength effect

The smallest filament inner aperture in laser produced plasmas has a scale of about one wavelength for the heating laser, resulting in the smallest holding energy in an individual filament obtained by adopting the $3\omega_0$ heating laser being a ninth of the energy of the $1\omega_0$ heating laser. In this sense, choosing the $3\omega_0$ heating laser will be absolutely necessary to avoid the essentially negative effect of the unstable SSSFI.

The other positive effect on the laser fusion physical process

The $3\omega_0/2$ filaments would be formed at the leading half-section of the self-focusing process, induced from the inhomogeneous corona plasma; a small fraction of laser light will be collected inside the filament with moderate laser power intensity, while most of the other fractions of the light will be diverged forward easily by the corona plasma inhomogeneity, bringing about a slightly more uniform illumination on the critical density surface—this effect could be described as the "first" physical uniformization, in order to differentiate from the "second" physical uniformization by the electron thermal conduction between the critical surface and the ablation surface.

As the laser power intensity increases, following the power intensity increase of the incident laser pulse leading edge, a larger fraction of the laser light will trap into the bunched filaments and later join the TPD process transiently, in a picosecond scale period. Moreover, there is still enough light, which can pass forward via the filament outer walls or ends, to diverge and to deposit the light energy randomly on the critical surface. Figure 10 shows that the laser produced a very smooth and flat shock front yielded by a seriously fluctuating irradiance less than around 5×10^{13} W/cm^2, and a rather low laser power intensity on the target surface. Although the laser irradiance distribution was just like that of figure 9(a) with the measured speckle contrast at 0.46 rms, or even larger actually, the laser produced shock front flatness was extremely good due to the two-step physical smoothing mentioned above.

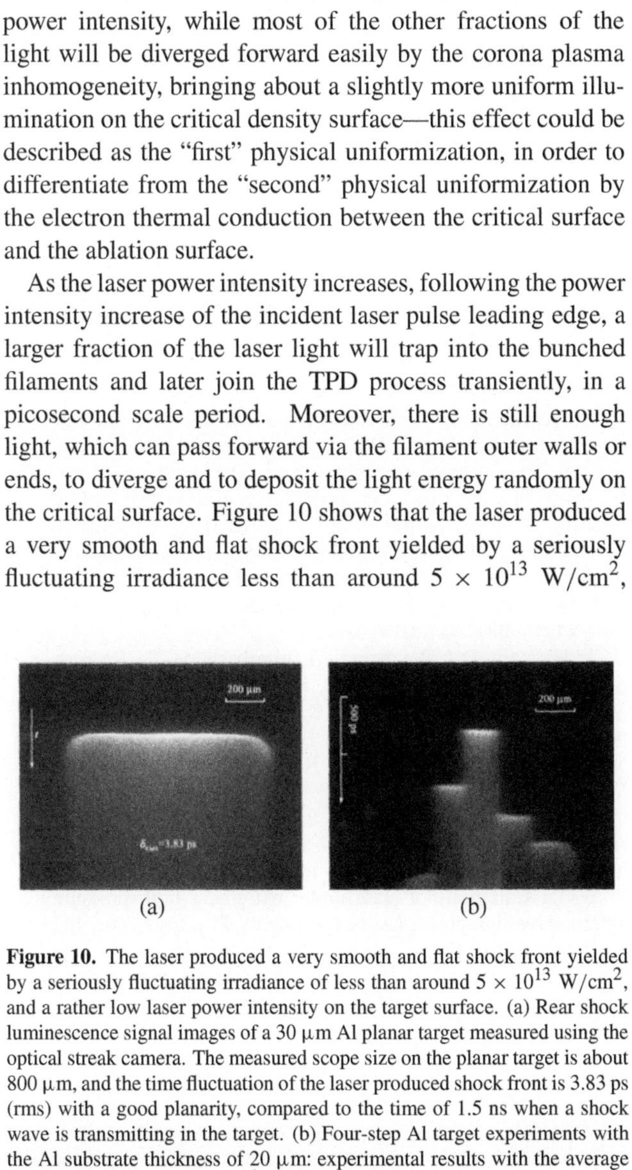

(a) (b)

Figure 10. The laser produced a very smooth and flat shock front yielded by a seriously fluctuating irradiance of less than around 5×10^{13} W/cm^2, and a rather low laser power intensity on the target surface. (a) Rear shock luminescence signal images of a 30 μm Al planar target measured using the optical streak camera. The measured scope size on the planar target is about 800 μm, and the time fluctuation of the laser produced shock front is 3.83 ps (rms) with a good planarity, compared to the time of 1.5 ns when a shock wave is transmitting in the target. (b) Four-step Al target experiments with the Al substrate thickness of 20 μm: experimental results with the average shock velocity of 20.9 km s^{-1}, with the velocity contrast at less than 0.5% and the linear correlation coefficient of 0.999,96. The two results show that planarity of the shock wave can still be maintained although there are many intensity modulations in the focal spot[28].

In the fast collapse phase of the bunch of individual filaments, some percentage of the light energy trapped inside the filaments, which did not take part in the TPD and the $3\omega_0/2$ composition process, will quickly escape outside and divergently deposit its energy on the critical surface.

Therefore, the unstable SSSFI process only has a moderate effect on the uniform target illumination except as regards the too bad boundary conditions for the corona plasma homogeneity.

4. Conclusions

We show that the SSSFI from either the SPMD or the PFD or both are almost the only origins of the TPD instability stimulated and collapsed experimentally in the leading fraction laser pulse produced plasmas in this paper. Also the other parametric instabilities, including SRS, SBS and RA ones, from the re-analysis of the early experimental results, were deduced to support an inseparable similar conclusion.

Actually there always are massive filaments generated and collapsed by the unstable SSSFI process unavoidably in the laser produced plasma corona for each filament lifetime of 10–20 ps. While the laser power intensities at the ends of the strong SPMD filaments reach the PFD self-focusing threshold, the strong filaments will be turned into PFD filaments, which will produce superhot electrons and hard X-rays, affecting the laser fusion physical process.

The SPMD unstable SSSFI itself on one hand ought to play a positive role in being the first laser intensity uniformization in the laser plasma corona, which is beneficial for laser fusion; on the other hand it plays a slightly negative role, due to the TPD collapse produced hot electrons and the 1.5–10 keV hard X-ray bursts in the duration of about 5 ps. The PFD unstable SSSFI process generated filaments will produce superhot electrons and superhard X-rays, and stir the high density plasma seriously, leading to the so called laser imprinting effect and a moderately harmful effect on the laser fusion physics.

To solve the laser irradiance uniformization problem, recently developed procedures including the CPP, DPP, RPP, and LA ones have been widely utilized and have partially improved the LPI physical boundary conditions. A further procedure should be carried out to break the laser field speckles into a quite large number of fine speckles, each with much less energy and intensity for all fine speckles. This smoothing of the laser irradiance will suppress the PFD unstable SSSFI generated filaments and superhot electrons and hard X-rays.

Adopting a very low Z material for laser target illumination is important for realizing the fluid balance more easily and quickly in laser produced plasmas in order to reduce the passively assigned fractions of the incident laser energy in each individual filament, resulting in a further suppression of the process of generation of PFD induced unstable filaments. Therefore a reasonable fusion target design will mitigate superhot electron and hard X-ray generation, bringing about a beneficial effect for laser fusion.

Acknowledgements

The authors would like to acknowledge fruitful discussions with R. Li, B. Shen, and several other colleagues abroad. We thank the referees for their helpful suggestions. We thank B. Chen and D. Huang for assistance with the technical management for this paper.

References

1. D. H. Chandal, Washington, DC 20585, Department of Energy, July 19 (2012).
2. C. S. Liu, M. N. Rosenbluth, and B. B. White, Phys. Fluids **17**, 1211 (1997) K. Estabrook, and W. L. Kruer, Phys. Fluids **126**, 1892 (1983).
3. D. W. Forslund, J. M. Kindel, and E. L. Lindman, Phys. Fluids **18**, 1002 (1975).
4. E. A. Jackson, Phys. Rev. **153**, 235 (1967) C. S. Liu, and M. N. Rosenbluth, Phys. Fluids **19**, 967 (1976); L. V. Power, and R. L. Berger, Phys. Fluids **27**, 242 (1984).
5. J. P. Freidberg, R. W. Mitchell, R. L. Morse, and L. I. Rudsinski, Phys. Rev. Lett. **28**, 795 (1972).
6. R. D. Jones, W. C. Mead, S. V. Coggeshall, C. H. Aldrich, J. L. Norton, G. D. Pollak, and J. M. Wallace, Phys. Fluids **31** (5), 1249 (1988).
7. K. Eidman, M. H. Key, and R. Sigel, J. Appl. Phys. **47**, 2402 (1976).
8. R. A. Hass, H. D. Shay, W. L. Kruer, M. J. Phillion, D. W. Rainer, V. C. Ruper, and H. V. Kornblum, Phys. Rev. Lett. **39**, 514 (1977).
9. H. A. Baldis, and P. B. Corkum, Phys. Rev. Lett. **45**, 1260 (1980).
10. Z. Lin, O. Will, and P. T. Rumsby, J. Phys. D **14**, L35 (1981).
11. O. Willi, P. T. Rumsby, C. Hooker, A. Raven, and Z. Q. Lin, Opt. Commun. **41**, 110 (1982).
12. Z. Lin, O. Willi, W. T. Toner, and A. F. Gibson, Proceedings of 15th ECLIM, Schliersee, January 18–22, D15-P, (1982).
13. Z. Lin, O. Willi, and P. T. Rumsby, Ratherford Report No RL-82-010.
14. Z. Lin, Opt. Commun. **42**, 351 (1982) C. E. Max, Ecole Polytechnique, UCRL-53107 (1981).
15. Z. Lin, W. Tan, M. Gu, G. Mei, C. Pan, W. Yu, and X. Deng, Laser and Particle Beams **4** (2), 223 (1986).
16. Z. Lin, H. Zhang, K. Lin, X. Wang, Y. Zhuang, X. Wei, Q. Lu, A. Shi, M. Dai, L. Tian, G. Fan, and J. Li, Phys. Rev. A **46** (8), 5123 (1992).
17. A. Giulietti, T. Afshar-Rad, S. Coe, O. Will, Z. Q. Lin, and W. Yu, Proceeding of 19th ECLIN–Madrid (1988).
18. E. Higson, R. Trines, J. Jiang, R. Bingham, K. L. Lancaster, J. R. Davies, and P. A. Norreys, New J. Phys. **15** (1), 015027 (2013).
19. F. H. Séguin, C. K. Li, M. J.-E. Manuel, H. G. Rinderknecht, N. Sinenian, J. A. Frenje, J. R. Rygg, D. G. Hicks, R. D. Petrasso, J. Delettrez, F. J. Marshall, and V. A. Smalyuk, Phys. Plasmas **19**, 012701 (2012).
20. M. J.-E. Manuel, C. K. Li, F. H. Séguin, N. Sinenian, J. A. Frenje, D. T. Casey, R. D. Petrasso, J. D. Hager, R. Betti, S. X. Hu, J. Delettrez, and D. D. Meyerhofer, Phys. Plasmas **20**, 056301 (2013).
21. H. Ahmed, M. E. Dieckmann, L. Romagnani, D. Doria, G. Sarri, M. Cerchez, E. Ianni, I. Kourakis, A. L. Giesecke, M. Notley, R. Prasad, O. Willi, and M. Borghesi, Phys. Rev. Lett. **110**, 205001 (2013).
22. J. H. Campbell, R. T. Maney, L. J. Atherton, R. C. Montesanti, J. J. DeYoreo, L. M. Sheehan, M. R. Kozlowski, and C. E. Barker, Large-aperture high-damage-threshold optics for beamlet, ICF Quarterly Report 5 (1), 1~85, LLNL, CA, UCRL-LR-105821-95-1 (1995).
23. R. Jedamzik, and F. Elsmann, Recent results on bulk laser damage threshold of optical glasses, Proc. SPIE 8603, High-Power Laser Materials Processing: Lasers, Beam Delivery, Diagnostics, and Applications II, 860305 Feb. 22, (2013), DOI: 10.1117/12.2000646.
24. W. Koechner, 1999 *Solid-state laser engineering*, 5th edn pp. 706. Springer-Verlag, Berlin, Heidelberg, and New York.
25. Shenlei Zhou, private communication. The small scale self-focusing filaments in rod amplifier are demonstrated at intensity around 5×10^9 W cm^{-2} at SG-II laser facility at the National Laboratory on High Power Laser and Physics, Shanghai Institute of Optics and Fine Mechanics.
26. Y. Jiang, S. Zhou, R. Wu, J. Li, X. Li, and Z. Lin, Chin. Opt. Lett. **11** (8), 081404 (2013).
27. C. A. Haynam, P. J. Wegner, J. M. Auerbach, M. W. Bowers, S. N. Dixit, G. V. Erbert, G. M. Heestand, M. A. Henesian, M. R. Hermann, K. S. Jancaitis, K. R. Manes, C. D. Marshall, N. C. Mehta, J. Menapace, E. Moses, J. R. Murray, M. C. Nostrand, C. D. Orth, R. Patterson, R. A. Sacks, M. J. Shaw, M. Spaeth, S. B. Sutton, W. H. Williams, C. C. Widmayer, R. K. White, S. T. Yang, and B. M. Van, Appl. Opt. **46** (16), 3276 (2007).
28. Y. Gu, S. Fu, X. Huang, J. Wu, J. Ye, H. Shu, M. Ma, J. He, and S. Wang, Equations of state under extremely high pressure obtained by high power laser experiments. Physics **36** (6), 465 (2007) (in Chinese).

22

Energy measurement system of a large-aperture high power laser experiment platform

Yanwen Xia[1], Yue Liang[1], Sen Li[1], Junpu Zhao[1], Zhitao Peng[1], Hongguang Li[2], Hua Liu[1], Zhihong Sun[1], Kuixing Zheng[1], and Xiaofeng Wei[1]

[1] *Research Center of Laser Fusion, CAEP, Mianyang 621900, China*

[2] *Xi'an Institute of Optics and Fine Mechanics, Chinese Academy of Sciences, Xi'an 710119, China*

Abstract

An energy measurement system in a large-aperture high power laser experiment platform is introduced. The entire measurement system includes five calorimeters, which carry out the energy measurement of the fundamental frequency before the frequency conversion unit, remaining fundamental frequency, remaining second-harmonics, third-harmonics, as well as the energy balance measurement after the frequency conversion unit. Combinational indirect calibration and direct calibration are employed to calibrate the sampling coefficients of the calorimeters. The analysis of the data showed that, regarding the energy balance coefficients, combinational calibration approach gives a higher precision, and leads to an energy balance with 1%; and regarding the energy sampling coefficients for the various wavelengths after the frequency conversion, the results from direct and combinational calibration are consistent. The uncertainties for all energy sampling coefficients are within 3%, which guarantees the reliability of the energy measurement for the laser facility.

Keywords: Laser diagnostics; pulse energy; calorimeter; high power laser

1. Introduction

The maximum output performance of the traditional high power laser facility[1–4] is often limited by the damage of the optical units, components and the entire system under high energy flux level. This is especially so for ultraviolet optical components. Various studies show that, not only the third-harmonics, but also the remaining fundamental frequency, and the second-harmonics also have a serious impact on the optical component damages[4–6]. Therefore, in order to systematically study the load capacity of the optical equipment, component and system under high intensity, it is necessary to have a high-precision measurement platform[7–9]. Not only does this platform have to have the sophisticated capabilities for laser parameter diagnosis, online damage detection for optical components, but it must also be able to comprehensively measure energy, which can obtain the energy of the remaining fundamental frequency, remaining second-harmonics, and third-harmonics entering the optical components.

This article introduces an energy measurement system for large-aperture High Power Laser Experiment Platform[10,11]. This system is composed of the sampling optical path and energy calorimeters, which are used to measure the energy of a single wavelength before and after the frequency conversion. We also show how to calibrate the calorimeter sampling coefficients.

Correspondence to: Yanwen Xia, Research Center of Laser Fusion, CAEP, P.O. Box 919-988, Mianyang 621900, China. Email: mengxiangfu07@163.com

2. System optical paths

Our large-aperture High Power Laser Experiment Platform is a comprehensive experiment verification platform that can output more than tens of thousands of joules. It can be used for comprehensively evaluating the load capacity of the optical equipment, component and system under high energy flux level. The optical paths of its energy measurement system are listed as in Figure 1. The laser from the last lens of the transmission spatial filter is divided into two paths: about 0.2% goes back the same optical path by sampling wedge as reflected light. Around the filtering holes in the spatial filter, the negative lens changes the beam aperture from 400 mm × 400 mm to 50 mm × 50 mm parallel light. Then it enters the calorimeter in the main amplifier diagnostic module, where the fundamental frequency laser energy is measured. After the main laser transmitted from the sampling wedge passes the vertically placed reflection mirrors ZM1 and ZM2, the height of the laser beam was lowered down to about 1.2 m above the ground. At the same time, the optical path is rotated for 90°. And then

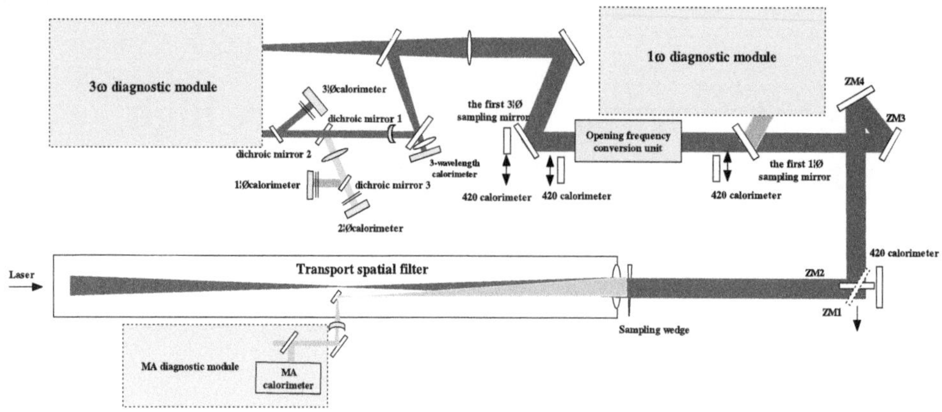

Figure 1. Schematic view of the energy diagnostic system of the large-aperture High Power Laser Experiment Platform.

it enters the ZM3 and ZM4 both at an angle of incidence of 22.5°, where the laser is turned to go parallel with the transport spatial filter. After being transmitted by the fundamental frequency sampling mirror 1, the light goes into the frequency conversion unit. The mixed parallel laser beams after the crystal conversion (remaining fundamental frequency, remaining second-harmonics, third-harmonics) enter the third-harmonics sampling mirror 1. After the reflected light is reflected and sampled by the third-harmonics sampling mirror 2, it enters the large-aperture beam shrink optical path, where the light beam is transformed into a 50 mm × 50 mm parallel light to complete the relevant parameter diagnosis of the third-harmonics. Its transmitted light is then absorbed by the absorber. An interface to a large-aperture calorimeter is placed in front of the absorber, so it can calibrate the total energy when needed.

There are two reflection mirrors in the third-harmonics system to turn the optical path. After being transmitted from the second reflection mirror, and being collimated by the three-wavelength collimation lens, it enters the three-wavelength calorimeter. That accomplishes the sampling of all three wavelengths.

The light through the third-harmonics beam shrink system is a mixture of fundamental frequency, second-harmonics and third-harmonics. After being reflected by dichroic mirror 1 and dichroic mirror 2, the light enters third-harmonics calorimeter. The purpose of dichroic mirror 1 is to only transmit third-harmonics light, and only reflect fundamental frequency and second-harmonics light; dichroic mirror 2 is a half-transmission-half-reflection mirror. At the same time, it serves as a wave aberration compensation for any measurement that follows.

After the mixed light is reflected by dichroic mirror 1, it becomes a mixture of fundamental frequency and second-harmonics. Due to the chromatic aberration, it needs to be collimated. After being converged by the energy convergence lens, the light was projected on dichroic mirror 3. Its reflected light enters the fundamental frequency calorimeter, and the transmitted light enters the second-harmonics calorimeter.

A filter was placed in front of each single-wavelength calorimeter, to ensure that the light entering the calorimeter is actually of a single wavelength.

Based on the modularization design requirement, all the calorimeters in the system use uniform calibers, and are composed of energy sensors, amplifiers, and display modules. The entire energy measurement system is centrally sampled and processed through remote control[12,13].

3. System calibration

For the fundamental frequency part of the main optical path, the energy values of two key points need to be known: the main amplifier output energy E (after the sampling wedge in Figure 1), and the energy before the crystal in the frequency conversion unit E_{in} . They can both be given by the main amplifier calorimeter. Therefore it is necessary to calibrate based on the sampling coefficients[14]. First, we take the light guide ZM1 out of the optical path, and replace it with a standard calorimeter with a caliber of 420 mm × 420 mm, (420 standard calorimeter), to calibrate the main amplifier output energy. The sampling coefficient of the main amplifier calorimeter is given by: $k_{MA} = E_{420}/W_{MA}$, where E_{420} is the energy given by the 420 calorimeter; W_{MA} is the main amplifier calorimeter reading. Then we place the 420 calorimeter in front of the frequency conversion unit. The same approach is used on the main amplifier calorimeter to calculate the energy sampling coefficient, k_{in}, before the frequency conversion unit. Therefore the main amplifier output energy, E_{MA}, as well as the energy before the crystal in the frequency conversion unit, E_{in}, are both given by the main amplifier calorimeter: $E_{MA} = k_{MA}\ W_{MA}$, $E_{in} = k_{in}\ E_{MA}$.

Here are four calorimeters after the frequency conversion unit. They are used to measure the energy of the remaining fundamental frequency, remaining second-harmonics, third-harmonics, and to sample the total energy for the purpose of energy balance. Their relationship can be shown in Figure 2. Here the single-wavelength calorimeter readings are from after the crystal. Energy balance refers to that the

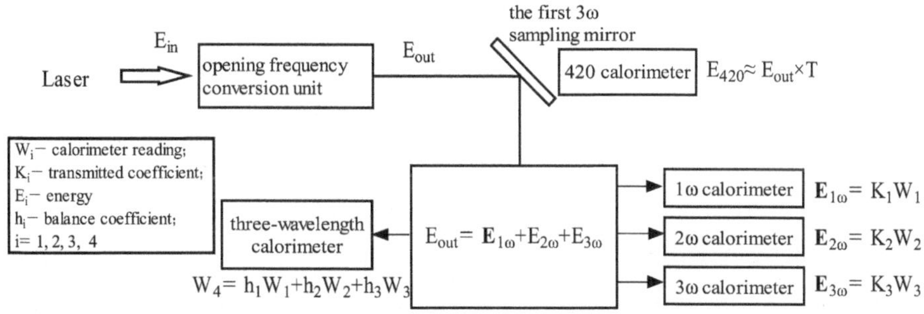

Figure 2. After-crystal energy relationship diagram.

sampling coefficients of the single-wavelength calorimeters maintain a constant quantitative relationship, which can be used to monitor the reliability of various sampling coefficients. Since there is no target focusing lens in the frequency conversion unit, it is difficult to achieve harmonic separation[3,4]. Therefore they cannot be calibrated using the traditional perforation method. Two different methods are employed here: direction and combinational calibration.

3.1. Direct calibration

We place the 420 standard calorimeter after the third-harmonics sampling mirror 1, and insert an absorbing glass element between the frequency conversion unit and the mirror 1, (two 1ω- and one 2ω- absorbing glass), to let only third-harmonics light pass for the main laser shot. That makes only third-harmonics calorimeter and the three-wavelength calorimeter have readings, and leads to the sampling coefficient of the third-harmonics $K_3'(E_{420} = K_3'W_3)$, as well as the energy balancing coefficient $h_3(W_4 = h_3W_3)$. Then we adjust the third-harmonics crystal, and only let fundamental frequency and second-harmonics pass the frequency conversion. Then we remove the 2ω absorbing glass, and only let second-harmonics pass. That leads to the sampling coefficient of the second-harmonics $K_2'(E_{420} = K_2'W_2)$, and its energy balancing coefficient, $h_2(W_4 = h_2W_2)$. Last we disable the second-harmonics, only the fundamental frequency can pass the frequency conversion unit. After taking down all the absorbing glass components, we use the main laser shot to obtain the sampling coefficient of the fundamental frequency, $K_1'(E_{420} = K_1'W_1)$, and the energy balancing coefficient, $h_1(W_4 = h_1W_1)$.

Then, based on the energy transmission–reflection rate of different wavelengths from the online calibration of the third-harmonics sampling mirror 1, we get the single-wavelength sampling coefficients, K_1, K_2, and K_3 after the third-harmonics crystal, and before the third-harmonics sampling mirror 1.

3.2. Combinational calibration

In the direct calibration approach, often we have a small number of calibrated laser shot missions. Also because the inserted absorbing glass element is difficult to guarantee a single-wavelength output, it results in a fair amount of dispersion in the sampling coefficient. Therefore we can also employ combinational calibration approach. There is no need to insert additional filters. The steps to calibrate the sampling coefficients are as follows: first, we place the 420 calorimeter after the third-harmonics sampling mirror 1. Within the range of $0.8 \sim 4$ kJ (the valid measurement range of the 420 calorimeter), we take incident fundamental frequency at different energy levels. We record the total energy from three wavelengths, and use the other four calorimeters to record the readings from the remaining fundamental frequency, remaining second-harmonics and third-harmonics. Then the relationship between the calorimeter readings can be expressed as follows:

$$E_{420i} = K_1'W_{1i} + K_2'W_{2i} + K_3'W_{3i} \tag{1}$$

$$W_{4i} = h_1W_{1i} + h_2W_{2i} + h_3W_{3i} \tag{2}$$

where $i = 1, 2, \ldots$ are the different laser shot index numbers. Then the total sampling coefficients and the energy balancing coefficients satisfy the following matrix relationship:

$$\begin{pmatrix} K_1' \\ K_2' \\ K_3' \end{pmatrix} = \begin{pmatrix} W_{1i} & W_{2i} & W_{3i} \\ W_{1j} & W_{2j} & W_{3j} \\ W_{1k} & W_{2k} & W_{3k} \end{pmatrix}^{-1} * \begin{pmatrix} E_{420i} \\ E_{420j} \\ E_{420k} \end{pmatrix} \quad \begin{pmatrix} h_1 \\ h_2 \\ h_3 \end{pmatrix} = \begin{pmatrix} W_{1i} & W_{2i} & W_{3i} \\ W_{1j} & W_{2j} & W_{3j} \\ W_{1k} & W_{2k} & W_{3k} \end{pmatrix}^{-1} * \begin{pmatrix} W_{4i} \\ W_{4j} \\ W_{4k} \end{pmatrix} \tag{3}$$

where i, j, and k are the numbering of any single laser shot. Then we use linear square to run linear regression, and give average coefficients and errors. Lastly, we process the obtained the energy sampling coefficient, K_1, K_2 and K_3 using the same approach.

4. Results and analysis

Table 1 shows the energy balancing coefficients from two calibration approaches. Figure 3 shows the corresponding energy balancing relationship. The hollow star in the figure means the energy balancing relationship obtained through direct calibration, and their relative errors. And the laser energy for only one wavelength exists after the frequency

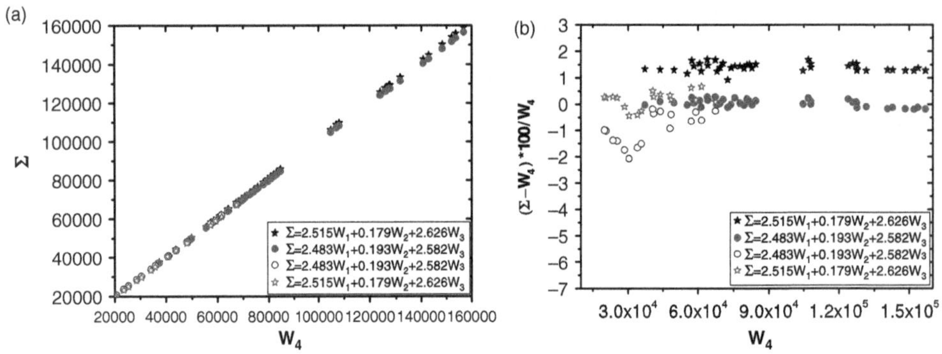

Figure 3. Comparison of the energy balancing relationship between the two approaches. (a) The relationship curve between $\Sigma = h_1W_1 + h_2W_2 + h_3W_3$ and energy reading from three-wavelength calorimeter reading W_4 and (b) their corresponding relative errors.

Table 1. Energy balancing coefficients from two approaches.

Items	h_1	h_2	h_3
Direct Calibration	2.51468	0.17886	2.62612
Combinational Calibration	2.48289	0.19274	2.58239

conversion at each laser shot. A solid star means the energy balancing coefficients are obtained through indirection calibration applied on mixed light. Here at each laser shot, the energy of all three wavelengths exists after the frequency conversion (remaining fundamental frequency, remaining second-harmonics and third-harmonics). A solid dot means to apply indirection combinational calibration on the raw data corresponding to the solid star to get the energy balancing relationship. A hollow dot means to apply the combinational calibration results on the raw data from hallow star to obtain the energy balancing relationship.

We can see from Figure 3 that, with each calibration shot mission, for the direct calibration marked by the hollow star and the combinational calibration marked by solid dot, their obtained energy balancing relationship $h_1W_1 + h_2W_2 + h_3W_3$ and their three-wavelength calorimeter reading W_4 match very well. All relative errors are within 1%. But if we apply the coefficients from the direct calibration on a normal laser shot mission, where all remaining fundamental frequency, second-harmonics and third-harmonics exist, the combinational result $h_1W_1 + h_2W_2 + h_3W_3$ and W_4 have a systematic error, and the relative error is about 1.4%. And if we apply the energy balancing relationship from the combinational calibration on the single-wavelength output, we also see systematic errors, in an opposite direction.

There are two areas of reasons for this situation. First there is a sensitivity difference in the calorimeters themselves. Since under the normal phase matching condition, the second-harmonics light from the harmonics crystal are almost all converted into third-harmonics with the crystal, the remaining second-harmonics light is far less than the remaining fundamental frequency and third-harmonics light. The design of the diagnosis system takes many factors into consideration, so that the energy sampling coefficients have a relatively smaller difference before entering the photosensitive surface in the calorimeter. Therefore, in order to guarantee the measurement precision of the second-harmonics energy, the sensitivity to the second-harmonics is about 10 times of that in the remaining fundamental frequency, third-harmonics, three-wavelength calorimeters, all of which have similar sensitivities. So during direct calibration, taking the measurement range of the 420 calorimeter into consideration, the actual energy value that enters the second-harmonics calorimeter is at the center of the linear region of this calorimeter[15]. But that is at the bottom of the three-wavelength calorimeter linear region. Therefore this leads to higher changes in h_2 and higher dispersions from different laser shots. Another reason is that even though absorbing glass is inserted into the optical path, it is difficult to get clean third-harmonics. Also we only have limited number of shot missions, these lead to a system error in the obtained average coefficients.

Based on the analysis above, the error from combinational calibration approach is smaller. The obtained energy balance relationship is with 1%. We can use it as the guidance on the accuracy of the energy measurement. Once the deviation exceeds this value, the reason is either the measured energy values are not within the linear regions of the calorimeters, or there is a change in the energy balancing relationship. For the latter, we need to check the optical component damage in optical path.

In the process of energy balancing coefficient calibration, only the readings from four calorimeters need to be recorded. There is no need to record the input energy levels. In order to obtain the input energy level, we need to calibrate the energy sampling coefficients of the optical path. Due to the limited number of laser shot missions, we only apply one group of calibration data, that from the direct calibration, in both groups of results. The result can be seen in Table 2 and Figure 4. Both approaches give out similar results.

In order to verify the accuracy of the calibration data, we accumulate the laser shot results that follow. More specifically, we use the main amplifier energy to backward calculate the total energy after the crystal, E_{out}. After comparing it to the after-crystal energy values from the sampling co-

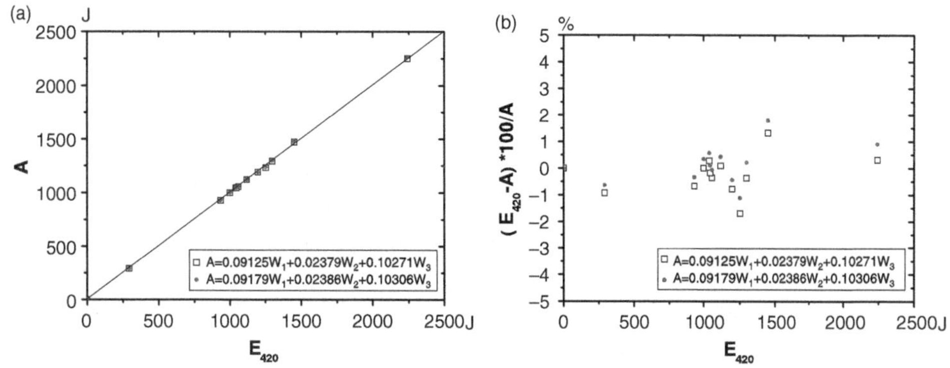

Figure 4. Comparison between the energy balancing relationships from two approaches. (a) The relationship curve between $A = K_1'W_1 + K_2'W_2 + K_3'W_3$ and energy reading from the 420 calorimeter, E_{420}, in back of the third-harmonics sampling mirror 1 and (b) their corresponding relative errors.

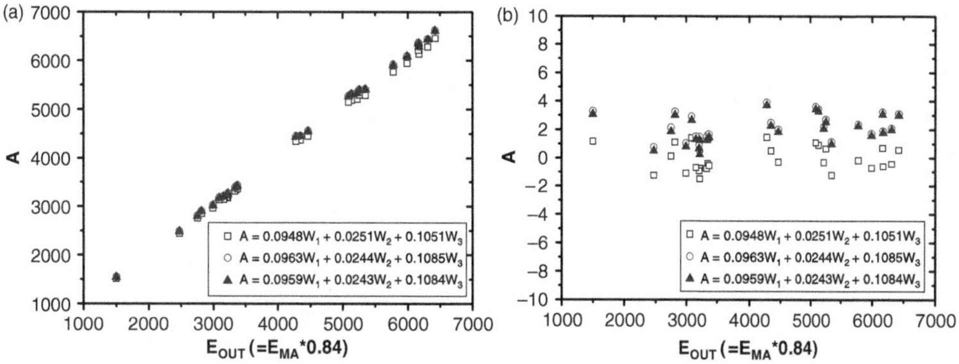

Figure 5. The relationship curve (a) between the total output energy from calibration and that derived from the main amplifier, and (b) their corresponding relative errors.

Table 2. The energy sampling coefficients from two approaches.

Items	K_1'	K_2'	K_2'	K_1	K_2	K_3
Direct Calibration	0.09179	0.02386	0.10306	0.0963	0.0244	0.1085
Combinational Calibration	0.09141	0.02381	0.10303	0.0959	0.0243	0.1084
Coefficients from MA calorimeter	–	–	–	0.0948	0.0251	0.1051

efficient combination, we observe the consistency between the two results. The energy transmission coefficient between the main amplifier and the frequency conversion unit is $k_{in} = 0.95$. The transmission coefficient of the frequency conversion unit is $T = 88\%$. We assume the transmission rate under different frequency conversion efficiencies stays the same, then the after-crystal energy, E_{out}, can be approximated as: $E_{out} = TE_{in} = Tk_{in}E_{MA} \approx 0.84E_{MA}$.

The experiment results can be seen in Figure 5. The hollow squares show that relationship between the derived energy from the main amplifier energy E_{out}, and combined energy from three-single wavelength calorimeter readings using the combination approach. Hollow circles represent the energy sampling coefficients from the direct calibration, and solid triangles represent the results from the combinational calibration.

We can see from Figure 5 that, there is a 2% system error between the output energy of the frequency conversion unit using the calibration, and the total output energy derived from the main amplifier. The dispersion of the average deviation is about 2%.

To explain such dispersions, besides the size of our data, also lies within the fact that the derived energy from main amplifier E_{out} is approximate. Since the crystal has a rather large absorption to the fundamental frequency light, there are some differences in the transmission rate from the crystal in the frequency conversion unit for different harmonics conversion efficiencies. Figure 6 shows the relationship curve between the main amplifier energy, E_{MA}, representing the total input energy E_{in}, and the actual measured total energy output, E_{out}, under various conversion efficiencies. The different marks in the graph represent only the measurement results at different shot periods. Apparently, under different conversion efficiencies, the transmission coefficients in the frequency conversion component have variations. The deviation is around ±3%. Therefore the dispersion in Figure 5

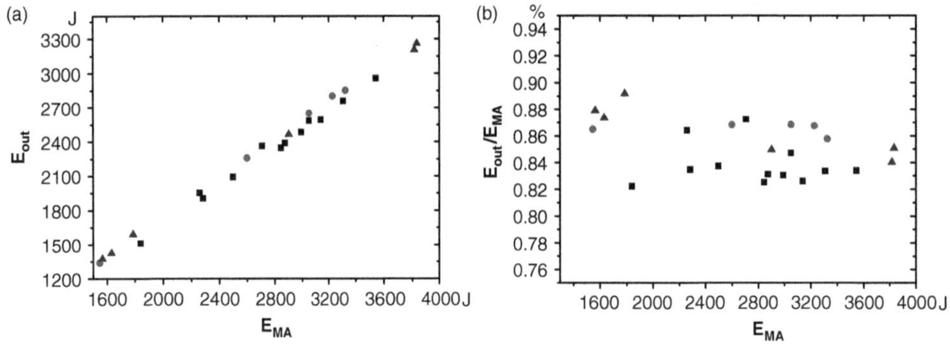

Figure 6. Under different harmonics conversion efficiencies, the relationship curve between (a) the main amplifier energy and the total output energy from the crystal, and (b) their corresponding relative errors.

only represents the deviations in E_{out}. It is also the error introduced by the transmission coefficients of the frequency conversion unit.

Regarding the system error, the main reason lies in that the fundamental frequency transmission rate in the frequency conversion unit is 88%, (no frequency conversion), in place of the total transmission rate of the frequency conversion unit. When there is significant harmonics conversion efficiency, the total transmission rate increases. That is also the reason why the third-harmonics sampling coefficients derived from the main amplifier is relatively smaller.

5. Conclusions

This article introduces the energy measurement system in a large-aperture high power laser experiment platform, and its calibration process. The data analysis shows that, regarding the energy balancing coefficients, results obtained from the combinational approach has a higher precision. The energy balance can be controlled within 1%. It can accurately monitor the reliability of the energy sampling efficiency. Regarding the energy sampling coefficients to all the wavelengths after the frequency conversion, the results from direct and the combinational calibration approaches are consistent. But due to the limited number of laser shot missions, there is about 2% system error in the total output energy, along with 2% dispersion. However, the results are guaranteed to ensure the single-wavelength energy measurement controlled within the 4.7% uncertainty range. To further improve the energy measurement accuracy of the third-harmonics, it is necessary to have a large number of laser shot statistics. This will be the focus of our future work.

Acknowledgement

This work was supported by the National Natural Science Foundation of China (No. 61377102).

References

1. S. C. Burkhart, W. C. Behrendt, and I. Smith, UCRL-LR-105821-95-1.
2. X. M. Zhang, W. G. Zheng, and X. F. Wei, Proc. SPIE **562**, 7 (2005).
3. N. Fleurot, C. Cavailler, and J. L. Bourgade, Fusion Eng. Des. **74**, (2005).
4. Z. Q. Lin, Chin. J. Lasers **37**, 9 (2010).
5. C. A. Haynam, P. J. Wegner, J. M. Auerbach, M. W. Bowers, S. N. Dixit, G. V. Eebert, G. M. Heestand, M. A. Henesian, M. R. Hermann, K. S. Jancaitis, K. R. Manes, C. D. Marshall, N. C. Mehta, J. Menapace, E. Moses, J. R. Murray, M. C. Nostrand, C. D. Orth, R. Patterson, R. A. Sacks, M. J. Shaw, M. Spaeth, S. B. Sutton, W. H. Williams, C. C. Widmayaer, R. K. White, W. T. Yang, and B. M. Van Wonterghem, Appl. Opt. **46**, 16 (2007).
6. P. Martin, A. Morono, and E. R. Hodgson, J. Nucl. Mater. 329 (2004).
7. A. Conde, T. Alger, S. Azevedo, J. Chang, S. Glenn, L. Kegelmeyer, J. Liebman, and P. Whitman, UCRL-PROC-236458 (2007).
8. Paul J. Wegner and Bruno M. Van Wonterghem, UCRL-JC-123070.
9. C. E. Thompson, D. E. Decker, and C. F. Knopp, UCRL-JC-130034 (1998).
10. Z. T. Peng, X. F. Wei, H. Y. Yuan, X. J. Fu, D. H. Chen, Z. H. Sun, H. Liu, and L. B. Xu, Infrared Laser Eng. **40**, 1111 (2011).
11. Y. W. Xia, Z. H. Sun, Z. S. Da, J. P. Zhao, Z. T. Peng, X. N. Dong, K. X. Zheng, and H. Liu, Acta Optica Sin. **33**, s112003 (2013).
12. M. Xia, X. Shan, J. Q. Gao, and W. Sun, Acta Meteorol. Sin. **30,6A**, 112 (2009).
13. T. Xu, J. Yu, Q. M. Fan, and Y. Q. Deng, Acta Meteorol. Sin. **30, 6A**, 116 (2009).
14. J. F. Wei, K. Zhang, S. S. Qian, X. Y. Gao, S. Zhou, and D. H. Zhou, High Power Laser Part. Beams **19,7**, 1103 (2007).
15. J. Yu, L. M. Xiong X Tao, Q. M. Fan, Y. Q. Deng, and Y. P. Zhang, Acta Meteorol. Sin. **30,6A**, 80 (2009).

23

Mechanistic study of continuous polishing

Xiang Jiao[1,2], Jianqiang Zhu[1], Quantang Fan[1], and Yangshuai Li[1,2]

[1] *Key Laboratory on High Power Laser and Physics, Shanghai Institute of Optics and Fine Mechanics, Chinese Academy of Sciences, Shanghai 201800, China*

[2] *University of Chinese Academy of Sciences, Beijing 100049, China*

Abstract

To establish the mechanism of surface change in a continuous polishing system, an ideal mathematical model is built based on Winkler's hypothesis and the Preston equation. The basic features of the model are the change rates in the surface peak–valley (PV) values of the workpiece, conditioning disk and pitch lap, rather than the specific surface shapes. In addition, an equilibrium state exists in the system, indicating that the surface change rates are all zero. Under equilibrium, the surface of the lap could remain flat, and it is insensitive to the surface error of the workpiece. These characteristics lay the theoretical foundations for high-efficiency and high-precision polishing. The methods to obtain an equilibrium state with flat surfaces are then proposed and confirmed experimentally. High-precision surfaces better than $\frac{\lambda}{10}$ ($\lambda = 632.8$ nm) are consistently produced experimentally.

Keywords: continuous polishing; equilibrium state; high accuracy; high efficiency; surface figure control

1. Introduction

Continuous polishing, which is widely used in optical processing, is very important to obtain high-precision and super-smooth surfaces. This process is also applied in high-power laser facilities[1, 2] such as the National Ignition Facility in America and the Ligne d'Integration Laser in France, which require numerous large-aperture slabs with minimal defects and low roughness. To date, the surface shape has been difficult to control accurately in continuous polishing. Polishing with polyurethane laps is also widely used, in which the interactions between the workpiece and the lap are similar to those in continuous polishing. This type of polishing method is limited by the different radial positions of the lap, which have different friction distances from the workpiece. This characteristic causes the lap to deteriorate regardless of the surface shape of the workpiece, even if it is an ideal plane. Scientists and engineers from the Lawrence Livermore National Laboratory have proposed a new polishing method called convergent polishing[3, 4]. In this new method, septa with specific shapes are applied to compensate for the uneven wear to the polishing lap so the surface that has been conditioned can be maintained for a long period. This study won the 2014 R&D100 Award. However, the new method has a drawback; that is, the lack of autorotation of the septum such that the polishing lap and the septum fit perfectly only if the lap is rotationally symmetric, which may cause the surfaces to become aspheric. Continuous polishing differs from polyurethane polishing because the former uses a conditioning disk. According to the study results, the shape of the pitch lap surface in continuous polishing can also be maintained for long periods with a particular position of the conditioning disk, indicating an equilibrium state. The ideal mathematical model of continuous polishing is built first in this paper. The system characteristics are discussed, and the theoretical foundations for high-efficiency and high-precision polishing are established. Continuous polishing is also established as an important step for a deterministic process.

Correspondence to: X. Jiao, No. 390, Qinghe Road, Jiading, Shanghai, CN 201800, China. Email: ziyoudeyunduo@126.com

2. Ideal continuous polishing model

The continuous polishing system consists of a massive rotating table with a polishing annulus, a conditioning disk and workpieces. Preston's equation[5], which is widely used to calculate the material removal rate for the polishing process, can be described as follows:

$$h = C \cdot \int_{t_1}^{t_2} P(t) \cdot V(t)\, \mathrm{d}t, \tag{1}$$

where C is a constant of proportionality; P, which changes over time t, is the local pressure; and V, which may also change over time, is the relative velocity.

In the continuous polishing process, the material removal equation is applied to the workpiece and the conditioning disk with different constants C. To a widely used pitch lap, this feature is similar to a series of independent springs because of the grooves on the surface. The lap is simplified to a one-dimensional Maxwell fluid based on the thickness, so the Preston constant considers lap wear, and viscous flow can be set for the pitch lap.

Winkler's hypothesis is applied to calculate the contact pressure for simplification and can be written as follows:

$$y = \frac{P}{k}, \tag{2}$$

where y is the deformation in the thickness of the lap, P is the pressure, and k is the elastic coefficient, which can be expressed as the ratio of the Young modulus to the lap thickness. Given the relative motion between the lap and workpiece/conditioning disk, as well as the considerable viscosity of the pitch, the contact pressure is determined by elastic mechanics rather than viscous elastic mechanics, even if the pitch lap is viscoelastic. A continuous polishing system can be described by Winkler's hypothesis because of the small thickness and the grooves in the lap surface.

The following assumptions should also be considered prior to modeling:

(1) The lap, workpiece and conditioning disk rotate at a synchronous angular velocity.

(2) Given the low velocity, hydromechanical effects are ignored, which suggests that the slurry film on the lap surface cannot influence the contact condition.

(3) The frictional force caused by the workpiece or the conditioning disk will not influence the deformation in the thickness of the lap.

(4) The workpiece and the conditioning disk will not deform, so they are considered rigid bodies. Similarly, the lower surface of the lap will also not deform.

(5) The surfaces of the workpiece, conditioning disk and lap are rotationally symmetric.

Given the force balance of the workpiece and the conditioning disk, the contact pressure distribution can be determined on the basis of Winkler's hypothesis and assumption (4). The changes in the pitch lap surface, workpiece and conditioning disk can then be established. To simplify the calculation, the contact pressure is assumed to remain unchanged in a cycle, and the surfaces at the end of the cycle are modified. A computer is used to control the surface according to the program flow chart in Figure 1, where H and θ correspond, respectively, to the relative position between the workpiece or conditioning disk and the lap surface in the thickness direction, and the dip angle of the workpiece or conditioning disk. F_N and M_o correspond to the sustaining

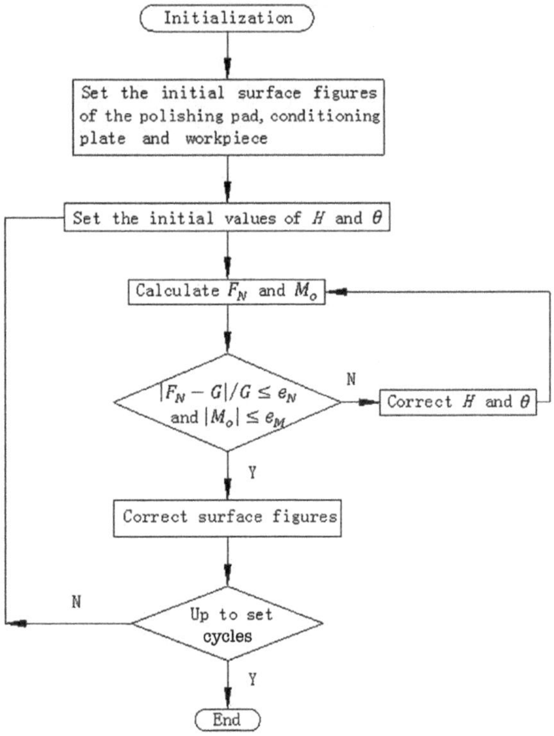

Figure 1. Program flow chart for surface calculation.

strength and sustaining moment to the workpiece or the conditioning disk determined by H and θ, which can be carried out though numerical integration by dividing the contact area into many elements. For convenient calculation, the grid is generated by the radial lines and concentric circles of the lap surface. The mesh density should be sufficiently high to guarantee computational accuracy, which can easily be verified by increasing the density and comparing the results. G is the weight of the workpiece or the conditioning disk. e_N and e_M are the given upper limits of calculation errors, which are positive numbers approximating to zero. The error criterion is given by

$$\frac{|F_N - G|}{G} \leqslant e_N, \quad |M_o| \leqslant e_M. \tag{3}$$

The initialization includes every size of the lap, conditioning disk and workpiece, the position and density of the conditioning disk and workpiece, Young's modulus of the lap, the Preston constants and rotation number, calculation errors, increments for H and θ, and some parameters for integral operation and output file path.

Setting the initial surface figure refers to setting the profiles along the radial directions of the workpiece, lap and conditioning disk. The profiles can be described as functions or height values at given radial positions in this program.

The workpiece and conditioning disk have different values of H and θ, but we consider that the computing method for both of them is identical, so H and θ refer to both the workpiece and conditioning disk. On the basis of the scanning and variable step method, values of H and θ that satisfy Equation (3) are easily found. For instance, we define an increasing value of H as standing for a decreasing amount of compression for the lap caused by the workpiece and the conditioning disk, with θ being positive standing for the outer part of the workpiece and the conditioning disk on the lap surface being lower than the inner part. If the initial given H and θ result in the total pressure being greater (smaller) than the weight G and the moment for the outer part of the workpiece or the conditioning disk being larger (smaller) than the inner part, then H should be increased (reduced) and θ should be reduced (increased) until the comparison reverses, following which H should be reduced (increased) and θ should be increased (reduced). If every time the comparison reverses, the increment applied to H or θ is multiplied by a constant smaller than unity, then Equation (3) will soon be true to a given accuracy. The initial values of H, θ and the increments are easy to set – for example, setting H to the thickness of the lap, θ to zero, the increment for H to 10^{-6} m and the increment for θ to 10^{-6} degrees, their values will soon be updated such that Equation (3) is satisfied.

A large number of numerical tests have been performed to verify the accuracy and consistency of the model. For example, it has been found that increasing the Preston constants causes faster changes in the surfaces, and changes in the surface figure from different initial shapes, such as concave, convex, 'M'-shape and 'W'-shape, are found to be consistent with experiment.

3. Study of the continuous polishing characteristics

3.1. Basic rules of surface figure adjustment

According to the model, two basic rules are found:

(1) Without moving the conditioning disk, the surface peak–valleys (PVs) of the pitch lap, conditioning disk and workpiece will change at constant speeds after their surfaces completely match.

This rule indicates that an inherent feature of the system is the rate of surface change rather than the specific surface. The rate of change in the workpiece surface PV based on the parameters in Table 1 is given in Figure 2. In the experiments, the material of the lap consists mainly of the Xi'an #66 polishing pitch produced in China with some additives such as beeswax and plastic powder. A square mesh is used in the lap. The squares have edges of 3.0 cm separated by grooves 3 mm in width and 2 mm in depth. A one-percent concentration, water-based CeO_2 polishing slurry is added every 5 min, 30 mL a time. The conditioning disk is made of 'Jinan Green' stone, sourced from Shandong, China. This is a kind of granite with high stability and low coefficient of thermal expansion. The temperature is controlled at 22 °C and the relative humidity is 60% in the polishing workshop. The horizontal axis in Figure 2 represents the distance between the center of the conditioning disk and the pitch lap, while the vertical axis represents the change rates, with a change to a convex surface when the axis is positive and a change to concave surface when it is negative. The Preston constants are all set to 0.7×10^{-12}/Pa[6, 7] according to the literature and polishing experience for years. Reference [6] indicates that the Preston constant of conventional polishing for BK7 is approximately 0.7×10^{-12}/Pa; the constant should be smaller for the fused silica, which is harder to polish. Reference [7] shows that the constant for conventional polishing of fused silica is 1.0×10^{-12}–1.2×10^{-12}/Pa. The value 0.7×10^{-12}/Pa may not fit the experiments, but very high accuracy is not necessary because the Preston constant does not influence the surface change rules, which were the main factors to be studied. The wear rate of the conditioning disk and the pitch lap are of the same order as the fused silica according to experience, so all the constants are set to 0.7×10^{-12}/Pa. The rotation speed is two revolutions per minute.

Some differences are found between the calculated and experimental results because the Preston equation is not completely satisfied, as a result of the nonlinear characteristic of the pitch lap. Errors exist in the experiments and calculation because of interference and inaccurate setting of the Preston constants. However, the basic rules shown by the model are accurate because the shapes of the two curves are similar.

(2) An equilibrium state exists in the system, indicating that the surface shape could remain unchanged, and the shape could be flat or spherical with a small curvature.

The conditioning disk position corresponding to the equilibrium state is the equilibrium position marked in Figure 2. The difference between the calculated and experimental result is 17 mm, and error is inevitable. Results for the surface maintained during the experiment are given in Figure 3, in which the fluctuation is caused by an unstable environment. The following analysis indicates that a polishing workpiece in equilibrium is the ideal work state of a continuous polishing machine. This characteristic lays the theoretical foundations for high-efficiency and high-precision polishing.

Table 1. Parameters for calculation and experiments.

	Outer diameter (mm)	Inner diameter (mm)	Thickness (mm)	Density (kg/m^3)
Lap	690	220	20	1800
Conditioning disk	350	—	30	3070
Workpiece	150	—	20	2500

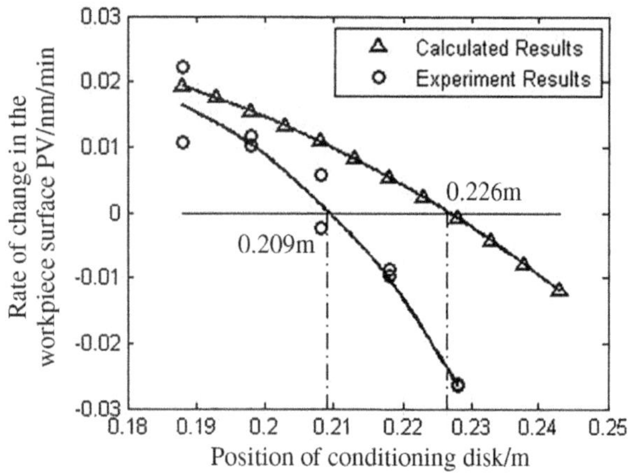

Figure 2. Rate of change in the workpiece surface PV.

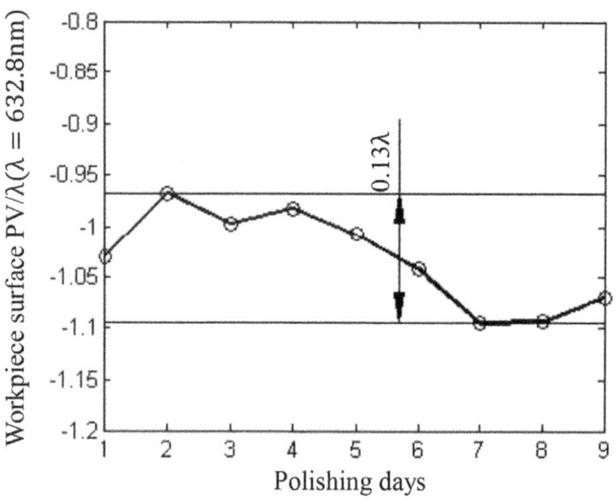

Figure 3. Maintained workpiece surface shape experiment.

3.2. Workpiece polishing in the equilibrium state

Different radial positions of the pitch lap have different friction distances from the workpiece. Thus, the lap deteriorates regardless of the shape of the workpiece surface, even though it may be an ideal plane. Figure 4 is a simulation of workpiece polishing without a conditioning disk according to the model. The initial shapes of the lap surface and the workpiece are both ideal planes. This figure is a sectional view of the lap, in which the fine lines stand for its worn shape and the dot-dash lines on the left side stand for the shape of the workpiece surface. The condition after 1st, 11th, 21st, 31st and 41st cycles are presented. The damage caused by a workpiece without surface error is illustrated. The damage to the lap surface contains two parts; one is the groove at the annulus center caused by the finite workpiece dimensions, called partial damage, and the other is departure of the bottom of the groove from the flat surface, called general damage.

If a conditioning disk is applied, the partial damage can be repaired naturally. Thus, the main task is to repair the general damage, which is most important in determining the workpiece surface figure. The workpiece surface can remain high-precision under equilibrium in flat cases, according to the above analysis. Further calculations show that the equilibrium state is insensitive to the surface error of the workpiece. If equilibrium with flat surfaces is obtained, exchanging the workpiece with another workpiece with surface error can also allow polishing to very high accuracy without moving the conditioning disk. This phenomenon indicates that in the process the curvature change is very small between the case of a mismatching workpiece and pitch lap and the perfectly matched case in the equilibrium state. These results can lead to a theoretical basis for high-efficiency and high-precision polishing. Figure 5 is the result of successive changes in workpieces according to the model. In the simulation, the system is first adjusted to the equilibrium state with flat surfaces, with the surface PV of the workpiece being approximately 1 nm. Then, the workpiece is replaced by other workpieces of the same material and size but with a concave surface with an error of 632.8 nm. Furthermore, a fresh workpiece is replaced each time the workpiece surface becomes stable. The process is repeated 20 times. The surface error of the final workpiece is found to be 8 nm, and the shape is concave. The parameters of the conditioner and pitch lap for the calculation are given in Table 1. The workpiece size is Ø20 × 30 mm with a

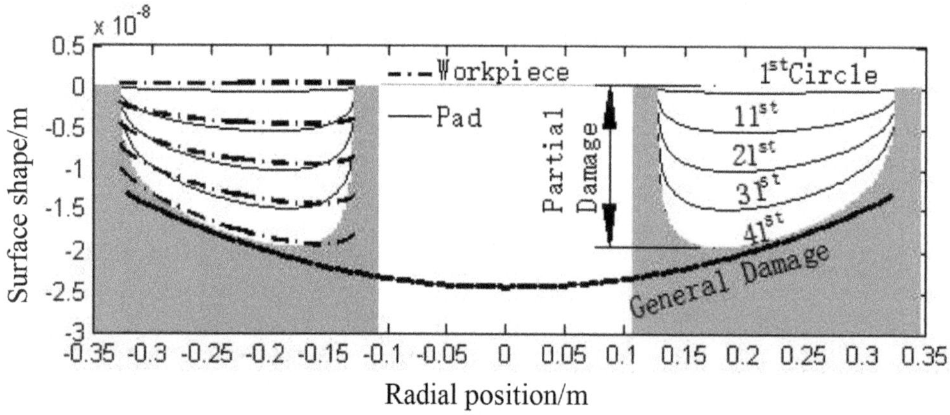

Figure 4. Lap surface damaged by the workpiece.

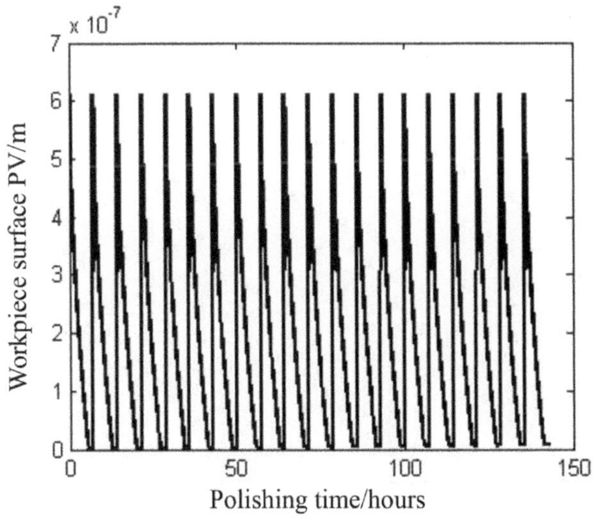

Figure 5. Change in the workpiece surface.

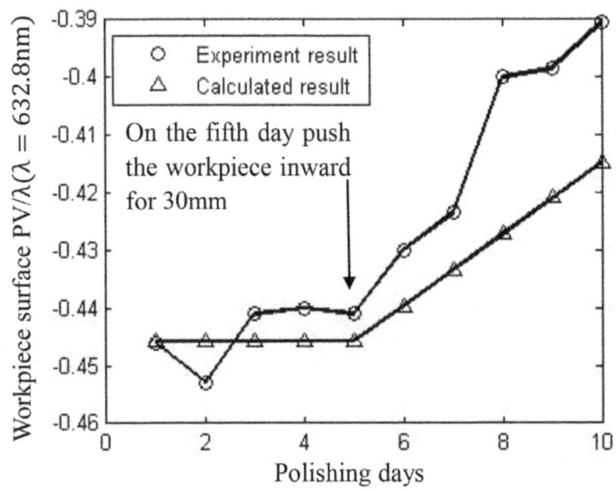

Figure 6. Change in the workpiece surface caused by motion in the radial direction of the lap.

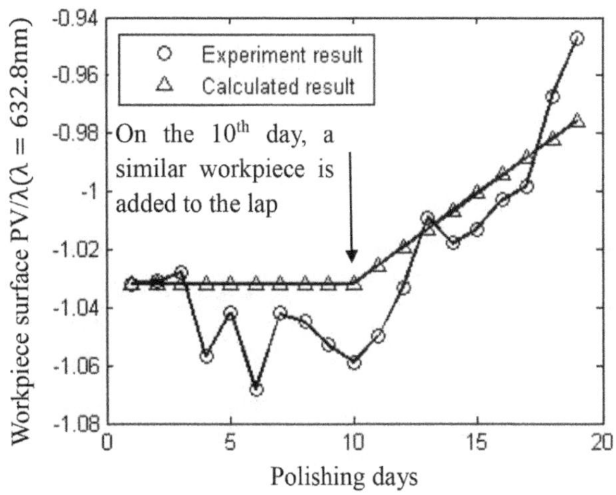

Figure 7. Change in the workpiece surface caused by a change in quantity.

density of 2500 kg/m^3. All Preston constants are set to 0.7×10^{-12}/Pa. When the concave workpieces are replaced with convex ones, the surfaces are still highly accurate.

Thus, analytical results show that polishing workpieces at equilibrium is efficient and highly precise. To obtain and maintain the equilibrium state, the following steps should be performed:

(1) the equilibrium position should be explored;

(2) an appropriate method to adjust the shape of the lap surface should be applied; and

(3) the conditioning disk position should be adjusted according to the polishing condition.

The equilibrium state can be found by measuring the rates of change of the workpiece surface under different conditioner positions (Figure 2). The equilibrium position is usually in the middle 1/3rd of the annulus; hence, experiments could be performed in this range. The equilibrium position is actually not fixed, even in a certain machine, and the machining parameters remain unchanged. The position will change according to the workpiece parameters, such as the workpiece position, size, quantity, pressures applied, and so on. Thus, the experiments should depend on the workpieces to be polished. Simulation results show that if the workpiece is placed at the center of the lap annulus, the extent of worn area in the inner side of the annulus is larger than that at the outer side. If the size, or the quantity, or the applied pressure of the workpiece is larger, the conditioning disk should be pushed outwards to maintain balance. If the workpiece is moved along the radial direction of the lap, the conditioning disk should be pushed towards the opposite side to maintain balance. Figure 6 shows that the surface changes to convex when the workpiece is pushed inwards, the original workpiece position is at the center of the annulus. The parameters are listed in Table 1. The workpiece material is fused silica. A single Preston constant is applied for all calculations, that is, 0.7×10^{-12}/Pa. Figure 7 demonstrates the changes in the workpiece surface caused by a change in the quantity. On the 10th day, a workpiece with the same material, size and surface figure is added to the other workstation on the pitch lap, and surface development into a convex shape is observed. The relationship between each change condition of the workpiece and the equilibrium position can be described mathematically, as will be discussed in future articles.

Given the delayed response of the pitch lap, measurement results beyond the equilibrium state cannot represent the true surface shape, which makes the lap surface difficult to control accurately. The model shows a delayed response in which the surface does not stop changing immediately after the conditioning disk is pushed to the equilibrium position.

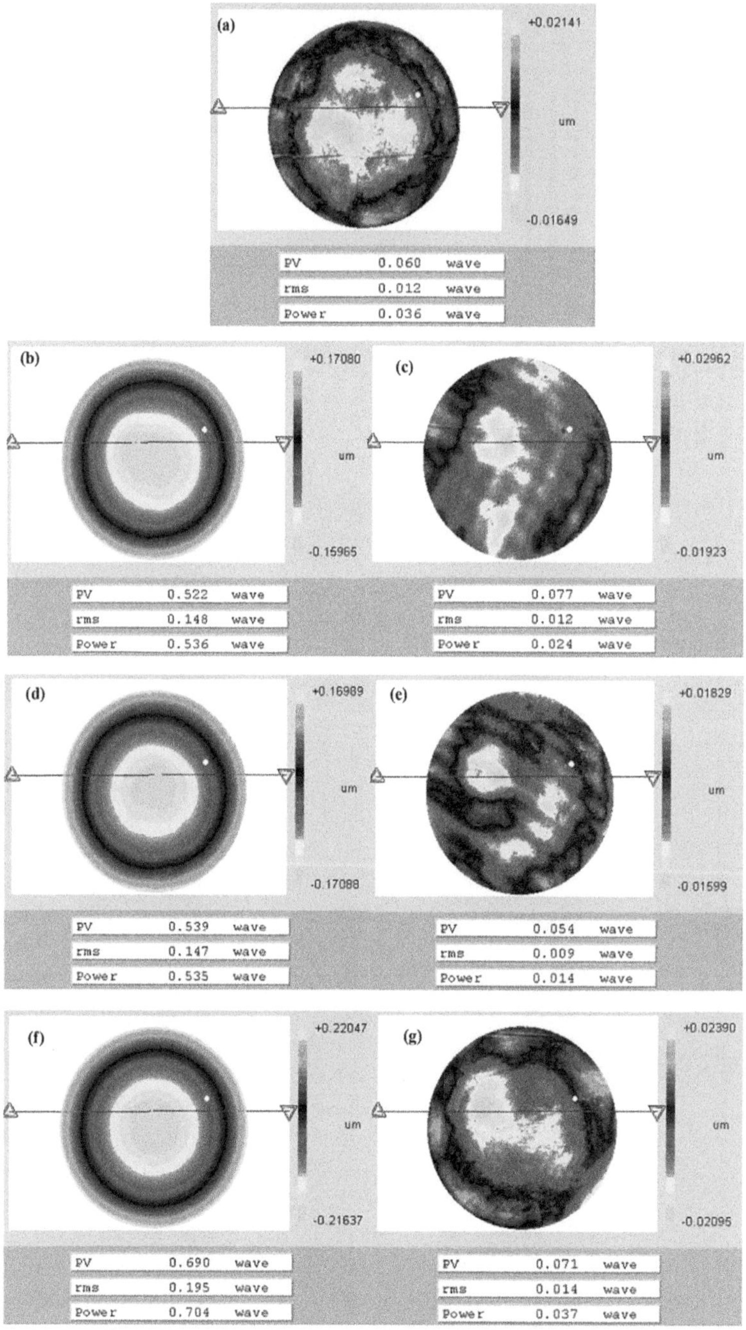

Figure 8. Workpieces continously polished in an equilibrium state.

To adjust the lap surface, the shape of the workpiece surface should be measured in the equilibrium state. Then, from the surface error and the change rates measured, the distance of movement for the conditioner and the conditioning time is determined. At the end of this calculated time, the conditioner is pushed back to the equilibrium position, and the surface figure is again checked when it is stable. If an error still exists, it may have been caused by an environmental disturbance or calculation errors. The disturbance and errors are eliminated, and the lap is reconditioned.

The key to stable machining is maintenance of the pitch lap surface. From the analyses above, a lap in the equilibrium state is insensitive to the surface error of the workpiece. However, when the position or size of the workpiece changes without moving the conditioning disk, the equilibrium state disappears, and the lap surface will deteriorate. Thus, the conditioner position should be adjusted according to the workpiece to prolong the ideal state.

Figure 8 shows experimental results with the system parameters in Table 1. The workpiece material is fused silica.

The shape of lap surface is adjusted in a similar way to the steps mentioned above. The equilibrium position is found after the change rate curve of the workpiece surface is measured, and the adjustment is in good numerical agreement with the measured rates. Figure 8(a) is the shape of the monitored workpiece surface after the adjustment process, at which point the workpieces are changed in succession. Figures 8(b), 8(d) and 8(f) are the workpieces before polishing. The workpieces have concave surfaces, and the testing wavelength is 632.8 nm. Figures 8(c), 8(e) and 8(g) are surface figures after polishing for 8 h without detection. The results clearly indicate the advantages of high efficiency and high accuracy when polishing at equilibrium. As a further test, a convex workpiece is polished. The convex workpiece always floats on the lap, because of hydromechanical effects[8–10], and the material removal rate is very low, so the edge of the workpiece is not polished after 8 h. When polishing concave workpieces the friction force is very large, which leads to a high polishing speed. Thus, the workpiece should be concave before polishing on a continuous machine.

4. Conclusion and discussion

An ideal mathematical model for a continuous polishing system is built. According to the model, the mechanism of surface changes in the system is examined, and the equilibrium state, which indicates that the surfaces remain unchanged, is found in the polishing system. The model shows that although the surface of the workpiece is an ideal plane, it will also damage the lap. Therefore, the damage caused by workpieces with surface errors may not be that serious – and this phenomenon has been proved theoretically. The change in curvature is very small between the case of workpiece–lap mismatch to that of a perfect match in the equilibrium state. This characteristic lays the theoretical foundations for high-efficiency and high-precision polishing. The procedures for obtaining an ideal equilibrium state are performed and verified experimentally.

Calculations show that the polishing process is actually deterministic, and the shapes of the workpiece surfaces can be precisely controlled. An unstable environment (such as fluctuations in temperature and humidity, and irregular airflows) and imperfect factors in polishing (such as the nonuniform temperature distribution of the lap and the workpieces, uneven polishing powders on the surface of the lap, unstable rotating speeds of the workpieces, and so on) will seriously impact the machining precision and make the polishing a nondeterministic process. However, with technological developments, such as a higher control accuracy of the environment temperature and humidity, enhanced performance of polishing machines, studies on the polishing materials and material removal mechanisms, as well as the application of computer simulations in conventional polishing, conventional polishing will become a deterministic process.

Acknowledgements

The study is supported by grants from the Chinese and Israeli Cooperation Project on High-power Laser Technology (2010DFB70490).

References

1. J. Néauporta, P. Cormonta, and N. Darboisa, Proc. SPIE **5965**, 59650N (2005).
2. R. Geyl, F. Houbre, G. Chabassier, and A. Roussel, Proc. SPIE **3739**, 63 (1999).
3. T. I. Suratwala, M. D. Feit, and W. A. Steele, J. Am. Ceram. Soc. **93**, 1326 (2010).
4. T. Suratwala, R. Steele, M. Feit, R. Desjardin, and D. Mason, Int. J. Appl. Glass Sci. **3**, 14 (2012).
5. F. Preston, J. Soc. Glass Technol. **9**, 214 (1927).
6. R. R. Berggren and R. A. Schmell, Proc. SPIE **3134**, 252 (1997).
7. A. A. Tesar and B. A. Fuchs, Proc. SPIE **1531**, 80 (1991).
8. R. Scoff, L. Runnels, and M. Eyman, J. Electrochem. Soc. **141**, 1698 (1994).
9. Y.-R. Jeng, P.-Y. Huang, and W.-C. Pan, J. Electrochem. Soc. **150**, G630 (2003).
10. E. J. Terrell and C. Fred Higgs, J. Electrochem. Soc. **153**, K15 (2006).

Conceptual design of the Kumgang laser: a high-power coherent beam combination laser using SC-SBS-PCMs towards a Dream laser

Hong Jin Kong[1], Sangwoo Park[1], Seongwoo Cha[1], Heekyung Ahn[1], Hwihyeong Lee[1], Jungsuk Oh[1], Bong Ju Lee[2], Soungwoong Choi[2], and Jom Sool Kim[3]

[1]*Department of Physics, Korea Advanced Institute of Science and Technology (KAIST), 291 Daehak-ro, Yuseong-gu, Daejeon 305-701, Republic of Korea*

[2]*Department of Advanced Green Energy and Environment, Handong Global University, Heunghae-eup, Buk-gu, Pohang-si, Gyeongbuk 791-708, Republic of Korea*

[3]*Laser Spectronix, 219 Gasan digital 1-ro, Geumcheon-gu, Seoul 153-704, Republic of Korea*

Abstract

In this review paper, we introduce a self-phase controlled stimulated Brillouin scattering phase conjugate mirror (SC-SBS-PCM) and the Kumgang laser. The SC-SBS-PCM was proposed and demonstrated its success at the academic low power level, ~100 mJ@10 Hz. The Kumgang laser is under development to verify whether the SC-SBS-PCM is operable at the kW level. It is a 4 kW beam combination laser combining four 1 kW beams using the SC-SBS-PCM. If the Kumgang laser functions successfully, it will be the most important step towards a Dream laser, a hypothetical laser with unlimited power and a high repetition rate.

Keywords: coherent beam combination; diode-pumped laser; high-power laser; self-phase controlled SBS-PCM

1. Introduction

1.1. High-power coherent beam combination laser

Pulsed lasers with high power and a high repetition rate are greatly needed in many application fields, such as laser machining/annealing, neutron/proton generation, particle acceleration, laser fusion energy development, etc.[1–4].

Beam combination can contribute to the development of these high-power, high-repetition-rate lasers. There are two main lines of research in beam combination, namely beam combination of fiber lasers and beam combination of bulk lasers. There are many research groups in the field of fiber beam combination lasers, including the ICAN Program[5–8]. Since the output power of an optical fiber laser is limited, a large number of fiber lasers need to be combined to get high enough power[9]. On the other hand, bulk lasers have higher output power and a relatively small number of lasers are needed to achieve high power and a high repetition rate, compared with fiber-based lasers. To build a coherent beam combination (CBC) laser, the wavefronts of the combining beams should be flat, and their piston errors, the relative phases between the laser beams, should be zero[10]. The M^2 factor of a single-mode fiber laser is very good, so it is greatly advantageous to combine fiber lasers because only the piston errors need to be considered. Meanwhile, high-power bulk lasers tend to have lower beam quality than fiber lasers and it is very difficult to combine the beams. The main sources of a bulk laser's low beam quality are the thermal lensing effect, pumping inhomogeneity and filamentation due to self-focusing.

To coherently combine bulk lasers, one should flatten the wavefront of the combining beams via active control or passive control. Active control utilizes an adaptive mirror[11], but it is difficult to make a perfect wavefront in this way. There are passive control methods using a phase conjugate mirror (PCM), and they can compensate for the distorted wavefront during amplification to give a lower M^2 factor[12]. Stimulated Brillouin scattering (SBS)[13] and degenerate four-wave mixing (DFWM)[14] are typical phase conjugation methods commonly used for bulk lasers. The SBS-PCM needs only one convex focusing lens, so it is optically very simple, but the phase of the reflected beam

Correspondence to: H. J. Kong, Department of Physics, Korea Advanced Institute of Science and Technology (KAIST), 291 Daehak-ro, Yuseong-gu, Daejeon, 305-701, Republic of Korea. Email: hjkong@kaist.ac.kr is inherently random because it originates from thermal noise[15]. The DFWM method maintains the phase of the reflected beam, but when it is applied to beam combination, the optical system becomes very complicated. Many studies have focused on phase locking of the reflected beam from an SBS-PCM[16–19]. They are mostly classified as the back-seeding method[16] and the focal point overlapping method[17, 18]. In the back-seeding method, the phase of an SBS wave follows the phase of a back-seeding beam so that the phase conjugation property disappears. On the other hand, in the overlapping method, the energy focused into a common focal point increases as the number of beams grows. Therefore, the number of combining beams is limited due to the energy limit at the focal spot, and the focal spot of each beam can move due to thermal effects, self-focusing effects, etc.

Thus, we need to develop a scalable beam combination technique using an SBS-PCM that produces a phase-controllable SBS wave.

1.2. Beam combination laser using an SBS-PCM

H. J. Kong, the main author of this paper, proposed the simplest technique for controlling the phase of an SBS-PCM[20]. A beam combination laser using this 'self-phase controlled' SBS-PCM (SC-SBS-PCM) is indefinitely scalable because the combining beams are not optically interconnected. The only requirements are a linewidth, a pulse width and the minimum energy of the beams for efficient SBS to occur[21]. The linewidth of the beams should be smaller than the Brillouin bandwidth, and the pulse width of the beams should be several times the acoustic lifetime, typically several nanoseconds. Moreover, the energy of the beams should be larger than several mJ because the reflectivity of an SBS-PCM is highly nonlinear to its input energy. These are general prerequisites for an SBS-PCM and not special conditions for the SC-SBS-PCM. The SBS medium, where the SBS process occurs, could be any phase of matter, solid, liquid, gas or plasma. A solid SBS material must be replaced when damage occurs, while a fluid SBS material like liquid, gas or plasma dissipates the damaged part and heals itself. In our research, fluorocarbonate liquids are utilized as the SBS material because they have good SBS properties[22].

The feasibility of a beam combination laser using fluorocarbonate SBS-PCMs has been proved experimentally in the low power range of several W[23–31]. Currently we are working on experimental verification of this beam combination laser with an average output power in the kW range. As mentioned above, the output power of a beam combination laser using an SC-SBS-PCM can be freely scaled. It should be noted that in this beam combination laser, the output power of each sub-beam is limited by the power limit of the SC-SBS-PCM. Moreover, the output power of the laser can be increased without additional

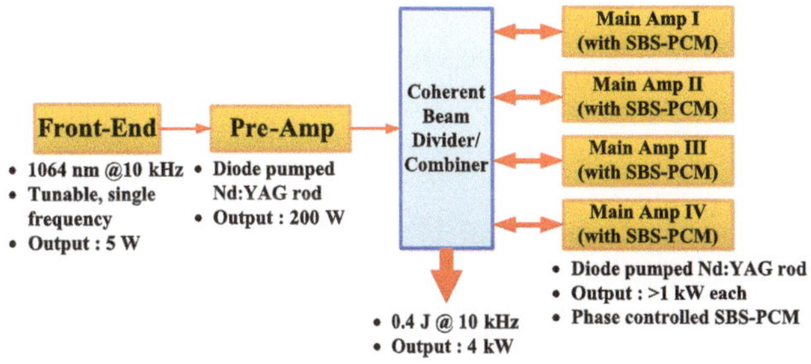

Figure 1. The schematic diagram of the Kumgang laser system.

problems by enlarging the size of the combined beam. Therefore, if this research is performed successfully, it will be an important step towards realization of the Dream laser, a hypothetical laser having unlimited power and a high repetition rate. We named the beam combination laser system employed in this research the 'Kumgang laser'.

1.3. Kumgang laser

A project for a 4 kW beam combination laser system using SC-SBS-PCMs began in 2012 and the laser is called 'Kumgang', meaning diamond (or very important) in Korean. Four sub-beams, each with an energy of 0.1 J/10 ns pulse width and a repetition rate of 10 kHz, are coherently combined to produce an output beam with an energy of 0.4 J. It is a master oscillator power amplifier (MOPA) system and all of its pre- and main amplifiers are diode-pumped Nd:YAG amplifiers. The project will be completed by the end of 2015 and its first industrial application will be patterned 2D laser cutting of a printed circuit board (PCB) substrate.

Figure 1 shows the schematic diagram of the Kumgang laser. The front end consists of a CW laser-diode master oscillator, an acousto-optic modulator (AOM) for 10 kHz pulse generation, fiber-based amplifier(s), and a regenerative amplifier for boosting the power. The average output power of the front end is 5 W [0.5 mJ/10 ns/10 kHz]. Diode-pumped Nd:YAG amplifiers in the pre-amplifier stage amplify the beam in a double-pass configuration. The output beam of the pre-amplifier stage has a power of 200 W [20 mJ/10 ns/10 kHz]. The coherent beam divider/combiner divides the beam from the pre-amplifier stage and combines the beams after amplification through the main amplifier stage. The volume Bragg grating (VBG) coherent beam divider/combiner is currently being researched[32, 33]. A polarizing beam splitter (PBS) will be used for the beam combiner before the VBG is developed. The main amplifier stage is composed of a diode-pumped Nd:YAG amplifier chain and an SC-SBS-PCM. When completed, each divided sub-beam from the coherent beam divider/combiner will be amplified to 1 kW with a double-pass configuration. After the coherent combination, the final output beam will have an energy of 400 mJ per pulse with a pulse width of 10 ns and a

Table 1. Beam combination methods.

Type	Subtype	Examples of combining device
IBC	Frequency combining	Dichroic mirror, prism, grating
	Side-by-side combining	Lens array, mirror, wedge
	Polarization combining	PBS
CBC	Wavefront combining	Lens array, mirror, wedge
	Amplitude combining	PBS, beam splitter, grating

Table 2. CBC methods.

	Wavefront combining	Amplitude combining
Beam positions of combined beams	Different (tiled)	Same (overlapping)
Pros	Output power is always the same	High beam quality (same as input beams)
Cons	Lower beam quality	Output power is degraded when the phase control is imperfect
Combining devices	Lens array, mirror, wedge, etc.	PBS, beam splitter, grating, etc.

repetition rate of 10 kHz, making the average output power 4 kW.

In Section 2, the general methods of beam combination are presented and two notable CBC methods are explained in detail. In Section 3, a beam combination laser system using SC-SBS-PCMs is described in detail. In Section 4, each subsystem of the Kumgang laser is shown with its prime application, the two-dimensional cutting of a PCB substrate. Finally, in Section 5, possible future projects for generating high-power fs/ps lasers are discussed.

2. CBC laser

2.1. Methods of beam combination

There are several methods of constructing beam combination lasers[10]. In general, they are divided into two categories, incoherent beam combination (IBC) and CBC. In IBC, the relative phases between the combining beams are ignored and only the directions of propagation are aligned, while in CBC, the relative phases are measured and controlled to be zero.

Table 1 summarizes the beam combination methods. There are three types of IBC. The frequency combining method combines beams with different wavelengths. The radiance of a frequency combined beam increases with the number of combining beams, but the spectral radiance is always the same. When building high-power beam combination lasers, frequency beam combination is less favorable, mainly due to the limited frequency range of high-power laser sources and additional difficulties in generating and handling beams with several frequencies.

The other two methods are monochromatic. The side-by-side beam combination method combines the beams blindly, only aligning them side by side, and does not consider their relative phases. The radiance does not scale with the number of combining beams and the focused beam pattern is irregular and random. This type of beam combination is only applicable when only the power of the beam is important. It is of no use for most applications where the focused beam intensity is critical.

The last type of IBC is the polarization combining method. Two beams with the same frequency and opposite polarization can be combined using a PBS or other polarization optics. Since the relative phase between the two beams is random, the output polarization is random and undetermined. Therefore, one cannot combine more than two beams using this method.

There are two types of CBC as shown in Table 2. The first one is wavefront combination/tiled-aperture combination. Like the side-by-side beam combination, it combines the beams in parallel, but by controlling the piston phases to be zero, the radiance scales as the number of beams combined. The advantage is that the output power of the wavefront combined beam is always the sum of the combining beams. Meanwhile, the beam quality can be degraded greatly compared with the individual combining beams. The arrangement of the combining beams causes the focused intensity pattern to scatter because the pattern is the product of the focused intensity pattern of one beam and the diffraction pattern of a two-dimensional multislit. If the piston phases are not exactly zero, the beam quality decreases and the focused intensity pattern becomes distorted. The wavefront beam combination can be performed using mirrors or wedges. The beams are directed to neighboring but non-overlapping positions and their propagation directions are aligned by tilting the optical devices. When many diverging beams are combined, e.g., when combining an array of laser diodes (LDs) or fiber lasers coupled to free space, a lens array can be used.

The other method of CBC is amplitude combination. It combines the beams to have the same position and the same direction. In this case, control of the relative phases is very important and it directly affects the output power by interference. The most beneficial property of amplitude beam combination is that the beam quality remains unchanged compared with the combining beams. This is because the shape of the combining beams is the same as that of the combined beam. On the other hand, the output power of the amplitude combined beam fluctuates as the piston phases deviate from zero. When all the piston phases are zero, the output power is maximal and is the same as the sum of the combining beams. Otherwise, the output power is lower than the sum. A beam splitter can combine two beams with the same polarization and power. The output power of the beam splitter is dependent on the phase difference between the two beams, and when the difference is zero, the output power is twice the power of the combining beams. One can combine 2^N beams by cascading the process N times. Amplitude beam combination using a PBS is a special case of beam splitting. Using two beams with opposite polarizations, the output power of the PBS is the sum of the two beams, but the polarization of the output is random and typically elliptical. By controlling the relative phase between the two beams, the output polarization is fixed to a known state and the combined output beam can be combined with another beam by the same method. A grating can also coherently combine the beams. This is done either by combining several high-order beams into a zeroth-order beam, or by combining several first-order beams of co-existing gratings into a common zeroth-order beam. In any case, the output power can be reduced without controlling the relative phases.

In beam combination using MOPA systems, the laser beam from the master oscillator is divided into several beams. After the amplification from the power amplifier, they are combined to make the output beam. Typically, the beam dividing optics and the beam combining optics are the same. When using the wavefront dividing method, division of the beam into sub-beams with sharp edges induces severe diffraction with very dangerous intensity spikes. To eliminate them, sophisticated methods like serrated apertures are needed and, therefore, the power of the divided beam is typically lower than that of the amplitude dividing method. Considering this limitation and other factors, the amplitude dividing/combining method will be employed for the Kumgang laser. Two possible methods of amplitude combination are presented below.

2.2. Amplitude beam combination using a PBS

Figure 2 shows a schematic diagram of a four-beam MOPA beam combination laser with the amplitude dividing method using a PBS. It employs a double-pass amplification scheme. PBS1 and quarter-wave plate (QWP1) serve as a beam diverter. When the laser passes through QWP1 and back, the polarization of the beam is rotated by 90° and the output beam is reflected on PBS1. PBS2 divides the input beam into two beams. With a proper angle of half-wave plate (HWP1), PBS2 can divide the beam into two beams with the same power. Likewise, HWP2, PBS3, HWP3 and PBS4 divide the two beams into four beams. Four power amplifier chains amplify each beam in a double-pass configuration. To measure the relative phases between the beams, at each combining stage, a small portion of the combined beam is directed towards a measurement stage. Each measurement stage consists of one QWP, one PBS and two energy meters. By measuring the energy on the two energy meters, the relative phase between two combining beams can be calculated[28]. If there are remaining phase errors between beams while the phase is being controlled, part of the beam leaks at PBS1 and PBS2, and the output power decreases.

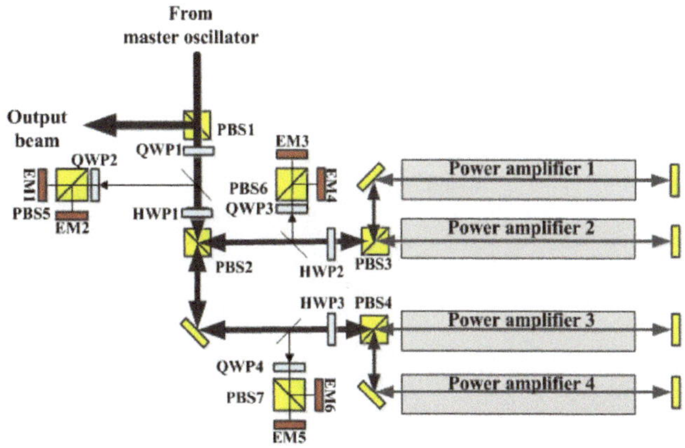

Figure 2. The MOPA beam combination laser with the amplitude combining method using PBSs: EM1–EM6, energy meters.

2.3. Amplitude beam combination using a VBG

Figure 3 shows a schematic diagram of a four-beam MOPA beam combination laser with the amplitude dividing method using a VBG. It employs a double-pass amplification scheme. The PBS and QWP serve as a beam diverter as before. The VBG is a grating in which variation in the refractive index is engraved within a material. When a beam passes the VBG, the input beam (zeroth-order beam) is gradually converted into a refracted beam (first-order beam) throughout the thickness of the VBG. Only slight modulation of the refractive index is needed to achieve 100% efficiency. By etching multiple gratings, a beam divider/combiner for many beams can be made. This multiplexed VBG divides the input beam (common zeroth-order beam) into multiple output beams (first-order beams from each grating). The recent development of photorefractive glass makes high power and high damage threshold application of the VBG possible. To measure the relative phases between the combining beams, an additional phase measurement device is needed. If there is a remaining phase error between the

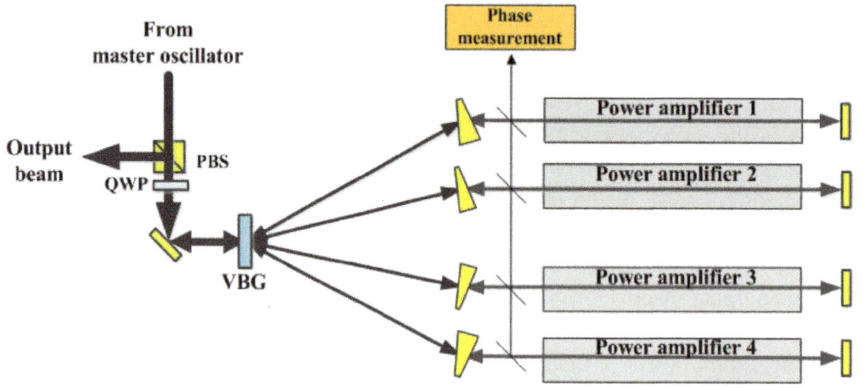

Figure 3. MOPA beam combination laser with the amplitude combining method using a VBG.

beams while the phase is being controlled, part of the beam leaks at the VBG, and the output power decreases.

3. CBC laser using an SBS-PCM

3.1. SBS-PCM

The SBS-PCM is one type of PCM. The SBS phenomenon can be explained as follows[34–37]. When a beam is focused into the SBS material, a small portion of the beam is spontaneously scattered by random thermal phonons, and the scattered beams interfere with the incoming beam and create weak optical gratings. These optical gratings generate acoustic gratings by an electrostriction effect. The acoustic gratings are generally weak and dissipate quickly. However, if a scattered beam happens to have the opposite direction to the incoming beam, the acoustic grating generated by it has the same direction and phase as the original thermal phonon that created that scattered beam, and the amplified phonon scatters the incoming beam even further. If the incoming beam has high peak power, the feedback loop is powerful enough that a significant portion of the incoming beam is scattered to make a 'reflected' beam. The reflectivity can be higher than 90%[21]. Since only a beam with a precisely opposite direction of the incoming beam is generated by the SBS process, the reflected beam is a phase conjugation of the incoming beam.

The main amplifiers in an MOPA beam combination laser are not exactly the same. Therefore, the wavefronts of the divided beams are distorted in different ways. Unless these distortions are compensated for, CBC cannot be achieved. By using a double-pass amplification scheme and utilizing the SBS-PCM as a back reflector of the main amplifier, the wavefronts are restored after the second pass due to the phase conjugation property of the SBS. In this case, CBC is possible if the piston phases are controlled. Unfortunately, since the reflected beam from the SBS-PCM originates from random thermal noise, the phases of the reflections are random, and the piston phases cannot be controlled.

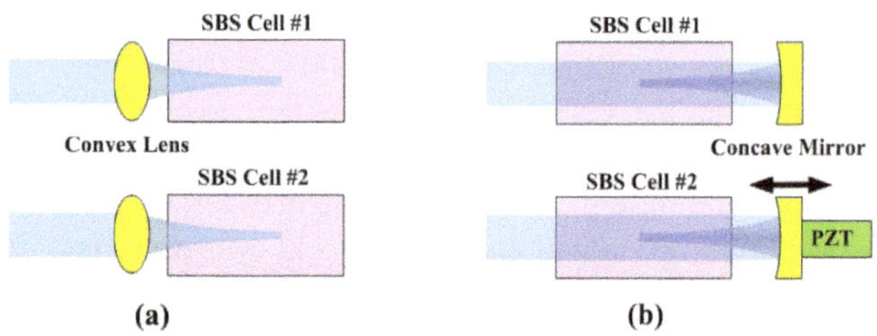

Figure 4. Types of SBS-PCM: (a) conventional SBS-PCM and (b) self-phase-locking SBS-PCM with PZT.

3.2. Self-phase control of an SBS-PCM

Figure 4(a) shows a conventional SBS-PCM. Convex lenses focus incoming beams into the SBS cells. As mentioned above, the relative phase between the reflected beams from SBS cell #1 and SBS cell #2 is random. Figure 4(b) shows one type of SC-SBS-PCM, in which concave mirrors focus the incoming beams. Unlike Figure 4(a), a standing wave from the incoming beam and the focused beam generates a grating and the grating initiates the SBS process. Since the grating is generated at a definite position and time, the relative phase between the incoming beam and the reflected beam is fixed, and the relative phase between the reflected beams from the two SBS cells is also fixed. By attaching a piezoelectric transducer (PZT) to one of the concave mirrors, the path length difference between the two beams can be controlled up to a wavelength. In practice, the path lengths of the beams fluctuate slowly due to air convection and other thermal effects. By measuring the piston phase, the path length fluctuation can be compensated for.

3.3. Experimental results for the beam combination laser using a SC-SBS-PCM

The principles of the SC-SBS-PCM were experimentally verified in 2005[20]. Since then, the results have improved and the method has been applied to four-beam coherent

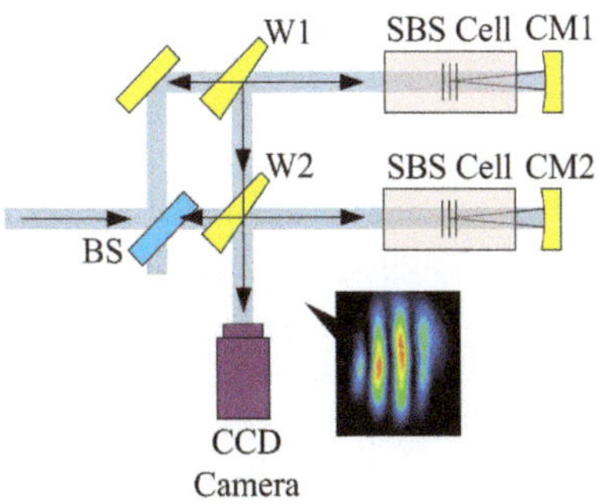

Figure 5. Experimental setup of the self-phase-locked SBS-PCM: BS, beam splitter; W1 and W2, wedges; CM1 and CM2, concave mirrors.

combination lasers with both the wavefront dividing method and the amplitude dividing method[23–31].

Figure 5 shows the experimental setup for verifying the principles of the SC-SBS-PCM[24]. A 1064 nm laser beam with a pulse width of 8 ns and a repetition rate of 10 Hz is divided into two beams. The input energy of each SBS-PCM is 10 mJ. After reflection from the self-phase-locked SBS-PCM, small parts of each beam are reflected by wedges. A CCD camera records the interference pattern between the beams. From the interference pattern, the relative phase between the beams is measured.

Figure 6(a) shows a mosaic intensity pattern. It is generated by lining up one line of the intensity profile pattern recorded at each CCD image. Figure 6(b) shows the measured relative phase. The standard deviation of the measured phase fluctuation is $\lambda/35.7$.

Figure 7 shows the experimental setup for controlling the phase of the SC-SBS-PCM[29]. In this case, a two-beam aperture with a beam diameter of 8 mm is used. After passing through the PBS, the prism divides the beam. A Faraday rotator (FR) rotates the polarization of the beam by 45°. After reflection from the SC-SBS-PCM and the second pass of the FR, the polarization of the beam is now perpendicular to the input beam and the combined beam is reflected at the PBS. The method for measuring the relative

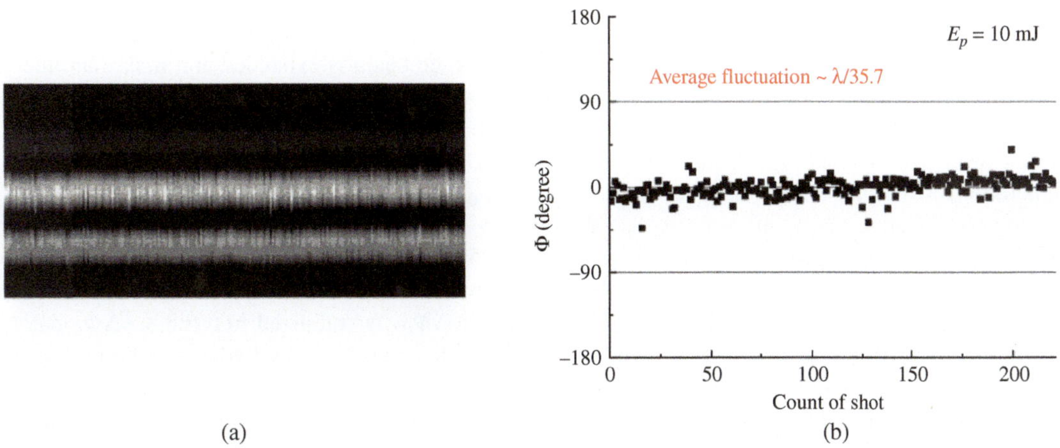

Figure 6. Experimental result of the self-phase-locked SBS-PCM. (a) A mosaic intensity pattern. It is generated by lining up one line of the intensity profile pattern recorded at each CCD image. (b) The measured relative phase. The standard deviation of the measured phase fluctuation is $\lambda/35.7$.

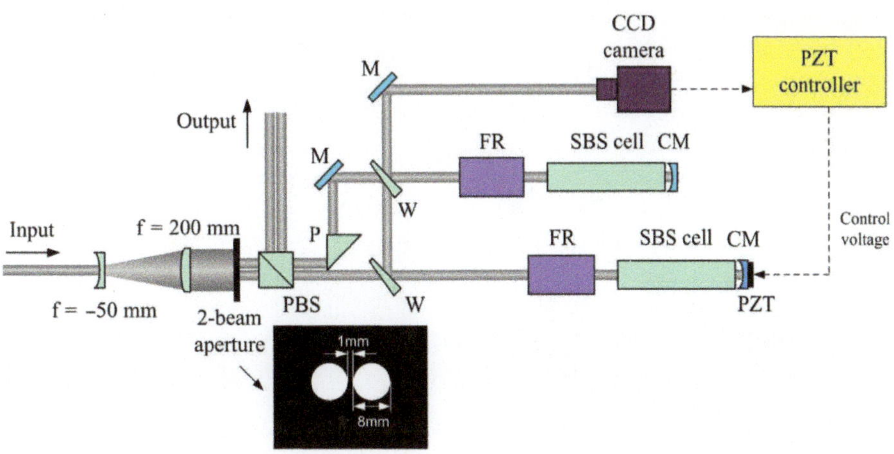

Figure 7. Experimental setup of beam combination using an SBS-PCM and a feedback loop: M, mirror; W, wedged window; CM, concave mirror.

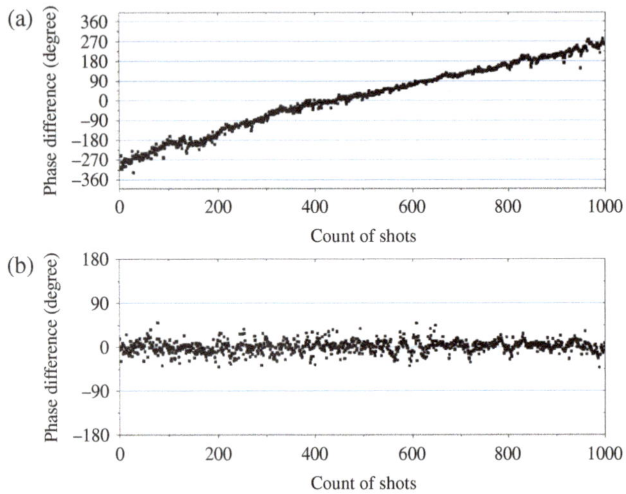

Figure 8. Experimental result of beam combination using an SBS-PCM and a feedback loop: (a) without operating the feedback loop and (b) with operation of the feedback loop.

phase is the same as in Figure 5. Using the measured relative phase, the PZT is operated to compensate for the path length fluctuation.

Figure 8(a) shows the relative phase between the two beams during 100 s when the PZT is not moving. Path length fluctuation causes the relative phase to drift. Figure 8(b) shows the same result when the feedback is on and the PZT is moving to compensate for the phase difference between the beams. The standard deviation of the measured phase fluctuation is $\lambda/28.8$.

Figure 9 shows the experimental setup of the four-beam combination laser with the amplitude dividing method[28] using a PBS. A detailed explanation of the measurement of the relative phases can be found in Ref. [28]. Using the measured phase between beam 1 and beam 2 ($\Delta\Phi_{12}$), PZT1 is operated to compensate for the phase difference. Likewise, using the measured phase between beam 3 and beam 4 ($\Delta\Phi_{34}$), PZT2 is operated. From the measured phase between the combined beam of 1&2 and the combined beam of 3&4 ($\Delta\Phi_{13}$) both PZT2 and PZT3 are operated to compensate for the path length difference.

Figure 10 shows the relative phases between beams during 200 s (2000 shots). $\Delta\Phi_{12}$, $\Delta\Phi_{34}$ and $\Delta\Phi_{13}$ are all stabilized and have standard deviations of $\lambda/34.3$, $\lambda/44.1$ and $\lambda/37.6$, respectively. The output energy is also stabilized to 6.16% of the average in standard deviation.

Figure 11 shows the experimental setup of the four-beam combination amplifier with the wavefront dividing method[30]. The laser beam from the oscillator is expanded four times by the beam expander consisting of a concave lens and a convex lens. The four-beam aperture shapes the input beam into four 6 mm beams. After separation, each divided beam passes through the amplifier twice in a double-pass configuration. The FR rotates the polarization of the beam by 90° and minimizes the thermal depolarization of the amplifier. At the end of each beam line, SC-SBS-PCMs reflect the beam. Inside one of the beam lines (beam 1), a wedge is installed to make a reference beam (beam 0). The reference beam is expanded and makes an interference pattern with the output beam. From the interference pattern, the relative phases between the reference beam and the four beams can be calculated ($\Delta\Phi_{01}$, $\Delta\Phi_{02}$, $\Delta\Phi_{03}$ and $\Delta\Phi_{04}$).

Figure 12 shows the relative phase between beams during 250 s (2500 shots). $\Delta\Phi_{01}$ is the interference between beam 1 and the reference beam. Both beams are from the same SBS-PCM and $\Delta\Phi_{01}$ is stabilized to $\lambda/68.8$. $\Delta\Phi_{02}$, $\Delta\Phi_{03}$ and $\Delta\Phi_{04}$ are the interferences between different beam lines, and they are all stabilized and have standard deviations of $\lambda/26.5$, $\lambda/28.0$ and $\lambda/26.1$, respectively. The input energy

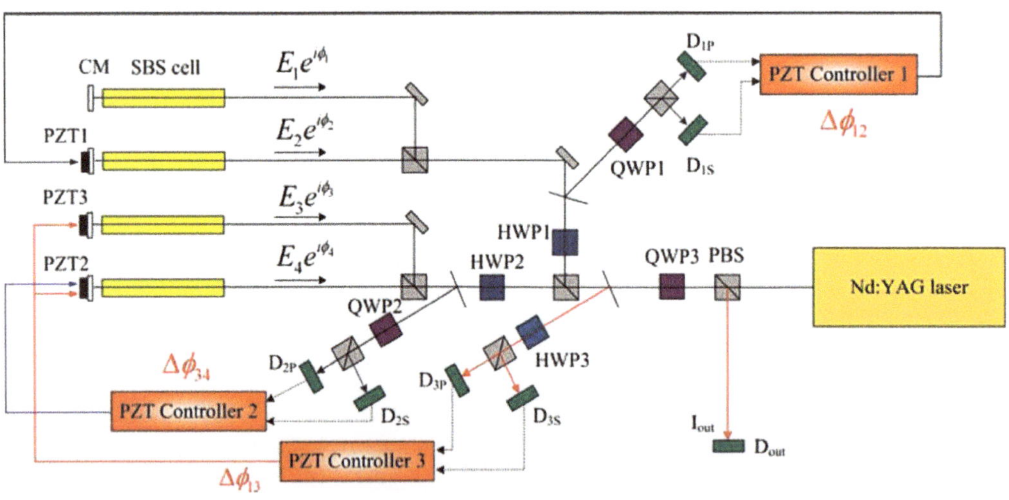

Figure 9. Experimental setup of the four-beam combination using an SBS-PCM with the amplitude dividing method: D_{1P}, D_{1S}, D_{2P}, D_{2S}, D_{3P}, D_{3S} and D_{out}, energy detectors; CM, concave mirror.

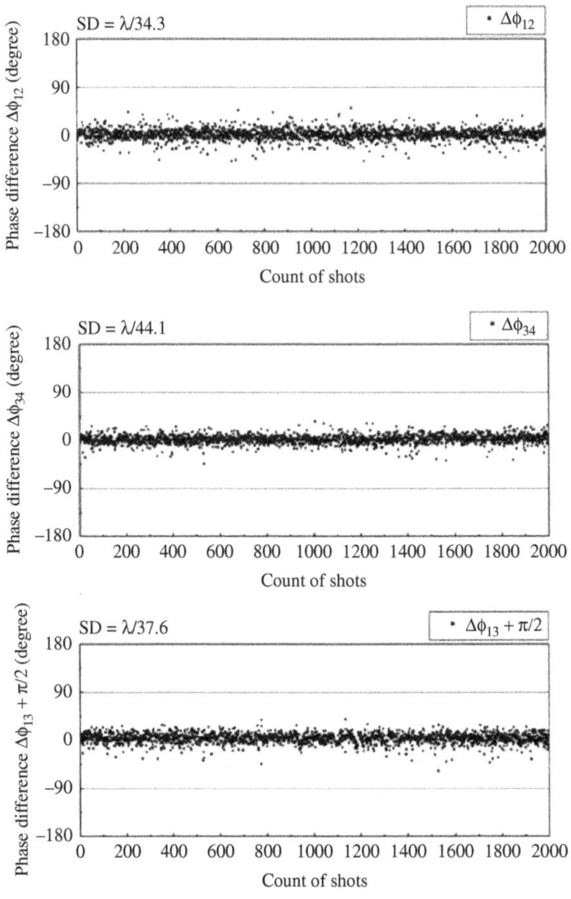

Figure 10. Measured phase fluctuation of four-beam combination using an SBS-PCM with the amplitude dividing method.

of the beam after the beam aperture is 32.2 ± 0.3 mJ and the output energy of the amplified combined beam is 169 ± 6 mJ.

4. Kumgang laser

The Kumgang laser, which is now under construction, will be a high-power, high-repetition-rate beam combination laser that uses SC-SBS-PCMs. Four sub-beams, each with a power of 1 kW, are combined to make a 10 kHz/10 ns laser beam with a pulse energy of 400 mJ. Therefore, the total output power is 4 kW. The research on the Kumgang laser and its application started in 2012 and will be completed by the end of 2015.

4.1. Front-end subsystem

Figure 13 shows the schematic diagram of the Kumgang laser's front-end subsystem. It consists of two parts, a 10 kHz master oscillator and a Nd:YAG regenerative amplifier. Recent technological developments of the LD enable stable and temperature tunable LDs with a narrow linewidth. By controlling the temperature of the LD, its frequency can be adjusted to match the frequency that the Nd:YAG amplifies most. The light from the laser diode passes through the AOM. The AOM is operated by a function generator to make a 10 kHz/10 ns pulse train. The pulse train has very stable timing and is almost free of jitter. Laser pulses now pass through an Yb fiber amplifier, which is pumped by another LD operating at 976 nm. To clear out unwanted frequency components generated by the fiber, the laser beam passes through a bandpass filter before it is coupled out of the fiber.

The Nd:YAG regenerative amplifier amplifies the beam further. The beam passes through FR1, HWP1 and PBS. The PBS acts as an output coupler. Next, the FR2 rotates the polarization of the beam by 45° and HWP2 reverses the effect of FR2. When the beam is injected into the regenerative amplifier through the thin-film polarizer (TFP1), the Pockels cell (PC) is in the 'open' state and does not change the polarization of the injected beam. The QWP rotates the polarization and the beam is reflected at TFP1. While the beam is being amplified at the gain module (GM) for the first time, the PC changes to a 'closed' state and reverses the effect of the QWP. Now the beam is trapped in the regenerative amplifier and passes between the high-reflectivity mirrors HR1 and HR2 many times. After sufficient amplification, the PC changes back to the 'open' state and the beam leaves the regenerative amplifier. The net effect on the polarization of the returning beam by FR2 and HWP2 is a 90° rotation and the beam is reflected at the PBS. The average output power of the front-end subsystem is 5 W (0.5 mJ/10 ns/10 kHz).

4.2. Pre-amplifier subsystem

The pre-amplifier subsystem is depicted in Figure 14. The beam from the front-end subsystem first enters an isolator (ISO). The ISO blocks the backward scattered beam and the spontaneously emitted CW background from going back to the front end. After passing through the HWP, the beam enters pre-amplifier stage I. There are two GMs in pre-amplifier stage I. The amplifying medium of the pre-amplifier GM is a Nd:YAG rod with a diameter of 4.5 mm and a length of 96 mm. The water-cooled GM is pumped by five 200 W LDs operated at 807 nm. To minimize the thermal depolarization effect, the polarization rotator rotates the polarization of the beam by 90° between two GMs. After passing through the FR twice, the beam is amplified again in the second pass. Relay lens pairs relay the image of the beam from the input position of the pre-amplifier to GM1, to GM2 and to the high-reflectivity mirror and back.

Pre-amplifier stage II has another two GMs with the same specifications as stage I. In this case, the beam passes through the amplifiers only once. An image of the beam at the output position of pre-amplifier stage I is relayed to GM3 and to GM4. The average output power of the pre-amplifier subsystem is 200 W (20 mJ/10 ns/10 kHz).

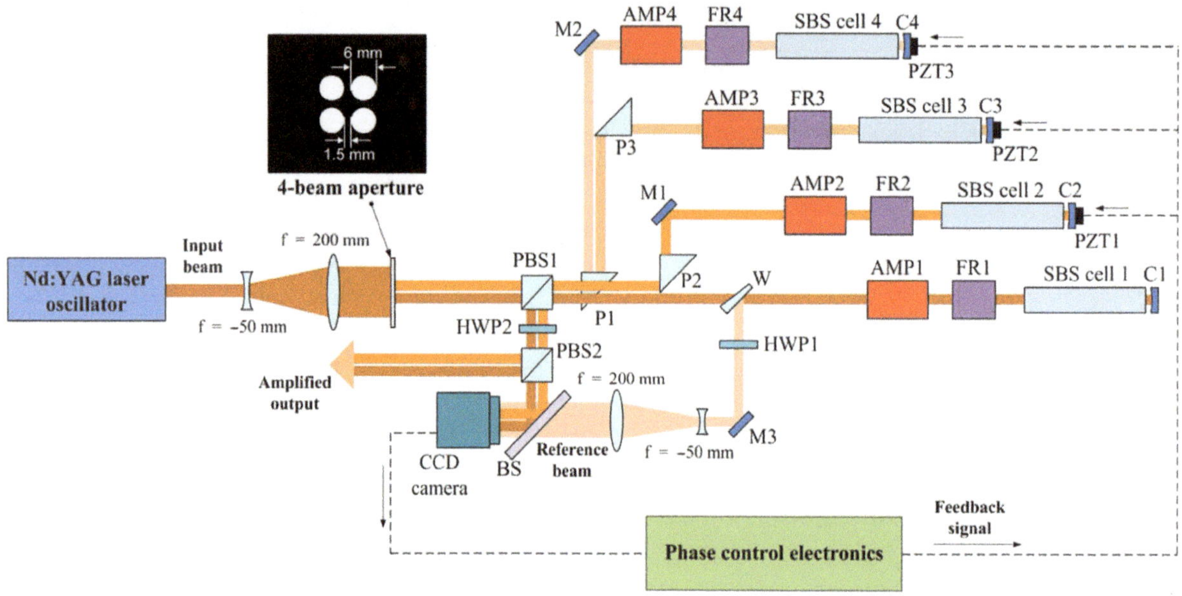

Figure 11. Experimental setup of the four-beam combination amplifier using an SBS-PCM with the wavefront dividing method: HWP1 and HWP2, half-wave plates; BS, beam splitter; P1–P3, Prisms; M1–M3, mirrors; AMP1–AMP4, amplifiers; C1–C4, concave mirrors; W, wedge.

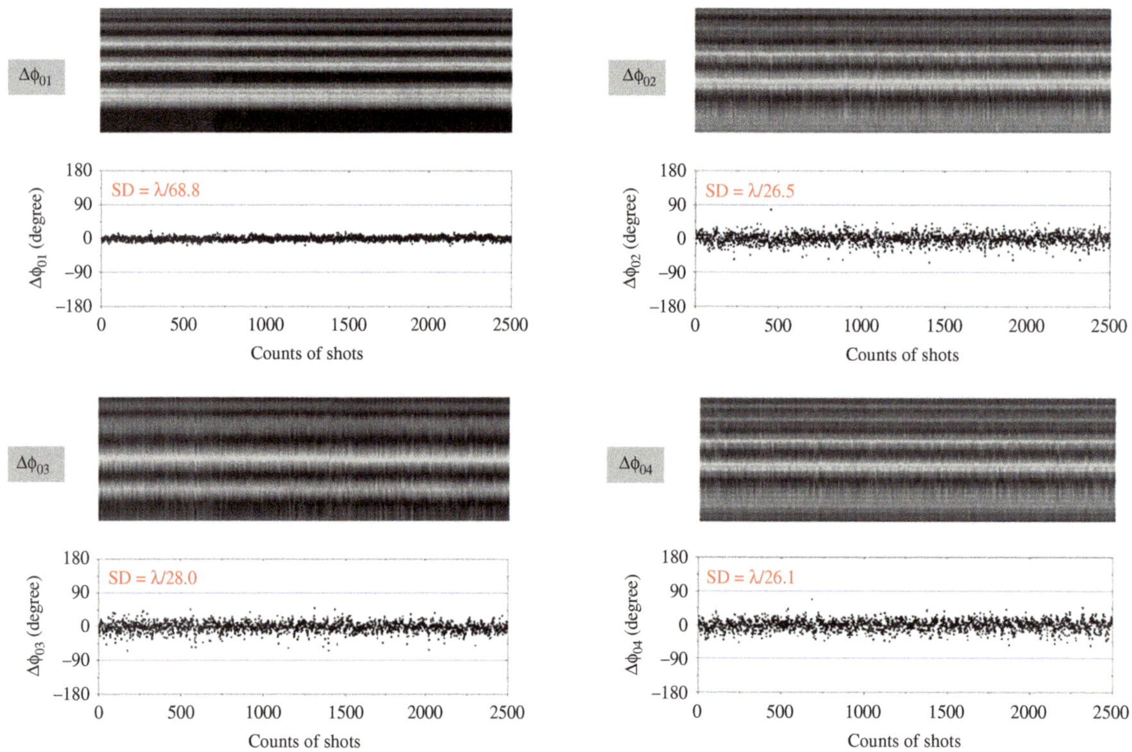

Figure 12. Measured phase fluctuation of the four-beam combination amplifier using an SBS-PCM with the wavefront dividing method.

4.3. Main amplifier subsystem

Figure 15 shows the main amplifier subsystem. The beam from the pre-amplifier subsystem is directed to the beam divider/combiner. The VBG coherent beam divider/combiner is currently under research. A PBS will be used for the beam combiner until the VBG is developed. Details of the proposed beam divider/combiner are explained in Section 2. After the division, the four sub-beams have a power of 50 W each. Four main amplifier modules amplify the sub-beams in the double-pass configuration. Each amplifier module has four main amplifier GMs, which are larger than

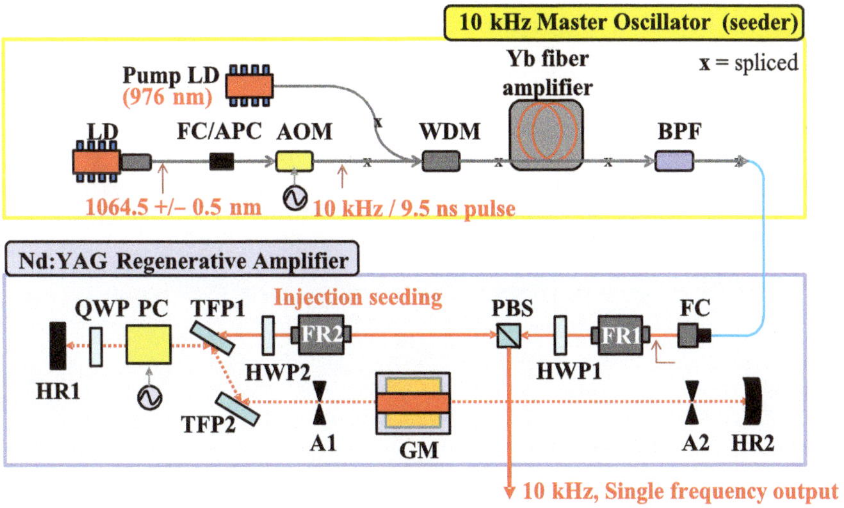

Figure 13. Schematic diagram of the front end: FC/APC, fiber connector/angled physical contact; WDM, wavelength division multiplexor; BPF, bandpass filter; FC, fiber collimator; HR1 and HR2; high-reflectivity mirrors; A1 and A2, apertures.

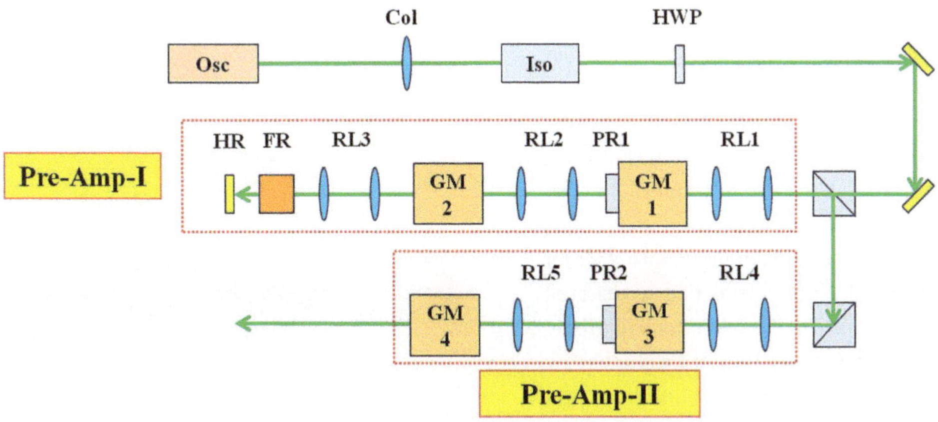

Figure 14. Schematic diagram of the pre-amplifier: RL1–RL5, relay lens pairs; PR1 and PR2, polarization rotators; HR, high-reflectivity mirror.

the pre-amplifier GMs. The amplifying medium of the main amplifier GM is a Nd:YAG rod with a diameter of 6.35 mm and a length of 144 mm. The water cooled GM is pumped by five 400 W LDs operated at 807 nm. The quartz rotators rotate the polarization of the beam by 90° after the first and the third GM to minimize thermal depolarization. Image relaying lenses relay the beam image from the output position of the pre-amplifier to the GMs one by one, and to the SBS-PCM and back.

After passing through the FR, the beam is reflected from the SC-SBS-PCM. The SC-SBS-PCM generates a phase conjugated wave, so any wavefront distortion in the first pass is compensated for automatically in the second pass. After amplification, each sub-beam has a power of 1 kW (100 mJ/10 ns/10 kHz), and four sub-beams are combined at the beam divider/combiner. A phase measurement device measures the relative phases between the four sub-beams. The exact nature of the phase measurement device needs to be studied and is dependent on the beam divider/combiner. Three of the four concave mirrors in the SC-SBS-PCMs are attached to the PZT. The relative phases are compensated for by the movement of these PZTs. The output beam of the Kumgang laser will be a 4 kW, 10 kHz pulse train. Each pulse will have a pulse width of 10 ns and an energy of 400 mJ.

4.4. Applications of the Kumgang laser

There are two distinct pulse width ranges among short-pulse lasers. For applications requiring high power and high peak power, ns pulsed lasers are suitable. The Kumgang laser falls in this category. On the other hand, for applications requiring high power and extremely high peak power, ps and/or fs pulsed lasers are suitable. This category will be discussed in Section 5.

There are many applications for high-power ns pulsed lasers like the Kumgang laser. For defense, high-power

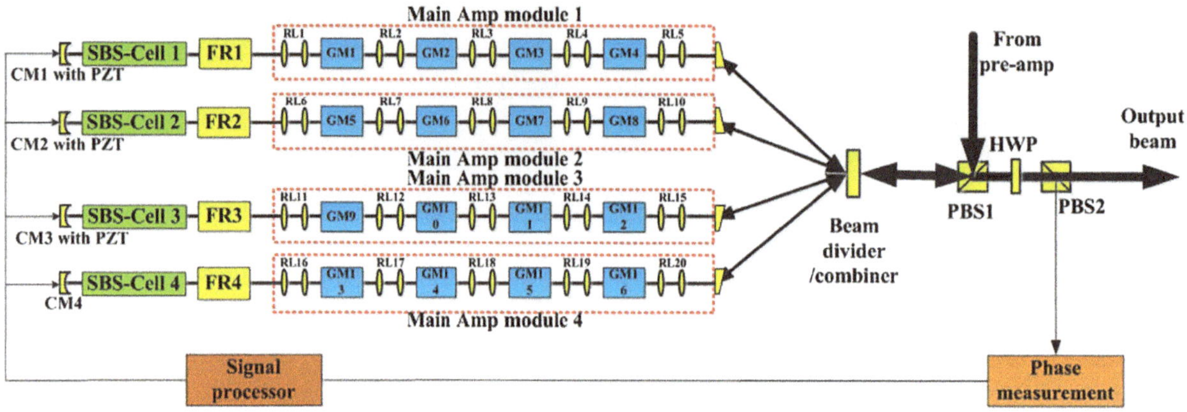

Figure 15. Schematic diagram of the main amplifier: RL1–RL20, relay lens pairs; CM1–CM4, concave mirrors.

lasers with powers of over 100 kW can be effective weapons for intercepting missiles, rockets and mortar shells with soft-kill/hard-kill methods. Further, using a neutron beam or x-ray generated by a laser, explosives and nuclear materials can be detected early. For medical applications, proton beams generated by lasers can be directed to cancer cells deep within the body and kill them while minimizing collateral damage. Neutron and proton beams can also generate radioactive isotopes that can be used in various fields including medical imaging. For manufacturing, laser peening technology can harden the surface of a material more than conventional peening. On the other hand, extreme UV (EUV) light can be generated with high brightness by high-power ns lasers. Using this EUV light, the linewidth of the lithography used in the semiconductor industry can be reduced severalfold. The laser fusion driver of inertial confinement fusion is also of great interest. To cost-effectively generate electricity, the laser fusion driver should operate with a repetition rate of at least 0.1 Hz and an energy of at least several kJ, and the whole fusion power plant will require one to two hundred such laser drivers.

The Kumgang laser aims for 2D cutting of a PCB substrate using holography. Figure 16 shows a schematic diagram of this application. The output beam of the Kumgang laser, with a power of 4 kW and a repetition rate of 10 kHz, passes through a glass hologram. The glass hologram deforms the wavefront of the laser beam and the desired cutting pattern is generated at the surface of the PCB. The conventional laser cutting industry uses CW or quasi-CW lasers. These have low peak power and need to scan a surface to operate. Since the output peak power of the Kumgang laser will be much higher than these CW lasers, the peak intensity at the PCB surface can be higher than the material processing threshold throughout the entire 2D pattern. With no moving parts, the processing time can be reduced significantly.

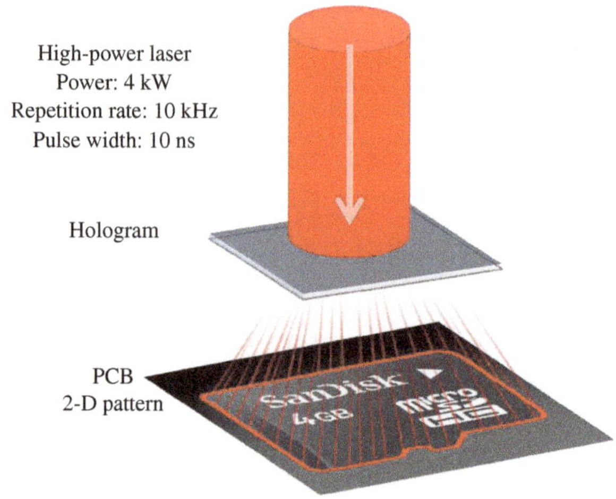

Figure 16. Schematic diagram of 2D laser processing using the Kumgang laser.

5. High-power fs/ps lasers generated by the Kumgang laser

5.1. Optical parametric chirped-pulse amplification

Since the peak power of fs/ps lasers is very high compared with ns lasers, chirped-pulse amplification (CPA) is usually employed to amplify fs/ps lasers to high power. It stretches the pulse width to a nanosecond range by dispersive optics. After amplification of the stretched and low-peak-power pulses, a compressor reverses the effect of the stretcher, creating a fs/ps pulse with higher energy. Due to many problems regarding the amplifying medium like thermal effects, amplified spontaneous emission and parasitic lasing, CPA has its limitations.

Optical parametric CPA (OPCPA) is a novel technology for building high-power fs/ps lasers. Instead of using a laser medium with a large amplification linewidth like CPA, a nonlinear crystal amplifies the chirped beam by an optical

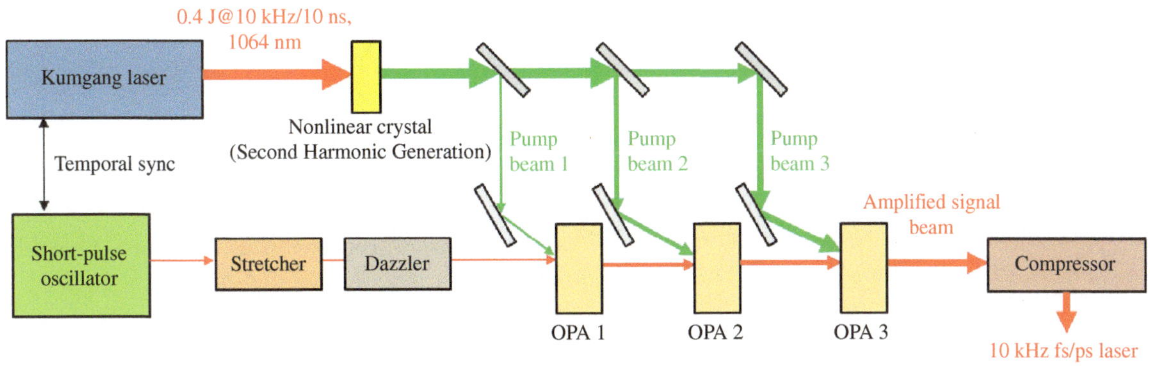

Figure 17. Schematic diagram of the generation of a fs/ps laser using the Kumgang laser as a pump source: OPA1–OPA3, optical parametric amplifiers.

parametric amplification (OPA) process. Since absorption and emission do not occur in the crystal, the thermal load is very low and the problems originating from spontaneous emission disappear. For efficient OPA to occur, the pump beam of the OPA should have a comparable pulse width to the amplifying beam. The Kumgang laser, a high-power, high-repetition-rate ns pulsed laser, will be a good pump beam for OPCPA of fs/ps lasers with a pulse energy of several tens of mJ and a repetition rate of 10 kHz.

Figure 17 shows the schematic diagram of a high-power, high-repetition-rate fs/ps laser using OPCPA. The Kumgang laser has a repetition rate of 10 kHz and a pulse energy of 400 mJ. Using a nonlinear crystal, the wavelength of the Kumgang laser is converted from 1064 to 532 nm. Meanwhile, a short-pulse oscillator generates a signal beam with a pulse width in the fs/ps range and a repetition rate of 10 kHz, which is in synchronization with the Kumgang laser optically or electronically. After passing through a stretcher, the signal beam is chirped and has a pulse width in the ns range. OPA crystals amplify the stretched pulse using the pump beam from the Kumgang laser. There are several OPA stages, and through the stages, the signal beam is amplified. After the final OPA stage, a compressor reduces the pulse width to the fs/ps range by inducing inverse chirp. A dazzler is inserted between the stretcher and the OPA stages to compensate for the residual chirping error. The final output beam is a 10 kHz pulse train with a pulse width in the fs/ps range.

5.2. Applications for a high-power, high-repetition-rate fs/ps laser

High-power, high-repetition-rate fs/ps lasers have many applications. The extremely high peak power and intensity enable various new fields. First of all, a high-brightness x-ray can be generated from the high-power fs/ps laser by various methods, including high-harmonic generation. High-brightness x-rays can probe very small and very fast processes in chemistry, materials science and medical science. On the other hand, a high-power fs/ps laser can generate a gamma-ray by Compton scattering. The properties of the laser-based gamma-ray can be adjusted easily and it is very useful in nuclear science. Furthermore, laser wakefield acceleration of protons and other ions has applications in cancer therapy and in the field of materials science. In addition, these accelerated ions can generate high-energy and high-density neutron beams by nuclear reaction. Since the neutron has no charge, it can penetrate matter at a long distance. Moreover, depending on the energy of the neutron beam, it can examine suspicious baggage non-destructively or treat cancer. Finally, the high electric field itself can probe linear and nonlinear relativistic effects in matter and in the vacuum.

6. Summary

Beam combination of lasers can contribute to the development of high-power, high-repetition-rate lasers. The SC-SBS-PCM used in a double-pass amplifier can compensate for the distorted wavefront from the amplifiers. Since the SC-SBS-PCM has a very simple and independently configured optical scheme, scalable beam combination of bulk pulsed lasers is possible.

An SC-SBS-PCM was demonstrated several times at the academic low power level of several watts. Using PZTs, the path length fluctuation between the beam lines is controlled. Four beam combination lasers using both the wavefront dividing method and the amplitude dividing method were proved experimentally.

The Kumgang laser, a beam combination laser combining four 1 kW beams to make 4 kW output (400 mJ/10 ns/10 kHz), is now under development to test whether this scheme is appropriate for high-power lasers. The Kumgang laser has four subsystems. The front-end subsystem generates the seed beam using a LD oscillator and an AOM. It amplifies the seed beam to 5 W (0.5 mJ/10 ns/10 kHz) using a fiber amplifier and a regenerative amplifier. The pre-amplifier subsystem amplifies the beam further to 200 W (20 mJ/10 ns/10 kHz) using Nd:YAG amplifiers. After dividing it into four sub-beams, the main amplifier amplifies each beam to 1 kW (100 mJ/10 ns/10 kHz) in a double-pass configuration. The wavefront combining method using a VHG or a PBS will be employed for the Kumgang laser.

There are many applications for high-power, high-repetition-rate lasers. One of the possibilities is to use a ns pulsed laser like the Kumgang as a pump source for ps/fs pulse generation by OPCPA. If the Kumgang laser functions successfully, it will be the most important step towards a Dream laser with ns, ps or fs pulse width, which is, a hypothetical laser with unlimited power and a high repetition rate.

Acknowledgements

This work was supported by the 'Dual Use Technology Program' at the Agency for Defense Development (ADD) of the Republic of Korea (UM12012RD1).

References

1. F. Bachmann, Appl. Surf. Sci. **208**, 125 (2003).
2. S. C. Wilks, A. B. Langdon, T. E. Cowan, M. Roth, M. Singh, S. Hatchett, M. H. Key, D. Pennington, A. MacKinnon, and R. A. Snavely, Phys. Plasmas **8**, 542 (2001).
3. A. Modena, Z. Najmudin, A. E. Dangor, C. E. Clayton, K. A. Marsh, C. Joshi, V. Malka, C. B. Darrow, C. Danson, D. Neely, and F. N. Walsh, Nature **377**, 606 (2002).
4. G. H. Miller, E. I. Moses, and C. R. Wuest, Opt. Eng. **43**, 2841 (2004).
5. X. Wang, P. Zhou, Y. Ma, J. Leng, X. Xu, and Z. Liu, Opt. Lett. **36**, 3121 (2011).
6. C. X. Yu, S. J. Augst, S. M. Redmond, K. C. Goldizen, D. V. Murphy, A. Sanchez, and T. Y. Fan, Opt. Lett. **36**, 2686 (2011).
7. G. D. Goodno, S. J. McNaught, J. E. Rothenberg, T. S. McComb, P. A. Thielen, M. G. Wickham, and M. E. Weber, Opt. Lett. **35**, 1542 (2010).
8. G. Mourou, B. Brocklesby, T. Tajima, and J. Limpert, Nat. Photonics **7**, 258 (2013).
9. R. L. Farrow, D. A. V. Kline, G. R. Hadley, and A. V. Smith, Opt. Lett. **31**, 3423 (2006).
10. Tso Yee Fan, IEEE J. Sel. Top. Quant. Electron. **11**, 567 (2005).
11. G. D. Goodno, H. Komine, S. J. McNaught, S. B. Weiss, S. Redmond, W. Long, R. Simpson, E. C. Cheung, D. Howland, P. Epp, M. Weber, M. McClellan, J. Sollee, and H. Injeyan, Opt. Lett. **31**, 1247 (2006).
12. R. A. Fisher, *Optical Phase Conjugation* (Academic Press, New York, 1983).
13. B. Y. Zel'dovich, V. I. Popovichev, V. V. Ragulsky, and F. S. Faizullov, Sov. Phys. JETP **15**, 109 (1972).
14. N. F. Andreev, E. A. Khazanov, S. V. Kuznetsov, G. A. Pasmanik, E. I. Shklovsky, and V. S. Sidorin, IEEE J. Quant. Electron. **27**, 135 (1991).
15. M. J. Damzen, V. I. Vlad, V. Babin, and A. Mocofanescu, *Stimulated Brillouin Scattering – Fundamentals and Applications* (Institue of Physics Publishing, Bristol and Philadelphia, 2003).
16. T. R. Loree, D. E. Watkins, T. M. Johnson, N. A. Kurnit, and R. A. Fisher, Opt. Lett. **12**, 178 (1987).
17. D. A. Rockwell and C. R. Giuliano, Opt. Lett. **11**, 147 (1986).
18. M. Valley, G. Lombardi, and R. Aprahamian, J. Opt. Soc. Am. B **3**, 1492 (1986).
19. N. F. Andreev, O. V. Palashov, G. A. Pasmanik, and E. A. Khazanov, Quant. Electron. **27**, 565 (1997).
20. H. J. Kong, S. K. Lee, and D. W. Lee, Laser Part. Beams **23**, 107 (2005).
21. H. J. Kong, S. K. Lee, and D. W. Lee, Laser Part. Beams **23**, 55 (2005).
22. W. L. J. Hasi, Z. W. Lu, S. Gong, S. J. Liu, Q. Li, and W. M. He, Appl. Opt. **47**, 1010 (2008).
23. H. J. Kong, S. K. Lee, D. W. Lee, and H. Guo, Appl. Phys. Lett. **86**, 051111 (2005).
24. S. K. Lee, H. J. Kong, and M. Nakatsuka, Appl. Phys. Lett. **87**, 161109 (2005).
25. H. J. Kong, J. W. Yoon, J. S. Shin, D. H. Beak, and B. J. Lee, Laser Part. Beams **24**, 519 (2006).
26. H. J. Kong, J. W. Yoon, J. S. Shin, and D. H. Beak, Appl. Phys. Lett. **92**, 021120 (2008).
27. J. W. Yoon, J. S. Shin, D. H. Beak, and H. J. Kong, Opt. Commun. **282**, 1000 (2009).
28. H. J. Kong, J. S. Shin, J. W. Yoon, and D. H. Beak, Laser Part. Beams **27**, 179 (2009).
29. H. J. Kong, J. S. Shin, J. W. Yoon, and D. H. Beak, Nucl. Fusion **49**, 125002 (2009).
30. J. S. Shin, S. Park, H. J. Kong, and J. W. Yoon, Appl. Phys. Lett. **96**, 131116 (2010).
31. H. J. Kong, J. S. Shin, and S. Park, J. Korean Phys. Soc. **57**, 316 (2010).
32. S. K. Case, J. Opt. Soc. Am. **65**, 724 (1975).
33. C. A. Lu, A. Flores, E. Bochove, W. P. Roach, V. Smirnov, and L. B. Glebov, Proc. SPIE **8601**, 86011A (2012).
34. Y. R. Shen, *The Principles of Nonlinear Optics* (John Wiley & Sons, Singapore, 1984).
35. S. Afshaarvahid, V. Devrelis, and J. Munch, Phys. Rev. A **57**, 3961 (1998).
36. H. J. Eichler and O. Mehl, J. Nonlinear Opt. Phys. **10**, 43 (2001).
37. R. Menzel and H. J. Eichler, Phys. Rev. A **46**, 7139 (1992).

Electromagnetic radiation from laser wakefields in underdense plasma

Yue Liu[1], Wei-Min Wang[2], and Zheng-Ming Sheng[1,3]

[1]Key Laboratory for Laser Plasmas (MoE) and Department of Physics and Astronomy, Shanghai Jiao Tong University, Shanghai, China

[2]Beijing National Laboratory of Condensed Matter Physics, Institute of Physics, CAS, Beijing, China

[3]Department of Physics, SUPA, Strathclyde University, Rottenrow 107, Glasgow, UK

Abstract

It is demonstrated by simulations and analysis that a wakefield driven by an ultrashort intense laser pulse in underdense plasma can emit tunable electromagnetic radiation along the laser propagation direction. The profile of such a kind of radiation is closely associated with the structure of the laser wakefield. In general, electromagnetic radiation in the terahertz range with its frequency a few times the electron plasma frequency can be generated in the moderate intensity regime. In the highly nonlinear case, a chain of radiation pulses is formed corresponding to the nonlinear structure of the wake. Study shows that the radiation is associated with the self-modulation process of the laser pulse in the wakefield and resulting transverse electron momenta from modulated asymmetric laser fields.

Keywords: laser wakefield; PIC simulation; tunable radiation; transverse momentum

1. Introduction

Radiation generation in relativistic laser–plasma interaction has been a hot issue in the last two decades due to its wide application in medicine, biology, chemistry, etc. Nowadays, radiation over a wide range of frequency, from terahertz (THz) to X/γ, can be produced. Meanwhile, various scenarios of radiation generation have been developed, such as high-order harmonics generation[1–3], betatron radiation[4, 5], Thomson scattering[6], Cerenkov radiation[7], and linear mode conversion[8–10]. Each scheme of producing radiation sources has its own advantages and disadvantages in terms of pulse duration, spectral range, collimating property, and energy conversion efficiency. It is well known that the laser wakefield driven in homogeneous underdense plasma generally cannot produce radiation by itself. Only when the wakefield is excited in inhomogeneous plasma[9] or in homogeneous plasma applied with an external DC magnetic field[11, 12], emission from laser wakefields can be generated. Recently, we showed that extremely ultraviolet (XUV) radiation can be produced from laser wakefields excited in homogeneous plasma[13]. The emission is caused by a transverse current sheet co-moving with the laser pulse, which is formed by an electron density spike with trapped electrons in the wakefields together with certain transverse residual momentum of electrons left behind the laser pulse.

In this work, we extend our previous work on the wakefield emission in homogeneous plasma[13] to a wider parameter range based upon numerical simulations. It is shown that electromagnetic radiation covering a wide frequency range from THz to XUV can be produced simultaneously with the laser wakefield excitation in underdense homogeneous plasmas. The strength and frequency of the radiation change significantly with the laser or plasma parameters. In the case with a weakly relativistic driving laser, the radiation produced turns out to be a multi-cycle THz pulse. If one increases the plasma density and the laser intensity, the radiation shows up in the form of a pulse chain with frequency around the laser frequency. As the laser intensity is enhanced further, wake wave-breaking occurs and XUV radiation is emitted[13, 14]. A simple theoretical analysis is presented for the physical mechanisms involved.

2. Wake radiation from one-dimensional PIC simulations

To illustrate the emission from laser wakefield excitation, a series of one-dimensional (1D) particle-in-cell (PIC) simulations were conducted with the code KLAP[15] and OSIRIS 2.0[16]. We take a laser pulse whose envelope in the form $a_L = a\sin^2(\pi t/\tau)$, where $a = \sqrt{I\lambda_L^2/(1.37\times10^{18}\ \mathrm{W\ \mu m^2 cm^2})}$,

Correspondence to: Zheng-Ming Sheng, Key Laboratory for Laser Plasmas (MoE) and Department of Physics and Astronomy, Shanghai Jiao Tong University, 200240, No. 800 Dongchuan Road, Minhang District, Shanghai, China. Email: zmsheng@sjtu.edu.cn

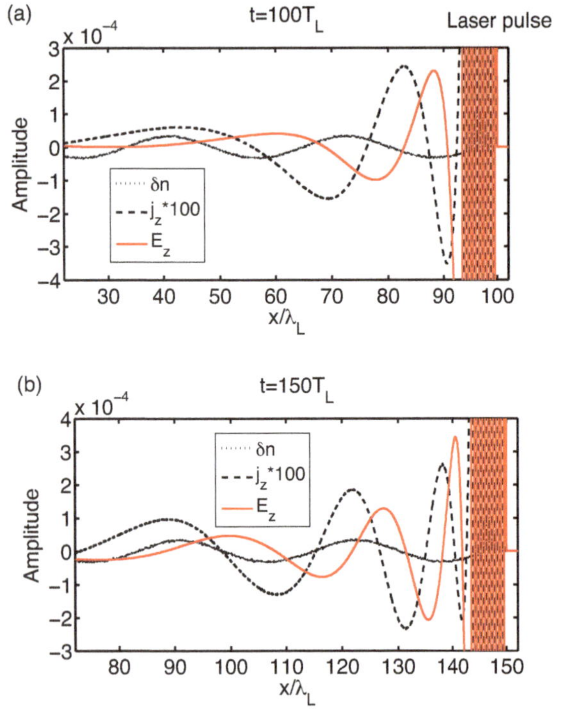

Figure 1. (Color online) The electromagnetic radiation waveform (red solid lines) in the wake at time $t = 100T_L$ and $t = 150T_L$; snapshots of the transverse current densities (blue dashed lines) and the electron densities (black dotted lines) are also plotted, where $a = 0.5$, $n_0 = 0.0001$. The filled red region indicates the incident laser.

with I the peak laser intensity. We also take $\lambda_L = 1\ \mu m$, and $\tau = 7T_L$ is the full laser duration, with T_L the laser oscillation period. It is assumed that the laser pulse is linearly polarized along the z-direction and is normally incident on a uniform plasma slab with density $n_0 = n_e/n_c = 0.001$ (corresponding to the electron plasma frequency $\omega_p/2\pi = 9.4$ THz) along the x-direction, where n_c is the critical density. The left vacuum–plasma boundary is set at $x = 20\lambda_L$. The length of the plasma slab depends on the laser propagation distance, and generally it varies between $100\lambda_L$ and $200\lambda_L$ in the simulation according to different cases. Simulations were conducted with different time and space resolutions to ensure that the results were not a numerical artefact.

Figure 1 displays the wake radiation waveform travelling forward in the wake right behind the laser pulse and the corresponding spatial distribution of the density perturbation (denoted by $\delta n = n - n_0$). The corresponding transverse current associated with the radiation is also shown. The frequency and the amplitude of the radiation behind the laser pulse are found to be different for each cycle, i.e., both the frequency and amplitude decrease with the distance from the driving laser pulse. At a later time $t = 150T_L$, there are more cycles of radiation formed. It is obvious that the amplitude of the wake radiation is larger at time $t = 150T_L$ than that at $t = 100T_L$. The wake radiation is found with the first cycle strongest. The radiation appears frequency chirped with

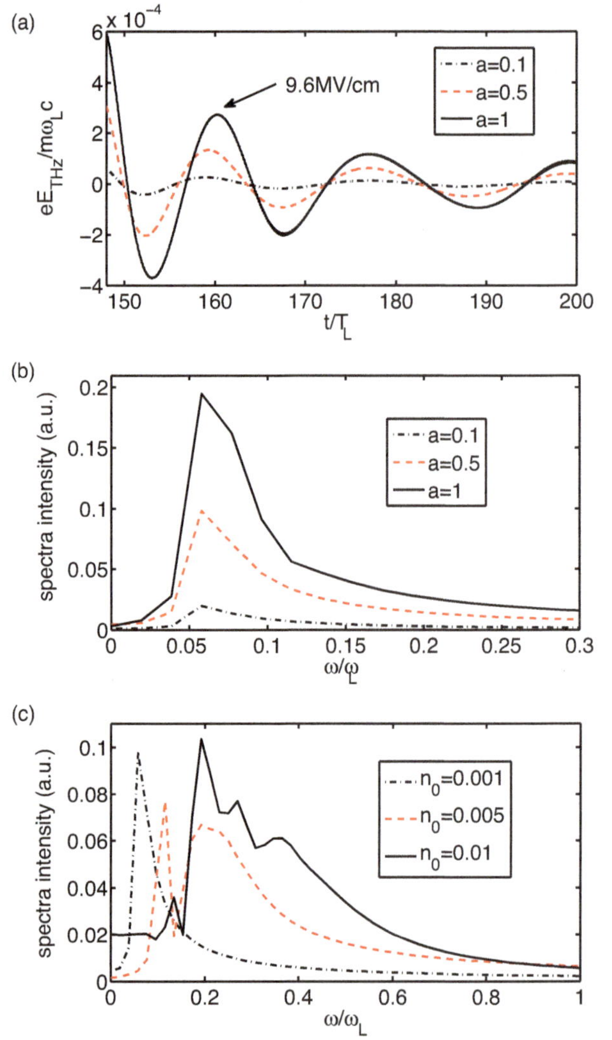

Figure 2. (Color online) The temporal waveforms of the radiation pulses observed at a fixed position at the right side of the simulation box in vacuum for three different laser intensities with initial plasma density $n_0 = 0.001$ (a) and the corresponding frequency spectra (b). Frequency spectra of the radiation produced with different plasma densities at a given laser amplitude $a = 0.5$ are also plotted in (c).

the high-frequency component at the front, which remains unchanged during the whole process. This can be understood partially as due to plasma dispersion, such that the high-frequency component propagates at a higher speed.

For plasmas with $n_0 = n_e/n_c = 0.001$, the wake radiation turns out to be in the THz frequency regime. We show the temporal evolution of the radiation observed at the right side of the plasma slab in vacuum for three typically different laser intensities and their corresponding spectra in Figures 2(a) and 2(b). The radiation penetrates through the plasma into the vacuum following the laser pulse and has a waveform with the first cycle strongest, as mentioned above. In Figure 2(a), the THz radiation amplitude is found to be up to 9.6 MV cm^{-1} for the incident laser at $\sim 10^{18}$ W cm^{-2}. However, the THz pulse frequency changes weakly with

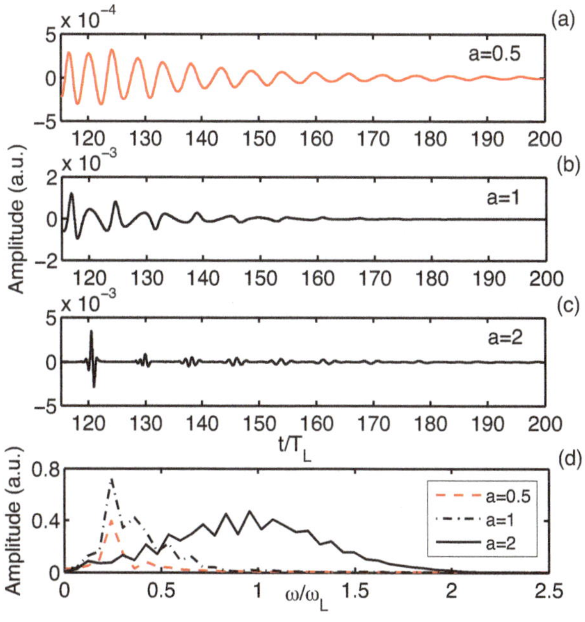

Figure 3. (Color online) The temporal evolution of the electromagnetic radiation emitted in the wake at three different laser intensities (a–c) and the corresponding frequency spectra (d), where $n_0 = 0.02$.

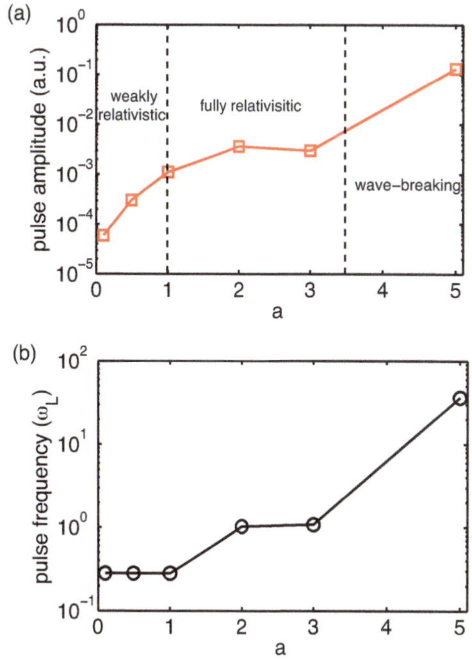

Figure 4. (Color online) The amplitude of the first pulse normalized by $m\omega_L c/e$ in the wake radiation (a) and its central frequency (b) versus the normalized laser strength a, where $n_0 = 0.02$.

the laser amplitude in Figure 2(b) in the moderate intensity regime. In this case the central frequency of the THz pulses is about 18.9 THz. In addition, if we increase the plasma density, the radiation frequency is also shown to increase. As plotted in Figure 2(c), by increasing the plasma density from $0.001n_c$ to $0.01n_c$, the frequency spectrum is broadened, and its central frequency grows from $0.0625\omega_L$ to $0.1875\omega_L$. This implies that the radiation frequency increases with increasing plasma density. Note that these frequencies are also larger than the corresponding electron plasma frequencies. In the $n_0 = 0.005$ and $n_0 = 0.01$ cases, in addition to the main peaks, there are also other subpeaks in the spectra.

Next, we increase laser intensity further. Figure 3 shows the radiation propagating forward in the right side of the simulation box in vacuum for three different incident laser intensities in plasma with initial density $n_0 = 0.02$, as well as the corresponding frequency spectra. At weakly relativistic laser intensity, the radiation displays similar temporal waveforms as shown above. But at higher intensity for $a \geqslant 1$, the radiation shows up in the form of a pulse train. This appears to be associated with the highly nonlinear features of the wakefield excitation. Note that, with the increase of the laser intensity, the separation between neighbouring subpulses also increases. This agrees with the fact that the wavelength of a nonlinear plasma wave increases with its amplitude due to the relativistic-mass increase of electrons. Similar to the case of $n_0 = 0.001$, the central frequency of the radiation is shown not to change a lot for different laser intensity in the weakly relativistic regime, even for plasma density $n_0 = 0.02$ in Figure 3(d). In the case of $a = 2$, the radiation frequency is approaching the laser frequency.

Figure 4 illustrates the amplitude of the wake radiation in its first cycle behind the laser pulse and its central frequency as a function of the normalized laser vector potential. In non-relativistic and weakly relativistic regimes, the radiation amplitude scales with the laser intensity following the power law, as shown in Figure 4(a), whereas it generally changes little with the laser intensity in the relativistic regime. When the laser intensity is high enough that wave-breaking occurs in the driven wakefield, electron injection and trapping in the first wave-bucket are found. Thus the intensity of the emitted radiation is strengthened remarkably. As shown in Figure 4(a), the field for $a = 5$ is about 20 times that for $a = 3$. This is the regime which we have identified before[13, 14]. Likewise, the frequency spectrum differs significantly for the weakly relativistic, fully relativistic, and the wave-breaking regimes, as illustrated in Figure 4(b). The central frequency of the radiation pulse remains nearly the same in the weakly relativistic regime, which confirms the previous results in Figure 2(b) under a different plasma density. In the fully relativistic regime, the radiation frequency becomes close to the laser frequency. In the wave-breaking regime, when the laser amplitude is enhanced to $a = 5$, the central frequency approaches $50\omega_L$ in the XUV regime.

3. Two-dimensional PIC simulations

To illustrate the multi-dimensional effects on the wake emission, we performed 2D PIC simulations. The 2D simulation box size is 50 μm × 50 μm, with 2500 cells along the laser propagation axis in the x-direction and 1000 cells along the

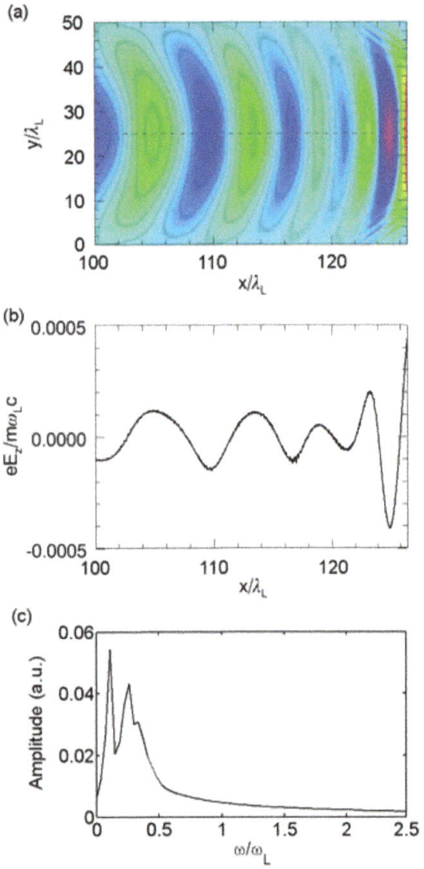

Figure 5. (Color online) 2D simulation results of wake radiation generation: (a) spatial distribution of electromagnetic radiation, (b) waveform of the radiation on the laser axis, and (c) frequency spectrum of the wake radiation, where $a = 0.5$ and $n_0 = 0.02$.

y-direction. The driving laser pulse has a Gaussian profile in the transverse direction with $10\lambda_L$ waist radius at the focal plane, and it is polarized in the z-direction with peak amplitude $a = 0.5$. The laser pulse enters plasmas with density $n_0 = 0.02$ at $t = 20T_L$, and we choose the same plasma parameters as the above 1D simulation in Figure 4. Figure 5 shows the wake radiation in the laser wakefields behind the driving laser and its transverse field on the laser axis, as well as the corresponding frequency spectrum. Study shows that the waveform of the radiation is consistent with the results found in the 1D simulation. In addition, the amplitude and the frequency of the radiation here are both similar to the results of 1D simulations (as shown in Figure 3). Generally, the wake radiation tends to decay more quickly in the 2D case. This is due to the spatial transverse spreading effect of the electromagnetic waves when the radiation source size is comparable to the radiation wavelength.

4. Theoretical model and discussions

In previous work, we proposed a simple model for XUV emission in the wakefields[13]. Here, we consider the case

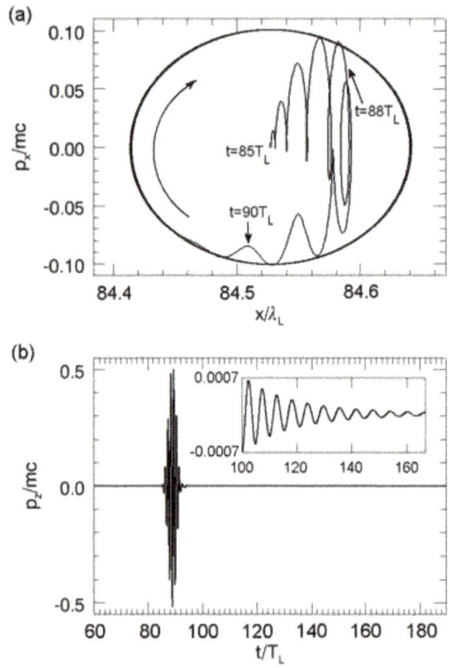

Figure 6. The trajectory of a typical electron during and after the laser interaction in phase space (a) and the temporal evolution of its transverse momentum (b), where $a = 0.5$, $n_0 = 0.02$. The inset in (b) shows the transverse momentum at a later time after the laser interaction.

for a weakly relativistic intense laser pulse passing through underdense plasmas. The key issue is to understand the mechanism of the transverse current formation, which is responsible for the wakefield emission. For the case with an ultrashort few cycle laser pulse with its waveform highly asymmetric, it has been shown that a slowly varying transverse current can be generated behind the laser pulse[17], which can produce radiation at the electron plasma frequency through the vacuum–plasma boundary. In the present case, however, the radiation is produced inside the homogeneous plasma, and its frequency is higher than the electron plasma frequency. Therefore, other physical processes may be involved.

Let us first check out a typical electron trajectory during and after the laser interaction. As illustrated in Figure 6(a), the electron is initially located at $x = 84.54\lambda_L$. During the laser interaction, it starts to move forward until it is at about $x = 84.59\lambda_L$, when it starts to move backward due to the charge separation field and the decreasing laser amplitude in its trailing edge. Finally, when the laser pulse passes by, it simply oscillates along the longitudinal direction in the established laser wakefield between $x = 84.42\lambda_L$ and $x = 84.64\lambda_L$. Figure 6(b) displays the transverse momentum of the electron during and after the laser interaction. After the laser interaction, a small transverse momentum is left. Note that the transverse momentum decays with time according to Figure 6(b). Correspondingly, the emitted radiation is also the strongest just behind the laser pulse, which agrees with the simulation results shown above. This transverse momentum oscillates much more slowly with time as compared with the laser frequency. Such oscillations occur after the laser pulse interacts with the electrons in the plasma. The net transverse momentum can be attributed to evolution of the laser pulse into slight asymmetry. Alternatively, the slowly varying transverse moment of electrons can also be attributed to self-modulation of the laser pulse. During this process, its carrier phase changes with time so that the electrons obtain net transverse acceleration at a certain time.

In the case when the driving laser is at fully relativistic intensity, the wakefield becomes highly nonlinear and has high-density peaks. The resulting radiation then appears in the form of a pulse chain with the subpulse separated by a plasma wavelength. This can explains the results shown in Figure 3(c). On either increasing the propagation distance or increasing the plasma density, simulation shows that the transverse momentum of electrons tends to increase because stronger modulation of the laser pulse may occur. This usually results in enhanced radiation, as shown in the simulation.

5. Conclusion

It is shown that, when an intense laser pulse propagates through homogeneous underdense plasma, tunable radiation in a wide frequency range can be produced in the wake of the laser pulse. Usually, it shows up with frequency higher than the electron plasma frequency and can transmit through plasma into vacuum following the driving laser pulse. Numerical simulations indicate that this kind of electromagnetic emission occurs over a wide range of parameters for laser intensities and plasma densities. In the weakly relativistic regime, the wake emission turns out to be THz radiation in extremely tenuous plasmas. With increasing plasma density and laser intensity, the radiation appears in the form of a pulse chain corresponding to the wake structure. If the laser intensity is increased further, then wave-breaking occurs in the first plasma bucket and XUV radiation is produced. Qualitatively, the self-modulation of the laser pulse and the laser wakefield excitation are responsible for the emission. To reach a quantitative understanding, further study is required. This mechanism may provide a new possibility to produce a tunable coherent radiation for various applications.

Acknowledgements

This research was supported by the National Science Foundation of China (Grant No. 11121504, 11374209, 11374210, and 11375261). The authors also would like to acknowledge the OSIRIS Consortium, consisting of UCLA and IST (Lisbon, Portugal), for providing access to the OSIRIS 2.0 framework. The computational resources utilized in this research were provided partially by Shanghai Supercomputer Center.

References

1. P. Salieres, A. L'Huillier, and M. Lewenstein, Phys. Rev. Lett. **74**, 3776 (1995).
2. E. J. Takahashi, Y. Nabekawa, and K. Midorikawa, Appl. Phys. Lett. **84**, 4 (2004).
3. B. Dromey, S. G. Rykovanov, D. Adams, R. Hörlein, Y. Nomura, D. C. Carroll, P. S. Foster, S. Kar, K. Markey, P. McKenna, D. Neely, M. Geissler, G. D. Tsakiris, and M. Zepf, Phys. Rev. Lett. **102**, 225002 (2009).
4. S. Kiselev, A. Pukhov, and I. Kostyukov, Phys. Rev. Lett. **93**, 135004 (2004).
5. S. Kneip, C. McGuffey, J. L. Martins, S. F. Martins, C. Bellei, V. Chvykov, F. Dollar, R. Fonseca, C. Huntington, G. Kalintchenko, A. Maksimchuk, S. P. D. Mangles, T. Matsuoka, S. R. Nagel, C. A. J. Palmer, J. Schreiber, K. Ta. Phuoc, A. G. R. Thomas, V. Yanovsky, L. O. Silva, K. Krushelnick, and Z. Najmudin, Nat. Phys. **6**, 980 (2010).
6. J. Meyer-ter Vehn and H. C. Wu, Eur. Phys. J. D **55**, 433 (2009).
7. C. D'Amico, A. Houard, M. Franco, B. Prade, A. Mysyrowicz, A. Couairon, and V. T. Tikhonchuk, Phys. Rev. Lett. **98**, 235002 (2007).
8. Z. M. Sheng, H. C. Wu, K. Li, and J. Zhang, Phys. Rev. E **69**, 025401 (2004).
9. Z. M. Sheng, K. Mima, J. Zhang, and H. Sanuki, Phys. Rev. Lett. **94**, 095003 (2005).
10. F. Quere, C. Thaury, P. Monot, S. Dobosz, Ph. Martin, J. P. Geindre, and P. Audebert, Phys. Rev. Lett. **96**, 125004 (2006).
11. Z. D. Hu, Z. M. Sheng, W. J. Ding, W. M. Wang, Q. L. Dong, and J. Zhang, J. Plasma Phys. **78**, 421 (2012).
12. Z. D. Hu, Z. M. Sheng, W. M. Wang, L. M. Chen, Y. T. Li, and J. Zhang, Phys. Plasmas **20**, 080702 (2013).
13. Y. Liu, Z. M. Sheng, J. Zheng, F. Y. Li, X. L. Xu, W. Lu, W. B. Mori, C. S. Liu, and J. Zhang, New J. Phys. **14**, 083031 (2012).
14. Y. Liu, F. Y. Li, M. Zeng, M. Chen, and Z. M. Sheng, Laser Part. Beams **31**, 233 (2013).
15. M. Chen, Z. M. Sheng, J. Zheng, Y. Y. Ma, and J. Zhang, Chin. J. Comput. Phys. **25**, 50 (2008).
16. R. A. Fonseca, L. O. Silva, F. Tsung, V. K. Decyk, W. Lu, C. Ren, W. B. Mori, S. Deng, S. Lee, T. Katsouleas, and J. C. Adam, Lect. Notes Comput. Sci. **2331**, 342 (2002).
17. W. M. Wang, S. Kawata, Z. M. Sheng, Y.-T. Li, and J. Zhang, Phys. Plasmas **18**, 073108 (2011).

Propagation characteristics of a high-power broadband laser beam passing through a nonlinear optical medium with defects

Xueqiong Chen[1], Xiaoyan Li[1], Ziyang Chen[1], Jixiong Pu[1], Guowen Zhang[2], and Jianqiang Zhu[2]

[1]*Fujian Provincial Key Laboratory of Light Propagation and Transformation, College of Information Science & Engineering, Huaqiao University, Xiamen 361021, China*

[2]*National Laboratory on High Power Laser and Physics Shanghai Institute of Optics and Fine Mechanics, Chinese Academy of Sciences, Qinghe Road, Jiading District 390, Shanghai 201800, China*

Abstract

The intensity distributions of a high-power broadband laser beam passing through a nonlinear optical medium with defects and then propagating in free space are investigated based on the general nonlinear Schrödinger equation and the split-step Fourier numerical method. The influences of the bandwidth of the laser beam, the thickness of the medium, and the defects on the light intensity distribution are revealed. We find that the nonlinear optical effect can be suppressed and that the uniformity of the beam can be improved for a high-power broadband laser beam with appropriate wide bandwidth. It is also found that, under the same incident light intensity, a thicker medium will lead to a stronger self-focusing intensity, and that the influence of defects in the optical elements on the intensity is stronger for a narrowband beam than for a broadband beam.

Keywords: broadband laser beam; defects; nonlinear medium; self-focusing

1. Introduction

The self-focusing of a laser beam is a basic physical phenomenon in nonlinear optics, and it has been the focus of several investigations[1–4]. It has been shown that nonlinear self-focusing is the fundamental cause that affects the laser output power. The nonlinear effect makes the whole intensity distribution of the laser change. Jokipii *et al.* presented solutions for reduction or elimination of catastrophic self-focusing in a medium[5]. Beckwitt *et al.*[6] used cascade quadratic nonlinearity to compensate self-focusing. Another factor influencing the intensity is defects in the optical elements[7,8]. The presence of defects may lead to a more complex intensity distribution and worse beam quality, and may even cause small-scale self-focusing which may damage optical elements[9,10].

The rapid development of laser techniques has promoted the generation and improvement of broadband lasers[11,12]. Increasing attention has been paid to broadband technology in high-power lasers: for example, the formation of a hot image from phase defects in high-power broadband laser beams[13], parametric characteristics of a broadband beam in free space[14], the smoothing effect through a dispersive wedge[15], and a broadband beam modulated by a hard-edged aperture[16] have been studied. However, little attention has been paid to high-power broadband lasers passing through a Kerr medium with defects, whose refractive index is dependent on the wavelength. In this paper, by treating a broadband laser as a superposition of several narrow-broadband lasers with different centre wavelengths, we study the propagation characteristics of a high-power broadband beam in a nonlinear medium, on the basis of the nonlinear Schrödinger equation. The influences of the bandwidth, the medium thickness, and the defects on the evolution of intensity are investigated. In particular, the role of the bandwidth in suppressing nonlinear self-focusing and improving uniformity is discussed. These results are important for high-power laser engineering.

2. Theory

In general, optical defects exist in optical materials, and when a high-power laser passes through optical materials, these optical defects may induce high-power laser self-focusing such that the optical materials could be severely damaged. The optical defects can be divided into two kinds: one is amplitude modulation defects, such as dirt, dust, and bubbles, and the other is phase modulation defects, such as

Correspondence to: Jixiong Pu, Fujian Provincial Key Laboratory of Light Propagation and Transformation, College of Information Science & Engineering, Huaqiao University, Jimei Road, Jimei District, Xiamen 361021, China. Email: jixiong@hqu.edu.cn

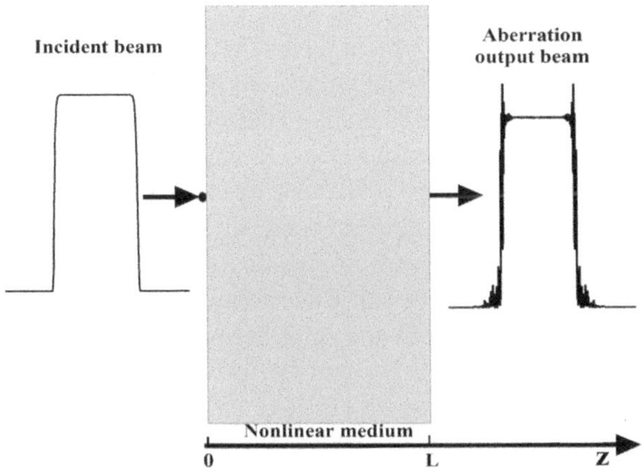

Figure 1. (Colour online) Schematic illustration of a super-Gaussian beam passing through a nonlinear medium.

protrusions, craters, and scratches. In a real situation, we assume that there is an amplitude defect with Gaussian-shaped amplitude transparency on the front surface of a Kerr medium. The mathematical description for this Gaussian-shaped amplitude modulation defect is

$$t(x, y) = 1 - A\exp\left[-\frac{(x-x_0)^2 + (y-y_0)^2}{a^2}\right], \quad (1)$$

where A is the modulation amplitude, a is the half-width of this defect, and (x_0, y_0) are the amplitude modulation (AM) centre coordinates.

It is known that the refractive index of a material depends on the wavelength of light. The widely used SF57 glass is taken as an example of a Kerr medium in this study; the dependence of the refractive index of SF57 glass on the wavelength can be expressed by dispersion formula (Sellmeier) as follows[17]:

$$n_0^2(\lambda) - 1 = \frac{B_1 \cdot \lambda^2}{(\lambda^2 - C_1)} + \frac{B_2 \cdot \lambda^2}{(\lambda^2 - C_2)} + \frac{B_3 \cdot \lambda^2}{(\lambda^2 - C_3)}, \quad (2)$$

where $B_1 = 1.81651371, B_2 = 0.428893641$, $B_3 = 1.07186278$, $C_1 = 0.0143704198$, $C_2 = 0.0592801172$, and $C_3 = 121.419942$. The unit of wavelength (λ) is μm.

The nonlinear refractive index n_2 of this glass in electrostatic units (*esu*) has the form[17]

$$n_2^{esu}(\lambda) = 6.57\frac{\left(n_0(\lambda)^2 - 1\right)^4}{n_0(\lambda)} \cdot 10^{-14} esu. \quad (3)$$

The propagation of a laser beam through a nonlinear medium is illustrated in Figure 1, in which a high-power broadband beam passes through a slab of a Kerr medium of thickness L, and then comes out of the medium and into free space. The propagation of the beam in the medium and in free space can be described by the nonlinear Schrödinger equation[18]

$$i\frac{\partial E}{\partial z} = -\frac{1}{2k}\nabla_\perp^2 E + \frac{\beta_2}{2}\frac{\partial^2}{\partial \tau^2}E - \frac{k\Delta n(\lambda)}{n_0(\lambda)}E, \quad (4)$$

and by the paraxial wave equation

$$i\frac{\partial E}{\partial z} = -\frac{1}{2k}\nabla_\perp^2 E. \quad (5)$$

The nonlinear refractive index arises from both self-phase and cross-phase modulation[18]; it is expressed as

$$\Delta n_j(\lambda_j) = n_2(\lambda_j)\left(|E_j|^2 + 2\sum_{\substack{jj=1 \\ jj\neq j}}^{n} |E_{jj}|^2\right). \quad (6)$$

Here, we divide the bandwidth $\Delta\lambda$ of the broadband laser beam into n elements, in which each bandwidth is $\Delta\lambda/n$, so that the transmission of the broadband laser can be considered as a superposition of monochromatic laser transmissions with different centre wavelengths λ_j. In this case, Equations (4) and (5) can be written as

$$i\frac{\partial E_j}{\partial z} = -\frac{1}{2k}\nabla_\perp^2 E_j + \frac{\beta_2}{2}\frac{\partial^2}{\partial \tau^2}E_j - \frac{k_j\Delta n_j(\lambda_j)}{n_0(\lambda_j)}E_j, \quad (7)$$

$$i\frac{\partial E_j}{\partial z} = -\frac{1}{2k_j}\nabla_\perp^2 E_j, \quad (8)$$

where $k_j = n_0(\lambda_j)\omega_0/c$ is the wavenumber of light, in which ω_0 is the circular frequency of the light, c is the light velocity in free space, β_2 is the group velocity dispersion coefficient of the medium, and $\Delta n(\lambda_j)$ is a additional change relating to the linear refractive index $n_0(\lambda_j)$ of the medium. In general, the laser pulse is quite long; in this case, the dependence of the light field E on τ in Equation (4) can be neglected.

It is assumed that the electric field of the incident beam is super-Gaussian, so the field can be written as

$$E_j(x, y, 0) = \exp\left[-0.5\left(\frac{x^2 + y^2}{w_0^2}\right)^N\right], \quad (9)$$

where w_0 denotes the beam radius of the incident beam and N is an integer representing the order of the super-Gaussian beam.

We can numerically calculate Equations (7) and (8) by employing the split-step Fourier method. Based on Equations Equation (1)–(9), the intensity distribution of the broadband laser beam can be obtained. In the following Section, we will discuss the influence of the bandwidth, the medium thickness and the surface defects on the intensity distribution of the laser beam.

3. Simulation results and discussion

3.1. The case of a nonlinear medium without defects

In this Section, unless other parameters are specified in the figure captions, we select the parameters as follows: the spatial sampling range is 1 cm × 1 cm, the transverse field is sampled over a grid of 1024 × 1024 points, for the incident super-Gaussian beam, $N = 4$, $w_0 = 3$ mm, and the central wavelength of broadband beam $\lambda_0 = 1053$ nm,

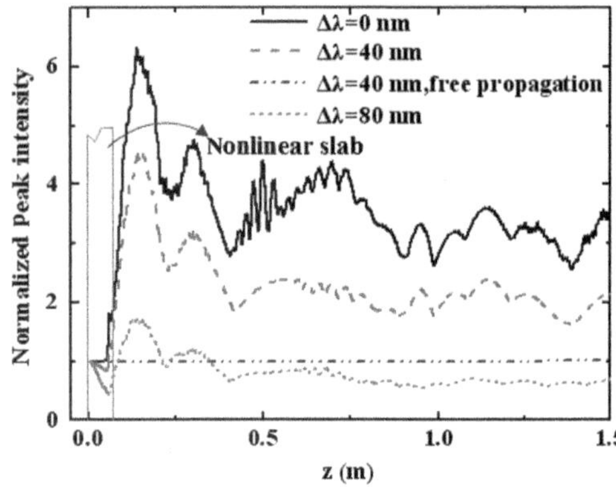

Figure 2. (Colour online) The normalized peak intensity of a super-Gaussian beam against the propagation distance for different bandwidths. $n_0(\lambda_0) = 1.812$, $n_2(\lambda_0) = 2.28 \times 10^{-6}$ cm^2/GW, $L = 6$ cm, the super-Gaussian beam $I_0 = 12$ GW/cm^2, $N = 4$, and $w_0 = 3$ mm.

the bandwidth $\Delta\lambda = 40$ nm, $n = 20$, and $\Delta\lambda_j = 2$ nm, $I_{0j} = I_0 = 12$ GW/cm^2. The intensity is normalized to the initial intensity I_0; the medium thickness $L = 6$ cm, $n_2(\lambda_0) = 2.28 \times 10^{-6}$ cm^2/GW, and $n_0(\lambda_0) = 1.812$.

To clearly demonstrate the effect of bandwidth on the nonlinear transmission, we first compare the properties of the peak intensity produced by a broadband laser system to that by a narrowband laser system. For a narrowband laser, we consider the limiting case of a monochromatic laser (i.e., $\Delta\lambda = 0$ nm). Figure 2 shows the variation of the normalized peak intensity of a laser beam with different bandwidths along the propagation distance. It is shown that when the incident laser is monochromatic, the maximum peak intensity reaches 6.3. We can find that the maximum peak intensity decreases with the increase of the bandwidth of the broadband beam; i.e., when $\Delta\lambda = 40$ nm, the maximum peak intensity is 4.6, and when $\Delta\lambda = 80$ nm, the maximum peak intensity decreases to 1.6. This result indicates that the increase of the bandwidth of the laser beam may effectively decrease the maximum peak intensity, so that decreases the risk of damage to the optical elements. The reason for decreasing the maximum peak intensity is due to the effect of dispersion for a broadband laser beam propagating in optical materials. The major effects of a monochromatic laser propagating in a nonlinear medium are diffraction and nonlinear effects: if the incident laser intensity is very high, the nonlinear effect is stronger than the diffraction effect, so the peak intensity of the laser beam increases with the propagation distance. Because a broadband beam propagating in a nonlinear medium is influenced by material dispersion, the self-focusing can be largely suppressed due to dispersion.

Figure 3 shows the evolution of the peak intensity of a broadband laser beam passing through a nonlinear medium with different thicknesses along the propagation direction. It is shown in Figure 3(a) that, under a certain incident intensity (such as $I_0 = 12$ GW/cm^2), a thicker medium will lead to a stronger self-focusing. This means that, for a thicker nonlinear medium, the focal position is closer to the left side. In Figure 3(b), we plot the evolution of the peak intensity of a broadband laser beam passing through a nonlinear medium with different thicknesses along the propagation direction for the case that the B integral is fixed. We can find that in this case the peak intensity is nearly constant.

It has been assumed that the incident laser beam is a super-Gaussian one. A previous reported study has shown that a larger order N of the super-Gaussian beam will result in stronger modulation on the laser beam; therefore, the uniformity of the light beam will be worse with a larger N. To clearly demonstrate the effect of the bandwidth of a laser beam on the beam uniformity, the order of the super-Gaussian laser beam is chosen as $N = 20$. Figure 4 shows the evolution of the transverse intensity for laser beams of two different bandwidths ($\Delta\lambda = 0$ nm and $\Delta\lambda = 40$ nm) propagating in a nonlinear medium and free space. It is found from Figure 4(a) that the intensity distribution of the narrowband beam has many spikes; i.e., the uniformity of the light intensity is very poor. These spikes of intensity may induce damage in the optical medium. In Figure 4(b), we plot the evolution of the transverse intensity distribution for the case of a broadband laser beam propagating in a nonlinear medium and free space. It is found that the spikes of the transverse intensity distribution are greatly decreased, and so the beam uniformity is improved. This phenomenon can be easily understood, because, for a different wavelength (or frequency), the transverse intensity distribution is slightly different, so the integrated intensity distribution for different wavelengths becomes uniform.

3.2. The case of a nonlinear medium with defects

All of the above studies focus on the cases of an ideal medium (i.e., a medium without defects). Now we consider the case of a laser beam passing through a nonlinear optical medium with defects. Figure 5 presents the evolution of the maximum intensity of a laser beam propagating through a medium in the presence of an amplitude defect on the front surface. It is found that, for a narrowband laser beam ($\Delta\lambda = 0$ nm), the defect on the medium surface makes the maximum intensity larger, and that accelerates the self-focusing process. By comparing Figure 2 with Figure 5, we find that the propagation characteristic of a broadband beam passing through a nonlinear medium with defects is different from that for a broadband beam passing through a nonlinear medium without defects. It is shown that the maximum intensity reaches 12 for a narrowband laser beam, which is nearly twice that for the case with the non-defect medium. This indicates that, for a laser beam passing through a nonlinear medium with defects, the maximum intensity is shown to be larger; therefore it may increase the risk of damaging optical systems. We also find from Figure 5 that a broadband laser beam can make the maximum intensity decrease, and that decreases the risk of damaging optical systems.

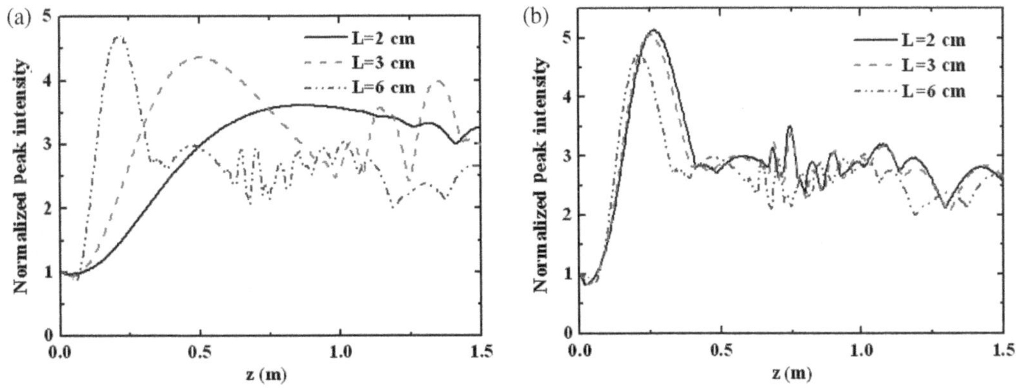

Figure 3. (Colour online) Evolution of the normalized peak intensity of a broadband super-Gaussian beam passing through different thicknesses of nonlinear medium along the propagation direction. (a) The incident intensity is fixed, $I_0 = 12$ GW/cm^2. (b) The *B* integral is fixed. $I_0 = 36$ GW/cm^2 when $L = 2$ cm; $I_0 = 24$ GW/cm^2 when $L = 3$ cm; $I_0 = 12$ GW/cm^2 when $L = 6$ cm. $N = 4$, $w_0 = 3$ mm, and $\Delta\lambda = 40$ nm.

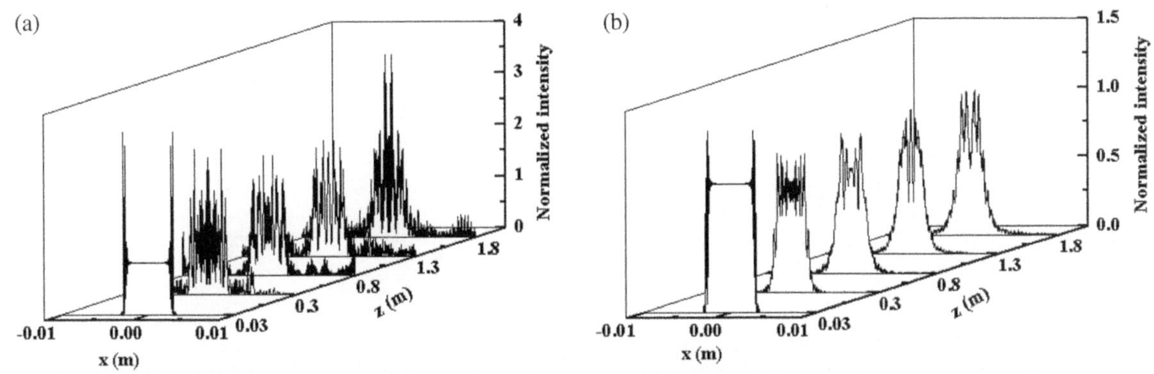

Figure 4. Intensity distribution of a narrowband beam, $\Delta\lambda = 0$ nm (a) and a broadband beam, $\Delta\lambda = 40$ nm (b) at different distances. $n_0(\lambda_0) = 1.812$, $n_2(\lambda_0) = 2.28 \times 10^{-6}$ cm^2/GW, $L = 6$ cm, $I_0 = 12$ GW/cm^2, $N = 20$, and $w_0 = 3$ mm.

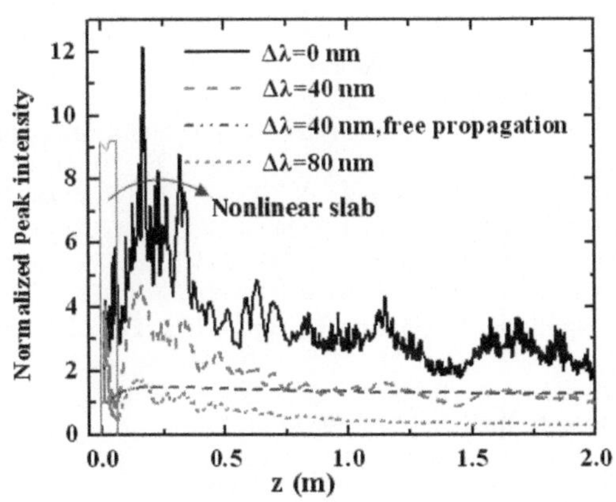

Figure 5. (Colour online) Evolution of the normalized peak intensity of a super-Gaussian beam for different bandwidths, considering the medium's front surface to have a defect. The defect size $a = 100$ μm, $A = 1$, $x_0 = 0$; $n_0(\lambda_0) = 1.812$, $n_2(\lambda_0) = 2.28 \times 10^{-6}$ cm^2/GW, $L = 6$ cm, $I_0 = 12$ GW/cm^2, $N = 4$, and $w_0 = 3$ mm.

We plot the evolution of the transverse intensity distribution of a laser beam with two different bandwidths ($\Delta\lambda = 0$ nm and $\Delta\lambda = 40$ nm) passing through a nonlinear medium with a defect in Figure 6, in which the parameters of the defect are the same as those in Figure 5. It is shown from Figure 6(a) that when the narrowband laser beam passes through a nonlinear medium with defects, the transverse intensity distribution has many spikes, being due to the high-frequency modulation by the defect. This high-frequency modulation seems to be an important reason for the acceleration of the nonlinear self-focusing effect. However, as shown in Figure 6(b), a broadband beam can significantly suppress small-scale self-focusing, partially erasing the self-focusing filaments and burrs caused by the defect and the diffraction effect.

In Figure 7, we plot the evolution of the transverse intensity distribution at several propagation positions for the case that the position of the defect modulation point is not in the centre of the beam. It is seen that the transverse intensity distribution is asymmetric, and when the incident laser beam is of narrowband type, there also exist many spikes. We can find that when the incident laser beam is of broadband type, the transverse intensity distribution is also asymmetric; however, the spikes in the transverse intensity distribution have been largely suppressed.

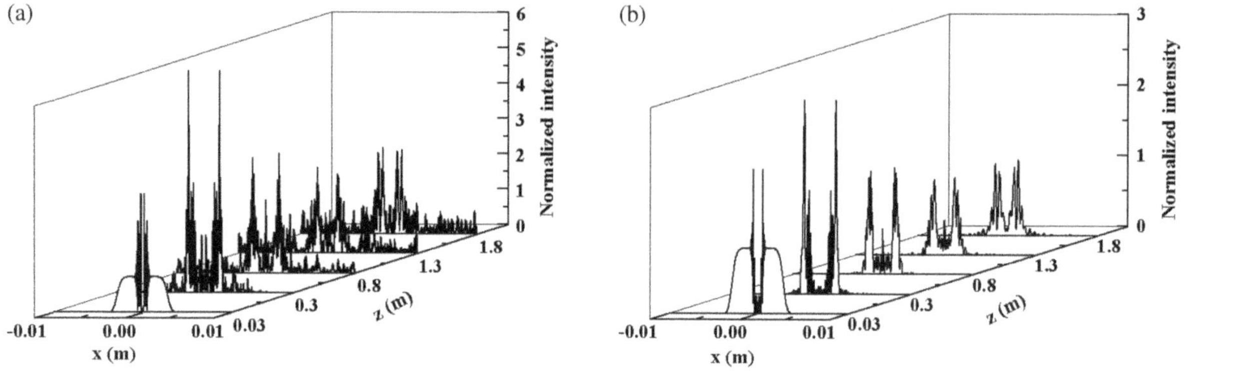

Figure 6. The propagation of a narrowband beam, $\Delta\lambda = 0$ nm (a) and a broadband beam, $\Delta\lambda = 40$ nm (b) through a nonlinear medium whose front surface contains a defect. Here, the parameters are the same as in Figure 5.

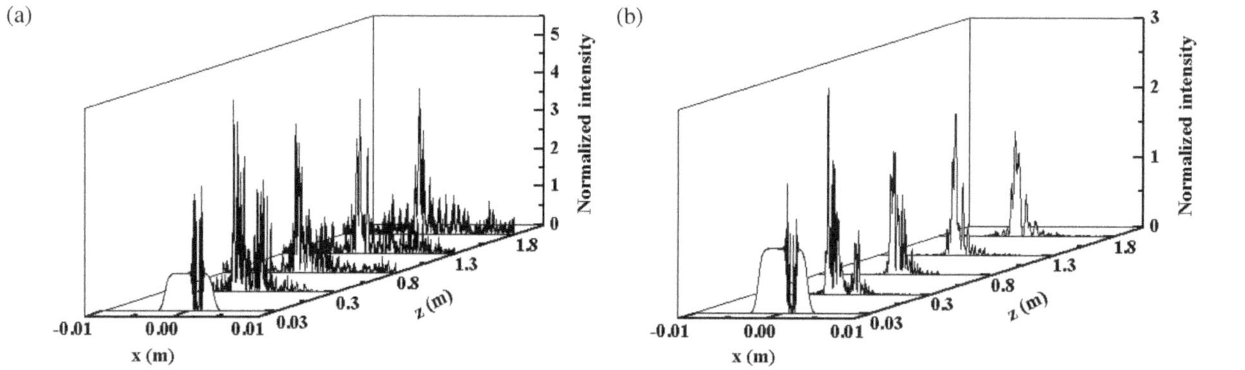

Figure 7. The propagation of a narrowband beam (a) and a broadband beam (b) when the defect is not in the centre, $x_0 = 0.001$ m; the other parameters are the same as in Figure 6.

4. Conclusions

In conclusion, we have studied the propagation characteristics of a high-power broadband laser beam in a nonlinear medium with defects or without defects. The effects of the bandwidth, the medium thickness, and defects on the evolution of the transverse intensity distribution have been studied. It has been found that, when the incident laser beam is of narrowband type, the transverse intensity distribution has many spikes, which may damage the optical medium. However when the incident laser beam is of broadband type, the spikes of the intensity distribution can be suppressed, decreasing the risk of the optical medium being damaged. The results of this paper are important for high-power laser engineering.

Acknowledgements

This research was supported by the National Natural Science Foundation of China (NSFC) under Grant Nos. 61178015 and 11304104, and the Open Research Fund of the National Laboratory on High Power Laser and Physics under Grant No. SG-001102.

References

1. A. M. Rubenchik, M. P. Fedoruk, and S. K. Turitsyn, Phys. Rev. Lett. **102**, 233902 (2009).
2. Fang Liu, Zhigang Liu, Liunian Zheng, Hongbiao Huang, and Jianqiang Zhu, High Power Laser Sci. Eng. **1**, 29 (2013).
3. M. Miclea, U. Skrzypczak, S. Faust, F. Fankhauser, H. Graener, and G. Seifert, Opt. Express **18**, 3700 (2010).
4. H. H. Lee, K. M. Chae, S. Y. YIM, and S. H. Park, Opt. Express **12**, 2603 (2004).
5. S. R. Jokipii, and J. Marbmger, Appl. Phys. Lett. **23**, 696 (1973).
6. K. Beckwitt, F. W. Wise, L. Qian, L. A. Wallker II, and E. Canto-Said, Opt. Lett. **26**, 1696 (2001).
7. J. Neauport, L. Lamaignere, H. Bercegol, F. Pilon, and J. C. Birolleau, Opt. Express **13**, 10163 (2005).
8. S. Mainguy, I. Tovena-Pecault, and B. Le Garrec, Proc. SPIE **5991**, 59910G (2005).
9. V. I. Bespalov, and V. I. Talanov, Jept. Lett. **3**, 307 (1966).
10. H. T. Jia, B. Xu, F. Wang, and L. D. Zhou, Appl. Opt. **51** 6089 (2012).
11. Y. Xu, J. Wang, Y. Huang, Y. Li, X. Lu, and Y. Leng, High Power Laser Sci. Eng. **1**, 98 (2013).
12. M. B. Johnston, D. M. Whittaker, A. Dowd, A. G. Davies, E. H. Linfield, and X. Li, Opt. Lett. **27**, 1935 (2011).
13. Y. H. Hu, Y. W. Wang, S. C. Wen, J. Q. Deng, and D. Y. Fan, Opt. Laser. Eng. **47**, 194 (2009).
14. H. Mao, and D. Zhao, Appl. Phys. B **100**, 611 (2010).

15. R. W. Peng, Y. X. Ye, Z. X. Tang, C. J. Zhao, S. C. Wen, and D. Y. Fan, Opt. Commun. **265**, 106 (2006).
16. R. W. Peng, Y. X. Ye, Z. X. Tang, and D. Y. Fan, J. Opt. Soc. Am. A **22**, 1903 (2005).
17. "Schott optical glass collection data sheets inquiry glass".
18. G. P. Agrawal, *Nonlinear Fiber Optics and Application of Nonlinear Fiber Optics* (Academic, 2001), p. 26.

Petawatt class lasers worldwide

Colin Danson[1], David Hillier[1], Nicholas Hopps[1], and David Neely[2]

[1]*AWE, Aldermaston, Reading RG7 4PR, UK*

[2]*STFC Rutherford Appleton Laboratory, Chilton, Didcot, Oxon OX11 0QX, UK*

Abstract
The use of ultra-high intensity laser beams to achieve extreme material states in the laboratory has become almost routine with the development of the petawatt laser. Petawatt class lasers have been constructed for specific research activities, including particle acceleration, inertial confinement fusion and radiation therapy, and for secondary source generation (x-rays, electrons, protons, neutrons and ions). They are also now routinely coupled, and synchronized, to other large scale facilities including megajoule scale lasers, ion and electron accelerators, x-ray sources and z-pinches. The authors of this paper have tried to compile a comprehensive overview of the current status of petawatt class lasers worldwide. The definition of 'petawatt class' in this context is a laser that delivers >200 TW.

Keywords: diode pumped; high intensity; high power lasers; megajoule; petawatt lasers

1. Motivation

The last published review of high power lasers was conducted by Backus *et al.*[1] in 1998. At this time there was only one petawatt class laser, the NOVA petawatt[2], in existence. The field has moved on a long way since then with over 50 petawatt class lasers currently operational, under construction or in the planning phase. The possibility of using focused high intensity laser beams to achieve previously unobtainable states of matter in the laboratory gained much attention after the demonstration of the first pulsed laser[3] in 1960. Potential applications, such as generating the conditions for fusion in the laboratory, became a major driver for the early development of high power lasers in the 1960s to 1980s. It was realized that although matter could be heated[4] to hundreds of electron volts using ~ns pulses and directly compressed using light pressure ($\sim I/c$, where I is the intensity and c is the speed of light), spherical compression using laser driven ablation could achieve much higher pressures and densities[5], suitable for the achievement of fusion conditions[6]. First estimates for laser driven fusion[7] proposed lasers delivering 20 ns shaped pulses of megajoule energies, operating at 100 Hz, eventually leading today to megajoule scale projects such as NIF[8] and LMJ[9].

The potential to interact with hot plasmas (greater than hundreds of electron volts) and probe the growth of instabilities and perturbations on timescales where hydrodynamic motion is small during the laser pulse ($\tau c_s \leqslant \lambda$, where c_s is the sound speed of the plasma, τ is the laser pulse length and λ is the laser wavelength) pushed the development of lasers with pulse durations τ of less than tens of picoseconds. The development and delivery of chirped pulse amplification (CPA)[10] in large aperture laser systems enabled a rapid push towards even shorter pulses (from picoseconds to attoseconds) and higher intensities, where relativistic and field effects associated with the laser pulse dominate the interaction physics[11]. Understanding and learning to control and manipulate the complex interactions taking place at the laser/matter interface led to a wide variety of experiments and potential new scientific[12] and industrial applications being pursued and necessitating the development of matching laser capability, some of which are outlined below.

The production of quasi-coherent VUV/soft x-ray sources for biological imaging or plasma probing was investigated using 'recombination pumping[13]' where, after heating a plasma, it was allowed to expand, usually into vacuum, and rapidly recombine, ideally creating a population inversion in an ionized state such a hydrogen-like carbon[14]. To achieve higher gain[15], shorter duration laser pulses were required, and by 1995, high power ~terawatt pulses of ~20 ps duration had been developed. Collisional excitation soft x-ray laser pumping using high (~kilojoule) energy, nanosecond pulses was first demonstrated at high gain with neon-like selenium[16] and subsequently and more efficiently

Correspondence to: C. Danson, AWE, Aldermaston, Reading RG7 4PR, UK. Email: colin.danson@awe.co.uk in neon-like germanium[17]. The highest possible brightness with a soft x-ray laser is obtained when it is operated in saturation, and this was initially achieved in 1992 using neon-like germanium[18] at 23 nm with a 500 ps pump, and later in 1997 with nickel-like samarium[19] at 7 nm using a 50 ps pump. To improve the efficiency of these devices[20], shorter pulse pumping using a mode of operation termed 'transient collisional excitation' helped to push the development of shorter pulse laser drivers. High transient collisional excitation gain was demonstrated using an 800 ps low energy pulse to pre-form a large plasma volume and then a 5 J ps pump to generate the transient collisional excitation to deliver gain at 14.7 nm[21] in nickel-like palladium.

The generation of quasi-coherent VUV/soft x-ray sources using high harmonics[22] rather than soft x-ray lasers was given a significant boost in the mid-1990s by the observation of the 68th harmonic of a 1.05 μm driving laser at 15.5 nm, using a 2.5 ps pulse focused to an intensity of 10^{19} W cm^{-2} on a solid target[23]. This was extended into the keV regime using petawatt power pulses focused to 10^{21} W cm^{-2} intensities[24] by 2007, and into the attosecond region[25] using even shorter few femtosecond optical driving pulses. The production of high currents of MeV electrons[26] and associated gamma-ray production[27] brought significant attention to the scale length of the interaction. At such intensities, any illumination of the target above the ionization threshold (10^{11}–10^{12} W cm^{-2}) can generate a pre-plasma which expands and dramatically changes the scale length of the interaction. The ability to control the scale length of the interaction[28] led to significant effort in improving the laser contrast and developing pre-pulse mitigation strategies (frequency doubling of high power short pulses[29, 30]; plasma mirrors[31]; saturable absorbers[32]; XPW techniques[33]; low gain OPA[34]; short pulse OPA[35]).

The concept of using the electric field associated with a laser driven plasma wave to accelerate electrons was given a major boost in the late 1970s when it was realized that GeV/cm fields could be potentially achieved[36]. Subsequent work utilizing the then available lasers investigated excitation of instabilities[37] and driving of the plasma at two different wavelengths using either 2 ns pulses from a carbon dioxide based laser[38] or Nd:glass lasers[39] to generate suitable beat-wave plasma modulations. By the mid 1990s, wave-breaking[40] generated electron beams with thermal-like spectra up to 45 MeV using a 25 TW picosecond driver were achieved. Using shorter 50 fs pulses at 10^{19} W cm^{-2} intensities, near mono-energetic beams of electrons were produced[41] and, currently, electrons of >GeV energies can be created with petawatt class sub-50 fs Ti:sapphire drivers[42].

Laser driven particle acceleration for applications such as ion driven fast ignition[43] (requiring <15 MeV protons) and medical purposes[44] (60–300 MeV protons or heavier ions) has attracted a significant amount of research effort since the turn of the century. 'Target normal sheath acceleration' using a petawatt class picosecond laser[45] was used to accelerate a population of electrons through a metallic foil, creating a large sheath field on the rear side which resulted in a highly laminar ion beam containing large fluxes (>10^{13}) of high energy (>10 MeV) ions. In subsequent experiments using tens of terawatt drivers, it was demonstrated that improved efficiency could be achieved by reusing the laser driven electrons[46] as they bounce back and forth in the foil target, termed 'recirculation'. As electron recirculation[47] experiments pushed to thinner and thinner targets (<50 nm thick foils by ~2007) at intensities of >10^{19} W cm^{-2}, laser system contrasts of >10^9 were routinely required. Currently, the new contrast enhancing techniques described earlier will need to used in combination with enhanced picosecond cleaning schemes to achieve picosecond intensity contrasts of >10^{11}, which are essential to explore new mechanisms[48] of ion acceleration at intensities of >10^{21} W cm^{-2}.

2. The road to petawatt class lasers

From the first demonstration of the laser, attempts have been made to increase the peak power and focused intensity in order to reach extreme conditions within the laboratory. Initial jumps in peak power came with the invention of Q-switching then mode locking, but progress slowed until the late 1980s and the dawn of CPA. The original use of CPA was in radar systems where short, powerful pulses that were beyond the capabilities of existing electrical circuits were needed. By stretching and amplifying the pulses prior to transmission, then compressing the reflected pulse, high peak powers within the amplifier circuitry could be avoided.

These ideas were first applied in a laser amplification scheme at the Laboratory for Laser Energetics at the University of Rochester, USA by Strickland and Mourou[10]. Here, the output from a mode-locked Nd:YAG oscillator was stretched and spectrally broadened by 1.4 km of optical fibre, amplified in a Nd:YAG regenerative amplifier then compressed using a Treacy grating pair[49] which compensated for the second order spectral phase imposed by the fibre.

Due to the limitations of mode-locked lasers operating at 1064 nm, early high power/energy CPA lasers[50–53] all relied on the use of self-phase modulation to generate enough bandwidth to support sub-few-picosecond pulses[54]. These systems generated large amounts of high order spectral phase and spectral modulations during the nonlinear process, making optimal compression hard to realize, and, moreover, these systems had poor stability due to the nonlinear process.

The development of Ti:sapphire mode-locked oscillators[55] allowed much shorter pulses to be produced. These systems could either directly seed Ti:sapphire amplifiers[56] or, if tuned to 1054 nm, be used to seed existing large aperture

Nd:glass systems. Other developments around this time included a neodymium based additive pulse mode locking system[57] which could generate pulses at under 0.5 ps at 1054 nm. These approaches were developed simultaneously in France[58, 59] and in the UK[60, 61], producing the first well defined, 100 TW class laser systems.

In the telecommunications industry, work was carried out on the use of prisms[62] and grating pairs[63] to compensate for the spectral phase distortions imposed on broad-bandwidth laser pulses by long lengths of optical fibre. By putting a telescope inside a grating pair Martinez produced a method to reverse the sign of the spectral phase that was imparted, thus creating a device that could stretch a pulse then exactly compress it. These systems were used in stretching pulses prior to propagation along the fibre then compressing them in order to reduce nonlinear effects. After the development of CPA various geometries of stretcher such as the Offner triplet[64] were developed, allowing longer stretches to be realized and more energy to be propagated for a constant stretched intensity.

The development of amplifiers capable of supporting broad bandwidths is also required to realize high peak powers. Early systems relied entirely on dye or Nd:glass amplifiers. While dye lasers could support very large bandwidths, their short lifetimes and low saturation fluences severely limited the amount of energy that could be extracted. Neodymium based lasers, on the other hand, could provide a large amount of energy but would support only a limited bandwidth.

This led to the search for a new laser material that could provide the energy and bandwidth required to support high energy short pulses. Ti:sapphire[65] and optical parametric amplification[66] provided the solution to these problems. These were initially used in the pre-amplification stages of multi-terawatt systems in conjunction with Nd:glass rod or disc amplifiers. They provided many orders of magnitude of gain at high bandwidth before larger amplifiers, generally Nd:glass, added the last few orders and reduced the bandwidth. As the quality and size of available Ti:sapphire and nonlinear crystals have improved, so has the energy that can be extracted from these systems.

3. Kilojoule glass systems

The first kilojoule glass system, or in fact the first laser configured to deliver a petawatt, was at the Nova Facility at Lawrence Livermore National Laboratory (LLNL)[2]. One beamline of the high energy Nova beamlines could be converted to operate in short pulse mode with a dedicated front end and vacuum compressor. The compressor was a pair of single pass in-house manufactured 940 mm diameter gold coated gratings, shown in Figure 1. These gratings would go on to be used in systems throughout the world. The beamline was capable of delivering 660 J in a 440 fs pulse giving 1.5 PW to the target and focused intensities of $>7 \times 10^{20}$ W cm^{-2}. All of the basic building blocks used on later systems were deployed on Nova, including broad-bandwidth pulse generation, optical pulse stretching, pulse amplification, deformable mirror, pulse compression and reflective focusing.

Figure 1. Inside the pulse compressor of the NOVA Petawatt – the first petawatt class laser worldwide (picture courtesy of LLNL).

Figure 2. The Vulcan Petawatt Target Hall (picture courtesy of STFC).

Vulcan was the first petawatt class laser to be used by the international plasma physics community as a dedicated user facility. It is a high power Nd:glass laser[67] which has been operational for over 30 years. It enables a broad range of experiments through a flexible geometry[68, 69]. It has two target areas: one with 6 × 300 J (1053 nm @1 ns) long pulses combined with two synchronized short pulse beams and a separate target area with high energy petawatt capability (500 J in 500 fs) synchronized with a single long pulse beamline, shown in Figure 2.

The concept of using an OPCPA (optical parametric chirped pulse amplification) system as a seed for the front end of a high power Nd:glass laser system was first proposed by Ross[70] from the Central Laser Facility. This allowed an ultra-short pulse to be amplified in a broad-bandwidth pre-amplifier before injection into the larger aperture Nd:glass chain, giving shorter pulses to the target and enabling higher contrasts to be realized[71]. The first OPCPA front-end system became operational on the Vulcan facility in 1998[72, 73]. In subsequent years, many facilities implemented these front-end systems[74–76].

In Asia, the first petawatt class laser was constructed as part of the high energy Nd:glass Gekko XII facility at Osaka University, Japan[77]. They started to implement ultra-short pulse lasers to couple up to Gekko XII[78] with a 30 TW GMII laser, initially for general ultra-high intensity research but lately more focused on the fast-ignition concept for ICF. The petawatt used an OPCPA front end with Nd:glass large aperture amplifiers and a double pass compressor to produce 420 J in a 470 fs pulse giving output powers of 0.9 PW. An F#7 off-axis parabola was used to focus to target, giving focused intensities of 2.5×10^{19} W cm^{-2} with contrast levels of 1.5×10^{-8}.

Titan[79] is one of the five lasers that make up the Jupiter Laser Facility at LLNL. It is a petawatt class laser coupled to a kJ beamline for a broad range of experiments. The short pulse beamline delivers up to 300 J in a sub-picosecond pulse and offers a 50 J high contrast green option.

An interesting development has been the coupling of petawatt beamlines to other sources, including ion beams and electron beams, and at Sandia National Laboratory coupled to the Z-pinch accelerator. The facility uses Beamlet[80], which was the original prototype facility for NIF at LLNL that was decommissioned in 1998 before being transferred to Sandia. Z-Beamlet[81] provides x-ray radiographic capability to the Z-pinch facility. The upgrading of the facility to Z-Petawatt[76] provides enhanced radiographic capability. The beamline, which consists of an OPCPA front end and Nd:phosphate glass amplifiers, delivers 500 J in 500 fs.

The Texas Petawatt Laser[82] based at the Texas Center for High Intensity Laser Science at the University of Texas at Austin uses a high energy OPCPA front end with optimized mixed glass to produce shorter pulses than traditional glass petawatt facilities. The OPCPA system amplifies pulses up to the joule level with broad bandwidth followed by a relatively modest final amplification factor of ~400 in mixed glass Nd:glass amplifiers. The first 64 mm rod is silicate with eight pass angular multiplexing then four pass through two pairs of phosphate disc amplifiers. The 1.1 PW beamline produces a bandwidth of 14.6 nm, delivering 186 J in 167 fs.

The PHELIX (Petawatt High Energy Laser for heavy Ion eXperiments) laser[83] was constructed at the Helmholtz Center GSI and is used in conjunction with a heavy ion accelerator. The laser can be switched between long and short pulse operation and in short pulse mode is designed to deliver 400 J in 400 fs.

The first petawatt laser in China was built as an auxiliary beamline to the Shenguang (Divine Light) II high energy facility at the Shanghai Institute of Optics and Fine Mechanics (SIOM)[84] and is still operational. SG-II was an eight-beam Nd:glass laser facility operating at a total of 6 kJ IR or 2 kJ 3ω. A ninth beam of 4.5 kJ was commissioned and made operational in 2005 and subsequently converted to the SG-II-U PW beamline. SG-II-U also included the building of a separate 24 kJ, 3ω, 3 ns eight-beam facility.

Figure 3. One of the Orion pulse compressor gratings (picture courtesy of AWE).

Orion is the latest facility to be built in the UK and became operational in April 2013[85]. It is a Nd:glass laser system which combines 10 long pulse beamlines (500 J, 1 ns @ 351 nm) with two synchronized infrared petawatt beams (500 J in 500 fs). One of the Orion large aperture compressor gratings is shown in Figure 3. An ultra-high contrast option is available by frequency doubling at sub-aperture (300 mm) one of the petawatt beamlines to operate in green, giving 100 J in <500 fs with nanosecond contrast levels of $<10^{-14}$[30].

4. Multi-kJ glass systems

The multi-kJ petawatt beamlines have all been primarily built to give advanced x-ray radiography capability to mega-joule class long pulse interaction facilities. They typically operate at a pulsewidth of ~10 ps with multi-kJ energy outputs. The beamlines are also used for fast-ignition experiments and as high intensity interaction beams in their own right[86].

The first of the multi-kJ petawatt facilities to be operational was built at the Laboratory for Laser Energetics (LLE) at the University of Rochester, USA. The laser is coupled with the well proven 30 kJ 60-beam long pulse Omega system. Omega EP (extended performance)[87], shown in Figure 4, is a four-beam system with an architecture very similar to that of NIF. Two of the beams can be operated in short pulse mode to add petawatt x-ray backlighting capability for ICF experiments plus options for fast-ignition investigations. The laser can operate between 1 and 100 ps,

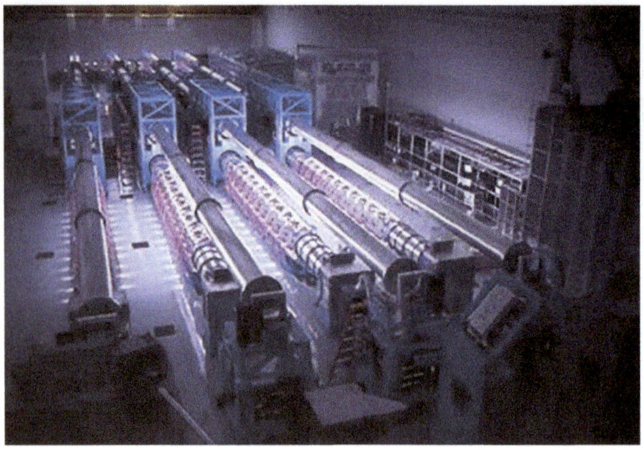

Figure 4. Omega EP beamlines (picture courtesy of LLE).

Figure 5. The Gekko XII and LFEX lasers at ILE, University of Osaka, Japan (picture courtesy of Osaka University).

delivering 1 PW performance at 1 ps and 2.6 kJ performance at pulsewidths >10 ps. It has driven the development of high damage threshold multi-layer dielectric gratings and their use in tiled geometry.

Laser Mégajoule (LMJ) is currently being commissioned by the CEA at a research establishment near Bordeaux, France. Short pulse capability is being added to LMJ through the PETAL beamline. PETAL was originally designed and built to be part of LIL (Laser Integration Line), the LMJ prototype beamline which was modified to incorporate CPA operation[88]. It uses four independent compressors with the beams phased together. In ~2009 it was decided to move the hardware into the LMJ facility where it will be used for high energy density physics and research on fast ignition. The beamline is specified to operate at 3.5 kJ and will be commissioned in 2016[89] at half of this energy while higher damage threshold transport optics are being produced.

Within the Gekko XII facility at the Institute of Laser Engineering (ILE), University of Osaka, Japan the LFEX facility, shown in Figure 5, is currently being commissioned as a fast ignitor[90] demonstrator for the FIREX project[91, 92]. The Laser for Fast Ignition EXperiment (LFEX) is designed to have a 1 ps rise time and 2 × 2 segmented dielectric gratings. Commissioning started in 2005 and delivered petawatt operation in 2010[93], with full operational capability expected by the end of 2014[94]. The beam is focused to target by a 4 m off-axis parabola, giving a spot of 30–60 μm in a 5 kJ beam in 1–20 ps, providing powers of 1–5 PW (although final specification is to deliver 10 kJ).

At LLNL, NIF ARC (Advanced Radiographic Capability)[95] is designed as an advanced x-ray radiography capability for NIF. NIF ARC uses four (one quad) of NIF's beams to obtain temporal resolution of tens of picoseconds. Each beam is split into two, producing 8 petawatt class beams delivering between 0.4 and 1.7 kJ at pulse lengths between 1 and 50 ps (0.5 PW each) in the infrared. First pulsed light during commissioning was achieved in September 2014, with commissioning planned to be completed by the end of 2015.

Figure 6. A technician inside the NIF target chamber (picture courtesy of LLNL).

5. Megajoule facilities

The megajoule class lasers, although designed to operate in the nanosecond regime, are true petawatt class facilities due to their enormous scale. The multi-pass technology allows close packing of the beamlines at large aperture, producing a multi-pass stacked laser architecture. They were originally designed jointly between the USA and France for use on NIF and LMJ and are now replicated throughout the world.

NIF (National Ignition Facility)[8], at LLNL, USA, is the first and currently the only megajoule scale facility to be operational. It has 192 40 × 40 beams delivering 1.8 MJ in 3 ns @ 3ω (0.6 PW) configured for indirect beam drive. Figure 6 is a photograph of inside the NIF target interaction chamber giving an idea of the scale of the facility. It became

Figure 7. The LMJ facility in Bordeaux, France (picture courtesy of CEA).

Figure 8. The first Ti:sapphire petawatt class laser facility J-KAREN, Japan (picture courtesy of JAEA).

operational and officially dedicated in March 2009. The facility has been operational for over five years and delivered data for both the NIC (National Ignition Campaign)[96] and its internal weapons programme.

LMJ (Laser Mégajoule)[9], shown in Figure 7, is a megajoule class laser currently under construction in Bordeaux, France by the French nuclear science directorate CEA. The facility is designed with 240 long pulse beams arranged in 30 lines of eight beams of 40 mm × 40 mm aperture. Initially only 176 beams will be commissioned, delivering a total energy of 1.4 MJ @ 3ω with a maximum power of 400 TW. The first beamlines will be operational in 2016 with two quads, eight beams, delivering long pulse energy combined with the PETAL short pulse facility[89, 97]. The rest of the beams will be commissioned during the following few years. Following an agreement between CEA and the Region Aquitaine, 20–30% of the time on LMJ/PETAL will be dedicated to academic access.

SG-IV (SG stands for Shenguang – Divine Light)[98] is to be built at CAEP (Chinese Academy of Engineering Physics) Research Center for Laser Fusion, Mianyang, China as an ignition demonstrator. The facility will be constructed following the successful commissioning of SG-III, which is designed to operate with 48 beams at 200 kJ. The initial specification of SG-IV is to be of similar scale to NIF and LMJ, although the design is yet to be finalized. Design options can be tested on SG-IIIP, a separate prototype beamline within the SG-III building.

In Russia, there are plans to construct a megajoule facility UFL-2M[99]. The facility is based on delivering 2.8 MJ of energy @ 2ω for ICF direct drive target illumination.

6. Ti:sapphire lasers

The introduction of Ti:sapphire lasers provided the opportunity to produce high-repetition-rate systems operating at relatively short pulses, typically 30 fs, due to the inherent broad bandwidth of the lasing medium. The lasers operate at 800 nm and are typically pumped by frequency doubled Nd:glass lasers at 527 nm. In recent years, the number of petawatt class Ti:sapphire lasers has grown significantly. The main reason for this is because the sub-components of the systems and/or the whole laser system itself have become commercially available. This takes away the need for the facility to be sited at a national laboratory and allows smaller research groups to enter the arena. It is also evident that these lasers are now being used for more specific research areas.

The J-KAREN (JAEA-Kansai Advanced Relativistic Engineering) laser system constructed at the APRC (Advanced Photon Research Center), JAEA (Japan Atomic Energy Agency), Kyoto, Japan was the world's first petawatt class Ti:sapphire facility and is shown in Figure 8. In 2003 the facility was generating 20 J @ 33 fs, giving 0.85 PW[100]. In 2010 the facility was upgraded with a high contrast technique, using an OPCPA front end to replace the conventional regenerative amplifier, producing contrasts below 10^{-10} [101]. The facility can operate at the 80 TW level at 10 Hz repetition rate and at the petawatt level with a reduced rate, once every 30 minutes, due to thermal considerations in the final booster amplifier.

SILEX-I was constructed at the CAEP (Chinese Academy of Engineering Physics) Research Center of Laser Fusion, Mianyang, China. The facility produced 9 J pulses at 30 fs, giving an output power of 286 TW at a repetition rate of 0.15 Hz[102]. The facility was able to produce focused intensities of 10^{21} W cm^{-2} without the need for deformable mirror corrections.

HERCULES (High Energy Repetitive CUos LasEr System) was constructed at the FOCUS Center and Center for Ultrafast Optical Science, University of Michigan, USA. In 2004 ultra-high intensities of up to 10^{22} W cm^{-2} in a 45 TW laser could be generated using wavefront correction

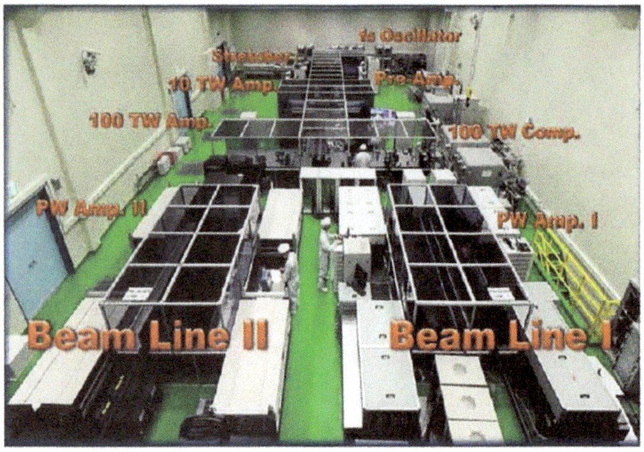

Figure 9. The APRI Petawatt Facility at GIST, South Korea (picture courtesy of GIST).

Figure 10. BELLA, the highest repetition rate petawatt class laser in the world (picture courtesy of Lawrence Berkeley National Laboratory).

and an F#0.6 off-axis parabola[103]. By adding a booster amplifier to the system 300 TW operation could be achieved at 0.1 Hz repetition rate[104]. When focused with an F#1 off-axis parabola this produced focused intensities of 2×10^{22} W cm^{-2}.

Astra-Gemini is a Ti:sapphire laser system[105] operated within the Central Laser Facility, STFC Rutherford Appleton Laboratory, UK. It is operated as an academic user facility. It has two ultra-high power beamlines each delivering 15 J in 30 fs pulses @ 800 nm, giving 500 TW beams to target, generating focused intensities $>10^{21}$ W cm^{-2} to target. Routine high contrast operation can be achieved with the use of a double plasma mirror assembly within the target chamber.

The LASERIX facility[106] at the University Paris Sud, France was designed to be a high-repetition-rate multi-beam laser to pump an XUV laser. The aim of this laser facility was to offer soft XRLs in the 7–30 nm range and an auxiliary IR beam, which could also be used to produce synchronized XUV sources. The laser was a combination of commercially supplied sub-systems primarily from Thales Laser for the front-end systems, Amplitude Technologies for the power amplification and Quantel for the Nd:glass pump laser. The laser performance was first demonstrated in 2006, delivering 36 J of energy although without full compression[107]. The facility is in the process of being moved to CILEX (Centre Interdisciplinaire Lumiere EXtreme).

A petawatt facility has been constructed at the Center of Femto-Science and Technology, Advanced Photonics Research Institute (APRI), Gwangju Institute of Science and Technology (GIST), South Korea. The facility, shown in Figure 9, first achieved petawatt capability in 2010 with a 33 J beam in 30 fs delivering 1.1 PW at a repetition rate of 0.1 Hz[108]. The facility was upgraded to deliver two petawatt beamlines operating at 1 PW and 1.5 PW which can be delivered into separate target chambers[109], and is claimed to be the very first 0.1 Hz Ti:sapphire petawatt laser in the world.

At the University of Quebec, the Advanced Laser Light Source (ALLS) is a commercial system built by Amplitude Technologies operating at 10 Hz delivering in excess of 150 TW[110].

The VEGA facility at the Center for Pulsed Lasers (CLPU) is based at the University of Salamanca, Spain. The facility has been operating with energies of 6 J at 30 fs giving output powers of 200 TW at a repetition rate of 10 Hz synchronized with a second 20 TW beamline. The system is currently being upgraded to provide a third beamline with amplifiers supplied by Amplitude Technologies to deliver 1 PW (30 J @ 30 fs) and will operate at a 1 Hz repetition rate[111].

Xtreme Light III (XL-III) operating at the Institute of Physics of the Chinese Academy of Sciences (IOP CAS) at Beijing National Laboratory for Condensed Matter, China generates 32 J in a 28 ps pulse delivering 1.16 PW to target at focused intensities $>10^{22}$ W cm^{-2} (Ref. [112]). The facility produces high fidelity pulses with contrasts of 10^{-10} @ 400 ps.

The BELLA (BErkeley Lab Laser Accelerator) project was launched in 2009 and is funded by the DOE for experiments on laser plasma acceleration at Lawrence Berkeley National Laboratory, USA. BELLA, shown in Figure 10, can operate at peak power levels of 1.3 PW with a record-setting repetition rate of 1 Hz for a petawatt laser[113]. The laser was commercially built by Thales and shipped to Berkeley in 2012.

The Diocles laser at the Extreme Light Laboratory, University of Nebraska – Lincoln, USA came online nominally at a power level of 100 TW and 10 Hz in 2008, and 1 PW at 0.1 Hz in 2012[114]. It has been modified since to have active feedback spectral phase control[115], and then more recently with a dual-compressor geometry[116]. Ref. [116] describes how it has been successfully used to generate 9 MeV x-rays via inverse Compton scattering.

The Scarlet laser facility at Ohio State University, USA[117] was built for studies in high energy density physics, in a purpose built building in 2007, and became operational in 2012. The front end was originally a Thales 40 TW system but has been upgraded to deliver 15 J in 40 fs, giving 400 TW with a shot every minute.

At SIOM the Qiangguang (Intense Light) Ti:S laser facility produces the highest powers ever achieved from a Ti:sapphire system (52 J @ 26 fs), giving output powers of 2 PW[118]. A high contrast front end gives contrasts to target of 1.5×10^{-11} @ 100 ps.

As part of the SLAC Linac Coherent Light Source (LCLS) at Stanford University, USA the MEC (Materials in Extreme Conditions instrument) femtosecond laser system has been operational at the 25 TW level in conjunction with the LCLS coherent x-ray beam. It is currently being upgraded to 200 TW to be operational in 2015.

DRACO (Dresden laser acceleration source)[119] at the Helmholtz-Zentrum Dresden-Rossendorf laboratory in Germany is a commercially sourced Ti:sapphire laser made by Amplitude Technologies. The facility is designed to investigate electron, ion and proton acceleration schemes for radiation therapy as part of ELBE (Electron Linac for beams with high Brilliance and low Emittance) – Center for High Power Radiation Sources. It is currently being commissioned to operate at 150 TW by the end of 2014, but will be operating at 1 PW by the end of 2015.

Two very similar systems are being constructed in France and Germany: Apollon[120] at CILEX (Centre Interdisciplinaire Lumiere EXtreme) and the Helmholtz Beamline[121] for the international accelerator project FAIR at GSI Helmholtz Centre for Heavy Ion Research, Darmstadt. Both lasers are mixed OPA and Ti:sapphire systems pumped by Nd:glass systems supplied by National Energetics, Texas, USA to realize short pulses at high energy. The systems are specified to deliver 150 J pulses at 15 fs, giving powers of 10 PW.

At the Centre for Advanced Laser Technologies INFLPR (National Institute for Laser, Plasma and Radiation Physics), Romania the CETAL petawatt laser (25 J in 25 fs) is currently being constructed[122]. The facility built commercially by Thales Optronique will operate at 0.1 Hz and be operational in 2015. Thales are also currently constructing a 200 TW system for Peking University, Beijing, China.

200 TW (5 J, 20 fs, 5–10 Hz PULSAR laser) systems from Amplitude Technologies, France have also been installed or are being installed at the following establishments:

- ETRI, Daejeon, Korea;
- INRD, Montreal, Canada;
- LLP, Shanghai Jiao Tong University, China[123];
- INFN, Frascati National Laboratories, Italy[124];

Figure 11. The SIOM OPCPA Qiangguang 10 PW laser facility (picture courtesy of SIOM).

- The Intense Laser Irradiation Laboratory (ILIL), CNR (Consiglio Nazionale delle Ricerche) National Institute of Optics, Pisa, Italy[124].

7. OPCPA systems

The OPCPA concept for large aperture systems was conceived at the Central Laser Facility, STFC Rutherford Appleton Laboratory by Ian Ross[73], with the first practical demonstration on Vulcan within the Central Laser Facility[125]. In this technique the frequency doubled light from a high energy Nd:glass laser facility is transferred to a chirped short pulse laser via parametric amplification in a KDP or LBO crystal at apertures of >100 mm.

The first operational OPCPA system was developed using a pump beam derived from the Luch Facility at the Institute of Applied Physics, Russian Academy of Science, Nizhny Novgorod. The laser delivered 0.2 PW in 2006[126] and was upgraded to 0.56 PW in 2007[127].

At SIOM (Shanghai Institute for Optics and Fine Mechanics), China the Qiangguang 10 PW (Intense Light) OPCPA system, shown in Figure 11, has been constructed with large aperture LBO crystals with a final aperture of 215 mm[128]. The current operating level is 28.7 J in a 33.8 fs pulse, giving output powers of 0.61 PW, the highest peak powers achieved anywhere in the world to date. In 2015, 5 PW (150 J in 30 fs) performance is planned; the final 10 PW (300 J in 30 fs) performance is currently delayed due to the availability of large aperture LBO but is due to be delivered in 2017.

Within the Central Laser Facility, STFC Rutherford Appleton Laboratory there are plans to upgrade the Vulcan facility with full aperture OPCPA following on from the first demonstrations[73, 125]. Two long pulse beamlines of Vulcan will be used to pump DKDP crystals to deliver 500 J in ~25 fs to deliver 20 PW[129].

PALS (Prague Asterix Laser System) is an iodine photodissociation laser. The Asterix facility was first built at MPQ Garching and completed in 1995. Asterix was moved to Prague and has been operational since September 2000[130]. PALS operates at 1315 nm and has extremely narrow linewidths, ~20 pm, making it unsuitable for direct short pulse operation. By frequency tripling the PALS beam it makes an ideal pump laser for an 800 nm seed. A design for a 1.4 PW interaction beam has been published[131] using the existing building geometry.

The Petawatt Field Synthesizer[132] is currently being constructed at the Max-Planck-Institute for Quantenoptik, Garching, Germany. It is a few-cycle petawatt system designed to produce isolated attosecond pulses for wakefield acceleration. The system is entirely OPCPA with ultrashort seed and pump pulses. The final specification of the system is 5 J in 5 fs and it will be operational in 2017.

At the Laboratory for Laser Energetics (LLE), University of Rochester, USA options are being investigated for an ultra-high energy OPCPA system using four OMEGA EP beamlines. The project is called OPAL[133] (Optical Parametric Amplifier Line) and would have available a total pump energy of 12 kJ @ 526 nm. Using this scheme it will be possible to generate 3 kJ, 15 fs pulses, giving peak powers of 200 PW and focused intensities of 10^{24} W cm^{-2}.

A similar planned project to that of ELI at the Institute of Applied Physics of the Russian Academy of Sciences in Nizhny Novgorod is the XCELS (Exawatt Centre for Extreme Light Studies). This megascience project in Russia is to produce an exawatt laser system for fundamental science. The system will use combined 15 PW OPCPA beamlines to reach >200 PW[134].

8. Diode pumped systems

Diode pumping has been identified as being on the critical path to the construction of ICF (inertial confinement fusion) power plants. Their high efficiency and low thermal deposition in the amplifier media make diode pumped systems ideal candidates for these developments. As the technology is developed it is being used in existing facilities to increase the repetition rates of amplifiers, in particular in their front ends. There are also an increasing number of entirely diode pumped petawatt class laser systems either operational or planned in the next few years.

It is proposed to use the Mercury laser facility at LLNL, USA, a diode pumped Yb:S-FAP laser, to pump a Ti:S laser to generate >1 PW powers at repetition rates of 10 Hz[135]. Mercury has been developed as a high average power laser (HAPL) using diode arrays and optimized gas cooling as a precursor to an advanced fusion driver[136].

Figure 12. The final amplifier of POLARIS (picture courtesy of Helmholtz Institute).

POLARIS (Petawatt Optical Laser Amplifier for Radiation Intensive experimentS) is based at the Helmholtz Institute Jena, Germany. It is designed as a fully diode pumped Yb:Glass petawatt class laser[137]. It operates at a central wavelength of 1030 nm and a bandwidth of ~10 nm. It is currently being upgraded from 4 J in 164 fs ~30 TW to 1 PW with the commissioning of the final amplifier to deliver 150 J in 150 fs in 2016. The final amplifier of the facility is shown in Figure 12.

PEnELOPE (Petawatt, Energy-Efficient Laser for Optical Plasma Experiments) is a high-repetition-rate diode pumped laser using broadband Yb-doped glass/CaF_2 under construction at the Helmholtz-Zentrum, Dresden-Rossendorf within the ELBE Centre for high power radiation sources[138]. It will be dedicated to the production of laser accelerated proton and ion beams with energies >100 MeV relevant to future cancer treatments. The facility, due to be commissioned in 2016, will deliver pulses of 150 J in 120 fs, giving >1 PW at 1 Hz. PEnELOPE and POLARIS are both programmes belonging to the German Helmholtz Society.

9. The next generation

Facilities that are changing the landscape of Petawatt class facilities are the three pillars of ELI (European Light

Infrastructure)[139], where three large scale laser user facilities are being built to exploit ultra-high intensity interactions in the Czech Republic (ELI-Beamlines), Hungary (ELI-Attosecond Light Pulse Source) and Romania (ELI-Nuclear Physics).

- ELI-Beamlines will provide a range of laser systems for the production of high brightness x-rays and accelerating particles. The beamlines use either OPCPA, Ti:sapphire or a combination of the two to produce pulses ranging from hundreds of millijoules at a kHz up to a kilojoule beamline firing once a minute. These will be coupled to separate interaction areas allowing a wide range of experiments to be performed. An artist's impression of the ELI-Beamlines building is shown in Figure 13.
- ELI-ALPS (Attosecond Light Pulse Source) will provide three high-repetition-rate OPCPA beamlines: 100 kHz, >5 mJ, <5 fs; a single cycle 1 kHz, >100 mJ, <5 fs; and a high intensity 5 Hz, >40 J, <15 fs. All the beamlines will be used to drive secondary sources (UV/XUV, x-rays, ions, etc.), which will be dedicated to extremely fast electron dynamics in atoms, molecules, plasmas and solids.
- ELI-NP (Nuclear Physics) will have two beamlines with OPCPA front ends and Ti:sapphire power amplifiers. The beamlines will either produce 1 PW at 1 Hz (20 J, <20 fs) or 10 PW at 1 shot per minute (220 J, <20 fs). The beamlines will be used to produce extremely high energy gamma rays for a wide range of nuclear physics applications.

During this review we have discussed stand alone flash-lamp pumped petawatt class lasers and also the megajoule class lasers currently operational or under construction. The next generation of these ICF demonstration facilities will use diode pumped technology to dramatically increase the repetition rate of the lasers. This will be a giant step on the road to building a commercial power plant using this technology. Large programmes have been examining the options for these systems both in the USA and Europe[140, 141].

Raman based plasma amplifiers have been the subject of speculation for many years[142] as a means of generating ultra-high powers. In these schemes, it is possible to transfer energy from multiple nanosecond laser pulses in a plasma to an ultra-short pulse seed. The benefit of this scheme is that it is not limited by the normal nonlinear propagation processes in laser amplifiers.

Systems based around VECSELs (vertical external cavity emitting lasers) have rapidly increased in output power in recent years. Thin disc lasers are currently used at facilities such as PEnELOPE[138] and mode-locked semiconductor VECSELs are surpassing what were believed to be their limitations[143].

Figure 13. Artist's impression of the ELI-Beamlines building (picture courtesy of ELI).

Femtosecond coherently combined fibre amplifiers have been demonstrated at the millijoule level[144] which show the potential for the construction of massively multiplexed short pulse lasers that could operate at high energy and repetition rates. Under IZEST (International Center for Zetta–Exawatt Science and Technology), based at Ecole Polytechnique, France, the ICAN (International Coherent Amplification Network) Project is looking to use thousands of fibre lasers coherently combined to build the next generation of particle accelerators. A demonstration system is aiming to coherently combine a fibre bundle to produce 10 J of energy in a 100–200 fs pulse[145]. When combined, the overall facility could produce >100 PW. This will potentially reach greater energies than are currently possible using conventional techniques in a vastly reduced footprint.

10. Conclusion

From national laboratories to university departments, the petawatt laser has evolved to become one of the most important tools in the scientific toolkit for the study of matter in extreme states. The first petawatt lasers were built at national laboratories by adapting beamlines from fusion laser systems. Over the last 20 years, as technologies have advanced, these systems have come down in size and cost such that they are commercially available and within the reach of university physics departments.

In this paper, we have noted over 50 petawatt class lasers (>200 TW) that are operational, under construction or in the planning phase. These range from kJ and even multi-kJ high energy systems to high-repetition table-top femtosecond devices.

Petawatt lasers are now being constructed for specific applications in fields ranging from proton therapy for the treatment of cancer to simulation of astrophysical phenomena, and many more besides. The next generation of lasers will approach exawatt power levels and allow us to reach conditions beyond those that naturally occur in the universe.

Acknowledgements

The authors are grateful for the cooperation of all of the managers of the petawatt class lasers who have contributed to making this paper as accurate as possible and who have allowed the figures to be used to illustrate their facilities.

The authors have tried to deliver a comprehensive review of petawatt class lasers worldwide and have conducted extensive research to ensure that all of the details in this paper are correct. Apologies if they have overlooked or misrepresented any facility within this exercise.

References

1. S. Backus, C. G. Durfee, III, M. M. Murnane, and H. C. Kapteyn, Rev. Sci. Instrum. **69, 3**, 1207 (1998).
2. M. D. Perry, D. Pennington, B. C. Stuart, G. Tietbohl, J. A. Britten, C. Brown, S. Herman, B. Golick, M. Kartz, J. Miller, H. T. Powell, M. Vergino, and V. Yanovsky, Opt. Lett. **24, 3** (1999).
3. T. H. Maiman, Nature **187**, 493 (1960).
4. J. M. Dawson, Phys. Fluids **7**, 981 (1964).
5. J. W. Daiber, A. Hertzberg, and C. Wittliff, Phys. Fluids **9**, 617 (1966).
6. J. D. Lawson, Proc. Phys. Soc. B **70**, 6 (1957).
7. J. Nuckolls, L. Wood, A. Thiessen, and G. Zimmerman, Nature **239**, 139 (1972).
8. G. H. Miller, E. I. Moses, and C. R. West, Opt. Eng. **43**, 2841 (2004).
9. J. Ebrardt and J. M. Chaput, J. Phys.: Conf. Ser. **244**, 032017 (2010).
10. D. Strickland and G. Mourou, Opt. Commun. **56, 3**, 219 (1985).
11. S. Bulanov, T. Zh. Esirkepov, Y. Hayashi, M. Kando, H. Kiriyama, J. K. Koga, K. Kondo, H. Kotaki, A. S. Pirozhkov, S. S. Bulanov, A. G. Zhidkov, N. N. Rosanov, P. Chen, D. Neely, Y. Kato, N. B. Narozhny, and G. Korn, Plasma Phys. Control. Fusion **53**, 124025 (2011).
12. S. P. Hatchet, C. G. Brown, T. E. Cowan, E. A. Henry, J. S. Johnson, M. H. Key, J. A. Koch, A. B. Langdon, B. F. Lasinski, R. W. Lee, A. J. Mackinnon, D. M. Pennington, M. D. Perry, T. W. Phillips, M. Roth, T. C. Sangster, M. S. Singh, R. A. Snavely, M. A. Stoyer, S. C. Wilks, and K. Yasuike, Phys. Plasmas **7, 5**, 2076 (2000).
13. R. C. Elton, *X-ray Lasers* (Academic, 1990), ISBN 0-12-238080-0.
14. M. A. Duguay and P. M. Rentzepi, Appl. Phys. Lett. **10, 12**, 350 (1967).
15. J. Zhang, M. H. Key, P. A. Norreys, G. J. Tallents, A. Behjat, C. Danson, A. Demir, L. Dwivedi, M. Holden, P. B. Holden, C. L. S. Lewis, A. G. MacPhee, D. Neely, G. J. Pert, S. A. Ramsden, S. J. Rose, Y. F. Shao, O. Thomas, F. Walsh, and Y. L. You, Phys. Rev. Lett. **74, 8**, 1335 (1995).
16. D. L Matthews, P. L. Hagelstein, M. D. Rosen, M. J. Eckart, N. M. Ceglio, A. U. Hazi, H. Medecki, B. J. MacGowan, J. E. Trebes, B. L. Whitten, E. M. Campbell, C. W. Hatcher, A. M. Hawryluk, R. L. Kauffman, L. D. Pleasance, G. Rambach, J. H. Scofield, G. Stone, and T. A. Weaver, Phys. Rev. Lett. **54, 2**, 110 (1985).
17. T. N. Lee, E. A. McLean, and R. C. Elton, Phys. Rev. Lett. **59, 11**, 1185 (1987).
18. A. Carillon, H. Z. Chen, P. Dhez, L. Dwivedi, J. Jacoby, P. Jaegle, G. Jamelot, J. Zhang, M. H. Key, A. Kidd, A. Klisnick, R. Kodama, J. Krishnan, C. L. S. Lewis, D. Neely, P. Norreys, D. O'Neill, G. J. Pert, S. A. Ramsden, J. P. Raucourt, G. J. Tallents, and J. Uhomoibhi, Phys. Rev. Lett. **68**, 19 (1992).
19. J. Zhang, A. G. MacPhee, J. Lin, E. Wolfrum, R. Smith, C. Danson, M. H. Key, C. L. S. Lewis, D. Neely, J. Nilsen, G. J. Pert, G. J. Tallents, and J. S. Wark, Science **276**, 1097 (1997).
20. H. Daido, Rep. Progr. Phys. **65, 10**, 1513 (2002).
21. J. Dunn, A. L. Osterheld, R. Shepherd, W. E. White, V. N. Shlyaptsev, and R. E. Stewart, Phys. Rev. Lett. **80, 13**, 2825 (1998).
22. R. Lichters, J. Meyer-ter-Vehn, and A. Pukhov, Phys. Plasmas **3, 9**, 3425 (1996).
23. P. A. Norreys, M. Zepf, S. Moustaizis, A. P. Fews, J. Zhang, P. Lee, M. Bakarezos, C. N. Danson, A. Dyson, P. Gibbon, P. Loukakos, D. Neely, F. N. Walsh, J. S. Wark, and A. E. Dangor, Phys. Rev. Lett. **76, 11**, 1832 (1996).
24. B. Dromey, S. Kar, C. Bellei, D. C. Carroll, R. J. Clarke, J. S. Green, S. Kneip, K. Markey, S. R. Nagel, P. T. Simpson, L. Willingale, P. McKenna, D. Neely, Z. Najmudin, K. Krushelnick, P. A. Norreys, and M. Zepf, Phys. Rev. Lett. **99, 8**, 085001 (2007).
25. P. B. Corkum and F. Krausz, Nat. Phys. **3**, 381 (2007).
26. F. N. Beg, A. R. Bell, A. E. Dangor, C. N. Danson, A. P. Fews, M. E. Glinsky, B. A. Hammel, P. Lee, P. A. Norreys, and M. Tatarakis, Phys. Plasmas **4, 2**, 447 (1997).
27. P. N. Norreys, M. Santala, E. Clark, M. Zepf, I. Watts, F. N. Beg, K. Krushelnick, M. Tatarakis, A. E. Dangor, X. Fang, P. Graham, T. McCanny, R. P. Singhal, K. W. D. Ledingham, A. Creswell, D. C. W. Sanderson, J. Magill, A. Machacek, J. S. Wark, R. Allott, B. Kennedy, and D. Neely, Phys. Plasmas **6, 5**, 2150 (1999).
28. A. J. Mackinnon, M. Borghesi, S. Hatchett, M. H. Key, P. K. Patel, H. Campbell, A. Schiavi, R. Snavely, S. C. Wilks, and O. Willi, Phys. Rev. Lett. **86, 9**, 1769 (2001).
29. D. Neely, C. N. Danson, R. Allott, F. Amiranoff, J. L. Collier, A. E. Dangor, C. B. Edwards, P. Flintoff, P. Hatton, M. Harman, M. H. R. Hutchinson, Z. Najmudin, D. A. Pepler, I. N. Ross, M. Salvati, and T. Winstone, Laser Part. Beams **17, 2**, 281 (1999).
30. D. I. Hillier, C. Danson, S. Duffield, D. Egan, S. Elsmere, M. Girling, E. Harvey, N. Hopps, M. Norman, S. Parker, P. Treadwell, D. Winter, and T. Bett, Appl. Opt. **52, 18** (2013).
31. C. Ziener, P. S. Foster, E. J. Divall, C. J. Hooker, M. H. R. Hutchinson, A. J. Langley, and D. Neely, J. Appl. Phys. **93, 1** (2003).
32. J. Itatani, J. Faure, M. Nantel, G. Mourou, and S. Watanabe, Opt. Commun. **148**, 70 (1998).
33. A. Jullien, O. Albert, F. Burgy, G. Hamoniaux, J. Rousseau, J. Chambaret, F. Augé-Rochereau, G. Chériaux, J. Etchepare, N. Minkovski, and S. M. Saltiel, Opt. Lett. **30**, 920 (2005).
34. R. C. Shah, R. P. Johnson, T. Shimada, K. A. Flippo, J. C. Fernandez, and B. M. Hegelich, Opt. Lett. **34**, 2273 (2009).
35. I. Musgrave, W. Shaikh, M. Galimberti, A. Boyle, C. Hernandez-Gomez, K. Lancaster, and R. Heathcote, Appl. Opt. **49**, 6558 (2010).
36. T. Tajima and J. M. Dawson, Phys. Rev. Lett. **43, 4**, 267 (1979).
37. C. Joshi, T. Tajima, J. M. Dawson, H. A. Baldis, and N. A. Ebrahim, Phys. Rev. Lett. **47, 18**, 1285 (1981).
38. C. E. Clayton, C. Joshi, C. Darrow, and D. Umstadter, Phys. Rev. Lett. **54**, 2343 (1985).
39. F. Amiranoff, D. Bernard, B. Cros, F. Jacquet, G. Matthieussent, J. R. Marques, P. Mora, A. Modena, J. Morillo, F. Moulin, Z. Najmudin, A. E. Specka, and C. Stenz, IEEE Trans. Plasma Sci. **24, 2**, 296 (1996).

40. A. Modena, Z. Najmudin, A. E. Dangor, C. E. Clayton, K. A. Marsh, C. Joshi, V. Malka, C. B. Darrow, C. Danson, D. Neely, and F. N. Walsh, Nature **377**, 606 (1994).
41. S. P. D. Mangles, C. D. Murphy, Z. Najmudin, A. G. Thomas, J. L. Collier, A. E. Dangor, E. J. Divall, P. S. Foster, J. G. Gallacher, C. J. Hooker, D. A. Jaroszynski, A. J. Langley, W. B. Mori, P. A. Norreys, F. S. Tsung, R. Viskup, B. R. Walton, and K. Krushelnick, Nature **431, 7008**, 535 (2004).
42. W. P. Leemans, R. Duarte, E. Esarey, S. Fournier, C. G. R. Geddes, D. Lockhart, C. B. Schroeder, C. Toth, J.-L. Vay, and S. Zimmermann, AIP Conf. Proc. **1299**, 3 (2010).
43. M. Roth, T. E. Cowan, M. H. Key, S. P. Hatchett, C. Brown, W. Fountain, J. Johnson, D. M. Pennington, R. A. Snavely, S. C. Wilks, K. Yasuike, H. Ruhl, F. Pegoraro, S. V. Bulanov, E. M. Campbell, M. D. Perry, and H. Powell, Phys. Rev. Lett. **86, 3** (2001).
44. K. W. D. Ledingham, P. McKenna, T. McCanny, S. Shimizu, J. M. Yang, L. Robson, J. Zweit, J. M. Gillies, J. Bailey, G. N. Chimon, R. J. Clarke, D. Neely, P. A. Norreys, J. L. Collier, R. P. Singhal, M. S. Wei, S. P. D. Mangles, P. Nilson, K. Krushelnick, and M. Zepf, J. Phys. D **37, 16**, 2341 (2004).
45. R. A. Snavely, M. H. Key, S. P. Hatchett, T. E. Cowan, M. Roth, T. W. Phillips, M. A. Stoyer, E. A. Henry, T. C. Sangster, M. S. Singh, S. C. Wilks, A. MacKinnon, A. Offenberger, D. M. Pennington, K. Yasuike, A. B. Langdon, B. F. Lasinski, J. Johnson, M. D. Perry, and E. M. Campbell, Phys. Rev. Lett. **85, 14**, 2945 (2000).
46. A. Mackinnon, Y. Sentoku, P. K. Patel, D. W. Price, S. Hatchett, M. H. Key, C. Andersen, R. Snavely, and R. R. Freeman, Phys. Rev. Lett. **88, 21**, 215006 (2002).
47. D. Neely, P. Foster, A. Robinson, F. Lindau, O. Lundh, A. Persson, C.-G. Wahlstrom, and P. McKenna, Appl. Phys. Lett. **89, 2**, 021502 (2006).
48. H. Daido, M. Nishiuchi, and A. S. Pirozhkov, Rep. Progr. Phys. **75, 5**, 056401 (2012).
49. E. B. Treacy, IEEE J. Quantum Electron. **5**, 454 (1969).
50. M. Ferray, L. A. Lompré, O. Gobert, A. L'huillier, G. Mainfray, C. Manus, A. Sanchez, and A. S. Gomes, Opt. Commun. **75**, 278 (1990).
51. F. G. Patterson and M. D. Perry, J. Opt. Soc. Am. B **8**, 2384 (1991).
52. K. Yamakawa, C. P. J. Barty, H. Shiraga, and Y. Kato, Opt. Lett. **16, 20**, 1593 (1991).
53. C. Sauteret, G. Mourou, D. Husson, G. Thiell, S. Seznec, S. Gary, and A. Migus, Opt. Lett. **16, 4**, 238 (1991).
54. B. Nikolaus, D. Grischkowsky, and A. C. Balant, Opt. Lett. **8, 3**, 189 (1983).
55. D. E. Spence, P. N. Kean, and W. Sibbet, Opt. Lett. **16, 1**, 42 (1991).
56. C. P. J. Barty, Opt. Lett. **19, 18**, 1442 (1994).
57. M. W. Phillips, Z. Change, C. N. Danson, J. R. M. Barr, D. W. Hughes, C. B. Edwards, and D. C. Hanna, Opt. Lett. **17**, 1453 (1992).
58. C. Rouyer, G. Mourou, A. Migus, É. Mazataud, I. Allais, A. Pierre, S. Seznec, and C. Sauteret, Opt. Lett. **18, 3**, 214 (1993).
59. N. Blanchot, C. Rouyer, C. Sauteret, and A. Migus, Opt. Lett. **20, 4**, 395 (1995).
60. C. N. Danson, L. J. Barzanti, Z. Chang, A. E. Damerell, C. B. Edwards, S. Hancock, M. H. R. Hutchinson, M. H. Key, S. Luan, R. R. Mahadeo, I. P. Mercer, P. Norreys, D. A. Pepler, D. A. Rodkiss, I. N. Ross, M. A. Smith, R. A. Smith, P. Taday, W. T. Toner, K. W. M. Wigmore, T. B. Winstone, R. W. W. Wyatt, and F. Zhou, Opt. Commun. **103, 5/6**, 392 (1993).
61. C. N. Danson, J. Collier, D. Neely, L. J. Barzanti, A. Damerell, C. B. Edwards, M. H. R. Hutchinson, M. H. Key, P. A. Norreys, D. A. Pepler, I. N. Ross, P. F. Taday, W. T. Toner, M. Trentelman, F. N. Walsh, T. B. Winstone, and R. W. W. Wyatt, J. Mod. Opt. **45, 8**, 1653 (1998).
62. R. L. Fork, O. E. Martinez, and J. P. Gordon, Opt. Lett. **9, 5**, 150 (1984).
63. O. E. Martinez, J. P. Gordon, and R. L. Fork, J. Opt. Soc. Am. **A1**, 1003 (1984).
64. G. Cheriaux, B. Walker, L. F. Rousseau, F. Salin, and J. P. Chambaret, Opt. Lett. **21, 6**, 414 (1996).
65. P. F. Moulton, J. Opt. Soc. Am. B **3**, 125 (1986).
66. A. Dubietis, G. Jonušauskas, and A. Piskarskas, Opt. Commun. **88**, 437 (1992).
67. C. N. Danson, P. A. Brummitt, R. J. Clarke, J. L. Collier, B. Fell, A. J. Frackiewicz, S. Hancock, S. Hawkes, C. Hernandez-Gomez, P. Holligan, M. H. R. Hutchinson, A. Kidd, W. J. Lester, I. O. Musgrave, D. Neely, D. R. Neville, P. A. Norreys, D. A. Pepler, C. J. Reason, W. Shaikh, T. B. Winstone, R. W. W. Wyatt, and B. E. Wyborn, IAEA J. Nucl. Fusion **44**, S239 (2004).
68. C. Hernandez-Gomez, P. A. Brummitt, D. J. Canny, R. J. Clarke, J. Collier, C. N. Danson, A. M. Dunne, B. Fell, A. J. Frackiewicz, S. Hancock, S. Hawkes, R. Heathcote, P. Holligan, M. H. R. Hutchinson, A. Kidd, W. J. Lester, I. O. Musgrave, D. Neely, D. R. Neville, P. A. Norreys, D. A. Pepler, C. J. Reason, W. Shaikh, T. B. Winstone, and B. E. Wyborn, J. Phys. IV **133**, 555 (2006).
69. I. O. Musgrave, A. Boyle, D. Carroll, R. Clarke, R. Heathcote, M. Galimberti, J. Green, D. Neely, M. Notley, B. Parry, W. Shaikh, T. Winstone, D. Pepler, A. Kidd, C. Hernandez-Gomez, and J. Collier, Proc. SPIE **8780**, 878003 (2013).
70. I. N. Ross, P. Matousek, M. Towrie, A. J. Langley, and J. L. Collier, Opt. Commun. **144**, 125 (1997).
71. C. N. Danson, D. Neely, and D. Hillier, High Power Laser Sci. Engng **2**, e34 (2014).
72. J. L. Collier, C. Hernandez-Gomez, I. N. Ross, P. Matousek, C. N. Danson, and J. Walczak, Appl. Opt. **38, 36**, 7486 (1999).
73. I. N. Ross, J. Collier, P. Matousek, C. N. Danson, D. Neely, R. M. Allott, D. A. Pepler, C. Hernandez-Gomez, and K. Osvay, Appl. Opt. **39, 15**, 2422 (2000).
74. V. Bagnoud, I. A. Begishev, M. J. Guardalben, J. Puth, and J. D. Zuegel, Opt. Lett. **30, 14**, 1843 (2005).
75. H. Kiriyama, Opt. Lett. **33, 7**, 645 (2008).
76. J. Schwarz, P. Rambo, M. Geissel, A. Edens, I. Smith, E. Brambrink, M. Kimmel, and B. Atherton, IFSA2007, J. Phys.: Conf. Ser. **112**, 032020 (2008).
77. Y. Kitagawa, H. Fujita, R. Kodama, H. Yoshida, S. Matsuo, T. Jitsuno, T. Kawasaki, H. Kitamura, T. Kanabe, S. Sakabe, K. Shigemori, N. Miyanaga, and Y. Izawa, IEEE J. Quantum Electron. **40**, 281 (2004).
78. M. Mori, Y. Kitagawa, R. Kodama, H. Habara, M. Iwata, S. Tsuji, K. Suzuki, K. Sawai, K. Tanaka, Y. Kato, and K. Mima, *Nuclear Instruments and Methods in Physics Research* p. 367 (Elsevier Science, vol. A410, 1998).
79. B. C. Stuart, J. D. Bonlie, J. A. Britten, J. A. Caird, R. Cross, C. A. Ebbers, M. J. Eckart, A. C. Erlandson, W. A. Molander, A. Ng, P. K. Patel, and D. Price, *CLEO Technical Digest* (Optical Society of America, 2006), JTuG3.
80. B. M. Van Wonterghem, J. R. Murray, J. H. Campbell, D. R. Speck, C. E. Barker, I. C. Smith, D. F. Browning, and W. C. Behrendt, Appl. Opt. **36**, 4932 (1997).
81. P. K. Rambo, I. C. Smith, J. L. Porter, Jr., Mi. J. Hurst, C. S. Speas, Ri. G. Adams, A. J. Garcia, E. Dawson, B. D. Thurston, C. Wakefield, J. W. Kellogg, M. J. Slattery, H. C. Ives, III, R. S. Broyles, J. A. Caird, A. C. Erlandson, J. E.

Murray, W. C. Behrendt, N. D. Neilsen, and J. M. Narduzzi, Appl. Opt. **44**, 2421 (2005).
82. E. W. Gaul, M. Martinez, J. Blakeney, A. Jochmann, M. Ringuette, D. Hammond, T. Borger, R. Escamilla, S. Douglas, W. Henderson, G. Dyer, A. Erlandson, R. Cross, J. Caird, C. Ebbers, and T. Ditmire, Appl. Opt. **49**, 9 (2010).
83. V. Bagnoud, B. Aurand, A. Blazevic, S. Borneis, C. Bruske, B. Ecker, U. Eisenbarth, J. Fils, A. Frank, E. Gaul, S. Goette, C. Haefner, T. Hahn, K. Harres, H.-M. Heuck, D. Hochhaus, D. H. H. Hoffmann, D. Javorková, H.-J. Kluge, T. Kuehl, S. Kunzer, M. Kreutz, T. Merz-Mantwill, P. Neumayer, E. Onkels, D. Reemts, O. Rosmej, M. Roth, T. Stoehlker, A. Tauschwitz, B. Zielbauer, D. Zimmer, and K. Witte, Appl. Phys. B **100**, 137 (2010).
84. G. Xu, T. Wang, Z. Li, Y. Dai, Z. Lin, Y. Gu, and J. Zhu, Rev. Laser Engng **36**, 1172 (2008).
85. N. W. Hopps, C. Danson, S. Duffield, D. Egan, S. Elsmere, M. Girling, E. Harvey, D. Hillier, M. Norman, S. Parker, Pl. Treadwell, D. Winter, and T. Bett, Appl. Opt. **52, 15**, 3597 (2013).
86. J. D. Zuegel, S. Borneis, C. Barty, B. Legarrec, C. Danson, N. Miyanaga, P. Rambo, C. Leblanc, T. J. Kessler, A. W. Schmid, L. Waxer, J. H. Kelly, B. Kruschwitz, R. Jungquist, E. Moses, J. Britten, I. Jovanovic, J. Dawson, and N. Blanchot, Fusion Sci. Technol. **49**, 453 (2006).
87. J. H. Kelly, L. J. Waxer, V. Bagnoud, I. A. Begishev, J. Bromage, B. E. Kruschwitz, T. J. Kessler, S. J. Loucks, D. N. Maywar, R. L. McCrory, D. D. Meyerhofer, S. F. B. Morse, J. B. Oliver, A. L. Rigatti, A. W. Schmid, C. Stoeckl, S. Dalton, L. Folnsbee, M. J. Guardalben, R. Jungquist, J. Puth, M. J. Shoup, III, D. Weiner, and J. D. Zuegel, J. Phys. IV **133**, 75 (2006).
88. N. Blanchot, G. Behar, T. Berthier, E. Bignon, F. Boubault, C. Chappuis, H. Coïc, C. Damiens-Dupont, J. Ebrardt, Y. Gautheron, P. Gibert, O. Hartmann, E. Hugonnot, F. Laborde, D. Lebeaux, J. Luce, S. Montant, S. Noailles, J. Néauport, D. Raffestin, B. Remy, A. Roques, F. Sautarel, M. Sautet, C. Sauteret, and C. Rouyer, Plasma Phys. Control. Fusion **50**, 1240045 (2008).
89. N. Blanchot, G. Behar, T. Berthier, B. Busserole, C. Chappuis, C. Damiens-Dupont, P. Garcia, F. Granet, C. Grosset-Grange, J.-P. Goossens, L. Hilsz, F. Laborde, T. Lacombe, F. Laniesse, E. Lavastre, J. Luce, F. Macias, E. Mazataud, J. L. Miquel, J. Néauport, S. Noailles, P. Patelli, E. Perrot-Minot, C. Present, D. Raffestin, B. Remy, C. Rouyer, and D. Valla, *EPJ Web of Conferences,* vol. 59, p. 07001 (2013).
90. M. Tabak, J. Hammer, M. E. Glinsky, W. L. Kruer, S. C. Wilks, J. Woodworth, E. M. Campbell, and M. D. Perry, Phys. Plasma **1**, 1626 (1994).
91. K. Mima, H. Azechi, Y. Johzaki, Y. Kitagawa, R. Kodama, Y. Kozaki, N. Miyanaga, K. Nagai, H. Nagatomo, M. Nakai, H. Nishimura, T. Norimatsu, H. Shiraga, K. A. Tanaka, and Y. Izawa, Fusion Sci. Technol. **47**, 662 (2005).
92. H. Shiraga, S. Fujioka, M. Nakai, T. Watari, H. Nakamura, Y. Arikawa, H. Hosoda, T. Nagai, M. Koga, H. Kikuchi, Y. Ishii, T. Sogo, K. Shigemori, H. Nishimura, Z. Zhang, M. Tanabe, S. Ohira, Y. Fujii, T. Namimoto, Y. Sakawa, O. Maegawa, T. Ozaki, K. A. Tanaka, H. Habara, T. Iwawaki, K. Shimada, H. Nagatomo, T. Johzaki, A. Sunahara, M. Murakami, H. Sakagami, T. Taguchi, T. Norimatsu, H. Homma, Y. Fujimoto, A. Iwamoto, N. Miyanaga, J. Kawanaka, T. Jitsuno, Y. Nakata, K. Tsubakimoto, K. Sueda, N. Morio, S. Matsuo, T. Kawasaki, K. Sawai, K. Tsuji, H. Murakami, T. Kanabe, K. Kondo, R. Kodama, N. Sarukura, T. Shimizu, K. Mima, and H. Azechi, High Energ. Dens. Phys. **8**, 227 (2012).
93. H. Azechi, K. Mima, Y. Fujimoto, S. Fujioka, H. Homma, M. Isobe, A. Iwamoto, T. Jitsuno, T. Johzaki, R. Kodama, M. Koga, K. Kondo, J. Kawanaka, T. Mito, N. Miyanaga, O. Motojima, M. Murakami, H. Nagatomo, K. Nagai, M. Nakai, H. Nakamura, T. Nakamura, T. Nakazato, Y. Nakao, K. Nishihara, H. Nishimura, T. Norimatsu, T. Ozaki, H. Sakagami, Y. Sakawa, N. Sarukura, K. Shigemori, T. Shimizu, H. Shiraga, A. Sunahara, T. Taguchi, K. A. Tanaka, and K. Tsubakimoto, Nucl. Fusion **49**, 104024 (2009).
94. H. Azechi, K. Mima, S. Shiraga, S. Fujioka, H. Nagatomo, T. Johzaki, T. Jitsuno, M. Key, R. Kodama, M. Koga, K. Kondo, J. Kawanaka, N. Miyanaga, M. Murakami, K. Nagai, M. Nakai, H. Nakamura, T. Nakamura, T. Nakazato, Y. Nakao, K. Nishihara, H. Nishimura, T. Norimatsu, P. Norreys, T. Ozaki, J. Pasley, H. Sakagami, Y. Sakawa, N. Sarukura, K. Shigemori, T. Shimizu, A. Sunahara, T. Taguchi, K. Tanaka, K. Tsubakimoto, Y. Fujimoto, H. Homma, and A. Iwamoto, Nucl. Fusion **53**, 104021 (2013).
95. C. P. J. Barty, M. Key, J. Britten, R. Beach, G. Beer, C. Brown, S. Bryan, J. Caird, T. Carlson, J. Crane, J. Dawson, A. C. Erlandson, D. Fittinghoff, M. Hermann, C. Hoaglan, A. Iyer, L. Jones, II, I. Jovanovic, A. Komashko, O. Landen, Z. Liao, W. Molander, S. Mitchell, E. Moses, N. Nielsen, H.-H. Nguyen, J. Nissen, S. Payne, D. Pennington, L. Risinger, M. Rushford, K. Skulina, M. Spaeth, B. Stuart, G. Tietbohl, and B. Wattellier, Nucl. Fusion **44**, S266 (2004).
96. J. D. Lindl, O. Landen, J. Edwards, E. Moses, and NIC Team, Phys. Plasmas **21**, 020501 (2014).
97. LMJ Website, (2014).
98. W. Y. Zhang and X. T. He, IFSA2007, J. Phys.: Conf. Ser. **112**, 032001 (2008).
99. V. B. Rozanov, S. Gus'kov, G. Vergunova, N. Demchenko, R. Stepanov, I. Doskoch, R. Yakhin, S. Bel'kov, S. Bondarenko, and N. Zmitrenko, IFSA-2013 (Nara, Japan, Sept. 8–13, 2013) O.Tu_A15 (2013).
100. M. Aoyama, K. Yamakawa, Y. Akahane, J. Ma, N. Inoue, H. Ueda, and H. Kiriyama, Opt. Lett. **28**, 1594 (2003).
101. H. Kiriyama, M. Michiaki, Y. Nakai, T. Shimomura, H. Sasao, M. Tanaka, Y. Ochi, M. Tanoue, H. Okada, S. Kondo, S. Kanazawa, A. Sagisaka, I. Daito, D. Wakai, F. Sasao, M. Suzuki, H. Kotakai, K. Kondo, A. Sugiyama, S. Bulanov, P. R. Bolton, H. Daido, S. Kawanishi, J. L. Collier, C. Hernandez-Gomez, C. J. Hooker, K. Ertel, T. Kimura, and T. Tajima, Appl. Opt. **49, 11** (2010).
102. H. S. Peng, W. Y. Zhang, X. M. Zhang, Y. J. Tang, W. G. Zheng, Z. J. Zheng, X. F. Wei, Y. K. Ding, Y. Gou, S. P. Zhou, and W. B. Pei, Lasers Part. Beams **23**, 205 (2005).
103. S. W. Bahk, P. Rousseau, T. A. Planchon, V. Chvykov, G. Kalintchenko, A. Maksimchuk, G. A. Mourou, and V. Yanovsky, Opt. Lett. **29, 24**, 2837 (2004).
104. V. Yanovsky, V. Chvykov, G. Kalinchenko, P. Rousseau, T. Planchon, T. Matsuoka, A. Maksimchuk, J. Nees, G. Cheriaux, G. Mourou, and K. Krushelnick, Opt. Express **16** (2008).
105. C. J. Hooker, J. L. Collier, O. Chekhlov, R. J. Clarke, E. J. Divall, K. Ertel, P. Foster, S. Hancock, S. J. Hawkes, P. Holligan, A. J. Langley, W. J. Lester, D. Neely, B. T. Parry, and B. E. Wyborn, Rev. Laser Engng **37, 6**, 443 (2009).
106. D. Ros, K. Cassou, B. Cros, S. Daboussi, J. Demailly, O. Guilbaud, S. Kazamias, J.-C. Lagron, G. Maynard, O. Neveu, M. Pittman, B. Zielbauer, D. Zimmer, T. Kuhl, S. Lacombe, E. Porcel, M.-A. duPenhoat, P. Zeitoun, and G. Mourou, Nucl. Instr. Meth. Phys. Res. A **653**, 76 (2011).

107. F. Ple, M. Pittman, G. Jamelot, and J. P. Chambaret, Opt. Lett. **32, 3** (2007).
108. J. H. Sung, S. K. Lee, T. J. Yu, T. M. Jeong, and J. Lee, Opt. Lett. **5**, 3021 (2010).
109. T. J. Yu, S. K. Lee, J. H. Sung, J. W. Yoon, T. M. Jeong, and J. Lee, Opt. Express **20**, 10807 (2012).
110. S. Formaux, S. Payeur, A. Alexandrov, C. Serbanescu, F. Martin, T. Ozaki, A. Kudryashov, and J. C. Kieffer, Opt. Express **16, 16**, 11987 (2008).
111. L. Roso, Proc. SPIE **8001**, 800113 (2011).
112. Z. Wang, C. Liu, Z. Shen, Q. Zhang, H. Teng, and Z. Wei, Opt. Lett. **36, 16** (2011).
113. W. P. Leemans, J. Daniels, A. Deshmukh, A. J. Gonsalves, A. Magana, H. S. Mao, D. E. Mittelberger, K. Nakamura, J. R. Riley, D. Syversrud, C. Toth, and N. Ybarrolaza, *Proceedings of PAC2013, Pasadena, CA, USA, THYAA1* (2013).
114. C. Liu, S. Banerjee, J. Zhang, S. Chen, K. Brown, J. Mills, N. Powers, B. Zhao, G. Golovin, I. Ghebregziabher, and D. Umstadter, Proc. SPIE **8599**, 859919 (2013).
115. C. Liu, J. Zhang, S. Chen, G. Golovin, S. Banerjee, B. Zhao, N. Powers, I. Ghebregziabher, and D. Umstadter, Opt. Lett. **39, 1** (2014).
116. C. Liu, G. Golovin, S. Chen, J. Zhang, B. Zhao, D. Haden, S. Banerjee, J. Silano, H. Karwowski, and D. Umstadter, Opt. Lett. **39, 14** (2014).
117. P. Poole, C. Willis, R. Daskalova, J. W. Sheng, D. L. Van Woerkom, R. Freeman, and E. Chowdhury, *Frontiers in Optics Conference, High Fields in Plasmas* (2011), ISBN: 978-1-55752-917-6.
118. Y. Chu, X. Liang, L. Yu, Y. Xu, Lu. Xu, L. Ma, X. Lu, Y. Liu, Y. Leng, R. Li, and Z. Xu, Opt. Express **21, 24** (2013).
119. K. Zeil, S. D. Kraft, S. Bock, M. Bussmann, T. E. Cowan, T. Kluge, J. Metzkes, T. Richter, R. Sauerbrey, and U. Schramm, New J. Phys. **12**, 045015 (2010).
120. J. P. Zou, *Invited talk at HPLSE 2014, Suzhou, China* (2014).
121. V. Bagnoud, *Invited talk at HPLSE 2014, Suzhou, China* (2014).
122. CETAL Website, (2014).
123. LLP Website, (2014).
124. L. A. Gizzi, C. Benedetti, C. Alberto Cecchetti, G. Di Pirro, A. Gamucci, G. Gatti, A. Giulietti, D. Giulietti, P. Koester, L. Labate, T. Levato, N. Pathak, and F. Piastra, Appl. Sci. **3**, 559 (2013).
125. O. V. Chekhlov, J. L. Collier, I. N. Ross, P. K. Bates, M. Notley, C. Hernandez-Gomez, W. Shaikh, C. N. Danson, D. Neely, P. Matousek, and S. Hancock, Opt. Lett. **31**, 3665 (2006).
126. V. V Lozhkarev, G. I. Freidman, V. N. Ginzburg, E. V. Katin, E. A. Khazanov, A. V. Kirsanov, G. A. Luchinin, A. N. Mal'shakov, M. A. Martyanov, O. V. Palashov, A. K. Poteomkin, A. M. Sergeev, A. A. Shaykin, and I. V. Yakovlev, Opt. Express **14**, 446 (2006).
127. V. V. Lozhkarev, G. I. Freidman, V. N. Ginzburg, E. V. Katin, E. A. Khazanov, A. V. Kirsanov, G. A. Luchinin, A. N. Mal'shakov, M. A. Martyanov, O. V. Palashov, A. K. Poteomkin, A. M. Sergeev, A. A. Shaykin, and I. V. Yakovlev, Laser Phys. Lett. **4**, 421 (2007).
128. L. Xu, L. Yu, X. Liang, Y. Chu, Z. Hu, L. Ma, Y. Xu, C. Wang, Xi. Lu, H. Lu, Y. Yue, Y. Zhao, F. Fan, H. Tu, Y. Leng, R. Li, and Z. Xu, Opt. Lett. **38, 22** (2013).
129. C. Hernandez-Gomez, S. P. Blake, O. Chekhlov, R. J. Clarke, A. M. Dunne, M. Galimberti, S. Hancock, R. Heathcote, P. Holligan, A. Lyachev, P. Matousek, I. O. Musgrave, D. Neely, P. A. Norreys, I. Ross, Y. Tang, T. B. Winstone, B. E. Wyborn, and J. Collier, J. Phys.: Conf. Ser. **244**, 032006 (2010).
130. B. Rus, K. Rohlena, J. Skála, B. Králiková, K. Jungwirth, J. Ullschmied, K. J. Witte, and H. Baumhacker, Lasers Part. Beams **17**, 179 (1999).
131. O. Novak, M. Divoký, H. Turčičová, and P. Straka, Lasers Part. Beams **31**, 211 (2013).
132. Z. Major, S. A. Trushin, I. Ahmad, M. Siebold, C. Wandt, S. Klingebiel, T. Wang, J. A. Fülöp, A. Henig, S. Kruber, R. Weingartner, A. Popp, J. Osterhoff, R. Hörlein, J. Hein, V. Pervak, A. Apolonski, F. Krausz, and S. Karsch, Laser Rev. **37, 6** (2009).
133. J. D. Zuegel, JTh4L.4, CLEO (2014).
134. XCELS Website, (2014).
135. Bayramian J. Armstrong, G. Beer, R. Campbell, B. Chai, R. Cross, A. Erlandson, Y. Fei, B. Freitas, R. Kent, J. Menapace, W. Molander, K. Schaffers, C. Siders, S. Sutton, J. Tassano, S. Telford, C. Ebbers, J. Caird, and C. Barty, J. Opt. Soc. Am. B **25**, B57 (2008).
136. A. Bayramian, P. Armstrong, E. Ault, R. Beach, C. Bibeau, J. Caird, R. Campbell, B. Chai, J. Dawson, C. Ebbers, A. Erlandson, Y. Fei, B. Freitas, R. Kent, Z. Liao, T. Ladran, J. Menapace, B. Molander, S. Payne, N. Peterson, M. Randles, K. Schaffers, S. Sutton, J. Tassano, S. Telford, and E. Utterback, Fusion Sci. Technol. **52**, 383 (2007).
137. M. Hornung, S. Keppler, R. Bödefeld, A. Kessler, H. Liebetrau, J. Körner, M. Hellwing, F. Schorcht, O. Jäckel, A. Sävert, J. Polz, A. K. Arunachalam, J. Hein, and M. C. Kaluza, Opt. Lett. **38, 5** (2013).
138. M. Siebold, F. Roeser, M. Loeser, D. Albach, and U. Schramm, Proc. SPIE **8780**, 878005 (2013).
139. ELI Website, (2014).
140. LIFE Website, (2014).
141. HiPER Website, (2014).
142. V. M. Malkin, G. Shvets, and N. J. Fisch, Phys. Rev. Lett. **82, 22**, 4448 (1999).
143. U. Keller, Proc. SPIE **8966**, 896602 (2014).
144. A. Klenke, S. Breitkopf, M. Kienel, T. Gottschall, T. Eidam, S. Hädrich, J. Rothhardt, J. Limpert, and A. Tünnermann, Opt. Lett. **38, 13**, 2283 (2013).
145. G. Mourou, B. Brocklesby, T. Tajima, and J. Limpert, Nat. Photon. **7** (2013).

28

Optimization of relativistic laser–ion acceleration

J. Schreiber[1,2], F. Bell[1], and Z. Najmudin[3]

[1] *Fakultät für Physik, Ludwig-Maximilians-Universität München, Am Coulombwall 1, D-85748 Garching, Germany*

[2] *Max-Planck-Institut für Quantenoptik, Hans-Kopfermann-Str. 1, D-85748 Garching, Germany*

[3] *The John Adams Institute, Blackett Laboratory, Imperial College London, London SW7 2AZ, United Kingdom*

Abstract

Experiments have shown that the ion energy obtained by laser–ion acceleration can be optimized by choosing either the appropriate pulse duration or the appropriate target thickness. We demonstrate that this behavior can be described either by the target normal sheath acceleration model of Schreiber *et al.* or by the radiation pressure acceleration model of Bulanov and coworkers. The starting point of our considerations is that the essential property of a laser system for ion acceleration is its pulse energy and not its intensity. Maybe surprisingly we show that higher ion energies can be reached with reduced intensities.

Keywords: laser–ion acceleration; relativistic laser plasma interaction

1. Introduction

Laser-driven ion acceleration has created enormous interest over the last few years[1, 2]. The rapid development of laser technology to intensities well beyond 10^{20} W cm^{-2} has enabled the generation of multi-MeV ion beams with exceptional characteristics[3]. Visions of reaching electron energies in the TeV[4] and even the PeV regime[5] have been published. While the ion beams in most experiments have exhibited extremely broad energy distributions, advanced target designs have been applied to achieve narrow energy distributions of protons and heavier ions[6–9]. The early, rapid developments have engendered much speculation about the use of laser-driven ions for fast ignition[10–12] and medical applications[13–17]. Laser-generated protons have already been successfully applied for time-resolved studies of the generation of electric and magnetic fields in the laser–plasma interaction on a ps timescale[18–20]. Moreover, the table-top generation of neutrons is discussed as a possible application and has been demonstrated in early and recent experiments[21–25].

The observation of ions emitted in laser–plasma interactions can first be traced to experiments employing high-intensity laser pulses with durations of a few ns to some hundreds of ps[26]. With the invention of chirped pulse amplification (CPA)[27], laser pulses with intensities well in excess of 10^{18} W cm^{-2} and fs durations were realized. The new era of relativistic laser–plasma interactions had begun, where the quiver velocity of electrons in the electromagnetic field of the laser approaches the speed of light. Moreover, the $v \times B$-term of the Lorentz force becomes dominant and pushes the electrons into the direction of laser propagation. The generation of relativistic electrons was the fundamental requisite for the acceleration of ions to high energy[28]. Foils of several tens of micrometer thickness which were irradiated by relativistically intense laser pulses were found to emit protons with energies of up to 60 MeV from the non-irradiated surface[29]. This behavior could be described by the target normal sheath acceleration (TNSA)[30] which has proved to be the dominant mechanism in most experiments performed until recent times. In recent years, competing mechanisms for ion acceleration have been introduced and discussed, the most prominent example being the radiation pressure acceleration (RPA)[31–40] or light-sail regime[41–44] which utilizes highly intense laser pulses with high contrast. Not too differently from the original ideas on the forces due to radiation[45–50], it seems feasible to accelerate the central part of an ultra-thin, nm scale foil to high energies according to the simple equation of motion[51].

Although a number of high-power, PW-class laser systems have been built around the world, the early record energy of 60 MeV[29] has improved little. Supported by analytical models, we demonstrate that this is due to the optimization

Correspondence to: J. Schreiber, Fakultät für Physik, Ludwig-Maximilians-Universität München, Am Coulombwall 1, D-85748 Garching, Germany. Email: Joerg.Schreiber@lmu.de problem that is encountered in both TNSA and RPA. We show that the key quantity of a laser system is its energy content in a single laser pulse. The laser energy simply is converted with a certain efficiency into a number and energy of ions and it is not the intensity that is most important as often intuitively assumed. This statement holds for both TNSA and RPA. As we will show, TNSA can be optimized by varying, even increasing, the duration of the laser pulse (i.e., decreasing intensity) and RPA by choosing an optimized target thickness. Although reduction of the pulse duration for a laser system of given energy means an increase in power and intensity, this also reduces the acceleration time so that inertial ions cannot reach the same final energy.

2. Target normal sheath acceleration

TNSA of ions has been extensively investigated over the last decade. In fact, until now it has proved to be the most effective method for ion acceleration when highly intense laser pulses are focused onto foils with thicknesses of several micrometers. TNSA relies on the efficient conversion of laser energy into hot, relativistic electrons. These electrons propagate through the target and set up fields at the target boundaries where they exit into vacuum. The electric field is created between the expelled electrons and the surface charge that they induce on the target. Since the electric field strength is of the same relative strength as the laser electric field amplitude (TV/m) which generates the hot electrons, most of the electrons return back into the target. Hence an electron cloud (sheath) is formed. Ions at the rear surface can be accelerated by the sheath fields to multi-MeV energies in only several tens of femtoseconds. Impurities such as hydrocarbon and water are present under most experimental conditions at the solid surface. Due to their higher charge-to-mass ratio, it is the protons from these contaminants that are most readily accelerated to high energies. However, by removing the contaminants by different means, the acceleration of heavier ions can also be optimized[52–55]. Due to strong spatial and temporal variations in the acceleration fields, the observed ions usually exhibit broad energy distributions extending from zero to a certain maximum cut-off energy. Reduction of this energy spread is a major challenge and some success has been achieved by micro-machining the targets in order to allow ions to be accelerated only in regions where the field is large and roughly uniform[6, 7], and by the use of mass-limited targets with an appropriate ion mixture[8, 56].

A large number of experimental results on laser-driven ion acceleration are now available[7, 29, 57–70], which can be used for a comparison with theoretical predictions. TNSA has been extensively studied numerically using particle-in-cell (PIC) simulations[71–75]. Several analytical models have also been developed which predict the dependence of the maximum ion energy on laser and target parameters[57, 76–82]. Substantial advance of the field is reported in a number of review papers available today[1, 2, 83]. In particular, the ion energy should be dependent on the laser irradiance, focal spot size and laser pulse duration.

2.1. Nonrelativistic TNSA

The original Schreiber model[57] is a nonrelativistic version of TNSA and calculates the energy gain of ions in an electrostatic field determined by the transfer of laser energy to the expanding surface by the divergent beam of hot electrons. This model has two main advantages. (i) Without choosing a distribution function for the laser-accelerated electrons, the potential that they set up depends only on the absorbed laser power ηP_L into those electrons and the transverse size of the electron cloud R_s. (ii) Due to the consideration of the transverse dimension R_s of the electron cloud when it exits the rear of the target, the resultant potential stays finite which is in contrast to most 1D models, where for infinite acceleration times the ion energies diverge[84]. Although the description of the potential appears to be based on ad hoc assumptions, the model is in excellent agreement (within a factor of 1.5 or so) with experiments performed in a wide range of parameters covering foil thicknesses from one to hundreds of micrometers, laser pulse energies in the sub-joule to kilojoule level, pulse durations from 50 to 5000 fs and ion species covering a major part of the periodic table ranging from protons to tungsten ions. The electric field at the rear side of the target is set up by fast electrons produced by the laser heating at the front side of the target. It should be noted that in general the targets are much thicker than the skin depth or hole-boring depth so that the laser does not interact with the rear of the target. Let us consider that the laser produces N_e electrons with an average energy E_e in a bunch of length $L = c\tau_L$, where τ_L is the laser pulse duration and the electrons are assumed to propagate with the speed of light c. On their way through the target, the electrons spread over a circular region with radius R_s. When exiting into vacuum a positive surface charge Qe is induced at the rear side of the target which yields a returning force $\propto Qe/(\pi R_s^2)$. Electrons run up the potential and eventually reverse their path at a distance z_u above the surface of the foil and re-enter the foil. In an equilibrium situation $2N_e z_u/L$ electrons are permanently outside the foil. To achieve global charge neutrality we identify this number with the number of positive surface charges Q. The potential of the corresponding charge density is

$$\Phi(r, z) = \frac{Qe}{4\pi\varepsilon_0\pi R_s^2}\int_0^{R_s}\int_0^{2\pi}\frac{r'\mathrm{d}r'\mathrm{d}\phi'}{\sqrt{r^2 + z^2 + r'^2 - 2rr'\cos\phi'}}. \tag{1}$$

In the following we will concentrate on the center of the acceleration region that is responsible for the most energetic ions, i.e., $r = 0$. In that case the integration of Equation (1) reads

$$-e(\Phi(0,z) - \Phi(0,0)) \equiv -e\Phi(\xi) = E_\infty\left(1 + \xi - \sqrt{1+\xi^2}\right) \equiv E_\infty s(\xi) \quad (2)$$

with

$$E_\infty = \frac{Qe^2}{2\pi\varepsilon_0 R_s} \quad (3)$$

and $\xi = z/R_s$. For short distances $\xi \ll 1$ one has $-e\Phi \cong E_\infty \xi$, which yields for the turning point $z_u = E_e R_s/E_\infty$. The total number of electrons is related to the laser energy E_L by $N_e E_e = \eta E_L$ where η is the efficiency by which the laser energy is converted into electron energy. The electrons are provided for the duration of the laser pulse τ_L. With $Q = 2N_e z_u/(c\tau_L) \propto P_L$ (the power of the laser pulse) we obtain the potential barrier

$$E_\infty = 2m_e c^2 \sqrt{\frac{\eta P_L}{P_{Re}}}. \quad (4)$$

We note that Equation (4) is independent of E_e because $E_\infty \propto Q \propto N_e z_u = (\eta E_L/E_e) \cdot (E_e R_s/E_\infty) = (\eta E_L R_s/E_\infty)$. The essential point is that laser energy is mainly converted into a large number of energetic electrons (fast enough to traverse the target) which in turn build up a dense electron sheath exhibiting extraordinary strong electric fields. It is thus not decisive what average energies E_e are gained by the laser. This is in contrast to the model proposed by Bulanov *et al.*[85] where ions are mainly accelerated by the longitudinal laser electric fields in a bored channel which acts as a waveguide with conducting walls. The potential of Equation (2) can be used to calculate the energy $E_i(\xi)$ an ion with charge q_i gains between $\xi = 0$ (the surface) and ξ:

$$E_i(\xi) = -q_i e\Phi(\xi). \quad (5)$$

The equation of motion yields

$$\tau_L = R_s \int_0^{\xi_m} \frac{d\xi}{v_i(\xi)} = R_s \int_0^{s_m} \frac{d\xi/ds}{v_i(s)} ds. \quad (6)$$

Now, nonrelativistically one has

$$v_i(s) = v_{i,\infty}\sqrt{s} \quad (7)$$

with $v_{i,\infty} = \sqrt{2E_{i,\infty}/m_i} = c\sqrt{2\varepsilon_{i,\infty}}$. Since

$$\frac{d\xi}{ds} = \frac{1+(1-s)^2}{2(1-s)^2}, \quad (8)$$

one obtains

$$\tau_L = \tau_0 \int_0^{s_m} \frac{1+(1-s)^2}{2(1-s)^2\sqrt{s}} ds = \tau_0\left[X + \int_0^X \frac{dx}{(1-x^2)^2}\right] \equiv \tau_0 F_{NR}(X) \quad (9)$$

with $\tau_0 = R_s/v_{i,\infty}$ and $X = \sqrt{s_m} = (E_{i,m}/E_{i,\infty})^{1/2}$. Finally, the integration yields

$$F_{NR}(X) = X + \frac{X}{2(1-X^2)} + \frac{1}{4}\ln\left(\frac{1+X}{1-X}\right) \quad (10)$$

with $X = (\varepsilon_{i,m}/\varepsilon_{i,\infty})^{1/2}$, where $\varepsilon_{i,m} = E_{i,m}/(m_i c^2)$ is the normalized maximum ion energy and $\varepsilon_{i,\infty}$ defines the normalized energy an ion could gain from the potential of the sheath if it were maintained stationary, and is given by $\varepsilon_{i,\infty}(\tau_L) = q_i 2m_e c^2\sqrt{\eta E_L/(\tau_L P_{Re})}/(m_i c^2) = 1.1 \times 10^{-3}\ (q_i/A_i)\sqrt{\eta E_L/(\tau_L P_{Re})}$. Here, τ_L and E_L are the laser pulse duration and energy respectively, q_i is the ion charge state, A_i is the nucleon number of the target, $P_{Re} = m_e c^3/r_e = 8.71$ GW is the relativistic power unit (r_e = classical electron radius, m_e = electron mass) and η is the absorption efficiency into hot electrons. Following Refs. [86, 87], this efficiency is evaluated as $\eta = 1.2 \times 10^{-15} I_L^{3/4}$, where I_L is in units of W cm^{-2}, up to a maximum $\eta = 0.5$. This scaling has been validated for a laser wavelength of $\lambda_L \sim 1$ µm and pulse durations of several hundreds of femtoseconds. We mention that a similar model has been developed by Bulanov *et al.*[15] who considered a conducting prolate ellipsoid which had been charged up by Q positive charges. The corresponding potential is given by Landau and Lifshitz[88]. However, there exists a major difference from our model: since it is assumed by the ellipsoid model that the surface is conducting, all transverse forces at $z = 0$ have to vanish, which can only be fulfilled by a specific distribution of the Q charges at the surface. However, that can be in contradiction to the laser-driven charge distribution of hot electrons which might be very different from any 'conducting' equilibrium.

One major result of the model presented is that for a given laser energy E_L, the shortest laser pulses and thus highest intensities are not necessarily optimal for TNSA. An important point is that the normalized maximum energy $\varepsilon_{i,\infty}$ an ion can gain depends on the pulse duration τ_L which has a strong consequence on the maximal ion energy $E_{i,m}$. This can be seen in Figure 1(a) where experimental proton energies $E_{i,m}$ (Ref. [57]) are plotted for a constant laser energy $E_L = 0.7$ J as a function of the laser pulse duration τ_L, showing explicitly that there exists an optimal duration τ_L^{opt}. This behavior has been verified on other laser systems as well[89] and can be understood by the following argument. The highest intensity (i.e., shortest pulse) produces the largest acceleration field, but only for a short time. On the other hand, a somewhat smaller field that is sustained over a longer duration of the laser pulse can result in higher ion

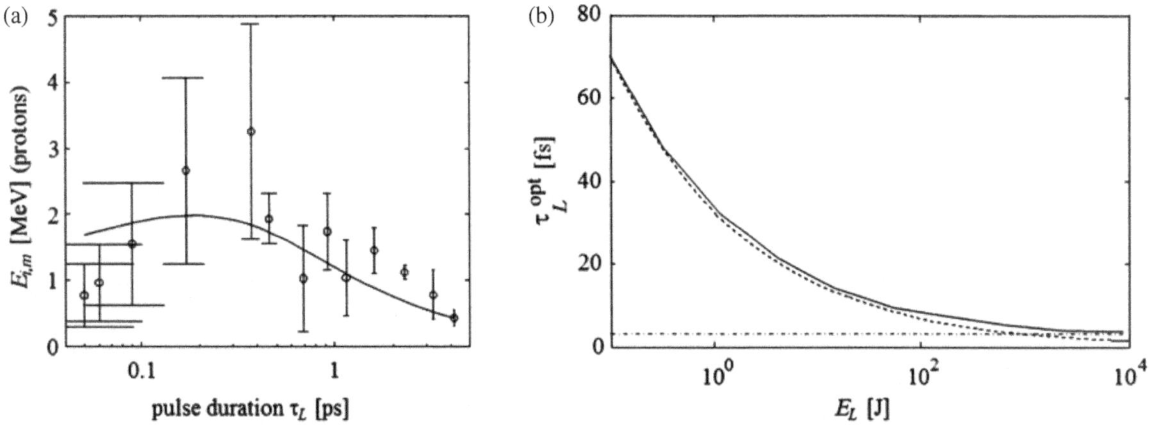

Figure 1. (a) Experimental data from Schreiber *et al.*[57] and the prediction of the nonrelativistic TNSA model. (b) Optimal pulse duration for the specific example of $R_s = 1$ μm, $\eta = 1$, $A_i/q_i = 2$ for the nonrelativistic consideration (Equation (12), dashed) and the relativistic consideration (Equation (29), solid). The ultra-relativistic limit is given by the dash-dotted line. For larger R_s the curves are globally shifted to respective larger optimal pulse durations.

energies. The acceleration is therefore optimized for some intermediate τ_L^{opt}. The solid curve in Figure 1(a) has been obtained from Equation (9) (for details, see Ref. [57]) and is in agreement with the experimental data, thus being able to describe an optimizing procedure. From Equation (9), for fixed laser energy and source radius, one can find the optimum using the condition $\mathrm{d}E_{i,m}/\mathrm{d}\tau_L|_{\tau_L^{\mathrm{opt}}} = 0$. For the nonrelativistic solution, Equation (10), this leads to

$$\left(\frac{\tau_L}{\tau_0}\right)^{\mathrm{opt}} = \frac{X_{\mathrm{opt}}}{3}\left(1 + \frac{1}{(1 - X_{\mathrm{opt}}^2)^2}\right), \tag{11}$$

where X_{opt} is evaluated at $\tau_L = \tau_L^{\mathrm{opt}}$. Insertion of Equations (11) in (10) yields the solution $X_{\mathrm{opt}} \cong 0.81$ with the corresponding value for $(\tau_L/\tau_0)^{\mathrm{opt}} \cong 2.55$. On inserting the nonrelativistic characteristic time τ_0, one obtains for the optimum pulse duration

$$\tau_L^{\mathrm{opt}}(E_L) = 195\left(\frac{R_s}{c}\right)^{4/3}\left(\frac{A_i}{q_i}\right)^{2/3}\left(\frac{P_{Re}}{\eta E_L}\right)^{1/3}. \tag{12}$$

Figure 1(b) shows the optimum pulse duration as a function of the laser energy for $A_i/q_i = 2$, $R_s = 1$ μm and $\eta = 1$. The dashed line is the nonrelativistic solution of Equation (12). The maximum ion energy then becomes

$$(\varepsilon_{i,m}^{\mathrm{opt}})_{\mathrm{TNSA}} = 1.3\frac{q_i}{A_i}\sqrt{\frac{\eta E_L}{\tau_L^{\mathrm{opt}} P_{Ri}}}, \tag{13}$$

where $P_{Ri} = (m_p/m_e)P_{Re} = (1836)^2 P_{Re} = 29.3$ PW is the relativistic power unit for a proton. From Equation (9) it follows that for a given laser system τ_0 should be minimized in order to obtain the largest maximum ion energy $E_{i,m}$. This is realized for the smallest possible source size $R_s = r_L$, where r_L is the radius of the beam spot. Therefore, it is convenient to use targets with thickness d much smaller than the radius of the focal spot, which is usually of the order of some micrometers.

2.2. *Relativistic TNSA*

The relativistic equation of motion is

$$\frac{\mathrm{d}p_i}{\mathrm{d}t} = q_i e E = -q_i e \frac{\mathrm{d}\Phi}{\mathrm{d}z} \tag{14}$$

with

$$-q_i e\Phi(\xi) = E_{i,\infty}\left(1 + \xi - \sqrt{1+\xi^2}\right) \equiv E_{i,\infty}s(\xi), \tag{15}$$

where $\xi = z/R_s$ and $E_{i,\infty} = q_i 2m_e c^2\sqrt{\eta P_L/P_{Re}}$, as before. The resulting two coupled first-order differential equations

$$\gamma_i^3\frac{\mathrm{d}\beta_i}{\mathrm{d}t^*} = \varepsilon_{i,\infty}\left(1 - \frac{\xi}{\sqrt{1+\xi^2}}\right) \tag{16}$$

and

$$\frac{\mathrm{d}\xi}{\mathrm{d}t^*} = \beta_i \tag{17}$$

have to be solved simultaneously with $t^* = t/t_{0,R} = ct/R_s$ and $\varepsilon_{i,\infty} = E_{i,\infty}/(m_i c^2)$. The initial condition is $t^* = 0$, $\xi = 0$ and $\beta_i = 0$. The ion energy is given by $\varepsilon_i = \gamma_i - 1$ with $\varepsilon_i = E_i/(m_i c^2)$. A first integration of Equation (16) yields

$$\int_0^{\beta_i}\gamma_i^3\beta_i\mathrm{d}\beta_i = \varepsilon_{i,\infty}\int_0^{\xi}\frac{\mathrm{d}s}{\mathrm{d}\xi'}\mathrm{d}\xi' \tag{18}$$

or

$$\gamma_i - 1 = \varepsilon_i = \varepsilon_{i,\infty}s(\xi). \tag{19}$$

Now, instead of Equation (7) one has

$$v_i(s) = c\sqrt{1-(1+\varepsilon_{i,\infty}s)^{-2}} \tag{20}$$

and thus

$$\frac{\tau_L}{\tau_{0,R}} = \int_0^{s_m} \frac{1+(1-s)^2}{2(1-s)^2\sqrt{1-(1+\varepsilon_{i,\infty}s)^{-2}}}\mathrm{d}s \tag{21}$$

with $\tau_{0,R} = R_s/c$ and $s_m = E_{i,m}/E_{i,\infty} = \varepsilon_{i,m}/\varepsilon_{i,\infty} = X^2$. It is immediately seen from Equation (21) that deviations from nonrelativistic TNSA occur for $\varepsilon_{i,\infty}s_m = \varepsilon_{i,m} \geqslant 1$ (compare Equation (9)).The essential difference is that nonrelativistically the final ion velocity $v_{i,\infty}$ might become arbitrarily large, but special relativity limits it to the speed of light. We note that Equation (21) can be written as

$$\begin{aligned}\frac{\tau_L}{\tau_0} &= \sqrt{2\varepsilon_{i,\infty}}\int_0^{s_m} \frac{1+(1-s)^2}{2(1-s)^2\sqrt{1-(1+\varepsilon_{i,\infty}s)^{-2}}}\mathrm{d}s \\ &\equiv F_R(X;\varepsilon_{i,\infty}).\end{aligned} \tag{22}$$

As a relativistic first-order correction to Equation (22) one obtains

$$\begin{aligned}\frac{\tau_L}{\tau_0} &= \int_0^X \frac{1+(1-x^2)^2}{(1-x^2)^2}\mathrm{d}x \\ &\quad - \frac{\varepsilon_{i,\infty}}{4}\int_0^X \frac{x^2(1+(1-x^2)^2)}{(1-x^2)^2}\mathrm{d}x \\ &= X + \frac{X}{2(1-X^2)} + \frac{1}{4}\ln\frac{1+X}{1-X} \\ &\quad - \frac{\varepsilon_{i,\infty}}{4}\left(\frac{X^3}{3} + \frac{X}{2(1-X^2)} - \frac{1}{4}\ln\frac{1+X}{1-X}\right).\end{aligned} \tag{23}$$

Unfortunately the simple scaling of Equation (9) is lost since now in F_R the additional parameter $\varepsilon_{i,\infty}$ accounts for relativistic effects. Using the value of $X_{\text{opt}} \cong 0.8$ of the nonrelativistic TNSA expression as a first-order solution we obtain a condition for the optimized pulse duration in the case of relativistic TNSA:

$$\left(\frac{\tau_L}{\tau_0}\right)^{\text{opt}} = 2.5 - 0.18\varepsilon_{i,\infty}^{\text{opt}} \tag{24}$$

or

$$\frac{c\sqrt{a}}{R_s}(\tau_L^{\text{opt}})^{3/4} = 2.5 - \frac{0.18a}{(\tau_L^{\text{opt}})^{1/2}} \tag{25}$$

with $a = 1.1\times10^{-3}(q_i/A_i)\sqrt{\eta E_L/P_{Re}}$. An analytic solution of Equation (25) reads

$$a = \frac{2.5}{\alpha_2} + \left(\frac{\alpha_1}{\alpha_2}\right)^2\left[1-\sqrt{1+\frac{5\alpha_2}{\alpha_1^2}}\right] \tag{26}$$

with $\alpha_1 = c(\tau_L^{\text{opt}})^{3/4}/R_s$ and $\alpha_2 = 0.18/(\tau_L^{\text{opt}})^{1/2}$. The nonrelativistic TNSA-limit Equation (11) is obtained for $\alpha_2 \to 0$:

$$a \cong \frac{25}{8\alpha_1^2} - \frac{125}{16\alpha_1^4}\alpha_2 = \frac{3.1R_s^2}{c^2(\tau_L^{\text{opt}})^{3/2}}\left(1-0.45\left(\frac{R_s}{c\tau_L^{\text{opt}}}\right)^2\right). \tag{27}$$

Again, the nonrelativistic optimal ion energy $(\varepsilon_{i,m}^{\text{opt}})_{\text{TNSA}}$ is obtained by Equation (13). The term within the bracket is the relativistic correction. A few things are worthy of note. (1) The correction due to relativistic ion motion becomes larger with decreasing optimal pulse duration, that is for larger laser energy E_L. This can be expected. However, (2) the correction decreases the value of a; in other words, the optimum pulse duration τ_L^{opt} is reduced even when the laser energy E_L remains constant. This in turn means that the optimum ion energy becomes larger if relativistic corrections are taken into account. (3) The correction increases with increasing R_s. Nonrelativistically τ_L^{opt} increases with $R_s^{4/3}$ (see Equation (12)), but relativistically this increase is less strong.

Writing Equation (21) as

$$\frac{\tau_L}{\tau_{0,R}} = \int_0^{s_m} H(s;\varepsilon_{i,\infty})\mathrm{d}s \tag{28}$$

and optimizing Equation (28) by the condition $\mathrm{d}\varepsilon_{i,m}/\mathrm{d}\tau_L|_{\tau_L^{\text{opt}}} = 0$ yields

$$\frac{1}{\tau_{0,R}} = \frac{\mathrm{d}s_m}{\mathrm{d}\tau_L}H(s_m;\varepsilon_{i,\infty}) + \int_0^{s_m}\frac{\partial}{\partial\tau_L}H(s,\varepsilon_{i,\infty})\mathrm{d}s. \tag{29}$$

It follows that

$$\frac{\mathrm{d}s_m}{\mathrm{d}\tau_L} = \frac{\mathrm{d}\varepsilon_{i,m}}{\mathrm{d}\tau_L}\frac{1}{\varepsilon_{i,\infty}} - \frac{\varepsilon_{i,m}}{\varepsilon_{i,\infty}^2}\frac{\mathrm{d}\varepsilon_{i,\infty}}{\mathrm{d}\tau_L} = \frac{s_m}{2\tau_L} \tag{30}$$

and

$$\begin{aligned}\frac{\partial H}{\partial\tau_L} &= \frac{1}{4\tau_L}\frac{1+(1-s)^2}{(1-s)^2} \\ &\quad\times\frac{\varepsilon_{i,\infty}s}{(1+\varepsilon_{i,\infty}s)^3(1-(1+\varepsilon_{i,\infty}s)^{-2})^{3/2}}.\end{aligned} \tag{31}$$

One therefore has the condition

$$\begin{aligned}4\frac{\tau_L^{\text{opt}}}{\tau_{0,R}} &= \frac{s_m^{\text{opt}}(1+(1-s_m^{\text{opt}})^2)}{(1-s_m^{\text{opt}})^2\sqrt{1-(1+\varepsilon_{i,\infty}^{\text{opt}}s_m^{\text{opt}})^{-2}}} \\ &\quad+\int_0^{s_m^{\text{opt}}}\frac{1+(1-s)^2}{(1-s)^2} \\ &\quad\times\frac{\varepsilon_{i,\infty}^{\text{opt}}s}{(1+\varepsilon_{i,\infty}^{\text{opt}}s)^3(1-(1+\varepsilon_{i,\infty}^{\text{opt}}s)^{-2})^{3/2}}\mathrm{d}s \\ &= 4\int_0^{s_m^{\text{opt}}}\frac{1+(1-s)^2}{2(1-s)^2(1-(1+\varepsilon_{i,\infty}^{\text{opt}}s)^{-2})^{1/2}}\mathrm{d}s,\end{aligned} \tag{32}$$

where $s_m^{\rm opt} = \varepsilon_{i,m}^{\rm opt}/\varepsilon_{i,\infty}^{\rm opt}$ and $\varepsilon_{i,\infty}^{\rm opt} = 1.1 \times 10^{-3}(q_i/A_i)$ $\sqrt{\eta E_L/(\tau_L^{\rm opt} P_{Re})}$. A numerical solution yields $s_m^{\rm opt}$ with $\varepsilon_{i,\infty}^{\rm opt}$ as a parameter. By inserting these solutions into Equation (29) one obtains the required $\tau_L^{\rm opt}/\tau_{0,R}$ (solid line in Figure 1(b)) and $\varepsilon_{i,m}^{\rm opt}$.

In the highly relativistic regime, the ion velocity approaches the speed of light for approximately all the acceleration time, which according to Equation (6) leads to

$$\frac{\tau_L}{\tau_{0,R}} = \xi_m \tag{33}$$

or

$$\varepsilon_{i,m} = 1.1 \times 10^{-3}(q_i/A_i)\sqrt{\frac{\eta E_L}{m_e c^2}\frac{R_s}{r_e}}G(\xi_m) \tag{34}$$

with

$$G(\xi_m) = \left(1 + \xi_m - \sqrt{1+\xi_m^2}\right)\Big/\sqrt{\xi_m}. \tag{35}$$

The function $G(\xi_m)$ has a rather broad maximum at $\xi_m = \tau_L^{\rm opt} c/R_s = 1$. Whereas in the nonrelativistic regime the optimum pulse duration decreases with increasing laser energy E_L, see Equation (12), it becomes a constant, $\tau_L^{\rm opt} = R_s/c$, for relativistic ion energies (dash-dotted line in Figure 1(b)). On inserting the value for G one obtains from Equation (34) the optimal ion energy

$$(\varepsilon_{i,m}^{\rm opt})_{\rm TNSA} = 0.64 \times 10^{-3}(q_i/A_i)\sqrt{\frac{\eta E_L}{m_e c^2}\frac{R_s}{r_e}}. \tag{36}$$

We conclude this section with two remarks. (i) The ion energy $E_{i,m}$ is a unique function of E_L and τ_L, i.e., for constant laser energy a function of τ_L only. Thus, $E_{i,m}$ might show a maximum with respect to τ_L or not. However, if there exists a maximum it is unique, i.e., there is a single-valued $E_{i,m}^{\rm opt} = E_{i,m}(\tau_L^{\rm opt})$ only. (ii) We propose a simple equation for the whole range,

$$\varepsilon_{i,m} = 1.3 \cdot \varepsilon_{i,\infty}\left(1 + \frac{\tau_L}{\tau^*} - \sqrt{1+\left(\frac{\tau_L}{\tau^*}\right)^2}\right), \tag{37}$$

with

$$\tau^* = \frac{R_s/c}{\sqrt{1 - \frac{1}{(1+\varepsilon_{i,\infty})^2}}}. \tag{38}$$

It might be that a better approximation is obtained if in the last equation $\varepsilon_{i,\infty}$ is replaced by $\varepsilon_{i,m}$. Bulanov *et al.*[85] have performed detailed PIC studies of the generation of high energy ions in overcritical targets. For strong focusing conditions ($R_s \cong 0.75$ μm), they obtained for $P_L = 1$ PW and a pulse duration of $\tau_L = 30$ fs (i.e., a laser energy of $E_L = 30$ J) a proton energy of 1.3 GeV. With $\varepsilon_{i,\infty} = 0.37$ and $\tau^* = 3.6$ fs we obtain from Equation (37) $E_{i,m} = 430$ MeV.

3. Radiation pressure acceleration

At the intensities available with present high-intensity lasers, it seems natural to consider RPA as a means of accelerating objects to high energy. RPA offers the most promising approach for the acceleration of plasma bunches with near-solid, or at least overcritical, density to relativistic velocities. The principle of RPA is the same as was proposed to use continuous wave lasers to drive interstellar vehicles to relativistic velocities[31, 50]. According to Simmons *et al.*[31], a body with rest mass comparable to the applied laser energy can be accelerated close to the speed of light. For a 5 nm thin carbon foil and a focal spot diameter of 2 μm, this would require only 5 J of energy. Unfortunately, the picture is not quite as simple due to the immense intensity of the applied laser. In Section 2, we have highlighted the importance of a high rate of absorption of laser energy into hot electrons for TNSA. For hole-boring or RPA to work efficiently, this heating must be suppressed, which may be achieved by the use of high-contrast systems with circular polarization[34]. In this way the ponderomotive force that acts on the plasma electrons is only composed of a secular term which can effectively expel the electrons from the focal region and push them into the target. The space charge separation in the focal region is then maintained over the duration of the laser pulse, or even longer when the reflection front starts to move. During this time, electrons stay cold, i.e., they do not gain a large longitudinal momentum spread, so they stay bound to this initial depletion zone.

The acceleration of an object with mass $M = m_i n_i {\rm d}\pi r_L^2$ by the radiation pressure, where n_i is the ion particle density, is described by Refs. [31, 34, 35]

$$\frac{{\rm d}(\gamma\beta)}{{\rm d}t} = \frac{1}{t_0}\frac{1-\beta}{1+\beta}, \tag{39}$$

where $t_0 = Mc^2/(2RP_L)$ and R denotes the reflectivity with which the laser is reflected. For a constant laser power P_L, a solution of Equation (39) in terms of the actual time t is not of much interest because of the retardation effect. More relevant is a solution in terms of the retarded time $t_{\rm ret} = t - \int_0^t \beta(t'){\rm d}t'$ (Refs. [31, 34]). Thus, Equation (39) becomes

$$\frac{{\rm d}\beta}{{\rm d}t_{\rm ret}} = \frac{1}{t_0}\frac{1-\beta}{\gamma}, \tag{40}$$

with the solution for the normalized ion energy

$$\varepsilon_{i,m}^{\rm RPA} = \gamma - 1 = \frac{1}{2}\left(1 + \zeta + \frac{1}{1+\zeta}\right) - 1, \tag{41}$$

where $\zeta = t_{\text{ret}}/t_0 = 2RE_L/(Mc^2)$, since the light pulse is terminated at the retarded time $t_{\text{ret}} = \tau_L$. We note that the interaction between the light and the plasma sheet may continue even when the laser has finished radiating: for relativistic velocities of the sheet the light pulse accompanies the bunch for interaction times $t = t_{\text{int}}$ much longer than the pulse duration

$$t_{\text{int}} = \tau_L \left(1 + \frac{\zeta}{2} + \frac{\zeta^2}{6}\right). \tag{42}$$

The essential point for the RPA mechanism is the assumption that the whole target of mass M is accelerated cooperatively as a charge-neutral plasma bunch. The physical mechanism behind this process is the following. Due to their small mass the electrons are accelerated by light pressure and then due to Coulomb forces drag the ions behind them (hence the name light-sail[43, 90]). Thus, one has to avoid the light pressure on the electrons exceeding the restoring forces due to the charge separation.

On inspecting Equation (41) it is readily seen that the ion energy $E_{i,m}^{\text{RPA}}$ increases with ζ and thus with decreasing target mass M, or with thinner target foils. However, a lower limit will be reached if the number of ions becomes so small that they cannot any longer retain the electrons by their Coulomb forces. One estimates the energy E_∞ to separate an electron from a sheet of ion charge density $q_i n_i = n_e$ to be $E_\infty = n_e \mathrm{d} r_L e^2/(4\varepsilon_0)$ (Ref. [57]). This estimate can also be derived from the 'capacitor' model for charge separation[37, 38, 91] assuming an effective cut-off of the induced electric field for distances larger than the lateral extension of the sheet. This assumption is equivalent to the demand to maintain the balance between charge separation and radiation pressure at least over a laser period[81, 92]. Evidently, the charge separating energy increases with the areal density $n_i d$ while the radiation pressure decreases like $(n_i d)^{-1}$, advocating once more the use of ultra-thin targets. The ion energy can be derived from Equation (41) in the nonrelativistic case $E_{i,m} \propto \zeta^2 \propto 1/M^2 \propto 1/d^2$. For a completely ionized carbon target with a thickness of 1 μg cm^{-2} one obtains $E_\infty \cong 10$ MeV. On the other hand, this estimate shows that there exists a lower limit on the electron mass M_e that can be accelerated by radiation pressure and which remains bound to the ions. The latter demand is essential for ion acceleration since it is the electrons that pull the ions behind them. If M_e becomes too small, the resulting fast electrons with energies $E_e = m_e c^2 \gamma \cong m_e c^2 \tau_L/(2t_0) = (m_e/M_e)E_L = E_L/N_e$ (note that at large electron energies the efficiency approaches 100%, i.e., all the laser energy is converted into the kinetic energy of N_e electrons) can surmount the potential barrier built up by the charge separation field (note that $M_e = m_e n_e \mathrm{d} r_L^2$). Thus, the minimum ion mass that prohibits charge separation and that can be accelerated as a charge-neutral plasma bunch

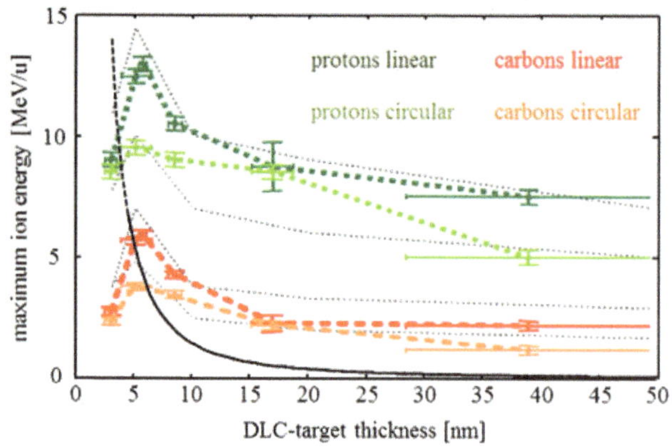

Figure 2. Maximum proton and carbon ion energies for varying thicknesses of nm-thin DLC foils reported in Henig *et al.*[51]. The solid curve represents the prediction for RPA, Equation (41), using the parameters $A_i/q_i = 2$, $E_L = 0.7$ J, $r_L = 1.8$ μm, $R = 1$, $\rho_{\text{DLC}} = 2.7$ g cm^{-3}. The optimum mass/thickness is indicated by the transition of the solid to a dashed curve.

becomes

$$M_{\min} = \frac{A_i}{q_i}\sqrt{\frac{4\varepsilon_0 E_L r_L}{e^2}}\, m_u \tag{43}$$

with the atomic mass unit m_u and A_i the mass number of the target. For a carbon target with $A_i/q_i = 2$, $E_L = 1$ J and $r_L = 3$ μm one obtains $M_{\min} = 2.2 \times 10^{-16}$ kg, corresponding to an optimum thickness of 4 nm. This is in strong contrast to a TNSA model of Andreev *et al.*[93] who predicted an optimum thickness of about 100 nm. In Refs. [51, 94] the maximum ion energy has been investigated as a function of the target mass M. As expected, the ion energy increases with $1/M$, but for masses close to $M_{\min}$ the energy starts to drop rather strongly with further decreasing target masses, thus confirming the estimate of Equation (43) quantitatively. Figure 2 reproduces the experimental results of Ref. [51] for the carbon energies obtained from nm-thin diamond-like carbon (DLC) foils. The solid curve has been obtained from Equation (41). Although the curve drops off faster towards thicker targets (due to the increasing domination of expansion), evidently there exists an optimal foil thickness, i.e., target mass. The estimate of Equation (43) is indicated where the solid line breaks into the dashed line. An estimate of the optimal and thus minimum foil thickness by Chen *et al.*[95] is in essence the same as that of Equation (43).

In a very recent theory of laser ion acceleration from thin foils[96] the dimensionless parameter

$$\xi = \frac{n_e \mathrm{d}}{n_c \lambda a_0} \tag{44}$$

has been introduced, which for small laser strength a_0 and thus $\xi \geqslant 1$ prohibits charge separation, i.e., allows collective ion acceleration induced by electrons riding ahead of the ions. Here, n_c is the critical electron density. It is easy to show that Equation (44) can also be written as $\xi =$

$M/M_{\min}$ with $M_{\min}$ of Equation (43). Thus, the condition $\xi \geqslant 1$ for an appropriate ion acceleration is equivalent to demanding $M \geqslant M_{\min}$. Identifying $M_{\min}$ with the optimal target mass M_{opt} for RPA and inserting Equations (43) into (41) one obtains $(\varepsilon_{i,m}^{\mathrm{opt}})_{\mathrm{RPA}}$, which can be compared with the equivalent expression for the case of optimal TNSA. In Figure 3 $\varepsilon_{i,m}^{\mathrm{opt}}$ has been plotted for either TNSA or RPA versus the laser energy E_L. The curves hold for $R_s = r_L = 1$ μm (solid lines) and $R_s = r_L = 10$ μm (dashed lines), $A_i/q_i = 2$ and $\eta = R = 1$. For a typical high-intensity laser with a wavelength of $\lambda_L = 1$ μm, a beam spot radius of 1 μm is close to the diffraction-limited minimum sized beam spot. The corresponding data in Figure 3 thus represent optimum ion energies also within this context. The extension of available laser energies up to 10 kJ might be optimistic but we note that in a recent paper by Tajima *et al.*[5] on electron acceleration up to PeV even GJ laser energies have been discussed. A laser with $E_L = 1$ GJ could accelerate by radiation pressure a carbon foil of optimal thickness of 300 μm with C ions in it up to energies of 60 TeV/ion ($\gamma = 5 \times 10^3$, which is close to the heavy-ion energies of CERN's Large Hadron Collider[97]). However, we also note that for a pulse duration of 1 ps[5], the interaction distance according to Equation (42) would become 5 km. The reflected radiation has a vanishingly small frequency and at the end of the acceleration process nearly all the laser pulse energy is converted into kinetic energy of the ions, $E_L \cong (\gamma - 1)M_{\mathrm{opt}}c^2$, see also Equation (45). It would not be easy to stop such an ion bunch. Its energy content is large enough to heat up and finally melt 15 tons of lead. For comparison, the 7.7 TeV protons stored in the LHC have an energy content of 0.6 GJ[98]. An elegant solution of such a 'beam-dump' problem has recently been proposed by Wu *et al.*[99], relying on the deceleration of energetic and dense particle bunches by collective electronic interactions in an underdense plasma, promising a compact and non-radioactive dump. However, two remarks might be made. (i) The main contribution to the energy loss of ultra-relativistic heavy ions is electron–positron pair creation and not electronic stopping ('Bethe–Bloch')[97]. According to the description of Ref. [97] we obtain for the case discussed – 60 TeV carbon ions in lead – an electronic energy loss of 0.07 TeV m^{-1} and one by pair production of 0.3 TeV m^{-1}. Energy losses due to bremsstrahlung can be neglected since the small impact parameters necessary to generate hard photon quanta lead unavoidably to a fragmentation of the projectile[97]. We thus estimate for the example mentioned above a stopping length of about 570 m. In contrast, a beam-dump length of several meters only is sufficient to break up the projectile by nuclear inelastic interactions, yielding a cascade of secondary particles with individual energies much less than the primary one[100]. It is mainly the electromagnetic component of the cascade that finally converts the energy into heat. (ii) The essential point is not to stop the bunch within a moderate distance and thus heat the absorber but to dissipate its enormous energy concentration without any further problem. The situation for a laser-generated ion bunch is quite different from that of the LHC where the particles are more or less homogeneously distributed around the 27 km long circumference of the storage ring: C ions within a bunch length of 300 μm compared with 5×10^{14} protons stored within 27 km. The corresponding pulse length of 90 μs is long enough to allow kicker magnets to sweep the beam across the absorber.

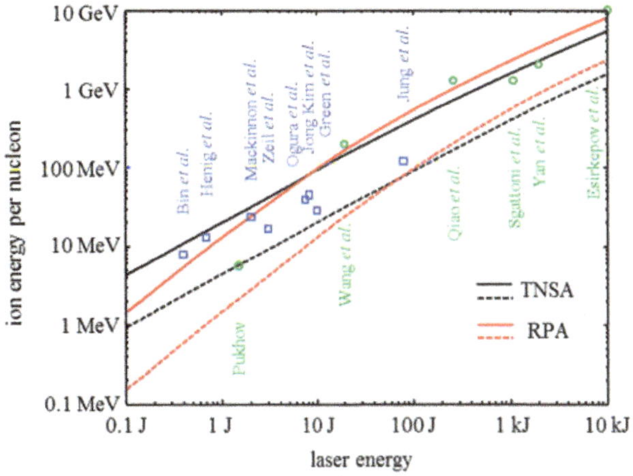

Figure 3. Optimum ion energies predicted by the models for TNSA, Equations (29)–(32) (black), and RPA, Equations (41)–(43) (red). The parameters are $A_i/q_i = 2$, $R = \eta = 1$, $r_L = R_s = 1$ μm (solid) and $r_L = R_s = 10$ μm (dashed). Some selected experimental results are represented by blue squares (Bin *et al.*[112], Henig *et al.*[51], Mackinnon *et al.*[108], Zeil *et al.*[113], Ogura *et al.*[68], Jong Kim *et al.*[114], Green *et al.*[115], Jung *et al.*[116]) and theoretical results obtained from PIC simulations are marked by green circles (Pukhov[71], Wang *et al.*[117], Qiao *et al.*[40], Sgattoni *et al.*[118], Yan *et al.*[75], Esirkepov *et al.*[35]). For details, see text.

It is readily seen that except for a minor factor over the whole range of laser energies both optimized theories yield the same maximum ion energies. It is therefore more a question of practicability what kind of optimization one chooses, τ_L^{opt} or M_{opt}. There might exist technical limits: it is certainly very difficult to obtain pulse durations shorter than say 1 fs. According to Equation (29) this would correspond to an upper limit of $E_L = 3 \times 10^4$ J, which in essence does not pose a strong limit. On the other hand, in the case of RPA a lower limit of M_{opt} is reached for a monolayer of atoms. Assuming a carbon foil as the target (the most popular choice) one obtains a lower limit of $M_{\mathrm{opt}} = 1.4 \times 10^{-18}$ kg, or from Equation (41) a lower limit of $E_L = 1.3 \times 10^{-4}$ J, which also does not have a practical influence. We emphasize that the optimizing procedure developed in this paper only works since the TNSA theory of Schreiber *et al.*[57] does not depend on the target thickness (at least to first order) and the RPA theory of Refs. [35, 101, 102] does not depend on the pulse duration.

In addition, experimental results are plotted in Figure 3 demonstrating that in most cases experiments are rather far away from being optimal. Multi-J laser systems especially suffer from having too long a pulse duration. We also note that normalized ion energies $\varepsilon_{i,m} \geqslant 1$ or $E_{i,m} \geqslant 1$ GeV/u, i.e., relativistic energies, are obtained for laser energies $E_L \geqslant$ 100 J only and so far have only been obtained in PIC simulations. Finally, we note that the efficiency of RPA can be written as

$$\eta_{\mathrm{RPA}} \equiv \frac{E_{i,m}}{E_L} = \frac{R\zeta}{1+\zeta}, \tag{45}$$

i.e., in the relativistic regime with $\gamma \gg 1$ or with Equation (41) $\zeta \gg 1$ one obtains that for $R = 1$, 100% of the laser energy is converted into ion energy.

Of course, if the transverse light intensity changes, different parts of the sail will be accelerated differently changing an initially plane sail to a convex one. The equation of motion of Equation (39) changes to

$$\frac{\mathrm{d}(\gamma\vec{\beta})}{\mathrm{d}t} = \frac{2RP_L}{Mc^2}\frac{1-\beta}{1+\beta}\vec{n}, \tag{46}$$

where both $P_L = P_L(y, z; t)$ and the unit vector $\vec{n} = \vec{n}(y, z; t)$ of the sail's surface normal depend on the transverse coordinates (y, z) and time t. Thus, a solution of Equation (46) has to account for a transverse expansion of the sail. An approximate solution of this self-consistent problem to obtain both the final velocity β and the shape of the sail, allowing for bowing while simultaneously submitting it to the condition of constant mass, has been given in Ref. [103]. The essential point is that instabilities like those of the Rayleigh–Taylor type lead to strong deviations from a uniform plasma front, forming cusps at an early stage of acceleration[104], resulting in emission of beamlets[105]. To avoid such instabilities Chen *et al.*[95] have proposed to use thin target foils shaped initially in the transverse direction to match the laser intensity profile. PIC simulations by Wilks *et al.*[106] performed as far back as in 1992 clearly demonstrated such instabilities for light intensities relevant to the light-sail regime. However, we remark that the ion energy differs for every beamlet. At a later stage the 'plasma foil' breaks into high-density clumps with diffuse lower density clouds between them[104], which due to its reduced mass may even accelerate ions to higher energies than the original foil[103]. A similar result has been obtained quite recently, demonstrating by PIC simulations that 'a relatively stable ion clump forms near the laser axis which is efficiently accelerated'[75]. A detailed study of Rayleigh–Taylor instabilities which yield filamentation and their optimization with respect to maximum ion energies has been published recently[85, 107].

4. Discussion

We are very much aware that analytic descriptions of the complex laser-assisted acceleration process of ions are hampered in many aspects. In contrast, PIC simulations describe in much more detail the complex processes described here. However, we also believe that an analytic description of the multi-parameter behavior of the process and its interdependences can give a more general overlook of the strategy to obtain the required outcomes such as, e.g., maximum ion energies. At the same time we remark that also PIC simulations which in essence are based on a mean field theory may rather severely suppress microscopic interactions. We also recognize a significant overshoot of PIC simulations, promising ion beams of great quality not verified by experiment hitherto. We cite a very recent paper: 'as pointed out recently in a number of papers circular polarized laser pulses can accelerate ions very efficiently and produce sharply peaked spectra'[75]. This hope results from PIC simulations only and still awaits experimental verification. In the following we will touch on some problems not considered in the analytic description. First, there is the question of the longitudinal and transverse extension of the laser beam. It has been assumed that the laser power is constant during its pulse duration, but even the use of a super-Gauss representation of the time dependence would merely change the conclusions. This might be rather different for the transverse extent which for a mono-mode laser system is Gaussian. In the case of TNSA this effect yields an electron sheet with transversely changing field strength which accelerates ions to different energies. Such broad spectra are not very favorable for specific applications. A similar problem arises in the case of RPA. The solution Equation (41) assumes that the light pressure is in essence transversely constant. RPA for a transverse Gaussian light beam has been investigated in detail by Bulanov *et al.*[92, 103] with PIC simulations showing in particular the buckling of an initially flat foil. Although the RPA process itself seems to remain stable, a rather broad ion spectrum results in this case also.

We note that such high energies as predicted by the analytical models have not been observed yet even though comparable laser conditions have been applied, for example by Mackinnon *et al.*[108]. Proton energies of up to 25 MeV have been obtained in the TNSA regime with micrometer thick targets. The discrepancy may be attributed to an absorption of laser energy into electrons much below 100%, especially when considering the high temporal contrast employed. The results could also be an indication that the electrons spread transversely over a size larger than $R_s \cong r_L$, even though the target thickness d was smaller than r_L in those experiments.

Hence, the studies presented here are encouraging in view of future applications that rely on high-repetition-rate laser systems. For example, for medical applications such as ion tumor therapy energies exceeding 100 MeV/u are envisioned. This energy range should be attainable even with sub-100 J laser systems, while relativistic energies can be achieved with energies slightly above 100 J. In order to increase the ion energy even further, i.e., above the multi-GeV level, kJ systems such as envisioned for the Extreme Light Infrastructure will be necessary. It may also become necessary to consider novel methods not discussed here. For example, once the ions move with the speed of light, staged acceleration possibly implementing plasma wake acceleration as used presently to accelerate electrons may provide a more effective means to reach higher energies[109–111]. In any case, even with lower output ion energies as compared with conventional accelerators, laser-accelerated ion bunches are still desirable, due to their bunch densities which may be close to solid density. Even with the low repetition rate of most high-power laser systems, the high number of reactions per unit volume that would be achievable with solid density bunches is perfectly suited for the exploration of the field of nonlinear nuclear physics, such as the production of exotic nuclei by fission–fusion reactions. Another important feature is the possible short ion bunch duration paired with the synchronism to other laser-driven radiation sources to allow for time- and space-resolved studies in pump–probe schemes.

5. Summary

Starting from experimental results, we have shown that current theories of laser–ion acceleration can and should be optimized in order to achieve maximum ion energies. Not all theories include such a possibility, but the TNSA theory of Schreiber *et al.*[57] can be optimized with respect to laser pulse duration and the RPA theory of Bulanov *et al.*[35, 101, 102] with respect to target thickness. It turns out that both the optimized TNSA and RPA theories yield approximately the same maximum ion energies over the range of laser energies $0.1\ \mathrm{J} < E_L < 10\ \mathrm{kJ}$. It is thus a matter of convenience whether one adjusts the optimal pulse duration or the optimal target thickness. For both theories, the decisive laser parameter is neither the power nor the intensity but solely the laser energy. Relativistic ion energies, i.e., energies beyond 1 GeV/u, can be obtained for systems with $E_L > 100$ J, where in addition a diffraction-limited small spot size has to be achieved.

Acknowledgements

This work was supported by the DFG Cluster of Excellence Munich Centre for Advanced Photonics (MAP) and the Transregio 18 as well as the Euratom/toIFE-project at MPQ. The authors thank Dr. Claudio Bellei for valuable and fruitful discussions.

References

1. H. Daido, M. Nishiuchi, and A. S. Pirozhkov, Rep. Progr. Phys. **75**, 056401 (2012).
2. A. Macchi, M. Borghesi, and M. Passoni, Rev. Modern Phys. **85**, 751 (2013).
3. T. E. Cowan, J. Fuchs, H. Ruhl, A. Kemp, P. Audebert, M. Roth, R. Stephens, I. Barton, A. Blazevic, E. Brambrink, J. Cobble, J. Fernández, J.-C. Gauthier, M. Geissel, M. Hegelich, J. Kaae, S. Karsch, G. P. Le Sage, S. Letzring, M. Manclossi, S. Meyroneinc, A. Newkirk, H. Pépin, and N. Renard-LeGalloudec, Phys. Rev. Lett. **92**, 204801 (2004).
4. M. Kando, H. Kiriyama, J. K. Koga, S. Bulanov (JAERI, Kyoto), A. W. Chao (SLAC), T. Esirkepov, R. Hajima, and T. Tajima, AIP Conf. Proc. **1024**, 197 (2008).
5. T. Tajima, M. Kando, and M. Teshima, Progr. Theoret. Phys. **125**, 617 (2011).
6. B. M. Hegelich, B. J. Albright, J. Cobble, K. Flippo, S. Letzring, M. Paffett, H. Ruhl, J. Schreiber, R. K. Schulze, and J. C. Fernández, Nature **439**, 441 (2006).
7. H. Schwoerer, S. Pfotenhauer, O. Jackel, K. U. Amthor, B. Liesfeld, W. Ziegler, R. Sauerbrey, K. W. Ledingham, and T. Esirkepov, Nature **439**, 445 (2006).
8. S. Ter-Avetisyan, M. Schnurer, P. V. Nickles, M. Kalashnikov, E. Risse, T. Sokollik, W. Sandner, A. Andreev, and V. Tikhonchuk, Phys. Rev. Lett. **96**, 145006 (2006).
9. S. M. Pfotenhauer, O. Jäckel, A. Sachtleben, J. Polz, W. Ziegler, H.-P. Schlenvoigt, K.-U. Amthor, M. C. Kaluza, K. W. D. Ledingham, R. Sauerbrey, P. Gibbon, A. P. L. Robinson, and H. Schwoerer, New J. Phys. **10**, 033034 (2008).
10. A. P. Fews, P. A. Norreys, F. N. Beg, A. R. Bell, A. E. Dangor, C. N. Danson, P. Lee, and S. J. Rose, Phys. Rev. Lett. **73**, 1801 (1994).
11. M. Roth, T. E. Cowan, M. H. Key, S. P. Hatchett, C. Brown, W. Fountain, J. Johnson, D. M. Pennington, R. A. Snavely, S. C. Wilks, K. Yasuike, H. Ruhl, F. Pegoraro, S. V. Bulanov, E. M. Campbell, M. D. Perry, and H. Powell, Phys. Rev. Lett. **86**, 436 (2001).
12. P. K. Patel, A. J. Mackinnon, M. H. Key, T. E. Cowan, M. E. Foord, M. Allen, D. F. Price, H. Ruhl, P. T. Springer, and R. Stephens, Phys. Rev. Lett. **91**, 125004 (2003).
13. S. V. Bulanov and V. S. Khoroshkov, Plasma Phys. Rep. **28**, 453 (2002).
14. E. Fourkal, B. Shahine, M. Ding, J. S. Li, T. Tajima, and C. M. Ma, Med. Phys. **29**, 2788 (2002).
15. S. V. Bulanov, H. Daido, T. Z. Esirkepov, V. S. Khoroshkov, J. Koga, K. Nishihara, F. Pegoraro, T. Tajima, and M. Yamagiwa, AIP Conf. Proc. **740**, 414 (2004).
16. V. Malka, S. Fritzler, E. Lefebvre, E. d'Humières, R. Ferrand, G. Grillon, C. Albaret, S. Meyroneinc, J. Chambaret, A. Antonetti, and D. Hulin, Med. Phys. **31**, 1587 (2004).
17. U. Linz and J. Alonso, Phys. Rev. ST Accel. Beams **10**, 094801 (2007).
18. L. Romagnani, J. Fuchs, M. Borghesi, P. Antici, P. Audebert, F. Ceccherini, T. Cowan, T. Grismayer, S. Kar, A. Macchi, P. Mora, G. Pretzler, A. Schiavi, T. Toncian, and O. Willi, Phys. Rev. Lett. **95**, 195001 (2005).
19. L. Romagnani, S. V. Bulanov, M. Borghesi, P. Audebert, J. C. Gauthier, K. Löwenbrück, A. J. Mackinnon, P. Patel, G. Pretzler, T. Toncian, and O. Willi, Phys. Rev. Lett. **101**, 025004 (2008).
20. T. Sokollik, M. Schnürer, S. Ter-Avetisyan, P. V. Nickles, E. Risse, M. Kalashnikov, W. Sandner, G. Priebe, M. Amin, T. Toncian, O. Willi, and A. A. Andreev, Appl. Phys. Lett. **92**, 091503 (2008).
21. P. A. Norreys, A. P. Fews, F. N. Beg, A. R. Bell, A. E. Dangor, P. Lee, M. B. Nelson, H. Schmidt, M. Tatarakis, and M. D. Cable, Plasma Phys. Control. Fusion **40**, 175 (1998).
22. S. Karsch, S. Düsterer, H. Schwoerer, F. Ewald, D. Habs, M. Hegelich, G. Pretzler, A. Pukhov, K. Witte, and R. Sauerbrey, Phys. Rev. Lett. **91**, 015001 (2003).

23. S. Ter-Avetisyan, M. Schnürer, D. Hilscher, U. Jahnke, S. Busch, P. V. Nickles, and W. Sandner, Phys. Plasmas **12**, 2702 (2005).
24. K. L. Lancaster, S. Karsch, H. Habara, F. N. Beg, E. L. Clark, R. Freeman, M. H. Key, J. A. King, R. Kodama, K. Krushelnick, K. W. D. Ledingham, P. McKenna, C. D. Murphy, P. A. Norreys, R. Stephens, C. Stöeckl, Y. Toyama, M. S. Wei, and M. Zepf, Phys. Plasmas **11**, 3404 (2004).
25. M. Roth, D. Jung, K. Falk, N. Guler, O. Deppert, M. Devlin, A. Favalli, J. Fernandez, D. Gautier, M. Geissel, R. Haight, C. E. Hamilton, B. M. Hegelich, R. P. Johnson, F. Merrill, G. Schaumann, K. Schoenberg, M. Schollmeier, T. Shimada, T. Taddeucci, J. L. Tybo, F. Wagner, S. A. Wender, C. H. Wilde, and G. A. Wurden, Phys. Rev. Lett. **110**, 044802 (2013).
26. S. J. Gitomer, R. D. Jones, F. Begay, A. W. Ehler, J. F. Kephart, and R. Kristal, Phys. Fluids **29**, 2679 (1986).
27. D. Strickland and G. Morou, Opt. Commun. **56**, 219 (1985).
28. F. N. Beg, A. R. Bell, A. E. Dangor, C. N. Danson, A. P. Fews, M. E. Glinsky, B. A. Hammel, P. Lee, P. A. Norreys, and M. Tatarakis, Phys. Plasmas **4**, 447 (1997).
29. R. A. Snavely, M. H. Key, S. P. Hatchett, T. E. Cowan, M. Roth, T. W. Phillips, M. A. Stoyer, E. A. Henry, T. C. Sangster, M. S. Singh, S. C. Wilks, A. MacKinnon, A. Offenberger, D. M. Pennington, K. Yasuike, A. B. Langdon, B. F. Lasinski, J. Johnson, M. D. Perry, and E. M. Campbell, Phys. Rev. Lett. **85**, 2945 (2000).
30. S. C. Wilks, A. B. Langdon, T. E. Cowan, M. Roth, M. Singh, S. Hatchett, M. H. Key, D. Pennington, A. MacKinnon, and R. A. Snavely, Phys. Plasmas **8**, 542 (2001).
31. J. F. L. Simmons and C. R. McInnes, Amer. J. Phys. **61**, 205 (1993).
32. L. O. Silva, M. Marti, J. R. Davies, R. A. Fonseca, C. Ren, F. S. Tsung, and W. B. Mori, Phys. Rev. Lett. **92**, 015002 (2004).
33. W. Yu, H. Xu, F. He, M. Y. Yu, S. Ishiguro, J. Zhang, and A. Y. Wong, Phys. Rev. E **72**, 046401 (2005).
34. A. Macchi, F. Cattani, T. V. Liseykina, and F. Cornolti, Phys. Rev. Lett. **94**, 165003 (2005).
35. T. Esirkepov, M. Borghesi, S. V. Bulanov, G. Mourou, and T. Tajima, Phys. Rev. Lett. **92**, 175003 (2004).
36. X. Zhang, B. Shen, X. Li, Z. Jin, F. Wang, and M. Wen, Phys. Plasmas **14**, 123108 (2007).
37. A. Robinson, M. Zepf, S. Kar, R. Evans, and C. Bellei, New J. Phys. **10**, 013021 (2008).
38. O. Klimo, J. Psikal, J. Limpouch, and V. T. Tikhonchuk, Phys. Rev. ST Accel. Beams **11**, 031301 (2008).
39. X. Q. Yan, C. Lin, Z. M. Sheng, Z. Y. Guo, B. C. Liu, Y. R. Lu, J. X. Fang, and J. E. Chen, Phys. Rev. Lett. **100**, 135003 (2008).
40. B. Qiao, S. Kar, M. Geissler, P. Gibbon, M. Zepf, and M. Borghesi, Phys. Rev. Lett. **108**, 115002 (2012).
41. B. Shen and Z. Xu, Phys. Rev. E **64**, 056406 (2001).
42. B. Qiao, M. Zepf, M. Borghesi, and M. Geissler, Phys. Rev. Lett. **102**, 145002 (2009).
43. A. Macchi, S. Veghini, and F. Pegoraro, Phys. Rev. Lett. **103**, 085003 (2009).
44. B. Qiao, M. Zepf, P. Gibbon, M. Borghesi, J. Schreiber, and M. Geissler, Proc. SPIE **8079**, 80790Q (2011).
45. P. Lebedew, Ann. Phys. **311**, 433 (1901).
46. E. F. Nichols and G. F. Hull, Astrophys. J. **17**, 315 (1903).
47. F. A. Zander, Tech. Zhizn **13**, 15 (1924).
48. A. S. Eddington, Mon. Not. R. Astron. Soc. **85**, 408 (1925).
49. V. I. Veksler, Sov. J. At. Energy **2**, 525 (1957).
50. G. Marx, Nature **211**, 22 (1966).
51. A. Henig, S. Steinke, M. Schnürer, T. Sokollik, R. Hörlein, D. Kiefer, D. Jung, J. Schreiber, B. M. Hegelich, X. Q. Yan, J. Meyer-ter-Vehn, T. Tajima, P. V. Nickles, W. Sandner, and D. Habs, Phys. Rev. Lett. **103**, 245003 (2009).
52. M. Hegelich, S. Karsch, G. Pretzler, D. Habs, K. Witte, W. Guenther, M. Allen, A. Blazevic, J. Fuchs, J. C. Gauthier, M. Geissel, P. Audebert, T. Cowan, and M. Roth, Phys. Rev. Lett. **89**, 085002 (2002).
53. J. Schreiber, M. Kaluza, F. Grüner, U. Schramm, B. M. Hegelich, J. Cobble, M. Geissler, E. Brambrink, J. Fuchs, P. Audebert, D. Habs, and K. Witte, Appl. Phys. B **79**, 1041 (2004).
54. P. McKenna, F. Lindau, O. Lundh, D. C. Carroll, R. J. Clarke, K. W. D. Ledingham, T. McCanny, D. Neely, A. P. L. Robinson, L. Robson, P. T. Simpson, C.-G. Wahlström, and M. Zepf, Plasma Phys. Control. Fusion **49**, B223 (2007).
55. A. Henig, D. Kiefer, K. Markey, D. C. Gautier, K. A. Flippo, S. Letzring, R. P. Johnson, T. Shimada, L. Yin, B. J. Albright, K. J. Bowers, J. C. Fernández, S. G. Rykovanov, H.-C. Wu, M. Zepf, D. Jung, V. Kh. Liechtenstein, J. Schreiber, D. Habs, and B. M. Hegelich, Phys. Rev. Lett. **103**, 045002 (2009).
56. A. V. Brantov, V. T. Tikhonchuk, O. Klimo, D. V. Romanov, S. Ter-Avetisyan, M. Schnürer, T. Sokollik, and P. V. Nickles, Phys. Plasmas **13**, 122705 (2006).
57. J. Schreiber, F. Bell, F. Grüner, U. Schramm, M. Geissler, M. Schnürer, S. Ter-Avetisyan, B. M. Hegelich, J. Cobble, E. Brambrink, J. Fuchs, P. Audebert, and D. Habs, Phys. Rev. Lett. **97**, 045005 (2006).
58. J. Fuchs, P. Antici, E. d'Humières, E. Lefebvre, M. Borghesi, E. Brambrink, C. A. Cecchetti, M. Kaluza, V. Malka, M. Manclossi, S. Meyroneinc, P. Mora, J. Schreiber, T. Toncian, H. Pépin, and P. Audebert, AIP Conf. Proc. **827**, 237 (2006).
59. P. McKenna, K. W. D. Ledingham, J. M. Yang, L. Robson, T. McCanny, S. Shimizu, R. J. Clarke, D. Neely, K. Spohr, R. Chapman, R. P. Singhal, K. Krushelnick, M. S. Wei, and P. A. Norreys, Phys. Rev. E **70**, 036405 (2004).
60. Y. Murakami, Y. Kitagawa, Y. Sentoku, M. Mori, R. Kodama, K. A. Tanaka, K. Mima, and T. Yamanaka, Phys. Plasmas **8**, 4138 (2001).
61. S. Fritzler, V. Malka, G. Grillon, J.-P. Rousseau, F. Burgy, E. Lefebvre, E. dHumieres, P. McKenna, and K. W. D. Ledingham, Appl. Phys. Lett. **83**, 3039 (2003).
62. L. Robson, P. T. Simpson, R. J. Clarke, K. W. D. Ledingham, F. Lindau, O. Lundh, T. McCanny, P. Mora, D. Neely, C.-G. Wahlström, M. Zepf, and P. McKenna, Nat. Phys. **3**, 58 (2006).
63. M. Nishiuchi, A. Fukumi, H. Daido, Z. Li, A. Sagisaka, K. Ogura, S. Orimo, M. Kado, Y. Hayashi, M. Mori, S. V. Bulanov, T. Esirkepov, K. Nemoto, Y. Oishi, T. Nayuki, T. Fujii, A. Noda, Y. Iwashita, T. Shirai, and S. Nakamura, Phys. Lett. A **357**, 339 (2006).
64. E. L. Clark, K. Krushelnick, M. Zepf, F. N. Beg, M. Tatarakis, A. Machacek, M. I. K. Santala, I. Watts, P. A. Norreys, and A. E. Dangor, Phys. Rev. Lett. **85**, 1654 (2000).
65. T. Fujii, Y. Oishi, T. Nayuki, and Y. Takizawa, Appl. Phys. Lett. **83**, 1524 (2003).
66. I. Spencer, K. W. D. Ledingham, P. McKenna, T. McCanny, R. P. Singhal, P. S. Foster, D. Neely, A. J. Langley, E. J. Divall, C. J. Hooker, R. J. Clarke, P. A. Norreys, E. L. Clark, K. Krushelnick, and J. R. Davies, Phys. Rev. E **67**, 046402 (2003).
67. S. Kar, K. F. Kakolee, B. Qiao, A. Macchi, M. Cerchez, D. Doria, M. Geissler, P. McKenna, D. Neely, J. Osterholz, R. Prasad, K. Quinn, B. Ramakrishna, G. Sarri, O. Willi, X. Y. Yuan, M. Zepf, and M. Borghesi, Phys. Rev. Lett. **109**, 185006 (2012).
68. K. Ogura, M. Nishiuchi, A. S. Pirozhkov, T. Tanimoto, A. Sagisaka, T. Zh. Esirkepov, M. Kando, T. Shizuma,

T. Hayakawa, H. Kiriyama, T. Shimomura, S. Kondo, S. Kanazawa, Y. Nakai, H. Sasao, F. Sasao, Y. Fukuda, H. Sakaki, M. Kanasaki, A. Yogo, S. V. Bulanov, P. R. Bolton, and K. Kondo, Opt. Lett. **37**, 2868 (2012).

69. D. Jung, L. Yin, B. J. Albright, D. C. Gautier, S. Letzring, B. Dromey, M. Yeung, R. Hörlein, R. Shah, and S. Palaniyappan, New J. Phys. **15**, 023007 (2013).
70. D. Jung, L. Yin, D. C. Gautier, H.-C. Wu, S. Letzring, B. Dromey, R. Shah, S. Palaniyappan, T. Shimada, R. P. Johnson, J. Schreiber, D. Habs, J. C. Fernández, B. M. Hegelich, and B. J. Albright, Phys. Plasmas **20**, 083103 (2013).
71. A. Pukhov, Phys. Rev. Lett. **86**, 3562 (2001).
72. Q. L. Dong, Z. M. Sheng, M. Y. Yu, and J. Zhang, Phys. Rev. E **68**, 026408 (2003).
73. T. Esirkepov, M. Yamagiwa, and T. Tajima, Phys. Rev. Lett. **96**, 105001 (2006).
74. L. Yin, B. J. Albright, B. M. Hegelich, K. J. Bowers, K. A. Flippo, T. J. T. Kwan, and J. C. Fernández, Phys. Plasmas **14**, 056706 (2007).
75. X. Q. Yan, H. C. Wu, Z. M. Sheng, J. E. Chen, and J. Meyer-ter-Vehn, Phys. Rev. Lett. **103**, 135001 (2009).
76. J. Fuchs, P. Antici, E. d'Humières, E. Lefebvre, M. Borghesi, E. Brambrink, C. A. Cecchetti, M. Kaluza, V. Malka, M. Manclossi, S. Meyroneinc, P. Mora, J. Schreiber, T. Toncian, H. Pépin, and P. Audebert, Nat. Phys. **2**, 48 (2006).
77. M. Passoni and M. Lontano, Laser Part. Beams **22**, 163 (2004).
78. M. Lontano and M. Passoni, Phys. Plasmas **13**, 042102 (2006).
79. A. P. Robinson, A. R. Bell, and R. J. Kingham, Phys. Rev. Lett. **96**, 035005 (2006).
80. M. M. Basko, Eur. Phys. J. D **41**, 641 (2007).
81. M. Passoni and M. Lontano, Phys. Rev. Lett. **101**, 115001 (2008).
82. A. V. Brantov, V. T. Tikhonchuk, V. Y. Bychenkov, and S. G. Bochkarev, Phys. Plasmas **16**, 043107 (2009).
83. G. A. Mourou, T. Tajima, and S. V. Bulanov, Rev. Modern Phys. **78**, 309 (2006).
84. P. Mora, Phys. Rev. Lett. **90**, 185002 (2003).
85. S. S. Bulanov, V. Yu. Bychenkov, V. Chvykov, G. Kalinchenko, D. W. Litzenberg, T. Matsuoka, A. G. R. Thomas, L. Willingale, V. Yanovsky, K. Krushelnick, and A. Maksimchuk, Phys. Plasmas **17**, 043105 (2010).
86. J. Yu, Z. Jiang, J. C. Kieffer, and A. Krol, Phys. Plasmas **6**, 1318 (1999).
87. M. H. Key, M. D. Cable, T. E. Cowan, K. G. Estabrook, B. A. Hammel, S. P. Hatchett, E. A. Henry, D. E. Hinkel, J. D. Kilkenny, J. A. Koch, W. L. Kruer, A. B. Langdon, B. F. Lasinski, R. W. Lee, B. J. MacGowan, A. MacKinnon, J. D. Moody, M. J. Moran, A. A. Offenberger, D. M. Pennington, M. D. Perry, T. J. Phillips, T. C. Sangster, M. S. Singh, M. A. Stoyer, M. Tabak, G. L. Tietbohl, M. Tsukamoto, K. Wharton, and S. C. Wilks, Phys. Plasmas **5**, 1966 (1998).
88. L. D. Landau, J. S. Bell, M. J. Kearsley, L. P. Pitaevskii, E. M. Lifshitz, and J. B. Sykes, *Electrodynamics of Continuous Media*. Vol. 8 (Elsevier, 1984).
89. T. Toncian, M. Swantusch, M. Toncian, O. Willi, A. A. Andreev, and K. Y. Platonov, Phys. Plasmas **18**, 043105 (2011).
90. S. Steinke, P. Hilz, M. Schnürer, G. Priebe, J. Bränzel, F. Abicht, D. Kiefer, C. Kreuzer, T. Ostermayr, J. Schreiber, A. A. Andreev, T. P. Yu, A. Pukhov, and W. Sandner, Phys. Rev. ST Accel. Beams **16**, 011303 (2013).
91. J. Meyer-ter-Vehn and H.-C. Wu, Eur. Phys. J. D **55**, 433 (2009).
92. F. Pegoraro and S. V. Bulanov, Eur. Phys. J. D **55**, 399 (2009).
93. A. Andreev, A. Levy, T. Ceccotti, C. Thaury, K. Platonov, R. A. Loch, and P. Martin, Phys. Rev. Lett. **101**, 155002 (2008).
94. S. Steinke, A. Henig, M. Schnürer, T. Sokollik, P.V. Nickles, D. Jung, D. Kiefer, R. Hörlein, J. Schreiber, T. Tajima, X. Q. Yan, M. Hegelich, J. Meyer-ter-Vehn, W. Sandner, and D. Habs, Laser Part. Beams **28**, 215 (2010).
95. M. Chen, A. Pukhov, T. P. Yu, and Z. M. Sheng, Phys. Rev. Lett. **103**, 024801 (2009).
96. X. Q. Yan, T. Tajima, M. Hegelich, L. Yin, and D. Habs, Appl. Phys. B **98**, 711 (2010).
97. A. H. Sørensen, Phys. Rev. A **81**, 022901 (2010).
98. T. L. S. Group, CERN Report CERN-91-03 (1991).
99. H.-C. Wu, T. Tajima, D. Habs, A. W. Chao, and J. Meyer-ter-Vehn, Phys. Rev. ST Accel. Beams **13**, 101303 (2010).
100. A. Ferrari, G. R. Stevenson, and E. Weisse, in *Proceedings of 3rd European Particle Accelerator Conference* 1545 (1992).
101. S. V. Bulanov, T. Esirkepov, P. Migliozzi, F. Pegoraro, T. Tajima, and F. Terranova, Nucl. Instrum. Methods Phys. Res. A **540**, 25 (2005).
102. F. Pegoraro, S. V. Bulanov, J. I. Sakai, and G. Tomassini, Phys. Rev. E **64**, 016415 (2001).
103. S. V. Bulanov, E. Y. Echkina, T. Z. Esirkepov, I. N. Inovenkov, M. Kando, F. Pegoraro, and G. Korn, Phys. Plasmas **17**, 063102 (2010).
104. C. A. J. Palmer, T. Kluge, K. Zeil, M. Bussmann, S. D. Kraft, T. E. Cowan, and U. Schramm, Phys. Rev. Lett. **108**, 225002 (2012).
105. F. Pegoraro and S. V. Bulanov, Phys. Rev. Lett. **99**, 065002 (2007).
106. S. C. Wilks, W. L. Kruer, M. Tabak, and A. B. Langdon, Phys. Rev. Lett. **69**, 1383 (1992).
107. G. Mourou, Z. Chang, A. Maksimchuk, J. Nees, S. V. Bulanov, V. Yu. Bychenkov, T. Zh. Esirkepov, N. M. Naumova, F. Pegoraro, and H. Ruhl, Plasma Phys. Rep. **28**, 12 (2002).
108. A. J. Mackinnon, Y. Sentoku, P. K. Patel, D. W. Price, S. Hatchett, M. H. Key, C. Andersen, R. Snavely, and R. R. Freeman, Phys. Rev. Lett. **88**, 215006 (2002).
109. B. Shen, X. Zhang, Z. Sheng, M. Y. Yu, and J. Cary, Phys. Rev. ST Accel. Beams **12**, 121301 (2009).
110. X. Zhang, B. Shen, L. Ji, F. Wang, M. Wen, W. Wang, J. Xu, and Y. Yu, Phys. Plasmas **17**, 123102 (2010).
111. F. L. Zheng, S. Z. Wu, H. C. Wu, C. T. Zhou, H. B. Cai, M. Y. Yu, T. Tajima, X. Q. Yan, and X. T. He, Phys. Plasmas **20**, 013107 (2013).
112. J. H. Bin, W. J. Ma, K. Allinger, H. Y. Wang, D. Kiefer, S. Reinhardt, P. Hilz, K. Khrennikov, S. Karsch, X. Yan, F. Krausz, T. Tajima, D. Habs, and J. Schreiber, Phys. Plasmas **20**, 073113 (2013).
113. K. Zeil, S. D. Kraft, S. Bock, M. Bussmann, T. E. Cowan, T. Kluge, J. Metzkes, T. Richter, R. Sauerbrey, and U. Schramm, New J. Phys. **12**, 045015 (2010).
114. I. J. Kim, K. H. Pae, C. M. Kim, H. T. Kim, J. H. Sung, S. K. Lee, T. J. Yu, I. W. Choi, C.-L. Lee, K. H. Nam, P. V. Nickles, T. M. Jeong, and J. Lee, Phys. Rev. Lett. **111**, 165003 (2013).
115. J. S. Green, A. P. L. Robinson, N. Booth, D. C. Carroll, R. J. Dance, R. J. Gray, D. A. MacLellan, P. McKenna, C. D. Murphy, D. Rusby, and L. Wilson, Appl. Phys. Lett. **104**, 214101 (2014).
116. D. Jung, *Ion acceleration from relativistic laser nano-target interaction* Dissertation (LMU München, 2012).
117. H. Y. Wang, C. Lin, F. L. Zheng, Y. R. Lu, Z. Y. Guo, X. T. He, J. E. Chen, and X. Q. Yan, Phys. Plasmas **18**, 093105 (2011).
118. A. Sgattoni, S. Sinigardi, and A. Macchi, Appl. Phys. Lett. **105**, 084105 (2014).

Defining the optimal gradient doped Yb^{3+}:YAG disk for room and low temperature diode pumped solid-state laser operations

J.-C. Chanteloup, M. Arzakantsyan, and S. Marrazzo
LULI, Ecole Polytechnique, CNRS, CEA, UPMC, Route de Saclay, 91128 Palaiseau, France

Abstract
We propose a general methodology to define the optimum doping ion volume distribution required for an efficient solid-state laser amplifier. This approach is illustrated in the context of two experimental diode pumped Yb:YAG amplifiers operating at 300 and 160 K. Processing of such tailored gain media is now possible through horizontal direct crystallization.

Keywords: diode solid-state laser; laser gain medium doping engineering; laser materials; ytterbium YAG laser

1. Introduction

The elaboration and processing of yttrium aluminum garnet $Y_3Al_5O_{12}$ (YAG) laser crystals with controlled spatial distribution of Yb^{3+} ions has been demonstrated with the horizontal direct crystallization (HDC) technique known as the Bagdasarov growth method[1]. HDC grown Yb:YAG disks as large as 77 mm have been produced, as illustrated in Figure 1. The advantages offered by such non-homogeneously doped structures in terms of both thermal and amplified spontaneous emission (ASE) management have been revealed in the context of disk amplifiers used in an active mirror architecture[2], where pumping and extraction take place on one side (AR coated) of the disk whereas cooling occurs on the other side (HR coated).

We detail here a methodology to define the optimum doping ion volume distribution required for two diode pumped Yb:YAG amplifiers. The proposed approach requires one first to define a convenient workspace to easily quantify the lasing ion distribution in an HDC engineered gain medium of thickness t (see Figure 1(a)). Since the doping distributions obtained with this growth technique are very linear[1, 2], we shall consider g_d [at.%], the doping linear ramp, as the first relevant quantity (horizontal axis). g_d/t is the actual gradient, expressed in at.% cm^{-1}. The second quantity is the average doping level d_0 [at.%] (vertical axis), which corresponds to the doping level at half the thickness of the considered disk, $d_0 = d(t/2)$. A disk is therefore defined in a two-dimensional $[g_d, d_0]$ workspace limited by the blue triangle in Figure 1.

We consider two experimental cases corresponding to the two power amplifiers of the Lucia laser project[3] at the Laboratoire pour l'Utilisation des Laser Intenses (LULI) of the Ecole Polytechnique, Palaiseau, France. In its current configuration this diode pumped solid-state laser (DPSSL) relies on two homogeneously doped Yb:YAG disks. Apart from having different dimensions and doping, the main differentiating feature is the operation temperature. Whereas the first amplifying stage operates at room temperature (300 K) and is water cooled[3], the second active mirror amplifier is designed to operate at a much colder temperature (160 K), cooled with an innovative helium cell[4].

Section 2 gives the respective cross sections used in the model derived to optimize the YAG disks' axial gradient in Yb concentration.

The 300 K amplifier disk is 60 mm in diameter and 7 mm thick, homogeneously doped at 2 at.%, defining a working point *WP* with coordinates ($g_d = 0$ at.%, $d_0 = 2$ at.%) in our workspace. The optimum point *OP* identification is detailed in Section 3, relying on stored energy and gain calculated through the Frantz and Nodvik formalism[5].

The 160 K amplifier disk is 77 mm in diameter and 10 mm thick, homogeneously doped at 1 at.%, defining a *WP* with coordinates ($g_d = 0$ at.%, $d_0 = 1$ at.%). The *OP* identification is detailed in Section 4.

Correspondence to: J.-C. Chanteloup, Laboratoire LULI, Ecole Polytechnique, Route de Saclay, 91128, Palaiseau CEDEX, France. Email: jean-christophe.chanteloup@polytechnique.edu

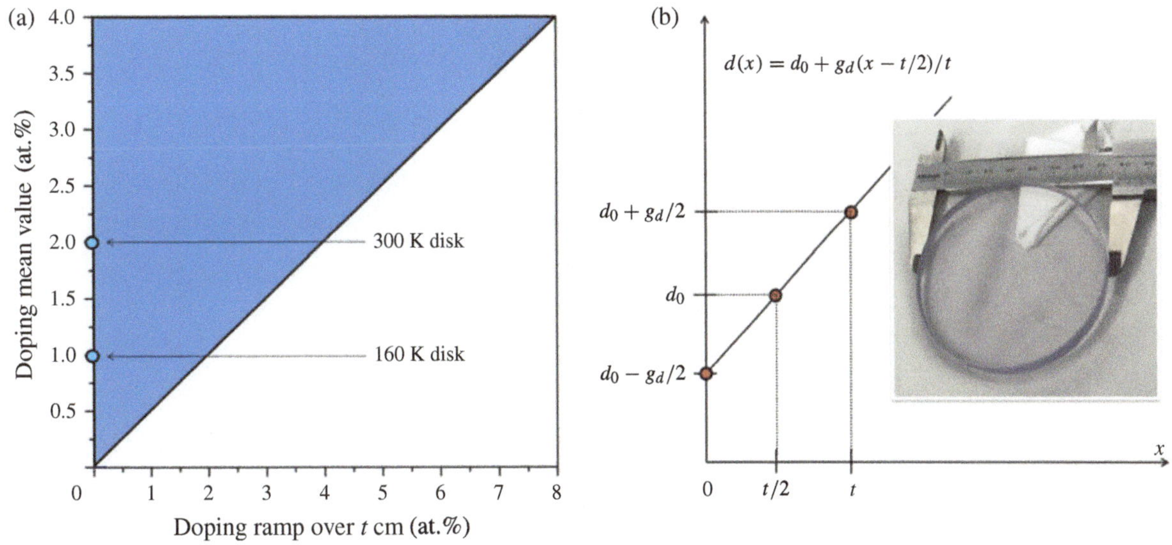

Figure 1. (a) The workspace defined by the disk doping ramp over its thickness and the doping mean value. (b) The linear doping equation $d(x) = d_0 + g_d(x - t/2)/t$. The inset shows the 77 mm disk used for the 160 K Lucia active mirror amplifier; it corresponds to the blue circle with coordinates (0, 1) in the workspace; the other circle (0, 2) refers to the 300 K disk. Both positions are qualified as *WP*.

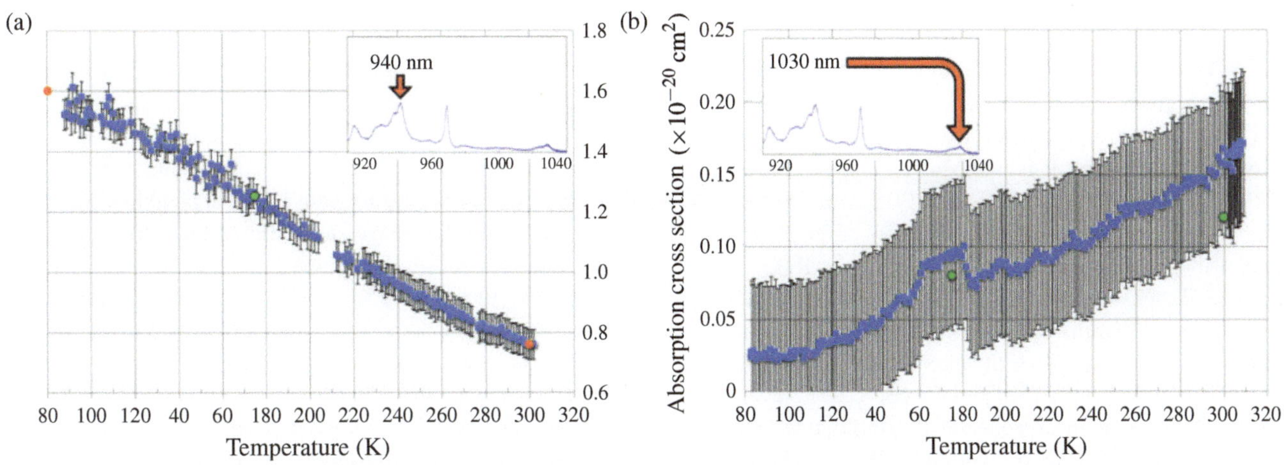

Figure 2. The thermal evolution of the absorption cross section between 90 and 300 K for an HDC grown homogeneously 1 at.% doped 1 cm thick Yb:YAG crystal. The data in (a) are collected at 940 nm whereas those in (b) correspond to the 1030 nm absorption peak. The red dots refer to data from Fan[7] whereas the green ones refer to data from Brown[8].

2. Absorption cross sections

The Lucia DPSSL currently operates two active mirror amplifiers at 300 and 160 K. In both cases, the gain medium is a YAG disk whose thickness and doping level have been optimized to maximize the laser efficiency considering a homogeneous volume distribution of Yb ions. The Lucia cryogenic amplifier relies on a thin (sub-mm) layer of helium for heat extraction[3, 6] through the HR coated surface of the active mirror. The helium layer thermal conductivity is adjusted through careful pressure control, allowing the absorption to be recorded over a 90 to 300 K temperature range, as illustrated in Figure 2. The cross sections considered in our model are reported in Table 1, the absorption cross sections being our measurements while the emission ones are taken from the literature[7, 8].

Table 1. Absorption and emission cross sections at 300 and 160 K.

Temperature	300 (K)	160 (K)
Absorption cross section at 940 nm	7.6×10^{-21} cm^2	1.3×10^{-20} cm^2
Absorption cross section at 1030 nm	1.6×10^{-21} cm^2	6.2×10^{-22} cm^2
Emission cross section at 1030 nm	2.4×10^{-20} cm^2	5.7×10^{-20} cm^2
Emission cross section at 940 nm	2.2×10^{-21} cm^2	9.7×10^{-22} cm^2

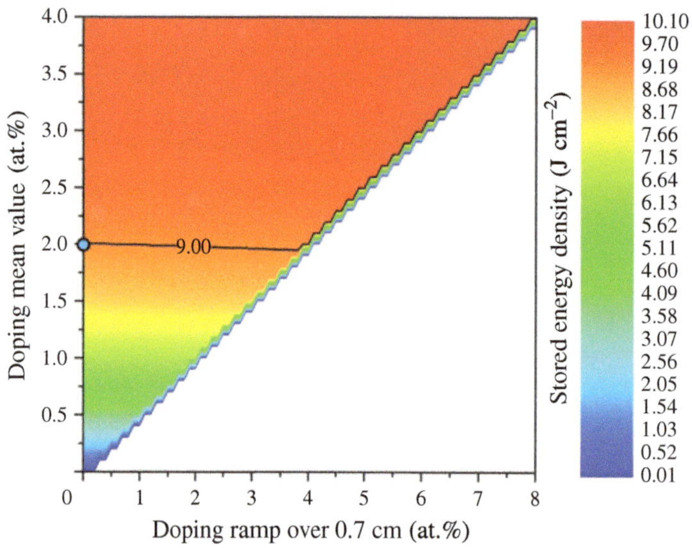

Figure 3. Stored energy density map versus doping ramp and mean value. For each specific doping ramp, a minimum average doping value exists below which this doping ramp is not defined; this explains the map's white triangular area. The iso-energy density 9 J cm^{-2} line above which the requested energetic performance will be reached is drawn.

3. Optimum Yb:YAG disk for a 300 K amplifier

The Lucia laser chain room temperature operated amplifier has previously been detailed[3]. The gain medium is an Yb:YAG disk 60 mm in diameter and 7 mm thick homogeneously doped at 2 at.%. It is pumped at a maximum level of 16 kW cm^{-2} over an area limited to 30 mm in diameter in order to take advantage of the 1030 nm absorption to mitigate ASE parasitic oscillations (Figure 16 of Ref. [3]). The relevant dimension to consider for ASE management is therefore $L_{\mathrm{ASE}} = 30$ mm. Considering the well-known limit for the constraint $g_0 L_{\mathrm{ASE}} \leqslant 4$, the maximum allowed value for the small signal gain is $g_{0,\max} = 1.33$ cm^{-1}. Wherever in the gain medium the gain exceeds this threshold, one enters into a regime where parasitic oscillations will start to deplete the gain.

Let us consider the workspace defined in Section 1 by the doping linear ramp g_d and the average doping level d_0. Extreme values of g_d are set from 0 to 8 at.% along the disk thickness of 7 mm. This means that the maximum explored gradient reaches 11.4 at.% cm^{-1}. Extreme values of d_0 are set from 0 to 4 at.%. The Lucia current room temperature operated crystal is then defined in this space by a *WP* whose coordinates are ($g_d = 0$ at.%, $d_0 = 2$ at.%). Let us first consider how the stored energy density (J cm^{-2}) is distributed in this workspace (Figure 3). The *WP* is materialized by the blue circle and we observe an energy density of 9 J cm^{-2}. The corresponding iso-energy density line is drawn on the map as well. It defines a boundary below which the Lucia room temperature energetic performance would not be satisfied. It is remarkable to observe that above this frontier, an increase in the average doping affects the energy density only very moderately, whereas below this line it decreases very rapidly. The *WP* has then been defined to be set on the edge of this plateau.

This map is obtained without considering ASE losses. In practice, increasing the doping mean value far above 2 at.% would actually not lead to a continuous increase of the stored energy density because at some point the gain medium would start seeing its gain being depleted by ASE related deleterious effects. We have defined such a threshold while introducing $g_{0,\max} = 1.33$ cm^{-1}. A maximum gain value map was then derived to identify the safe operation area within the workspace (Figure 4(a)). To derive this map, for each (g_d, d_0) point, the gain axial variation $g_0(z)$ is first computed. In Figure 4(b), two of these curves are given, the black one for *WP*. The maximum value of each curve, Max[$g_0(z)$], is then simply reported on the map.

On this map, the 1.33 cm^{-1} threshold iso-gain curve (solid back) is reported together with the 9 J cm^{-2} iso-energy density line, the first defining an upper boundary and the second the lower frontier of a small optimum region (dashes). The inset gives an enlarged view of the area of interest (AoI) within which this optimum region falls. The Lucia current *WP* is marked with the blue circle (1.73 cm^{-1}) whereas the *OP* (offering the lower $g_{0,\max} = 1.29$ cm^{-1}) is marked in yellow.

Obtaining an axial gain distribution as homogeneous as possible helps in staying below the parasitic oscillation threshold whatever the depth into the gain medium and reduces the gain medium internal stresses induced by the thermal load[2].

In order to identify the area of best homogeneity, it is convenient to study the gain contrast, defined as $Cg_0 = (g_0^{\max} - g_0^{\min})/(g_0^{\max} + g_0^{\min})$. The map of Figure 5 gives the Cg_0 distribution. As a reference, the AoI rectangle is

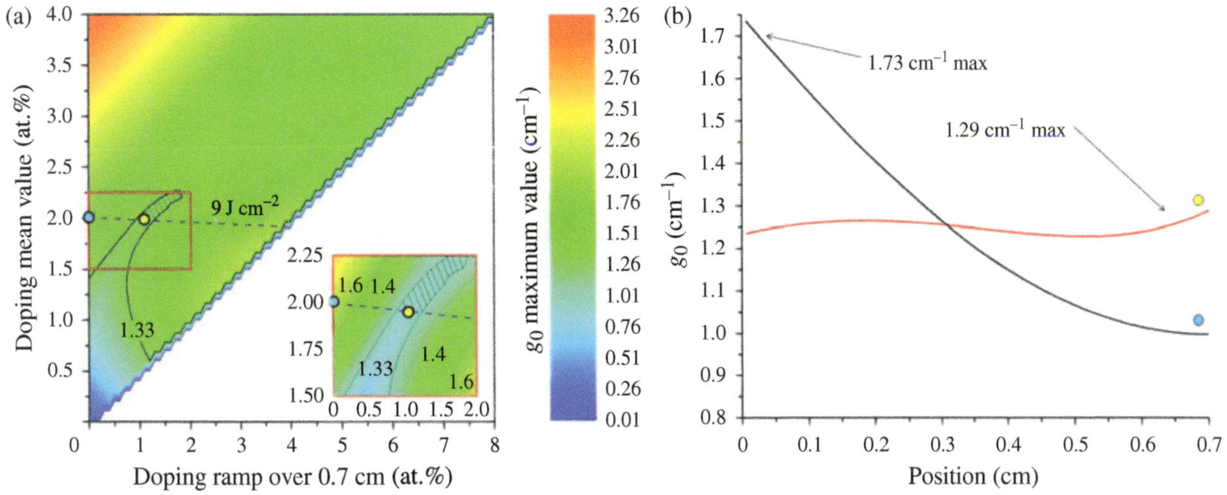

Figure 4. (a) Maximum gain map versus doping ramp and mean value. The 1.33 cm^{-1} threshold iso-gain curve is drawn in solid back. Below this value, ASE triggered parasitic oscillations are expected to be severely limited, guaranteeing a satisfactory efficiency for the amplifier. The iso-energy density 9 J cm^{-2} line is also drawn (dashed line). The bottom right inset gives an enlarged view of the AoI. (b) Small signal gain distribution over the 7 mm thick crystal for the current 2 at.% constant doped crystal (black) and optimum 1.9 at.% average/0.9 at.% doping ramp crystal (red).

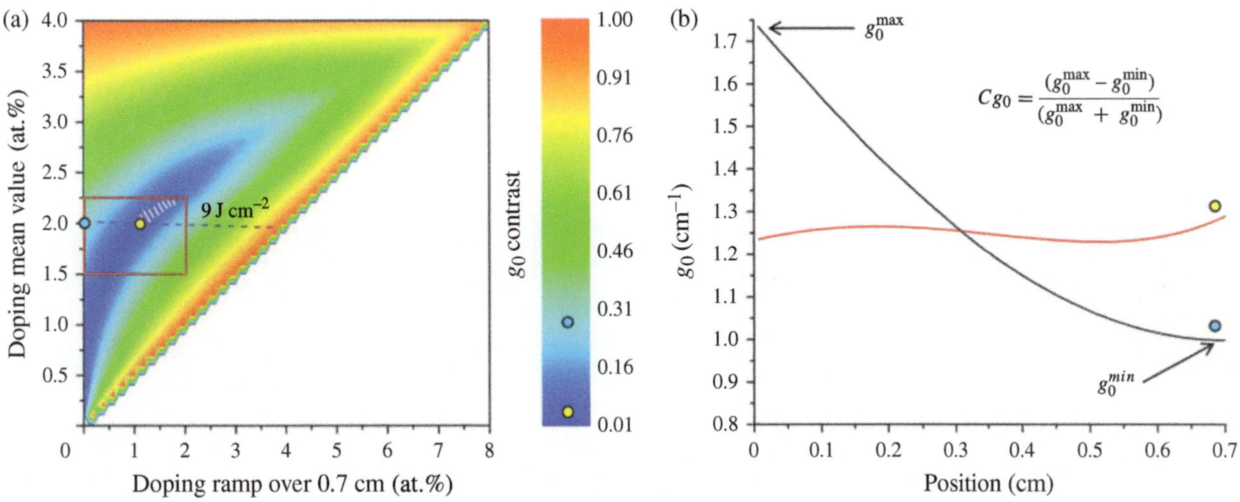

Figure 5. (a) Gain contrast Cg_0 map versus doping ramp and mean value. The red rectangle gives the AoI. The iso-energy density 9 J cm^{-2} line is dashed. The optimum region is white dashed. The Lucia current operating point *WP* is marked with the blue circle (27%) and the optimum one is marked in yellow (2.4%). (b) The gain profiles for the blue and yellow points.

also displayed as well as the 9 J cm^{-2} iso-energy density line and both the *WP* and the *OP*. We observe a very low 2.4% contrast for the *OP*, a situation noticeably better than the 27% value for the *WP*.

4. Optimum Yb:YAG disk for a 160 K amplifier

The Lucia low temperature amplifier hosts a 10 mm thick, 1 at.% doped Yb:YAG disk. It is 77 mm wide but pumped over a 23.4 mm × 30.3 mm elliptical surface, leading to a $g_{0,\max} = 1.32$ cm^{-1} similar to the previous 300 K case. At Lucia working point coordinates (0 at.%, 1 at.%), the stored energy density is 3.3 J cm^{-2}. Figure 6(a) gives the stored energy density (J cm^{-2}) map for the explored region. The iso-energy density 3.3 J cm^{-2} line is shown on the map. It defines a boundary below which the Lucia low temperature energetic performance would not be satisfied. We observe here also that this boundary is set on the edge of an energy density plateau culminating at a 3.5 J cm^{-2} energy density as shown by the 3.4, 3.42, and 3.44 J cm^{-2} iso-energy density curves.

Like for the 300 K case, a maximum gain value map is derived to identify the ASE-compatible area (Figure 6(b)). The 1.32 cm^{-1} threshold iso-gain curve (solid back) is reported together with the 3.3 J cm^{-2} iso-energy density line, the first one defining an upper boundary and the second one the lower frontier of the optimum region (dashes).

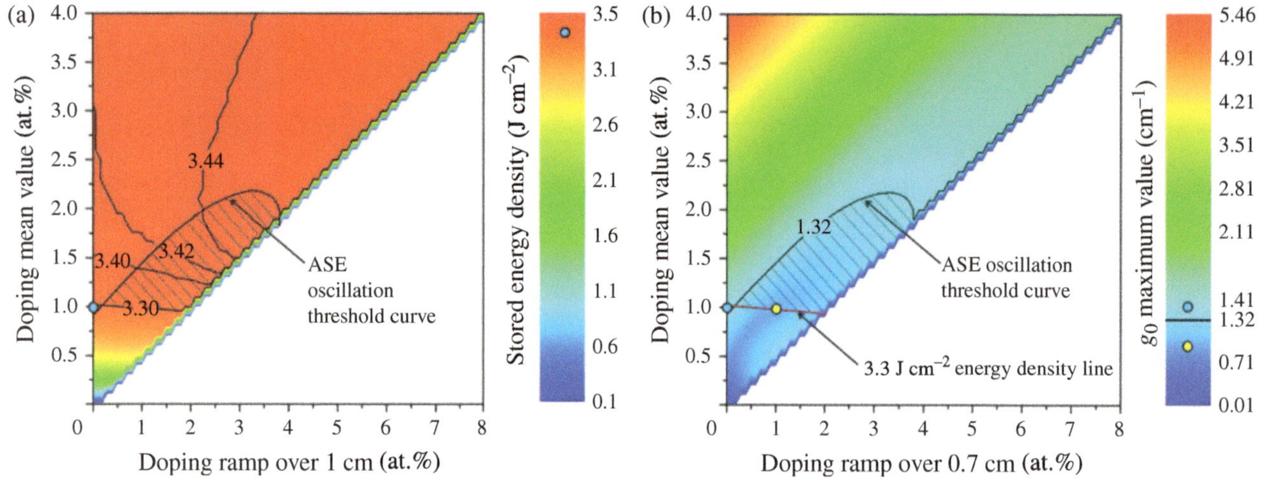

Figure 6. (a) Stored energy density map versus doping ramp and mean value. (b) Maximum gain map versus doping ramp and mean value. The optimum region (dashed) is defined by an upper boundary: the 1.32 cm^{-1} threshold iso-gain curve. Below this value, ASE triggered parasitic oscillations are expected to be severely limited, guaranteeing a satisfactory efficiency for the amplifier. The iso-energy density 3.3 J cm^{-2} line defines the lower boundary. The Lucia low temperature amplifier operating point is marked with the blue circle (0, 1). The yellow circle is located at the lowest gain position along the 3.3 J cm^{-2} energy density line.

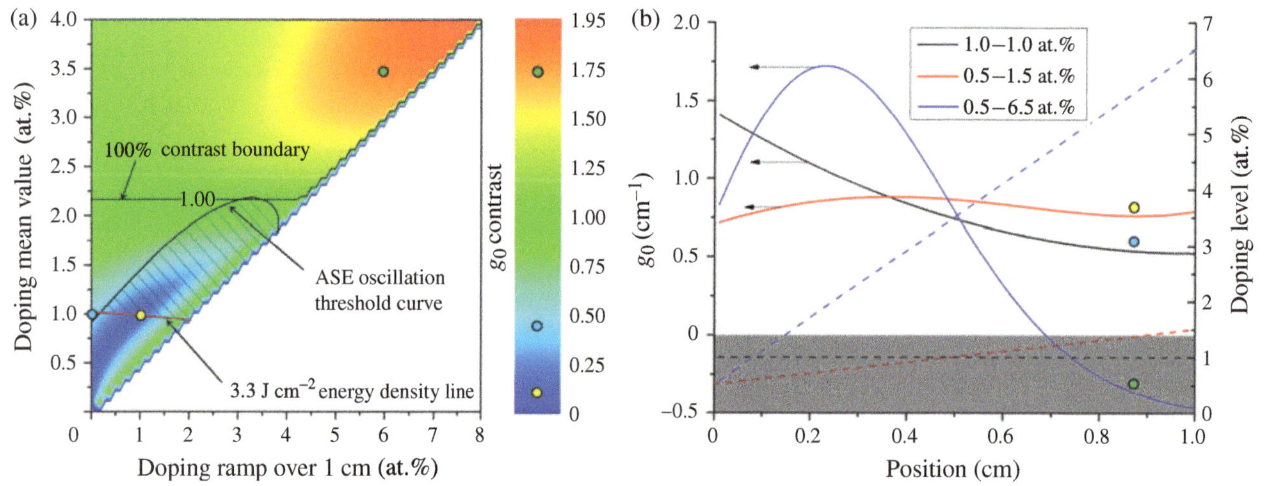

Figure 7. (a) Gain contrast Cg_0 versus doping ramp and mean value. The unity contrast line defines the frontier of doping domains where the minimum gain becomes negative due to strong 1030 nm absorption ($d_0 > 2.2$ at.%). (b) Gain (bold curves, left scale) and Yb^{3+} concentration (dashed lines, right scale) distributions within three 1 cm thick crystals. The respective extreme doping values are given in the inset. The grayed-out bottom area defines the negative gain location.

Since the Lucia low temperature amplifier was designed to satisfy the $g_0 L_{\mathrm{ASE}} \leqslant 4$ criterion as closely as possible, the associated *WP* (blue circle) obviously falls in the vicinity of the 1.32 cm^{-1} curve: its value is 1.41 cm^{-1}. With respect to ASE, a more optimum point marked in yellow can be identified where the maximum gain falls to 0.88 cm^{-1}. At the other extremity of the optimum region, it is possible to slightly improve the energy storage to reach 3.5 J cm^{-2} if we accept an increased level of ASE losses and, more critically, to the extent of a four times stronger doping gradient and two times higher average doping level! Finally, Figure 7 reveals that in this area the gain contrast Cg_0 is extremely bad, of the order of 100%, a value that would cause severe issues in terms of heat load thermal management. The value of Cg_0 is 10% for the *OP* (yellow circle), while it is 46% for the current configuration (blue circle).

At first sight, it can be surprising to observe that, above a 2.2 at.% average Yb^{3+} concentration, we enter into a *greater-than-100%-contrast* regime for the gain. In order to explain this phenomenon, let us consider a specific point located where Cg_0 reaches a value above 150%: the green circle of (6 at.%, 3.5 at.%) coordinates. Such a 1 cm crystal would present a 0.5 at.% doped entrance face to the pumping beam while the cooled side would exhibit a 6.5 at.% concentration, as illustrated on the graph of Figure 7 (blue dashed line with the right scale).

Table 2. Energetic improvement for the low temperature amplifier.

Pump intensity (kW cm^{-2})	5.5	6	6.5	7	8	9
Stored energy density (J cm^{-2})	3.29	3.76	4.06	4.34	4.78	4.9
Storing efficiency, η_S (%)	59.8	62.7	62.5	62.0	59.8	54.4
Doping mean value, d_0 (at.%)	1.0	1.8	1.6	1.4	1.0	0.7
Doping gradient (at.% cm^{-1})	0	3.0	2.4	2.0	1.1	0.3

As a reference, the 1 at.% cm^{-1} yellow *OP* crystal gradient is also displayed (0.5 to 1.5 at.%, red dashed line) as well as the 1 at.% constant doping distribution (*WP*, black dashed line). The three bold curves give the associated gain distributions within the 1 cm crystals' thicknesses. As expected, the red gain curve is very flat with a 10% contrast and the black Lucia low temperature amplifier curve illustrates the 46% contrast resulting from the stronger population inversion located at the pumped face of a constant doped crystal. The blue gain distribution shows that a strong population inversion takes place 2 to 3 mm after the pump entrance face of the (6 at.% cm^{-1}, 3.5 at.%) crystal. For such a high average doping level (3.5 at.%), this crystal is too thick. Indeed, the last 3 mm of the crystal are actually absorbing (negative gain) the 1030 nm light. This negative gain area is grayed out on the graph. Having such a negative minimum value for the gain explains why the contrast Cg_0 (as defined) exceeds the 100% value.

5. Parametric study

Higher values of stored energies might be achieved by choosing the appropriate combination of pump power, doping mean value d_0 and doping ramp g_d. This obviously influences also the storage efficiency η_s. Table 2 compares the current configuration (first column) based on a 1 at.% constant doping gain medium (*WP*) with alternative solutions offered by gradient doped gain media. To obtain the values presented in the table, the $g_0 < 1.33$ cm^{-1} ASE limitation was imposed and the crystal thickness was kept at 1 cm. By increasing the pump power, one can increase the stored energy density, while never exceeding a local gain above 1.33 cm^{-1}, by choosing the appropriate doping distribution. Increasing the pump power above 9 kW cm^{-2} does not significantly improve the stored energy density due to the imposed limitation ($g_0 < 1.33$ cm^{-1}). The stored energy efficiency reaches a 62.7% optimum for 6 kW cm^{-2} pump power and a crystal of 0.3–3.3 at.% linear doping distribution. The amount of stored energy can be significantly increased from 3.29 up to 4.78 J cm^{-2} by increasing the pump power to 8 kW cm^{-2} while keeping the 59.8% efficiency of the constant doped case. The doping gradient necessary for this case is only $g = 1.1$ at.% cm^{-1}, a value already achieved with HDC and easier to obtain than the 3 at.% cm^{-1} of the previous case. It should be kept in mind that any pump power increase is related to major investment in laser diodes. The thickness of the crystal could also be reduced, but such optimization would only make sense in order to achieve better cooling efficiency. Considering that the current low temperature cooling system is highly effective[4, 6], this optimization does not really seem useful.

6. Conclusion and outlook

A methodology to define the optimum doping ion volume distribution was derived in the context of the Lucia DPSSL room and low temperature active mirror amplifiers. After delimiting an optimum region whose boundaries are defined by minimum energy storage and maximum ASE-compatible gain considerations, the optimum point is selected considering the axial gain contrast. Minimizing this gain modulation will help in homogenizing the thermal load, therefore reducing the internal stresses and associated losses (like depolarization) the gain medium is submitted to Ref. [9].

This approach is valid for other DPSSL amplifier architectures like the DIPOLE[10] and PEnELOPE[11] projects, where pumping takes place from both sides through the coolant.

Acknowledgement

The research leading to these results has received funding from LASERLAB-EUROPE (grant agreement no 284464, EC's Seventh Framework Programme).

References

1. M. Arzakantsyan, D. Albach, N. Ananyan, V. Gevorgyan, and J.-C. Chanteloup, J. Crystal Growth **329**, 39 (2011).
2. M. Arzakantsyan, D. Albach, N. Ananyan, V. Gevorgyan, and J.-C. Chanteloup, Opt. Mater. Express **2**, 1219 (2012).
3. T. Gonçalvès-Novo, D. Albach, B. Vincent, M. Arzakantsyan, and J.-C. Chanteloup, Opt. Express **21**, 855 (2013).
4. A. Lucianetti, D. Albach, and J.-C. Chanteloup, Opt. Express **19**, 12766 (2011).
5. L. M. Frantz and J. S. Nodvik, J. Appl. Phys. **34**, 2346 (1963).
6. J.-C. Chanteloup, A. Lucianetti, D. Albach, and T. Gonçalvès-Novo, Plasma Fusion Res. **8**, 3404043 (2013).
7. T. Y. Fan, D. J. Ripin, R. L. Aggarwal, J. R. Ochoa, B. Chan, M. Tilleman, and J. Spitzberg, IEEE J. Sel. Top. Quantum Electron. **13**, 448 (2007).
8. D. C. Brown, R. L. Cone, Y. Sun, and R. W. Equall, IEEE J. Sel. Top. Quantum Electron. **11**, 604 (2003).
9. D. Albach, T. Gonçalvès-Novo, and J.-C. Chanteloup, Plasma Fusion Res. **8**, 3405049 (2013).
10. S. Banerjee, K. Ertel, P. D. Mason, P. J. Phillips, M. Siebold, M. Loeser, C. Hernandez-Gomez, and J. Collier, Opt. Lett. **37**, 2175 (2012).
11. M. Siebold, F. Roeser, M. Loeser, D. Albach, and U. Schramm, Proc. SPIE **8780**, 878005 (2013).

EUV ablation of organic polymers at a high fluence

Chiara Liberatore[1,2], Klaus Mann[3], Matthias Müller[3], Ladislav Pina[2], Libor Juha[1], Jorge J. Rocca[4], Akira Endo[1], and Tomas Mocek[1]

[1] *HiLASE Centre, Institute of Physics ASCR, v.v.i., Za Radnicí 825, 25241 Dolní Břežany, Czech Republic*

[2] *Czech Technical University, Thákurova 2077/7, 160 00 Prague 6, Czech Republic*

[3] *Laser Laboratorium Göttingen (LLG), 37077 Göttingen, Germany*

[4] *Department of Electrical and Computer Engineering, Colorado State University, Fort Collins, CO 80523, USA*

Abstract

A preliminary investigation on short-wavelength ablation mechanisms of poly(methyl methacrylate) (PMMA) and poly (1,4-phenylene ether ether-sulfone) (PPEES) by extreme ultraviolet (EUV) radiation at 13.5 nm using a table-top laser-produced plasma from a gas-puff target at LLG (Göttingen) and at 46.9 nm by a 10 Hz desktop capillary discharge laser operated at the Institute of Physics (Prague) is presented. Ablation of polymer materials is initiated by photo-induced polymer chain scissions. The ablation occurs due to the formation of volatile products by the EUV radiolysis removed as an ablation plume from the irradiated material into the vacuum. In general, cross-linking of polymer molecules can compete with the chain decomposition. Both processes may influence the efficiency and quality of micro(nano)structuring in polymer materials. Wavelength is a critical parameter to be taken into account when an EUV ablation process occurs, because different wavelengths result in different energy densities in the near-surface region of the polymer exposed to nanosecond pulses of intense EUV radiation.

Keywords: EUV ablation; organic polymer; photo-erosion mechanisms; wavelength effect

1. Introduction

Laser ablation is an efficient removal of a material from its surface and near-surface region irradiated by an intense laser beam at a fluence above the single-shot ablation threshold[1, 2]. Due to its potential use in surface patterning/structuring and other applications (e.g., pulsed laser deposition (PLD)), ablation has attracted attention from the advent of laser technology in the sixties. Among the lasers widely utilized to initiate material ablation, ultraviolet (UV) lasers are often applied because of their limited penetration depth and ability to induce photochemical ablation while avoiding thermal artefacts[3]. It is known that photochemical, non-thermal decomposition plays an important role when the wavelengths decrease. Some ablation studies at shorter wavelengths in the vacuum UV (VUV) region have been conducted, for example, with F_2 excimer lasers ($\lambda = 157$ nm[4]) and radiation generated by four-wave-sum-frequency mixing of a frequency-doubled Nd:YAG laser ($\lambda = 125$ nm[5]). Laboratory-scaled laser sources operating at even shorter wavelengths have recently become operational in the extreme UV (EUV) spectral range ($\lambda < 100$ nm)[6–8].

The EUV sources utilized to induce material erosion emit at both low peak power and high peak power. In principle, this results in different modes of material removal[9, 10]. With low-peak-power sources, materials are removed by photo-induced desorption of their volatile fragments from the irradiated sample surface. Each EUV photon carries enough energy to break any chemical bond. This energy is also usually higher than the cohesive energy of any solids. Therefore, the energetic photons absorbed by the photo-effect in atoms located in a near-surface region may create small fragments of a sample material, which are desorbed into the vacuum.

Quite a different situation is expected if a high-peak-power source delivers a single high-energy pulse to the material. The sample is then exposed to a high local dose of radiation (given by the energy content of the pulse and the absorption length of the radiation in the irradiated material) in a short period of time (given by the pulse duration), i.e., a very high dose rate. This means that a large number of events that cause radiation-induced structural decomposition (i.e., organic polymer chain scissions, etc.) occur almost simultaneously in a relatively thick layer of irradiated material.

Correspondence to: C. Liberatore, HiLASE Centre, Institute of Physics ASCR, v.v.i., Za Radnicí 825, 25241 Dolní Břežany, Czech Republic. Email: liberatore@fzu.cz

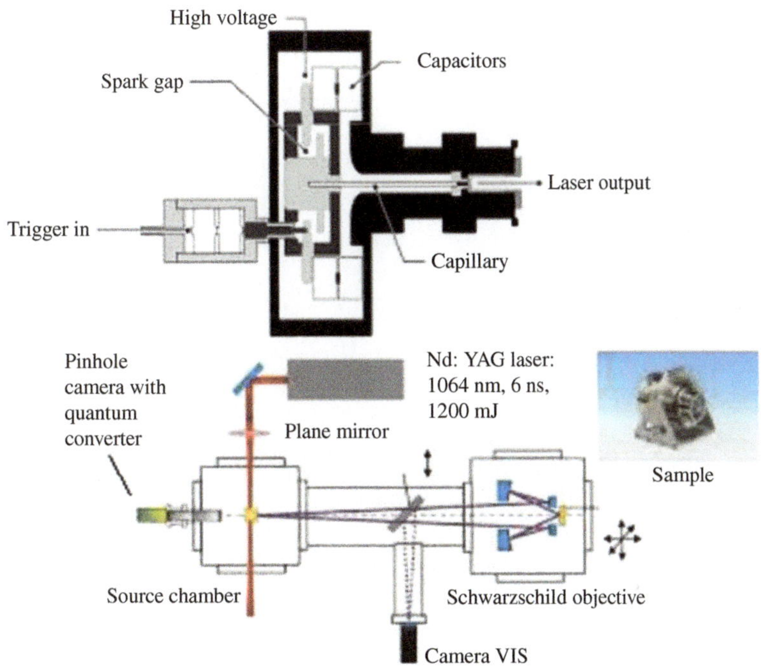

Figure 1. Experimental systems: scheme of the CDL system operated in Prague[15, 16] and scheme of the LPP system operated at LLG in Göttingen[17].

Since a portion of the radiation energy absorbed in the material will be thermalized, the sudden heating of the layer, which is also heavily chemically altered by the radiation, must be taken into account when the fluence is above the single-shot ablation threshold. The overheated, fragmented region of the irradiated material represents a new phase, which tends to blow off into the vacuum. These particular processes, as well as specific features of short-wavelength ablation with respect to ablation induced by conventional, long-wavelength (UV–Vis–IR) sources, together with their potential applications in micro(nano)structuring, represent a main goal of our research program.

In this contribution, a comparison of ablation results obtained in poly(1,4-phenylene ether ether-sulfone) PPEES at 13.5 nm and 46.9 nm is presented.

2. Materials and methods

The ablation experiment has been performed at LLG using poly(methyl methacrylate) (PMMA) as a reference material and then, with the same setup, on a sample material, i.e., PPEES. The material chosen as the target to be tested is PPEES[11] because poly(olefin sulfone)s (POSs) are well-known radiation sensitive polymers[12–14]. Among various POS materials, PPEES was selected because its molecule contains π electrons, in contrast to the aliphatic molecules of PMMA.

The experimental system used in the EUV irradiation experiments is shown in Figure 1. The capillary-discharge Ne-like Ar EUV laser (CDL)[15] developed and built at CSU in Fort Collins, installed and operated in Prague, was used to ablate PPEES by intense EUV radiation. Its output characteristics are as follows: wavelength 46.9 nm, pulse length 1.5 ns (FWHM), maximum pulse energy approximately 10 μJ, repetition rate 5 Hz (typical) and 12 Hz (maximum), capillary lifetime (2–3) $\times$ 10^4 pulses, current $\sim$21 kA. The capillary is an Al_2O_3 (inner diameter: 3.2 mm, length: 210 mm), Ar-filled (50 Pa). A spherical multilayer Sc/Si mirror with a focal length of 0.25 m was used to focus the CDL beam. The mirror was designed and manufactured to have a maximum reflectivity of 46.9 nm radiation for a 6° angle between the incident and reflected beams[16].

The second experimental system, built and operated at LLG, is based on a driving Nd:YAG laser (wavelength 1064 nm, pulse length 7 ns, jitter <1 ns, divergence <0.5 mrad – Gaussian profile at far field, pulse energy stability <1% for 90% of pulses, maximum pulse energy 1.2 J, i.e., twice that used by Barkusky *et al.* in earlier EUV ablation experiments[17]) focused into a Xe gas-puff target. EUV emission of the laser-produced Xe plasma is collected and focused by a Schwarzschild objective with a Mo/Si multilayer coating. The source provides EUV surface energy densities (i.e., fluences) exceeding 0.1 J cm^{-2} at 13.5 nm, 2% bw.

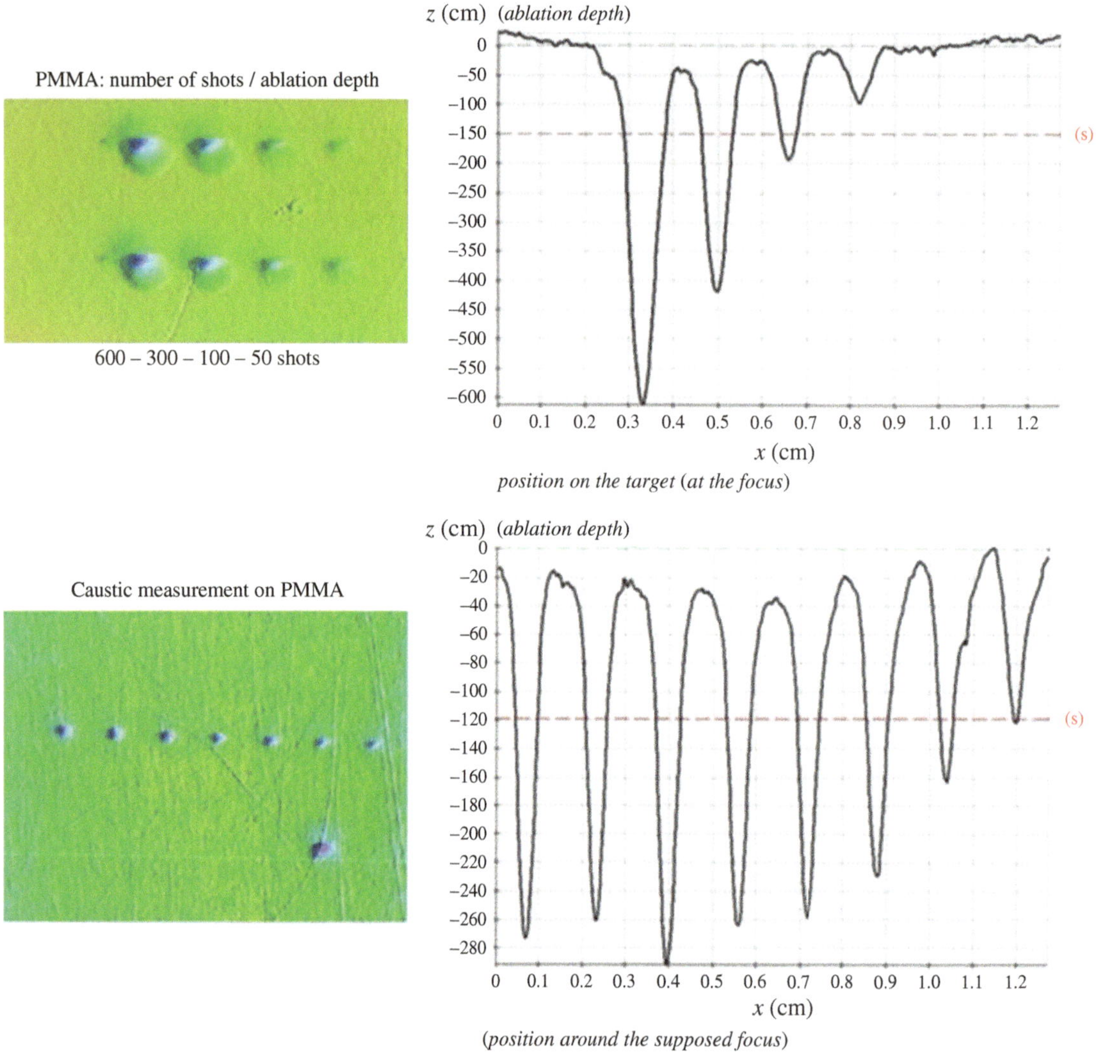

Figure 2. LLG source at a wavelength of 13.5 nm: example of caustic (bottom) and dependence of ablation depth on the number of shots (top). Measurements on PMMA.

This second wavelength is particularly promising because it allows one to modify smaller areas of the material, which is an advantage for many applications (i.e., lithography, nanostructuring).

Typical ablation tests were performed in the tight focus, varying the number of shots. Procedures making it possible to find the focus (i.e., a measurement of caustics, z scan) to determine the EUV energy distribution in the focus and to obtain other irradiation characteristics have been explained elsewhere[18].

The caustics (Figures 2 and 3) are lines of ablated spots, obtained by moving the target in fixed steps of a few microns around the supposed best focal position. At each new position the target was irradiated by a greater number of laser shots. This procedure allows one to verify the focal position of the beam on the sample and, at the same time, to demonstrate that the polymer under test can be ablated by EUV radiation.

3. Results

PMMA was efficiently ablated by 13.5 nm radiation[18], as shown in Figure 2. The ablation process is very stable and the craters maintain the same morphology with an increasing number of shots. Under very similar irradiation conditions, the measurement has been repeated with the PPEES sample. Surprisingly, the ablation pattern has not been observed.

This can be explained with respect to our previous positive PPEES ablation results obtained at 46.9 nm (Figures 3 and 4)[18] by recalling that the absorption of EUV photons on polymers is expressed by Lambert's law:

$$\frac{\mathrm{d}I(z)}{\mathrm{d}z} = -\alpha I(z), \qquad (1)$$

where I is the intensity, z is the depth below the surface, and α is the absorption coefficient of the radiation in the material. The attenuation length is then $1/\alpha$. For PPEES (density:

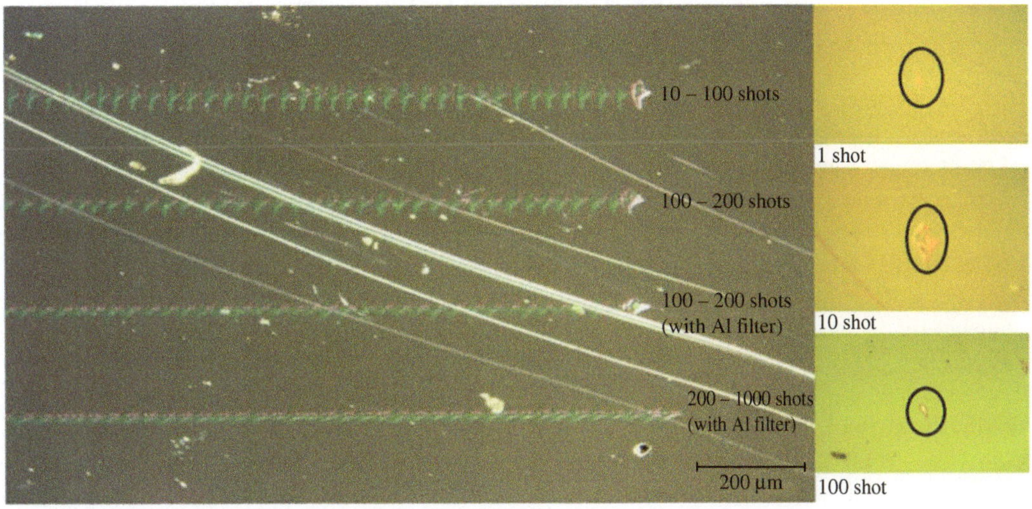

Figure 3. CDL at a wavelength of 46.9 nm: caustic measurement wavelength 46.9 nm (below) and ablation measurement (top). The shape of the crater is almost the same along the line of the caustic. Looking at the image on the right it can be seen that the annular form is preserved although, on increasing the number of shots, its dimensions increase.

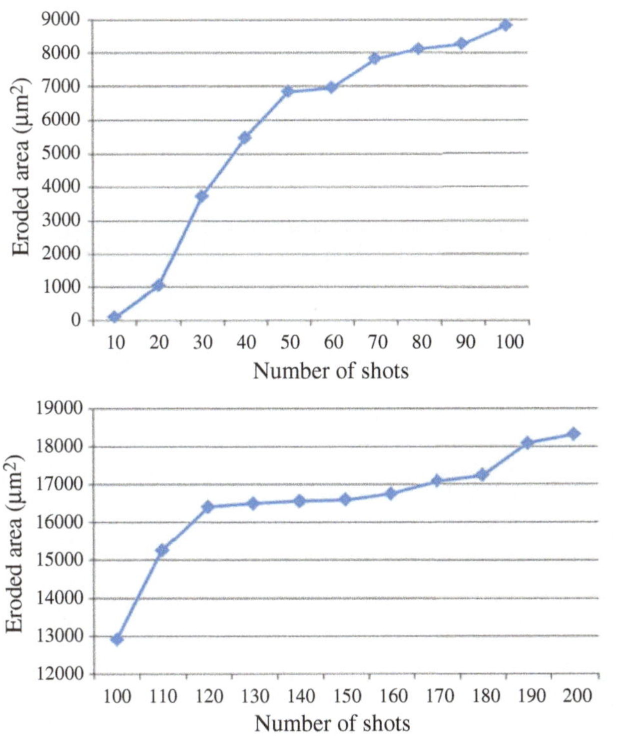

Figure 4. CDL at a wavelength of 46.9 nm: eroded area on number of shots (100–200 shots, bottom; 10–100 shots, top).

1.24 g cm^{-3}; elemental composition: C18SO4H12), the attenuation length at a wavelength of 13.5 nm (photon energy: 91.8 eV) is 215 nm[19]. The attenuation length at a wavelength of 46.9 nm (photon energy: 26.4 eV) is much shorter, i.e., approximately 20 nm[19].

Therefore, there is at least one order of magnitude higher energy density in the PPEES near-surface region when irradiated at 46.9 nm than in the case of irradiation of the material at 13.5 nm at the same surface energy density. Two conclusions follow from this fact:

- at 46.9 nm the ablation threshold should be much lower than at 13.5 nm;
- during the exposure at 13.5 nm, the near-surface region is not so 'overexposed/overheated' as in the previous case (i.e., at 46.9 nm), so that single-photon radiolytical processes would play an important role in material ablation, making visible a difference in the radiation stabilities of PPEES and PMMA. PPEES is much more radiation stable than PMMA because of the π electrons in the PPEES molecules.

The differences in the efficiency of the process explained thanks to the differences in the ablation thresholds and in the efficiency of the radiolytic process mentioned above explain why no ablation was registered at 13.5 nm for fluences achieving a relatively high value, i.e., 127 mJ cm^{-2}. It also makes reasonable the finding of ablation craters on the PPEES surface illuminated by 46.9 nm at a lower fluence, i.e., around 50 mJ cm^{-2}.

These considerations provide evidence of the importance of the chemical structure of the material. Indeed, considering just the scaling law related to the optical systems used in this experimental campaign, results at a lower fluence would have been expected. Indeed, the resolution σ is proportional to the wavelength:

$$\sigma \propto \frac{\lambda}{\mathrm{NA}}, \qquad (2)$$

where NA is the numerical aperture of the optics; while for the penetration depth:

$$d \propto \frac{1}{\lambda}. \qquad (3)$$

Considering the parameters of the LLG experimental system (NA $= 0,4$; $E > 0.1$ J cm^{-2} and $\lambda = 13.5$ nm) from the relation:

$$\text{fluence} \propto \frac{E}{\sigma^2}, \tag{4}$$

a fluence of 87 mJ cm^{-2} must have been sufficient to ablate. Indeed, this is the case for the less radiation resistant PMMA[17]. The difference in the study between PMMA and PPEES ablation using the EUV wavelength, in particular when it is shorter, comes directly from the impossibility of using the current scaling law for the second polymer, because it is no longer valid, being affected by the chemical structure of the polymer.

Repeating the experiment on PPEES, which presents a higher radiation stability, is interesting because use of a better resistor would provide an improvement in the industrial application of EUV ablation processes. It guarantees a longer lifetime of such systems before damage.

For this reason, in the next phase of the project, we are going to increase the output energy of the LLG source and focus its EUV emission onto a smaller spot to exceed the EUV ablation threshold of PPEES at 13.5 nm and to evaluate the required fluence experimentally.

However, in future industrial EUV polymer ablation, the wavelength-dependent ablation depth $alpha$ is the guiding parameter; therefore multiple EUV wavelengths may be required, necessitating the use of different elements (gas or solid).

4. Conclusions

In this contribution, we have demonstrated the possibility of creating well-developed ablation craters in PPEES exposed to focused 46.9 nm CDL radiation. However, the more penetrating 13.5 nm radiation (its attenuation length in PPEES exceeds 0.2 μm) was not able to ablate PPEES, although PMMA was ablated quite effectively under these irradiation conditions. The higher radiation stability of PPEES is expected because of the presence of π electrons in the PPEES molecule, in contrast to the less radiation resistant polymer structures (PMMA). The increased attenuation length at the shorter wavelength is likely responsible for the observed difference in the response of PPEES to radiation emitted by the CDL source ($\lambda = 46.9$ nm) and the LLG source ($\lambda = 13.5$ nm).

Acknowledgements

This work benefitted from: the support of the Ministry of Education, Youth and Sports to the HiLASE (CZ.1.05/2.1.00/01.0027) and DPSSLasers (CZ.1.07/2.3.00/20.0143) projects, co-financed from the European Regional Development-Fund, this research has been supported by the grant RVO 68407700; financial support by the 'Deutsche Forschungsgemeinschaft' within the Sonderforschungsbereich 755 'Nanoscale Photonic Imaging' and the Czech Science Foundation (14-29772S) and from EU COST MPO1203, from the Ministry of Education, Youth and Sports, COST CZ VES14 LD14032 are also appreciated.

References

1. C. Phipps, *Laser Ablation and its Applications* (Springer, New York, 2006).
2. D. Bäuerle, *Laser Processing and Chemistry* 3rd edn (Springer-Verlag, Berlin, 2000).
3. T. Lippert, Plasma Process. Polym. **2**, 525 (2005).
4. M. Lapczyna and M. Stuke, Appl. Phys. A **66**, 473 (1998).
5. D. Riedel and M. Claude Castex, Appl. Phys. A **69**, 375 (1999).
6. B. R. Benware, C. D. Macchietto, C. H. Moreno, and J. J. Rocca, Phys. Rev. Lett. **81**, 5804 (1998).
7. S. Kranzusch and K. Mann, Opt. Commun. **200**, 223 (2001).
8. A. Bartnik, H. Fiedorowicz, R. Jarocki, L. Juha, J. Kostecki, R. Rakowski, and M. Szczurek, Microelectron. Engng **78–79**, 452 (2005).
9. L. Juha, A. Andrejczuk, R. Sobierajski, L. Ryc, F. P. Boody, H. Fiedorowicz, J. Feldhaus, V. Létal, J. Krzywinski, M. E. Grisham, P. Kubát, L. Pína, M. Horváth, G. O. Vaschenko, R. Nietubyc, Z. Otcenasek, A. Bartnik, M. Bittner, D. Chvostová, J. Mikolajczyk, J. B. Pelka, C. S. Menoni, J. J. Rocca, R. Rakowski, J. Krása, M. Kozlová, M. Pfeifer, J. Polan, A. R. Präg, B. Rus, and M. Stupka, J. Micro/Nanolith. MEMS MOEMS **4**, 033007 (2005).
10. J. Chalupský, L. Juha, V. Hájková, J. Cihelka, L. Vyvsín, J. Gautier, J. Hajdu, S. P. Hau-Riege, M. Jurek, J. Krzywinski, R. A. London, E. Papalazarou, J. B. Pelka, G. Rey, S. Sebban, R. Sobierajski, N. Stojanovic, K. Tiedtke, S. Toleikis, T. Tschentscher, C. Valentin, H. Wabnitz, and P. Zeitoun, Opt. Express **17**, 208 (2009).
11. http://www.sigmaaldrich.com/catalog/product/aldrich/44096 5?lang=en&region=CZ.
12. M. J. Bowden and J. H. O'Donnell, Dev. Polym. Degrad. **6**, 21 (1985).
13. B. W. Yates and D. M. Shinozaki, J. Mater. Res. **7**, 520 (1992).
14. F. Iacona and G. Marletta, Nucl. Instrum. Meth. Phys. Res. B **166–7**, 676 (2000).
15. S. Heinbuch, M. Grisham, D. Martz, and J. J. Rocca, Opt. Express **13**, 4050 (2005).
16. L. Vyvsín, T. Burian, J. Chalupský, M. Grisham, V. Hájková, S. Heinbuch, K. Jakubczak, D. Martz, T. Mocek, P. Pira, J. Polan, J. J. Rocca, B. Rus, J. Sobota, and L. Juha, Proc. SPIE **7361**, 73610O (2009).
17. F. Barkusky, C. Peth, A. Bayer, and K. Mann, J. Appl. Phys. **101**, 124908 (2007).
18. C. Liberatore, K. Mann, M. Müller, L. Pina, L. Juha, L. Vyšín, J. J. Rocca, A. Endo, and T. Mocek, Phys. Scr. T **161**, 014066 (2014).
19. http://henke.lbl.gov/optical_constants/filter2.html.

High energy density physics at the Atomic Weapons Establishment

Andrew Randewich and Colin Danson

AWE, Aldermaston, Reading RG7 4PR, UK

Abstract

The Atomic Weapons Establishment (AWE) is tasked with supporting Continuous At Sea Deterrence (CASD) by certifying the performance and safety of the national deterrent in the Comprehensive Test Ban Treaty (CTBT) era. This means that recourse to further underground testing is not possible, and certification must be achieved by supplementing the historical data with the use of computer calculation. In order to facilitate this, AWE operates some of the largest supercomputers in the UK. To validate the computer codes, and indeed the designers who are using them, it is necessary to carry out further experiments in the right regimes. An excellent way to meet many of the requirements for material property data and to provide confidence in the validity of the algorithms is through experiments carried out on high power laser facilities.

Keywords: high energy density physics; HEDP; high performance computing; HPC; high power lasers

1. High performance computing

Certification of the national deterrent in the Comprehensive Test Ban Treaty (CTBT) era must be achieved through the use of computer calculation[1]. The large supercomputers used at Atomic Weapons Establishment (AWE) are supplemented by high resilience data storage, high speed networks, visualization clusters and high end desktops that enable the supercomputers to be exploited effectively. The current systems have recently been upgraded to provide around two and a half petaflop/s (2.5 million billion floating point operations per second), and AWE has also recently procured several advanced architecture platforms in order to ensure that the most efficient (in terms of both cost and energy) hardware possible can be deployed in future years. Figure 1 shows AWE's high performance computer 'Blackthorn'.

This high performance computing (HPC) environment is of course useless without computer programmes or 'codes' to run on it. Multi-physics algorithms are therefore developed to solve the equations of compressible fluid flow coupled to transport algorithms and other required physics such as fusion burn. These state-of-the-art algorithms are in turn worthless without material property data. Perhaps unsurprisingly, it is particularly challenging to obtain the required material properties such as opacity, strength and equation of state in the extreme conditions pertinent to nuclear warheads. The requirements span time scales of many orders of magnitude and pressures up to gigabars, with temperatures ranging up to thousands of electron volts (keV, or tens of millions of degrees).

The regimes of hot dense matter (HDM) and warm dense matter (WDM) are pertinent, representing material that is heated while remaining at a density close to solid; similar conditions exist in the cores of large planets and in the Sun, see Figure 2. In some cases it is possible to harness the power of HPC to calculate the material properties in question, although this is non-trivial; the conditions of WDM sit between well understood regimes, in particular because the thermal energy of WDM ions is typically comparable with their potential energy. The physics is therefore intermediate to solid-state physics (dominated by interatomic potentials) and traditional plasma physics (dominated by thermal energy), so simplifying assumptions required for modelling will not be valid across all the required parameter space. In any case, it is always important to underpin calculated properties with measurements in appropriate conditions, and especially so in such a difficult regime. Once the material property data have been incorporated into the multi-physics codes, meaningful calculations become possible.

2. High power lasers for high energy density physics

High power lasers have been used at AWE for the study of high energy density conditions for almost forty years[2].

Correspondence to: A. Randewich, AWE, Aldermaston, Reading RG7 4PR, UK. Email: Andrew.randewich@awe.co.uk

Figure 1. AWE's high performance computer 'Blackthorn'.

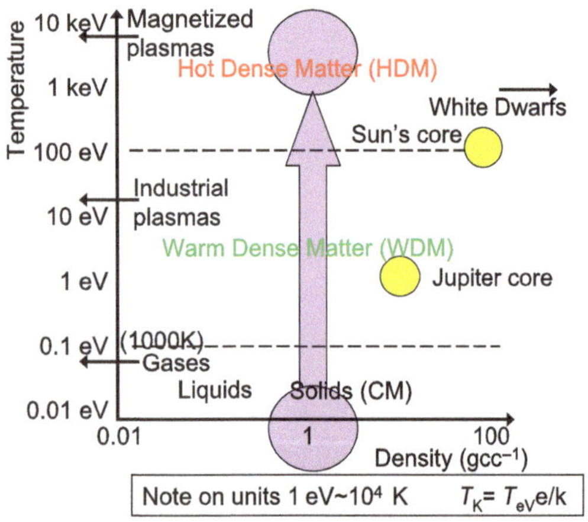

Figure 2. WDM and HDM are generated in planetary and solar cores as well as within nuclear weapons.

Following the laser fusion concept of Nuckolls[3], AWE began modelling laser plasmas in the early 1970s. The concept of using a hohlraum of high atomic number material to convert laser energy to thermal x-rays rather than directly driving a capsule was developed and within ten years the HELEN laser[4] was opened, see Figure 3. HELEN was initially a two-beam system, each delivering a terawatt of power (a kilojoule in a nanosecond). In parallel with HELEN experiments, the much higher energy drive available from underground nuclear tests was used as a platform to drive fusion experiments. While these experiments were successful in demonstrating the validity of the key science of laser fusion, this research was effectively terminated when nuclear testing ceased in 1991. Experiments in HELEN's target chamber (see Figure 4) continued, and the laser was upgraded at regular intervals, with the two beams being converted to green light, and a third beamline being added as a backlighter for diagnostic purposes.

By the year 2000 it was realized that HELEN had almost exhausted the phase space that it could access, and yet requirements remained for further data on material properties in hot dense plasmas and experiments to validate calculations. The construction of a laser system with sufficient energy to compress and heat to the conditions of interest using nanosecond-class laser pulses was a daunting prospect, but nevertheless one that was tackled in the US at the National Ignition Facility (NIF)[5] and in France at Laser Mega-Joule (LMJ)[6].

Figure 3. The HELEN laser was decommissioned in 2009 after demonstrating the feasibility of carrying out relevant high energy density physics (HEDP) experiments on laser systems.

Figure 4. The HELEN target chamber.

In the UK a novel approach was undertaken to break the problem into two parts; it was realized that, if it is possible to use one set of laser beams to compress a target and then a separate, much shorter, laser pulse at higher power to heat, then extreme conditions might be accessed with a much smaller laser than NIF or LMJ. Furthermore, such a laser system would be an ideal staging platform to larger lasers for collaboratively carrying out those experiments that do require megajoule class energies. This concept for the Orion laser facility combining short and long pulse lasers[7] was endorsed, and building work commenced in 2006.

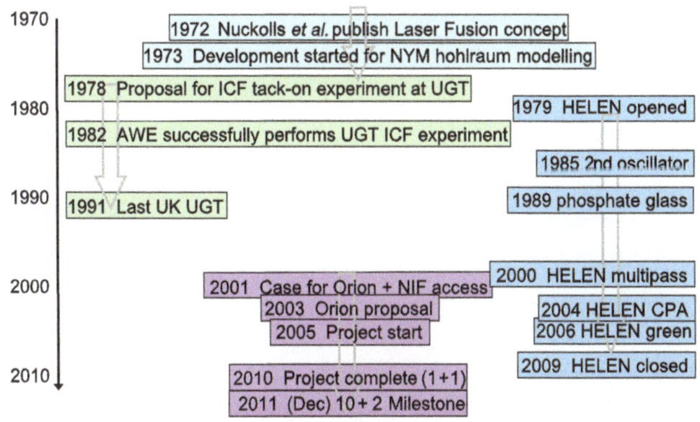

Figure 5. HEDP experiments at AWE followed dual tracks (albeit with a very limited number of experiments in the green underground testing (UGT) line) until nuclear testing stopped. HELEN had several upgrades, but has now been replaced by the Orion facility (NYM is a 2D hydrocode used for modelling and ICF is inertial confinement fusion).

Meanwhile, the HELEN laser undertook a final upgrade to prototype a chirped pulse amplification (CPA)[8] short pulse laser system. This system, similar to that demonstrated at the STFC Rutherford Appleton Laboratory's Central Laser Facility (CLF) Vulcan petawatt laser[9], would be required for deployment on Orion. Experiments were very successful, and ultimately HELEN demonstrated opacity data at 500 eV (around five million kelvin)[10] for material at solid density. HELEN was closed in 2009 as it came time for Orion to enter its final commissioning phase. The high energy density physics timeline at AWE is illustrated schematically in Figure 5.

3. The Orion laser/plasma interaction facility

Orion effectively consists of two neodymium-doped glass laser systems. The 'long pulse' system (see Figure 6) comprises ten beamlines, each delivering a nominal 500 J of energy in a nanosecond with highly configurable pulse shape and relative timing. The beams are 300 mm in diameter and are generated and amplified at 1053 nm in the infrared before being frequency tripled to 351 nm (ultraviolet light) and then delivered to the target. The light enters the 4 m diameter target chamber in two cones of five beams at an angle of 50° to the axis, allowing targets to be driven from either or both sides.

The 'short pulse' laser system consists of two beams each with an output beam diameter of 600 mm. The beams start out at as low energy sub-picosecond pulses and are temporally stretched and amplified using the CPA technique which makes use of the small amount of spectral bandwidth available in the pulse. The stretched pulses can then undergo recompression using a pair of large diffraction gratings (940 mm diameter) and are delivered to the target at petawatt powers (500 J in 0.5 ps) and up to 10^{21} W cm^{-2} focused intensity.

Figure 6. Orion's long pulse amplifier chain.

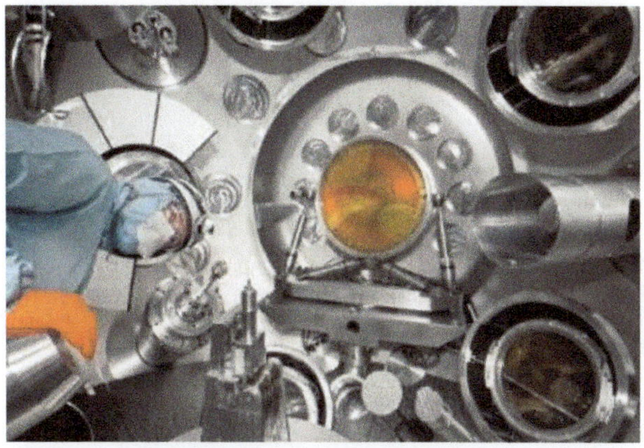

Figure 7. Inside the Orion target chamber.

The target chamber is shown in Figure 7, to which the two CPA beams are delivered orthogonally and can be used for heating or diagnostic purposes (a short pulse focusing parabola can be seen at the centre of Figure 7, coloured orange). One of the short pulse beamlines has the option of being converted to the second harmonic (527 nm) at a 300 mm sub-aperture, and thereby provides up to 100 J with a significantly reduced pre-pulse (i.e. higher contrast $\sim 10^{-14}$)[11]. A further upgrade providing a high contrast front-end to Orion has also recently been commissioned, providing unimaginably high contrast levels $\sim 10^{-18}$ when used in combination with the second harmonic option[12].

In order to carry out experiments, the laser needs to be provided with a series of targets on which to focus its energy. These targets may be relatively simple flat layered structures or complex 3D assemblies depending on the experiment in question, see Figure 8. They are always very small, with typical dimensions from millimetres to microns, and with very close tolerance, typically microns to nanometres, and require the integration of material science, precision engineering, assembly, characterization and metrology for successful delivery. Some campaigns will require large numbers of these targets.

Figure 8. Targets range from 3D assemblies such as the hohlraum shown as a cartoon (top left) and as an actual target (top right) to flat foils (bottom right). They range in size, but the tolerances are always very tight. The image at the bottom left is a demonstration of target manufacturing capability on a miniature scale.

A laser facility also requires a range of diagnostics to provide useful data on the high energy density plasmas generated by the system. In principle, there is interest in measuring both the flux and the spectrum of any electromagnetic (optical, x-ray, etc.) or particle (electron, neutron, etc.) output from the target. The requirements, in each case, may be for time integrated or temporally resolved data, and spatial (or directional) resolution may also be required. The resolutions vary from experiment to experiment, but time scales as short as picoseconds and length scales of microns are not uncommon. For this reason, the Orion target chamber is equipped with many diagnostic ports, see Figure 9; some of these are committed to permanent diagnostics, but great flexibility is also provided by the use of six ten-inch manipulators (TIMs). Experimental teams can have access to instruments housed therein without needing to break the vacuum on the target chamber, and their use permits varied combinations of diagnostics to be fielded in different relative configurations.

The optical diagnostics on Orion include streak cameras, active and passive shock breakout systems, pyrometry and Velocity Interferometer System for Any Reflector (VISAR). The particle systems include electron spectrometers, Thomson parabolas, Faraday cups and Neutron Time of Flight (nToF) systems. There are also x-ray microscopes, a Dante soft x-ray system, a filter fluorescer (FFLEX), a transmission grating spectrometer, a hard x-ray spectrometer and thermoluminescent dosimetry.

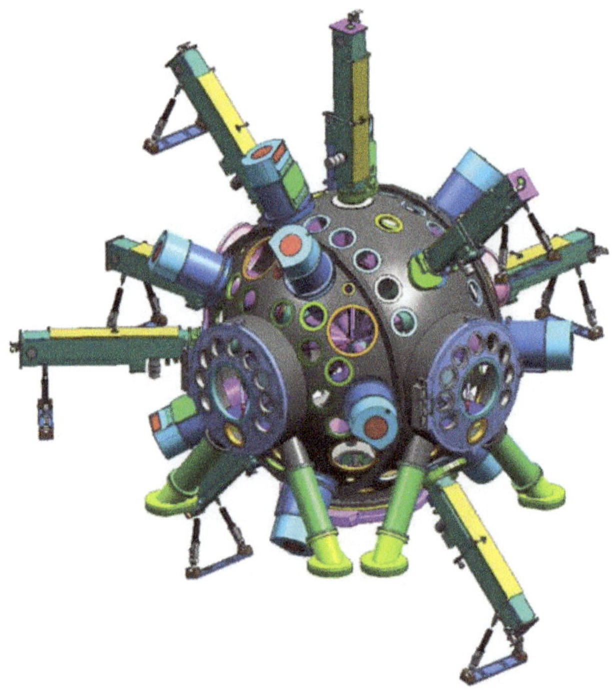

Figure 9. The Orion target chamber is peppered with ports, the majority being for diagnostic access. The TIMs are shown as the large oblong devices protruding from the chamber and the target inserter is mounted vertically at the top of the chamber.

The Orion laser was handed over from construction in December 2010 to enter a two-year commissioning programme. At the time of handover a mid-energy synchronization demonstration of one long pulse and one short pulse beam to the target had been carried out. By the end of the first year, the facility demonstrated all twelve beamlines delivering synchronized energy to the target chamber, and during the final commissioning period carried out the first opacity experiments on the facility as well as commissioning many auxiliary systems such as diagnostics.

The aim of the opacity experiments was to extend the experiments on HELEN, as described earlier, where aluminium was heated to 500 eV; by use of the combination of short and long pulse systems together, Orion is able to exceed these conditions. The experiment consisted of an aluminium foil sandwiched between two layers of plastic. When this foil is irradiated by the short pulse laser it is heated rapidly, and conditions are then diagnosed by x-ray spectroscopy. With much increased power and excellent contrast in the green, the Orion short pulse beam allows higher temperatures to be attained than on HELEN, but in particular it is possible to use the high energy long pulse laser beams to generate compression by launching a shock timed such that, after short pulse heating, the aluminium is at increased density as well as temperature[13, 14].

4. Summary

While we have emphasized the importance of Orion in certification under the CTBT, it will also be a very powerful and important tool to the wider UK scientific community and their international collaborators. It has already been stated that the physics of material properties in extreme conditions is an area of active interest in order to improve our understanding of stellar and planetary interiors. Similar conditions also exist in the inertial confinement fusion (ICF) capsules that could one day lead us to a world of clean energy, and research on Orion will be a useful staging platform for experiments in this context. To this end the facility is available for access by the scientific community for up to 15% of the available time, with access being prioritized through the Central Laser Facility's established processes.

At the point of writing, Orion had completed its first 18 months of operation, delivering a highly successful internal programme and having completed two academic experiments[15]. These two campaigns were both from the University of Oxford;the first was an experiment to investigate materials under multi-megabar pressures led by Dr. Andrew Higginbotham and the second was to study the generation of plasma shock waves relevant to the conditions found in binary stars led by Prof. Gianluca Gregori. Preliminary analysis of the data obtained for both of these campaigns is very promising. Two further Orion academic access experiments have also been scheduled on Orion during 2014/2015: the investigation of colliding shocks relevant to astrophysics led by Dr. Francisco Suzuki-Vidal, Imperial College London, and a study of the effects of strong magnetic fields in plasmas using proton radiography led by Prof. Nigel Woolsey, University of York.

Acknowledgements

The work referred to in this paper has been carried out by members of the Plasma Physics and Design Physics departments at AWE, and they are therefore credited with the achievements described. The former heads of Plasma Physics, Peter Roberts, Daryl Landeg and Steve Rose, were central to the delivery of the Orion project.

References

1. K. O'Nions, R. Pitman, and C. Marsh, Nature **415**, 853 (2002).
2. P. D. Roberts, AIP Conf. Proc. **1438**, 35 (2012).
3. J. Nuckolls, L. Wood, A. Thiessen, and G. Zimmerman, Nature **239**, 139 (1972).
4. M. J. Norman, J. E. Andrew, T. H. Bett, R. K. Clifford, J. E. England, N. W. Hopps, K. W. Parker, K. Porter, and M. Stevenson, Appl. Opt. **41**, 3497 (2002).
5. G. H. Miller, E. I. Moses, and C. R. Wuest, Opt. Eng. **43**, 2841 (2004).
6. J. Ebrardt and J. M. Chaput, J. Phys. Conf. Ser. **244**, 032017 (2010).
7. N. W. Hopps, C. Danson, S. Duffield, D. Egan, S. Elsmere, M. Girling, E. Harvey, D. Hillier, M. Norman, S. Parker, P. Treadwell, D. Winter, and T. Bett, Appl. Opt. **52**, 3597 (2013).
8. D. Strickland and G. Mourou, Opt. Commun. **56**, 219 (1985).
9. C. N. Danson, P. A. Brummitt, R. J. Clarke, J. L. Collier, B. Fell, A. J. Frackiewicz, S. Hancock, S. Hawkes, C. Hernandez-Gomez, P. Holligan, M. H. R. Hutchinson, A. Kidd, W. J. Lester, I. O. Musgrave, D. Neely, D. R. Neville, P. A. Norreys, D. A. Pepler, C. J. Reason, W. Shaikh, T. B. Winstone, R. W. W. Wyatt, and B. E. Wyborn, Nucl. Fusion **44**, S239 (2004).
10. C. R. D. Brown, D. J. Hoarty, S. F. James, D. Swatton, S. J. Hughes, J. W. Morton, T. M. Guymer, M. P. Hill, D. A. Chapman, J. E. Andrew, A. J. Comley, R. Shepherd, J. Dunn, H. Chen, M. Schneider, G. Brown, P. Beiersdorfer, and J. Emig, Phys. Rev. Lett. **106**, 185003 (2011).
11. D. I. Hillier, C. N. Danson, S. J. Duffield, D. A. Egan, S. P. Elsmere, M. T. Girling, E. J. Harvey, N. W. Hopps, M. J. Norman, S. J. F. Parker, P. A. Treadwell, D. N. Winter, and T. H. Bett, Appl. Opt. **52**, 18 (2013).
12. D. I. Hillier, S. Elsmere, M. Girling, N. Hopps, D. Hussey, S. Parker, P. Treadwell, D. Winter, and T. Bett, Appl. Opt. **53** (2014).
13. D. J. Hoarty, P. Allan, S. F. James, C. R. D. Brown, L. Hobbs, M. P. Hill, J. W. O. Harris, J. Morton, M. G. Brookes, R. Shepherd, J. Dunn, H. Chen, E. Von Marley, P. Beiersdorfer, G. Brown, and J. Emig, Phys. Rev. Lett. **110**, 265003 (2013).
14. D. J. Hoarty, P. Allan, S. F. James, C. R. D. Brown, L. M. R. Hobbs, M. P. Hill, J. W. O. Harris, J. Morton, M. G. Brookes, R. Shepherd, J. Dunn, H. Chen, E. Von Marley, P. Beiersdorfer, H. K. Chung, R. W. Lee, G. Brown, and J. Emig, High Energy Density Physics **9**, 661 (2013).
15. C. N. Danson, in *Proceedings of 33rd ECLIM* MoIO2 (2014).

32

Scaling and design of high-energy laser plasma electron acceleration

Kazuhisa Nakajima, Hyung Taek Kim, Tae Moon Jeong, and Chang Hee Nam

Center for Relativistic Laser Science, Institute for Basic Science (IBS), Gwangju 500-712, Korea

Abstract

Recently there has been great progress in laser-driven plasma-based accelerators by exploiting high-power lasers, where electron beams can be accelerated to multi-GeV energy in a centimeter-scale plasma due to the laser wakefield acceleration mechanism. While, to date, worldwide research on laser plasma accelerators has been focused on the creation of compact particle and radiation sources for basic sciences, medical and industrial applications, there is great interest in applications for high-energy physics and astrophysics, exploring unprecedented high-energy frontier phenomena. In this context, we present an overview of experimental achievements in laser plasma acceleration from the perspective of the production of GeV-level electron beams, and deduce the scaling formulas capable of predicting experimental results self-consistently, taking into account the propagation of a relativistic laser pulse through plasma and the accelerating field reduction due to beam loading. Finally, we present design examples for 10-GeV-level laser plasma acceleration, which is expected in near-term experiments by means of petawatt-class lasers.

Keywords: electron beam loading; GeV-level electron beam acceleration; laser plasma (wakefield) accelerators; petawatt-class lasers; propagation of relativistic laser pulses in plasma

1. Introduction

In this decade, active research has been carried out on the laser plasma acceleration concept[1] in order to achieve high-energy, high-quality electron beams with GeV energy in a cm-scale plasma[2–6], 1%-level energy spread[7], 1-mm-mrad-level transverse emittance[8] and 1-fs-level bunch duration[9], ensuring that the stability of reproduction is as high as that of present high-power ultrashort-pulse lasers[10]. Recently, staged laser plasma acceleration[11, 12] has been successfully demonstrated in conjunction with ionization-induced injection[13–15]. Based on recent results on laser plasma acceleration experiments and large-scale particle-in-cell (PIC) simulations[16–18], design consideration and feasibility studies on applications for high-energy frontier colliders with TeV-range center-of-mass energy have been carried out[19, 20]. In this context, state-of-the-art PW-class lasers allow us to study the feasibility of laser plasma accelerators toward the 10–100-GeV range in a full-scale experiment. For example, a large-scale experiment[21] is proposed to implement the demonstration of 100-GeV electron beam acceleration by means of a laser plasma accelerator driven with a multi-PW laser capable of delivering 3.5-kJ, 500-fs pulses[22]. This capability allows us to explore laser plasma acceleration operated in the entire laser wakefield regime, from the linear regime to the nonlinear bubble regime. Such a large-scale laser plasma accelerator may comprise a gas jet or a short gas cell, which acts as an injector, followed by a long, uniform, low-density plasma or preformed plasma channel (plasma waveguide) that acts as an accelerating medium. In order to implement this project, we need to strongly corroborate design formulas that scale experimental results properly.

We organize the remainder of this paper as follows. Section 2 presents an overview of progress and recent achievements in laser plasma electron acceleration. Section 3 considers the energy scaling of laser wakefield acceleration for the self-guided and channel-guided cases in both the quasi-linear and the nonlinear bubble regimes as well as the comparison of such scaling formulas with recent experimental results on multi-GeV electron acceleration. Section 4 presents the design formulas and examples of 10-GeV-level laser wakefield accelerators as well as a comparison with the result from a three-dimensional full-scale PIC simulation[16].

Correspondence to: K. Nakajima, Center for Relativistic Laser Science, Institute for Basic Science (IBS), Gwangju 500-712, Korea.
Email: naka115@dia-net.ne.jp

2. Overview of laser plasma electron acceleration

Laser-driven plasma-based accelerators have evolved from a groundbreaking concept by Tajima and Dawson[1] into the reality of next-generation particle accelerator technologies. Relativistic electron beams from ultraintense laser plasma interactions can be conceived to be compact particle accelerators, inspiring a wide range of applications of unique particle beam and radiation sources such as THz[23] and betatron x-ray radiation[24]. Furthermore, it is envisaged that laser plasma accelerators will downsize large-scale particle accelerators such as x-ray free-electron lasers[25–28] and high-energy frontier colliders[19, 20] to a realistic extent in both dimensions and costs.

In fact, there has been significant experimental progress in laser wakefield acceleration of electron beams since the incipient experiments on laser wakefield accelerators successfully demonstrated ultrahigh gradient acceleration of the order of 100 GeV m^{-1}, using chirped pulse amplification lasers with 10-TW class peak power and 1 ps pulse duration[29, 30]. Such experiments are characterized in terms of the self-modulated wakefield regime[31], where the laser power should be higher than the critical power for relativistic self-focusing and the laser pulse duration is longer than the plasma period. In this regime, the laser pulse undergoes temporal intensity modulation and self-guiding through nonlinear interactions with the plasma, so that large-amplitude plasma waves are resonantly excited. Ultimately, wave breakings occur, generating relativistic electrons to be randomly trapped and accelerated by wakefields throughout the acceleration distance. Therefore, electron beams produced from single-stage experiments showed energy spectra with 100% energy spread, as characterized by a Maxwellian distribution with the highest energy tail reaching at most 100 MeV[32]. The energy gain of accelerated electrons should be determined by the acceleration distance, which is restricted due to dephasing of electrons with respect to the correct acceleration phase of the wakefield and due to depletion of the laser pulse energy. For most experiments using a supersonic gas jet, the acceleration distance extends only to a few mm, so that the energy gain is limited to the order of 200 MeV[33].

For many practical applications of electron beams, quality, stability and controllability of the beam performance such as energy, energy spread, emittance and charge are indispensable in addition to compact and robust features of accelerators. In this context, breakthrough experiments[34–36] have succeeded in producing high-quality electron beams, so-called quasi-monoenergetic beams, with ultrashort pulses, small energy spread and low emittance. Quasi-monoenergetic electron beams have been obtained from the use of ultrashort laser pulses with durations of the order of several tens of femtoseconds and by controlling the plasma density precisely to make the dephasing length long enough to exceed the acceleration distance. Under these conditions, once the plasma electrons expelled by the ponderomotive force (radiation pressure) of the laser pulse form a plasma cavity called a 'bubble', then some of them are self-injected into the wakefield by a wave-breaking or restoring force exerted by an ion channel remaining unshielded behind the laser pulse. As a result of beam loading of trapped electrons, the nonlinear wakefield amplitude inside the bubble is reduced below the trapping threshold. Consequently, electrons trapped in the wakefield undergo the processes of acceleration and bunching toward the wave crest to increase their energy and brightness unless the acceleration distance exceeds the dephasing length. This is a scenario of quasi-monoenergetic electron beam acceleration, based on the self-injection mechanism in the bubble regime, which is visually shown by multi-dimensional PIC simulations[37, 38].

Although self-injection is a robust method relying on self-focusing, self-compression of the laser pulse and expansion of the bubble[39] which occur during the propagation of relativistic laser pulses, initially heated (accelerated) electrons with large transverse momentum are injected into nonlinear wakefields that excite betatron oscillation of accelerated electrons due to the strong focusing field. Hence, if the self-injection and the deterioration of beam quality are suppressed, high-quality electron beams can be produced by controlled injection schemes such as colliding optical injection[40, 41], density-transition injection[42] or density down-ramp[43] and ionization-induced injection[13–15] in the quasi-linear regime of wakefields driven by a laser pulse with a moderate intensity. These injection schemes provide us with high-quality electron beam injectors for the front end of multi-stage high-energy accelerators. As a simplest case, two-stage laser plasma acceleration has been successfully demonstrated in combination with ionization-induced injection[11, 12].

For laser plasma acceleration reaching GeV-level energies, it is essential to propagate intense laser pulses over a centimeter-scale distance in underdense plasma. For this purpose, a preformed plasma density channel with a parabolic radial distribution[44] has been developed for guiding a laser beam over many Rayleigh lengths without diffraction which limits the acceleration distance to a few mm in a uniform plasma. Plasma density channels stabilize the propagation of relativistically intense laser pulses under a matched condition, preventing laser plasma nonlinear instabilities, such as filamentation and hosing which often occur in self-guiding[45, 46]. Therefore, the concept of channel-guided laser wakefield accelerators has been a long-standing proposal[47]. Employing a centimeter-scale plasma waveguide, experiments on GeV-level electron acceleration have been carried out with a gas-filled or ablative discharge capillary[2, 7, 8] to demonstrate quasi-monoenergetic electron beams. Experimental progress beyond 1 GeV has been reported in experiments making use of a 100-TW class laser and a centimeter-scale capillary discharge plasma waveguide[4] or a gas cell relying on self-guiding[3]. During the last two decades, a number of laser plasma accelerator experiments have been carried out under various conditions. Comparing these data with theoretical laser wakefield acceleration models, it may be useful to find a correct scaling law capable of predicting energy gain, accelerated electron charge and the required laser and plasma conditions.

3. Energy scaling of laser wakefield acceleration in the relativistic regime

3.1. Propagation of relativistic laser pulses in plasma

The wave equation for the normalized vector potential describing the evolution of a laser pulse with laser wavelength λ_L and duration τ_L (full width at half maximum, FWHM) in a plasma channel can be written as[48]

$$\left(\nabla^2 - \frac{\partial^2}{c^2\partial t^2}\right)\mathbf{a} = k^2\left(1-\eta^2\right)\mathbf{a}, \qquad (1)$$

where $\mathbf{a} \equiv e\mathbf{A}/m_ec^2$ is the vector potential $\mathbf{A}$ of the laser pulse normalized with respect to the electron rest energy m_ec^2, satisfying the Coulomb gauge $\nabla \cdot \mathbf{a} = 0$, and $k = \omega/c = 2\pi/\lambda_L$ is the free-space wavenumber along the propagation direction. The (squared) refractive index for linearly polarized electromagnetic waves in the long-pulse limit ($ck_p\tau_L \gg 1$) is given by[49]

$$\eta^2(r,z) = 1 - \frac{k_p^2}{k^2\gamma_L}\left[1 + \frac{1}{k_p^2}\nabla_\perp^2\gamma_L + \frac{\Delta n}{n_0}\frac{r^2}{r_0^2}\right], \qquad (2)$$

where $k_p = \omega_p/c = (4\pi r_e n_0)^{1/2}$ is the plasma wavenumber evaluated with the unperturbed on-axis density n_0 and the classical electron radius $r_e = e^2/mc^2$, and $\gamma_L = (1 + a^2/2)^{1/2}$ is the relativistic factor of the laser intensity for linear polarization. In Equation (2), the first term represents free-space propagation, and the three terms in the square brackets correspond to relativistic self-focusing, ponderomotive channeling and a preformed plasma channel with a parabolic density distribution of the form $n(r) = n_0[1 + (\Delta n/n_0)(r^2/r_L^2)]$, where r_L is the laser spot radius and Δn is the channel depth. Analysis of the wave equation with the standard paraxial form provides the matched spot radius r_m under the condition of a beam propagating with a constant spot size r_L, i.e., $k_pr_m = k_pr_L = R_m$, given by[21]

$$\begin{aligned} k_p^2r_m^2 &\equiv R_m^2(a_0^2, \Delta n/n_0) \\ &= 2\ln(\gamma_{L0})\{\gamma_{L0} - 1 - 2\ln((1+\gamma_{L0})/2) \\ &\quad + (\Delta n/n_0)[1 - \gamma_{L0} + \ln((1+\gamma_{L0})/2) \\ &\quad + (a_0^2/4)\,{}_4F_3(1/2,1,1,1;2,2,2;-a_0^2/2)]\}^{-1}, \end{aligned} \qquad (3)$$

where $\gamma_{L0} = (1 + a_0^2/2)^{1/2}$ is the relativistic factor for a Gaussian laser beam with peak amplitude a_0 and ${}_pF_q$ denotes the generalized hypergeometric series of order q and class $q - p + 1$.

Under the matched condition that no phase shift of the laser pulse occurs, the group velocity is written as $\beta_g^2 \approx 1 - k_p^2/(\kappa_{ch}k^2)$, where a correction factor for the group velocity is defined as[21]

$$\begin{aligned} \kappa_{ch}(a_0^2, \Delta n/n_0) &= \frac{a_0^2}{8}[\gamma_{L0} - 1 - \ln((1+\gamma_{L0})/2) \\ &\quad + \frac{a_0^2}{8}\frac{\Delta n}{n_0}{}_4F_3(1/2,1,1,1;2,2,2;-a_0^2/2)]^{-1}. \end{aligned} \qquad (4)$$

Under the matched condition for self-guiding, the spot radius and the group velocity correction factor are given by Equations (3) and (4), respectively, where the channel depth is set to be $\Delta n = 0$.

3.2. Laser plasma acceleration in the quasi-linear regime

In the linear laser wakefield with the accelerating field $E_z = E_{z0}\cos\Psi$, the equations of the longitudinal motion of an electron with the normalized velocity $\beta_z = v_z/c \approx 1$ and electron energy $\gamma = E_e/m_ec^2$ are given by[50]

$$\begin{aligned} \frac{d\gamma}{dz} &= k_p\frac{E_{z0}}{E_0}\cos\Psi \quad \text{and} \\ \frac{d\Psi}{dz} &= k_p\left(1 - \frac{\beta_p}{\beta_z}\right) \approx \frac{k_p}{2\gamma_g^2}, \end{aligned} \qquad (5)$$

where $\Psi = k_p(z - v_pt) + \Psi_0$ is the phase of the plasma wave, $E_0 = mc\omega_p/e$ is the non-relativistic wave-breaking field approximately given by $E_0 \approx 96$ [GV m^{-1}] $(n_e/10^{18}$ [cm^{-3}]$)^{1/2}$, $\beta_p = v_p/c \approx v_g/c = \beta_g$ is the phase velocity v_p of the plasma wave normalized to c, and $\gamma_g = (1-\beta_g^2)^{-1/2} \gg 1$ is assumed. By integrating Equations (5), the energy and phase of the electron can be calculated as

$$\begin{aligned} \gamma(z) &= \gamma_0 + 2\gamma_g^2\frac{E_{z0}}{E_0}[\sin\Psi(z) - \sin\Psi_0] \quad \text{and} \\ \Psi(z) &\approx \frac{k_pz}{2\gamma_g^2} + \Psi_0. \end{aligned} \qquad (6)$$

Setting the initial electron phase $\Psi_0 = 0$ at $z = 0$, the maximum energy gain is given by

$$\Delta\gamma_{max} = \gamma_{max} - \gamma_0 = 2\gamma_g^2\frac{E_{z0}}{E_0}, \qquad (7)$$

at $k_pz = \pi\gamma_g^2$ or $z = \lambda_p\gamma_g^2/2$. As shown from Equations (6), setting $\Psi_0 = -\pi/2$, the maximum energy gain reaches $\Delta\gamma_{max} = 4\gamma_g^2E_{z0}/E_0$ at $k_pz = 2\pi\gamma_g^2$ or $z = \lambda_p\gamma_g^2$. However, electrons undergo both acceleration and focusing only for the phase $0 \leqslant \Psi \leqslant \pi/2$. Hence, we define the dephasing length as $L_{dp} = \lambda_p \gamma_g^2/2$. Considering a driving laser pulse of normalized intensity a_0^2 moving in a plasma channel with channel depth Δn at the group velocity $\beta_g = v_g/c$ with the corresponding relativistic factor $\gamma_g^2 = (1-\beta_g^2)^{-1} \approx \kappa_{ch}(\omega^2/\omega_p^2) = \kappa_{ch}(n_c/n_e) = \kappa_{ch}\gamma_{g0}^2$, where $\gamma_{g0} = \omega/\omega_p$ is the relativistic factor for the group velocity in a uniform plasma and its correction factor κ_{ch} is given by Equation (4), the maximum energy gain and the dephasing length are written as $\Delta\gamma_{max} = 2\kappa_{ch}\gamma_{g0}^2(E_{z0}/E_0)$ and $L_{dp} = (\lambda_p/2)\kappa_{ch}\gamma_{go}^2$, respectively. In the limit of $a_0^2 \ll 1$, $\kappa_{ch} \approx (1+\Delta n/n_0)^{-1}$. In the quasi-linear regime, taking the beam loading effect into account, the maximum accelerating field driven by a Gaussian laser pulse is given by[20]

$$\frac{E_{z0}}{E_0} \simeq \sqrt{\pi}\alpha a_0^2\left(\frac{k_p\sigma_L}{4}\right)\exp\left(-\frac{k_p^2\sigma_L^2}{4}\right) \approx 0.38\alpha a_0^2, \quad (8)$$

where α denotes a factor of accelerating field reduction due to the beam loading effect and σ_L is the rms pulse length of the Gaussian temporal profile with the FWHM length $c\tau_L \sim 0.375\lambda_p$ for $k_p\sigma_L = \sqrt{2}$.

3.3. Laser plasma acceleration in the bubble regime

Previous laser plasma acceleration experiments that successfully demonstrated the production of quasi-monoenergetic electron beams with narrow energy spread have been elucidated in terms of self-injection and an acceleration mechanism in the bubble regime[37, 38]. In these experiments, electrons are self-injected into a nonlinear wake, often referred to as a bubble, i.e., a cavity void of plasma electrons consisting of a spherical ion column surrounded by a narrow electron sheath, formed behind the laser pulse instead of a periodic plasma wave in the linear regime. The phenomenological theory of a nonlinear wakefield in the bubble regime describes the accelerating wakefield $E_z(\xi)/E_0 \approx (1/2)k_p\xi$ in the bubble frame moving in the plasma with velocity v_B, i.e., $\xi = z - v_Bt$. In the bubble (blowout) regime for $a_0 \geqslant 2$, since the electron-evacuated cavity shape is determined by balancing the Lorentz force of the ion sphere exerted on the electron sheath with the ponderomotive force of the laser pulse, the bubble radius R_B is approximately given as $k_pR_B \approx 2\sqrt{a_0}$ (Ref. [38]). Thus, the maximum accelerating field is obtained as $E_{z0}/E_0 = (1/2)\alpha k_pR_B$, where α represents a factor that takes into account the difference between the simulation and theoretical estimation, and more significantly the accelerating field reduction due to the beam loading effects.

In self-guided laser wakefield acceleration, where a driving laser pulse propagates by means of self-channeling, the equations of longitudinal motion of an electron are approximately written as[21]

$$\begin{aligned} \frac{d\gamma}{dz} &= k_p\frac{E_{z0}}{E_0}\left(1-\frac{\xi}{R_B}\right) = \frac{1}{2}\alpha k_p^2R_B\left(1-\frac{\xi}{R_B}\right) \\ \text{and}\quad \frac{d\xi}{dz} &= 1-\frac{\beta_B}{\beta_z} \approx 1-\beta_B \approx \frac{3}{2\gamma_g^2}, \end{aligned} \quad (9)$$

where $\xi = z - v_Bt$ $(0 \leqslant \xi \leqslant R_B)$ is the longitudinal coordinate of the bubble frame moving at the velocity $v_B = c\beta_B \approx v_g - v_{etch}$ and taking into account the diffraction at the laser front that etches back at the velocity $v_{etch} \simeq c(\omega_p/\omega)^2$ (Ref. [38]). Integrating Equations (5), the energy and phase of the electron can be calculated as

$$\begin{aligned} \gamma(z) &= \gamma_0 + \frac{1}{3}\alpha\gamma_g^2k_p^2R_B\xi(z)\left(1-\frac{1}{2}\frac{\xi(z)}{R_B}\right) \\ \text{and}\quad \xi(z) &= \frac{3}{2}\frac{z}{\gamma_g^2}, \end{aligned} \quad (10)$$

where $\gamma_0 = \gamma(0)$ is the injection energy. Hence, the maximum energy gain is obtained at $\xi = R_B$ as

$$\begin{aligned} \Delta\gamma_{max} &= \gamma_{max} - \gamma_0 \approx \frac{1}{6}\alpha\gamma_g^2k_p^2R_B^2 \approx \frac{2}{3}\alpha a_0\gamma_g^2 \\ &= \frac{2}{3}\alpha\kappa_{self}a_0\frac{n_c}{n_e}. \end{aligned} \quad (11)$$

The dephasing length L_{dp} for the self-guided bubble regime is given by

$$k_pL_{dp} \approx \frac{2}{3}k_pR_B\gamma_g^2 = \frac{4}{3}\sqrt{a_0}\kappa_{self}\frac{n_c}{n_e}, \quad (12)$$

i.e.,

$$\begin{aligned} L_{dp} &\simeq \frac{4}{3\pi}\lambda_L\sqrt{a_0}\kappa_{self}\left(\frac{n_c}{n_e}\right)^{3/2} \\ &\approx 24.7\,[\text{mm}]\sqrt{a_0}\kappa_{self}\left(\frac{0.8\,\mu\text{m}}{\lambda_L}\right)^2\left(\frac{10^{18}\,\text{cm}^{-3}}{n_e}\right)^{3/2}, \end{aligned} \quad (13)$$

while the pump depletion length due to pulse-front erosion is given by

$$L_{pd} \approx c\tau_L\frac{n_c}{n_e}, \quad (14)$$

where $\kappa_{self} \equiv \kappa_{ch}(a_0^2, 0)$ is the correction factor of the group velocity for the self-guided pulse in Equation (4).

For a driving laser pulse propagating in a plasma channel, the equations of electron motion are given by setting $v_B = c\beta_B \approx v_g$ in Equations (5), i.e., $d\xi/dz \approx 1-\beta_B \approx 1/2\gamma_g^2$.

Table 1. Parameters of experiments on GeV-class laser wakefield acceleration.

Ref.	$n_e/10^{18}$ (cm^{-3})	$\Delta n/n_e$ (%)	L_{acc} (mm)	P_L (TW)	P/P_c	r_L (μm)	τ_L (fs)	a_0	E_{ex} (GeV)	$\Delta E/E$ (%)
[51]	5.7	0	8.7	182	35	19	55	3.9	0.8	12
[52]	3	0	8	65	6.6	15	60	2.8	0.72	14
[3]	1.3	0	13	110	4.9	15	60	3.8	1.45	~100
[11]	5.7 + 2.5	0	1 + 3	45	3.8	16	40	2.3	0.8	25
[12]	3	0	3 + 5	40	4.1	15	60	2.3	0.46	5
[5]	0.48	0	67	625	18	50	160	3	2	10
[6]	2.1	0	4	212	15	21	60	3.7	0.35	~25
[6]	1.3	0	10	212	8	21	60	3.7	0.87	~30
[6]	2 + 0.8	0	4 + 10	212	5.8	21	60	3.7	3	~30
[2]	4.3	<5	33	40	5.8	25	37	1.4	1	5.9
[2]	3.5	<5	33	12	1.4	25	80	0.75	0.5	13
[8]	8.4	<5	15	18	5.1	23	42	0.84	0.5	2.5
[7]	1.9	<5	40	24	1.55	17	27	1.7	0.56	2.8
[53]	1.8	<5	30	32	1.96	22	80	1.4	0.52	5
[4]	3.1	<5	40	130	13.7	21	55	3	1.8	~50

Hence, the maximum energy gain is

$$\Delta\gamma_{\max} = \gamma_{\max} - \gamma_0 \approx \frac{1}{2}\alpha\gamma_g^2 k_p^2 R_B^2 \approx 2\alpha a_0 \gamma_g^2 = 2\alpha\kappa_{\text{ch}} a_0 \frac{n_c}{n_e}, \tag{15}$$

and the dephasing length L_{dp} is

$$k_p L_{\text{dp}} \approx 2k_p R_B \gamma_g^2 = 4\sqrt{a_0}\kappa_{\text{ch}}\frac{n_c}{n_e}, \tag{16}$$

where κ_{ch} is given by Equation (4). For channel-guided laser plasma acceleration, the pump depletion length, at which the total field energy becomes half of the initial laser energy, is given by

$$L_{\text{pd}} \approx \frac{\sqrt{\pi}}{2\alpha^2} a_0 \sigma_L \frac{n_c}{n_e} = \frac{1}{4}\sqrt{\frac{\pi}{\ln 2}\frac{a_0 c\tau_L}{\alpha^2}\frac{n_c}{n_e}}, \tag{17}$$

where $\sigma_L = c\tau_L/(2\sqrt{\ln 2}) \approx 0.6c\tau_L$ is the rms pulse length.

The matched power P_m corresponding to the matched spot size r_L is calculated as

$$P_L = \frac{k_p^2 r_L^2 a_0^2}{32} P_c, \tag{18}$$

where $P_c = 17 n_c/n_0$ [GW] is the critical power for the relativistic self-focusing at the plasma density n_0 and the required pulse energy is $U_L = P_L \tau_L$.

3.4. Beam loading effects

In laser wakefield acceleration, an accelerated electron beam induces its own wakefield and cancels the laser-driven wakefield. Assuming the beam loading efficiency $\eta_b \equiv 1 - E_z^2/E_M^2$ defined by the fraction of the plasma wave energy absorbed by particles of the bunch with the rms radius σ_b, the beam-loaded field is given by $E_z = \sqrt{1-\eta_b}E_M = \alpha E_M$, where E_M is the accelerating field without beam loading, given by $E_M \approx a_0^{1/2} E_0$ for the bubble regime $a_0 \gtrsim 2$. Thus, the loaded charge is calculated as[20]

$$Q_b \simeq \frac{e}{4k_L r_e}\frac{\eta_b k_p^2 \sigma_b^2}{1-\eta_b}\frac{E_z}{E_0}\left(\frac{n_c}{n_e}\right)^{1/2} \approx 76\,[\text{pC}]\frac{(1-\alpha^2)a_0^{1/2}k_p^2\sigma_b^2}{\alpha}\left(\frac{n_e}{10^{18}\,\text{cm}^{-3}}\right)^{-1/2}. \tag{19}$$

Here, the field reduction factor α for accelerating charge Q_b in the operating plasma density n_e is obtained by solving the equation

$$\alpha^2 + B\alpha - 1 = 0, \tag{20}$$

where the coefficient B is defined as

$$B \equiv \frac{1}{a_0^{1/2}k_p^2\sigma_b^2}\left(\frac{Q_b}{76\,\text{pC}}\right)\left(\frac{n_e}{10^{18}\,\text{cm}^{-3}}\right)^{1/2}. \tag{21}$$

Thus, the field reduction factor is obtained by

$$\alpha = \frac{B}{2}\left[\left(1+\frac{4}{B^2}\right)^{1/2} - 1\right]. \tag{22}$$

For $B \ll 1$, the field reduction factor becomes $\alpha \approx 1$ and contrarily, for $B \gg 1$, $\alpha \approx 0$.

3.5. Comparison with experimental results on GeV-class electron beams

Table 1 summarizes the parameters for experiments on laser wakefield acceleration driven by a self-guided laser pulse with channel depth $\Delta n/n_e = 0$ (Refs. [3, 5, 6,

11, 12, 51, 52]) and a channel-guided laser pulse with $\Delta n/n_e \neq 0$ (Refs. [2, 4, 7, 8, 53]). Since the maximum energy gain scales as $\Delta\gamma_{\max} \propto n_c \propto \lambda_L^{-2}$ for a given a_0, most previous experiments have employed chirped pulse amplification lasers with wavelength $\lambda_L = 800$ nm and pulse duration $\tau_L \leqslant 80$ fs, except for the case of Ref. [5], where a PW-class laser with wavelength $\lambda_L = 1057$ nm and $\tau_L \sim 150$ fs was employed. The validity of the energy scaling formulas (Equations (7), (11) and (15)) based on the present analytical methods may be verified by comparison with these experimental results. Figure 1 shows the comparison of measured electron beam energies in laser wakefield acceleration with the energy scaling on the operating plasma density for the self-guided case (Figure 1(a)) in the bubble regime and the channel-guided case (Figure 1(b)) in both the quasi-linear regime and the bubble regime. In the energy scaling formulas, we assume that the field reduction due to beam loading and the group velocity correction due to relativistic effects cancel each other out, i.e., $\alpha\kappa_{\rm self} \sim 1$ or $\alpha\kappa_{\rm ch} \sim 1$.

For self-guided laser wakefield acceleration, the multi-GeV acceleration results reported in Refs. [5, 6] provide us with informative examples for testing the energy scaling formula (Equation (11)). In Ref. [5], a 625-TW, 160-fs laser pulse with wavelength $\lambda_L = 1057$ nm is focused onto a $1/e^2$ spot radius of $r_L = 50\,\mu$m, producing $a_0 = 3.6$ at the entrance of a 7-cm long gas cell with $n_e = 4.8 \times 10^{17}$ cm^{-3}. The accelerated electron beam has a quasi-monoenergetic peak at 2.0 GeV with a relative energy spread of 10% (FWHM), containing a total charge of 540 pC in a bunch. In this case, from Equations (3) and (4), the dimensionless matched spot radius $R_m = k_p r_m$ and the correction factor of the group velocity $\kappa_{\rm self} = \kappa_{\rm ch}(a_0^2, 0)$ are $R_m = 2.0$ and $\kappa_{\rm self} = 1.46$, respectively, at $a_0 = 3.6$. Thus, since the matched spot radius is $r_m = 15\,\mu$m, the laser pulse undergoes self-focusing after propagating through the gas cell. Using Equations (20) and (21) and assuming the electron beam size $k_p\sigma_b = 1$, the field reduction factor α is calculated as $\alpha \approx 0.32$ ($B \approx 2.84$). From Equation (13), the dephasing length is $L_{\rm dp} \approx 118$ mm, while from Equation (14), the pump depletion length due to pulse-front erosion is $L_{\rm pd} \approx 100$ mm. From Equation (10), the electron beam energy at the accelerator length $z = L_{\rm acc}$ is estimated as

$$
\begin{aligned}
E_b(L_{\rm acc}) &= E_{b0} + m_e c^2 \alpha \sqrt{a_0} k_p L_{\rm acc}\left(1 - \frac{L_{\rm acc}}{2L_{\rm dp}}\right) \\
&\approx E_{b0} + 96\,[\text{MeV}]\,\alpha\sqrt{a_0}\left(1 - \frac{L_{\rm acc}}{2L_{\rm dp}}\right) \\
&\quad \times \left(\frac{L_{\rm acc}}{1\,\text{mm}}\right)\left(\frac{n_e}{10^{18}\,\text{cm}^{-3}}\right)^{1/2}, \qquad (23)
\end{aligned}
$$

where E_{b0} is the injection beam energy. For $L_{\rm acc} = 70$ mm, the beam energy is evaluated to be $E_b = 2.0$ GeV. This estimate is in good agreement with the measured beam energy of 2.0 ± 0.1 GeV[5], taking into account the field reduction factor α due to beam loading.

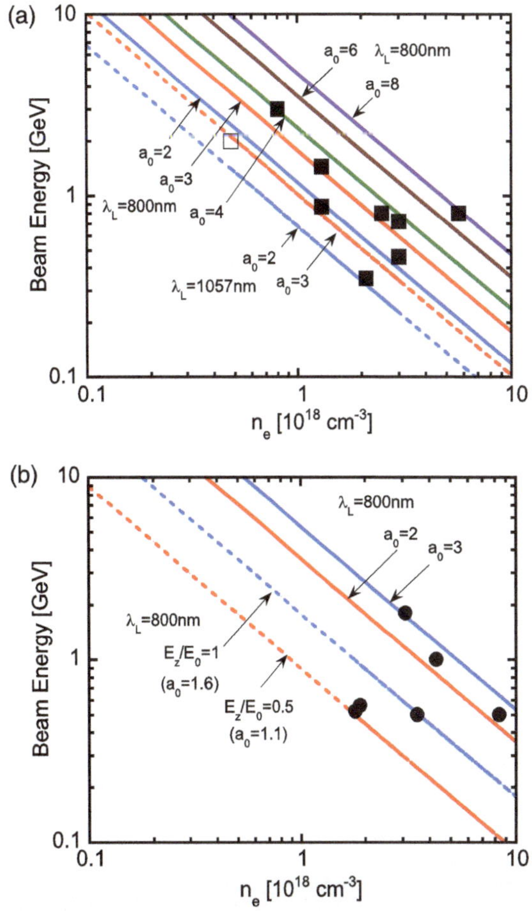

Figure 1. A comparison of measured electron beam energies in laser wakefield acceleration with the energy scaling as a function of the operating plasma density for (a) the self-guided case in the bubble regime at laser wavelengths of 800 nm (solid line) and 1057 nm (dashed line) and (b) the channel-guided case in both the quasi-linear regime (dashed line) and the bubble regime (solid line). The experimental data are plotted with filled squares for $\lambda_L = 800$ nm and the open square for $\lambda_L = 1057$ nm in (a), and with filled circles for $\lambda_L = 800$ nm in (b).

In Ref. [6], a 212-TW, 60-fs laser pulse is focused on a $1/e^2$ spot radius of $r_L = 21\,\mu$m, producing $a_0 = 3.7$ at the entrance of a gas jet for three cases consisting of a 4-mm long single stage with $n_e = 2.1 \times 10^{18}$ cm^{-3}, a 10-mm long single stage with $n_e = 1.3 \times 10^{18}$ cm^{-3}, and two stages comprising a 4-mm long injector with $n_e = 2 \times 10^{18}$ cm^{-3} and a 10-mm long accelerator with $n_e = 0.8 \times 10^{18}$ cm^{-3}. As shown in Figure 2, the accelerated electron beams for the three cases have peak energies of 0.35, 0.87 and 3 GeV, respectively, and total loaded charges of 88, 110 and 80 pC, respectively. From Equations (3) and (4), the dimensionless matched spot radius $R_m = k_p r_m$ and the correction factor of the group velocity $\kappa_{\rm self} = \kappa_{\rm ch}(a_0^2, 0)$ are $R_m = 2.0$ and $\kappa_{\rm self} = 1.48$, respectively, at $a_0 = 3.7$. Since the matched spot radius is $r_m = 7.3\,\mu$m for the 4-mm single-stage case, $r_m = 9.3\,\mu$m for the 10-mm single-stage case and $r_m = 7.5\,\mu$m in the injector jet and $r_m = 12\,\mu$m in the accelerator jet for the two-stage case, respectively, it is inferred that a laser pulse with

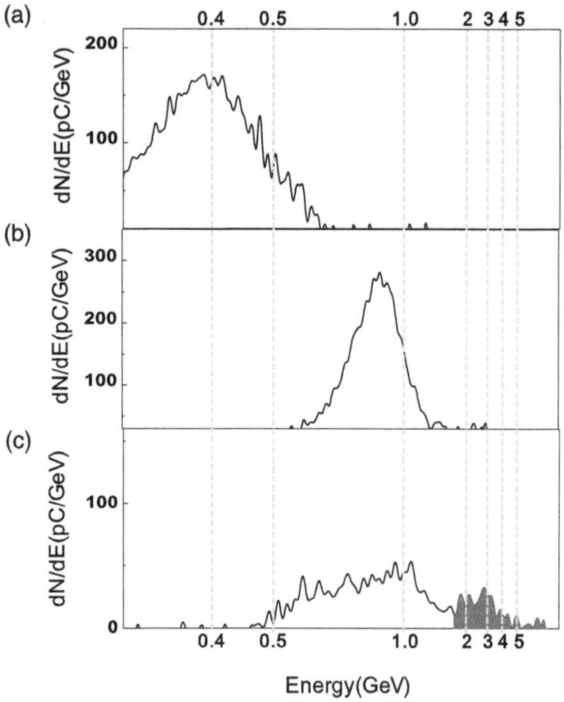

Figure 2. Electron beam energy spectra obtained from the experiment[6], where a 212-TW, 60-fs laser pulse is focused on a $1/e^2$ spot radius of $r_L = 21\,\mu\text{m}$ producing $a_0 = 3.7$ at the entrance of a gas jet for three cases consisting of (a) a 4-mm long single stage with $n_e = 2.1 \times 10^{18}\,\text{cm}^{-3}$, (b) a 10-mm long single stage with $n_e = 1.3 \times 10^{18}\,\text{cm}^{-3}$ and (c) two stages comprising a 4-mm long injector with $n_e = 2 \times 10^{18}\,\text{cm}^{-3}$ and a 10-mm long accelerator with $n_e = 0.8 \times 10^{18}\,\text{cm}^{-3}$.

a focused spot radius of $r_L = 21\,\mu\text{m}$ is initially self-focused down to the matched spot radius. Using Equations (20) and (21) and assuming the electron beam size $k_p\sigma_b = 1$, the field reduction factor α is calculated as $\alpha \approx 0.66\,(B = 0.86)$ for the 4-mm single-stage case, $\alpha = 0.66\,(B = 0.86)$ for the 10-mm single-stage case and $\alpha = 0.66\,(B = 0.86)$ in the injector/$\alpha \approx 0.78\,(B = 0.49)$ in the accelerator for the two-stage case. From Equation (13), the dephasing length is $L_{\text{dp}} \approx 23.1$ mm for the 4-mm single-stage case, $L_{\text{dp}} \approx 47.4$ mm for the 10-mm single-stage case and $L_{\text{dp}} \approx 24.9$ mm in the injector/$L_{\text{dp}} \approx 98.3$ mm in the accelerator for the two-stage case. The pump depletion length due to pulse-front erosion is $L_{\text{pd}} \approx 14.9$ mm for the 4-mm single-stage case, $L_{\text{pd}} \approx 24.1$ mm for the 10-mm single-stage case and $L_{\text{pd}} \approx 15.7$ mm in the injector/$L_{\text{pd}} \approx 39.2$ mm in the accelerator for the two-stage case. From Equation (23), the beam energy is estimated to be $E_b \approx 353$ MeV with an effective acceleration length of $L_{\text{acc}} = 2$ mm for the 4-mm single-stage case and $E_b \approx 1.02$ GeV with $L_{\text{acc}} = 8$ mm for the 10-mm single-stage case. For the two-stage case, the output energy of the injector is $E_b \approx 409$ MeV with $L_{\text{acc}} \approx 2.5$ mm and the output energy of the accelerator stage reaches $E_b \approx 2.46$ GeV with $L_{\text{acc}} \approx 10$ mm, assuming that the injection energy is $E_{b0} \approx 409$ MeV and the focused spot size at the entrance of the accelerator stage is decreased to $r_L = 7.5\,\mu\text{m}$ due to self-focusing in the injector stage, increasing the normalized vector potential up to $a_0 \approx 10.4$. In this experiment, most of the charge produced in the injector is injected into the accelerator stage, while the large energy spread is attributed to the fact that the accelerator length is shorter than the dephasing length at which the energy compression takes place in the phase space as well as the maximum energy.

4. Design of 10-GeV-level laser plasma accelerators

At present, the most near-term prospects for 10-GeV-level laser plasma acceleration are confidently given by the scaling and methods described in the previous section. Here, we consider design examples of laser plasma accelerators capable of delivering 10-GeV electron beams with bunch charges of 160 pC (10^9 electrons per bunch) for three cases: a self-guided laser plasma accelerator in the bubble regime with $a_0 = 3$, a channel-guided laser plasma accelerator in the bubble regime with $a_0 = 2$ and a channel-guided laser plasma accelerator in the quasi-linear regime with $a_0 = 1.5$. For all three cases, we present design parameters for the laser and plasma for a driving laser wavelength of 800 nm. Table 2 shows the design parameters of the 10-GeV laser plasma accelerators for the abovementioned three cases and a 40-GeV laser plasma accelerator with $a_0 = 2$ and a beam loading charge of 300 pC by comparison with the results of 3D PIC simulation from the Lorentz-boosted frame OSIRIS code[16]. The design parameters for the 40-GeV laser plasma accelerator are in good agreement with the PIC simulation results in terms of the operating plasma density, the accelerator length, i.e., the dephasing length, the matched spot radius and the matched power. The design formulas we used to evaluate these parameters are described in the following.

4.1. Self-guided laser plasma accelerator in the bubble regime

For a given energy gain W, the operating plasma density is determined from Equation (11) as

$$n_e = \frac{2}{3}\alpha\kappa_{\text{self}}a_0\frac{n_c}{\Delta\gamma_{\max}}$$
$$\approx 5.94 \times 10^{16}\,[\text{cm}^{-3}]\,\kappa_{\text{self}}a_0\left(\frac{0.8\,\mu\text{m}}{\lambda_L}\right)^2\left(\frac{10\,\text{GeV}}{W/\alpha}\right). \quad (24)$$

The accelerator length, equal to the dephasing length, becomes

$$L_{\text{acc}} = L_{\text{dp}} \approx \sqrt{\frac{3}{2}}\frac{(\Delta\gamma_{\max}/\alpha)^{3/2}}{\pi a_0\kappa_{\text{self}}^{1/2}}\lambda_L$$
$$\approx \frac{0.9\,[\text{m}]}{a_0\kappa_{\text{self}}^{1/2}}\left(\frac{\lambda_L}{0.8\,\mu\text{m}}\right)\left(\frac{W/\alpha}{10\,\text{GeV}}\right)^{3/2}, \quad (25)$$

Table 2. Design parameters for 10-GeV-level laser plasma accelerators in comparison with the results of the 3D PIC simulation[16]. Case A stands for the self-guided case in the bubble regime, designed by the formulas given in Section 4.1, case B for the self-guided case in the bubble regime, designed by the formulas given in Section 4.2, case C for the channel-guided case in the quasi-linear regime, designed by the formulas given in Section 4.3, and case D for the self-guided case in the bubble regime at 40 GeV, designed by the formulas given in Section 4.1.

Case	A	B	C	D	Ref. [16]
W (GeV)	10	10	10	40	38
n_e (10^{17} cm^{-3})	1.9	2.7	1.1	0.28	0.22
$\Delta n/n_0$ (%)	0	5	5	0	
L_{dp} (m)	0.38	0.42	0.87	4.7	5
λ_L (nm)	800	800	800	800	800
a_0	3	2	1.5	2	2
κ_{ch}	1.35	1.13	1.06	1.19	
R_m	2.3	2.9	3.6	3.2	
r_L (μm)	29	29	57	103	100
τ_L (fs)	128	95	127	238	160
P_L (TW)	238	114	250	1483	1400
P_L/P_c	1.5	1.1	0.91	1.3	1.04
U_L (J)	30	11	32	353	220
Q_b (pC)	160	160	160	300	300
C	0.596	0.945	0.989	1.05	
α	0.77	0.68	0.67	0.79	

while the pump depletion length due to pulse-front erosion is given by $L_{\mathrm{pd}} \approx c\tau_L n_c/n_e$. The dephasing length should be less than the pump depletion length, i.e., $L_{\mathrm{pd}} \geqslant L_{\mathrm{dp}}$. Therefore, the pulse length is set to be $c\tau_L \geqslant (2/3\pi)\sqrt{a_0\kappa_{\mathrm{self}}}\lambda_p$.

$$\tau_L \geqslant \sqrt{\frac{2}{3}\frac{\lambda_L}{\pi c}\kappa_{\mathrm{self}}^{1/2}}\left(\frac{\Delta\gamma_{\max}}{\alpha}\right)^{1/2}$$
$$\geqslant 97\,[\mathrm{fs}]\,\kappa_{\mathrm{self}}^{1/2}\left(\frac{\lambda_L}{0.8\,\mu\mathrm{m}}\right)\left(\frac{W/\alpha}{10\,\mathrm{GeV}}\right)^{1/2}. \quad (26)$$

The matched spot radius becomes

$$r_L = \frac{1}{2\pi}\sqrt{\frac{3}{2}}\lambda_L\frac{R_m}{\sqrt{a_0\kappa_{\mathrm{self}}}}\left(\frac{\Delta\gamma_{\max}}{\alpha}\right)^{1/2}$$
$$\approx 22\,[\mu\mathrm{m}]\frac{R_m}{\sqrt{a_0\kappa_{\mathrm{self}}}}\left(\frac{\lambda_L}{0.8\,\mu\mathrm{m}}\right)\left(\frac{W/\alpha}{10\,\mathrm{GeV}}\right)^{1/2}, \quad (27)$$

where $R_m(a_0^2, 0) = k_p r_L$ is the dimensionless matched spot radius given by Equation (3) for self-guiding, $\Delta n/n_0 = 0$. The matched power is calculated as

$$P_L = \frac{k_p^2 r_L^2 a_0^2}{32}P_c = \frac{51}{64}\,[\mathrm{GW}]\frac{a_0 R_m^2}{\kappa_{\mathrm{self}}}\frac{\Delta\gamma_{\max}}{\alpha}$$
$$\approx 15.6\,[\mathrm{TW}]\frac{a_0 R_m^2}{\kappa_{\mathrm{self}}}\left(\frac{W/\alpha}{10\,\mathrm{GeV}}\right). \quad (28)$$

The required pulse energy is

$$U_L = P_L\tau_L \geqslant 1.51\,[\mathrm{J}]\frac{a_0 R_m^2}{\kappa_{\mathrm{self}}^{1/2}}\left(\frac{\lambda_L}{0.8\,\mu\mathrm{m}}\right)\left(\frac{W/\alpha}{10\,\mathrm{GeV}}\right)^{3/2}. \quad (29)$$

The field reduction factor α for loading charge Q_b up to a given energy W is obtained by solving the equation

$$\alpha^2 + C\alpha^{3/2} - 1 = 0, \quad (30)$$

where the coefficient C is given as

$$C = \frac{Q_b}{312\,[\mathrm{pC}]}\frac{\kappa_{\mathrm{self}}^{1/2}}{k_p^2\sigma_b^2}\left(\frac{0.8\,\mu\mathrm{m}}{\lambda_L}\right)\left(\frac{10\,\mathrm{GeV}}{W}\right)^{1/2}. \quad (31)$$

4.2. Channel-guided laser plasma accelerator in the bubble regime

The operating plasma density is determined by

$$n_e = 2\alpha\kappa_{\mathrm{ch}}a_0\frac{n_c}{\Delta\gamma_{\max}}$$
$$\approx 1.78\times 10^{17}\,[\mathrm{cm}^{-3}]\,\kappa_{\mathrm{ch}}a_0\left(\frac{0.8\,\mu\mathrm{m}}{\lambda_L}\right)^2\left(\frac{10\,\mathrm{GeV}}{W/\alpha}\right), \quad (32)$$

and the accelerator length becomes

$$L_{\mathrm{stage}} = L_{\mathrm{dp}} \approx \frac{\lambda_L}{\sqrt{2}\pi\kappa_{\mathrm{ch}}^{1/2}a_0}\left(\frac{\Delta\gamma_{\max}}{\alpha}\right)^{3/2}$$
$$\approx \frac{0.5\,[\mathrm{m}]}{\kappa_{\mathrm{ch}}^{1/2}a_0}\left(\frac{\lambda_L}{0.8\,\mu\mathrm{m}}\right)\left(\frac{W/\alpha}{10\,\mathrm{GeV}}\right)^{3/2}. \quad (33)$$

The pump depletion length, at which the total field energy becomes half of the initial laser energy, is given by

$$L_{\rm pd} \approx \frac{\sqrt{\pi}}{2\alpha^2} a_0 \sigma_L \frac{n_c}{n_e} = \frac{1}{4}\sqrt{\frac{\pi}{\ln 2}} \frac{a_0 c\tau_L}{\alpha^2} \frac{n_c}{n_e} \approx \frac{0.156\,[\mathrm{m}]}{\alpha^2 \kappa_{\rm ch}} \left(\frac{W/\alpha}{10\,\mathrm{GeV}}\right)\left(\frac{\tau_L}{100\,\mathrm{fs}}\right), \tag{34}$$

where $\sigma_L = c\tau_L/(2\sqrt{\ln 2}) \approx 0.6 c\tau_L$ is the rms pulse length. The requirement for the accelerator length $L_{\rm acc} = L_{\rm dp} \leqslant L_{\rm pd}$ bounds the minimum pulse duration as

$$c\tau_L \geqslant \frac{8}{\pi}\left(\frac{\ln 2}{\pi}\right)^{1/2} \frac{\alpha^2 \kappa_{\rm ch}}{\sqrt{a_0}} \lambda_p \quad \text{or} \quad \tau_L \geqslant 320\,[\mathrm{fs}] \frac{\alpha^2 \kappa_{\rm ch}^{1/2}}{a_0}\left(\frac{\lambda_L}{0.8\,\mu\mathrm{m}}\right)\left(\frac{W/\alpha}{10\,\mathrm{GeV}}\right)^{1/2}. \tag{35}$$

The matched spot radius becomes

$$r_L = \frac{\lambda_L}{2\pi}\left(\frac{n_c}{n_e}\right)^{1/2} R_m \approx 12.6\,[\mu\mathrm{m}] \frac{R_m}{\sqrt{\kappa_{\rm ch} a_0}}\left(\frac{\lambda_L}{0.8\,\mu\mathrm{m}}\right)\left(\frac{W/\alpha}{10\,\mathrm{GeV}}\right)^{1/2}, \tag{36}$$

where $R_m(a_0^2, \Delta n/n_0) = k_p r_L$ is the dimensionless matched spot radius given by Equation (3). The matched power is calculated as

$$P_L = \frac{k_p^2 r_L^2 a_0^2}{32} P_c \approx 5.2\,[\mathrm{TW}] \frac{a_0 R_m^2}{\kappa_{\rm ch}}\left(\frac{W/\alpha}{10\,\mathrm{GeV}}\right), \tag{37}$$

and the required pulse energy is

$$U_L = P_L \tau_L \geqslant 1.66\,[\mathrm{J}] \frac{\alpha^2 R_m^2}{\kappa_{\rm ch}^{1/2}}\left(\frac{\lambda_L}{0.8\,\mu\mathrm{m}}\right)\left(\frac{W/\alpha}{10\,\mathrm{GeV}}\right)^{3/2}. \tag{38}$$

The field reduction factor α is determined from the coefficient C, given by

$$C \equiv \frac{Q_b}{180\,[\mathrm{pC}]} \frac{\kappa_{\rm self}^{1/2}}{k_p^2 \sigma_b^2}\left(\frac{0.8\,\mu\mathrm{m}}{\lambda_L}\right)\left(\frac{10\,\mathrm{GeV}}{W}\right)^{1/2}. \tag{39}$$

4.3. Channel-guided laser plasma accelerator in the quasi-linear regime

For a given $a_0 \leqslant 1.7$, the pulse duration is given by

$$\tau_L \approx \frac{160\,[\mathrm{fs}]}{a_0\sqrt{\kappa_{\rm ch}}}\left(\frac{\lambda_L}{0.8\,\mu\mathrm{m}}\right)\left(\frac{W/\alpha}{10\,\mathrm{GeV}}\right)^{1/2}. \tag{40}$$

For this pulse duration, the operating plasma density becomes

$$n_e \approx 6.8 \times 10^{16}\,[\mathrm{cm}^{-3}]\, \kappa_{\rm ch} a_0^2 \left(\frac{0.8\,\mu\mathrm{m}}{\lambda_L}\right)^2 \left(\frac{10\,\mathrm{GeV}}{W/\alpha}\right). \tag{41}$$

Thus, the required accelerator length $L_{\rm acc}$ can be set to be

$$L_{\rm acc} = L_{\rm dp} \approx \frac{1.65\,[\mathrm{m}]}{a_0^3 \kappa_{\rm ch}^{1/2}}\left(\frac{\lambda_L}{0.8\,\mu\mathrm{m}}\right)\left(\frac{W/\alpha}{10\,\mathrm{GeV}}\right)^{3/2}, \tag{42}$$

where n_c/n_e is given by Equation (9). The matched spot radius is calculated from

$$r_L \approx 20\,[\mu\mathrm{m}] \frac{R_m}{\kappa_{\rm ch}^{1/2} a_0}\left(\frac{\lambda_L}{0.8\,\mu\mathrm{m}}\right)\left(\frac{W/\alpha}{10\,\mathrm{GeV}}\right)^{1/2}, \tag{43}$$

where $R_m = k_p r_L$ is the dimensionless matched spot radius for a given a_0 and $\Delta n/n_0$, given by Equation (3). The required peak power of the laser pulse is given by

$$P_L = \frac{k_p^2 r_L^2 a_0^2}{32} P_c \approx 13.7\,[\mathrm{TW}] \frac{R_m^2}{\kappa_{\rm ch}}\left(\frac{W/\alpha}{10\,\mathrm{GeV}}\right), \tag{44}$$

and the required pulse energy becomes

$$U_L = P_L \tau_L \approx 2.2\,[\mathrm{J}] \frac{R_m^2}{a_0 \kappa_{\rm ch}^{3/2}}\left(\frac{\lambda_L}{0.8\,\mu\mathrm{m}}\right)\left(\frac{W/\alpha}{10\,\mathrm{GeV}}\right)^{3/2}. \tag{45}$$

The field reduction factor α is determined from the coefficient C, given by

$$C \equiv \frac{Q_b}{111\,[\mathrm{pC}]} \frac{\kappa_{\rm self}^{1/2}}{a_0 k_p^2 \sigma_b^2}\left(\frac{0.8\,\mu\mathrm{m}}{\lambda_L}\right)\left(\frac{10\,\mathrm{GeV}}{W}\right)^{1/2}. \tag{46}$$

5. Conclusion

We have provided an overview of recent progress in laser plasma accelerators from the perspective of experiments on the production of GeV-level electron beams, and scaling formulas to describe energy gain for a self-guided laser plasma accelerator in the bubble regime ($a_0 \geqslant 2$), a channel-guided laser plasma accelerator in the bubble regime ($a_0 \geqslant 2$) and a channel-guided laser plasma accelerator in the quasi-linear regime ($a_0 < 2$). Although most previous experiments have been focused on electron injection into the plasma bubble and the production of high-quality electron beams with small energy spread and emittance, employing a millimeter-scale gas jet, recent experimental results beyond 1-GeV acceleration allow us to test the scaling formulas in depth, which are necessary for the design of the operating parameters of laser plasma accelerators to satisfy requirements such as energy gain and beam charge. Taking account of the group velocity correction factor $\kappa_{\rm ch}$ in the propagation of laser pulses with relativistic intensity, characterized by a_0, through plasma channels, including initially uniform plasma with $\Delta n/n_0 = 0$ and a preformed plasma channel with $\Delta n/n_0 \neq 0$, provides the correct accelerator length equal to the dephasing length as well as the proper operating plasma density. Meanwhile, we found that the accelerating field reduction factor α due to beam loading can be properly evaluated by applying the resultant scaling formulas to recent experimental results for multi-GeV laser plasma accelerators[5, 6], leading to self-consistent design parameters for the driving laser pulse achieving the requirement of beam energy and charge. Independently, we confirmed that the accelerating field reduction factor α can be obtained from the simple analytic formulas Equations (20)–(22), indicating good agreement with that obtained from 3D PIC simulation results[16, 38]. A further detailed analysis on the beam loading effects such as bunch length and shape that affect the accelerating fields, energy gain and energy spread will be pursued in future work.

Acknowledgements

This work was supported by Project Code IBS-R012-D1. K. Nakajima was also supported by the National Natural Science Foundation of China (Project No. 51175324).

References

1. T. Tajima and J. M. Dawson, Phys. Rev. Lett. **43**, 267 (1979).
2. W. P. Leemans, B. Nagler, A. J. Gonsalves, C. Toth, K. Nakamura, C. G. R. Geddes, E. Esarey, C. B. Schroeder, and S. M. Hooker, Nat. Phys. **2**, 696 (2006).
3. C. E. Clayton, J. E. Ralph, F. Albert, R. A. Fonseca, S. H. Glenzer, C. Joshi, W. Lu, K. A. Marsh, S. F. Martins, W. B. Mori, A. Pak, F. S. Tsung, B. B. Pollock, J. S. Ross, L. O. Silva, and D. H. Froula, Phys. Rev. Lett. **105**, 105003 (2010).
4. H. Lu, M. Liu, W. Wang, C. Wang, J. Liu, A. Deng, J. Xu, C. Xia, W. Li, H. Zhang, X. Lu, C. Wang, J. Wang, X. Liang, Y. Leng, B. Shen, K. Nakajima, R. Li, and Z. Xu, Appl. Phys. Lett. **99**, 091502 (2011).
5. X. Wang, R. Zgadzaj, N. Fazel, Z. Li, S. A. Yi, X. Zhang, W. Henderson, Y. Y. Chang, R. Korzekwa, H. E. Tsai, C. H. Pai, H. Quevedo, G. Dyer, E. Gaul, M. Martinez, A.C. Bernstein, T. Borger, M. Spinks, M. Donovan, V. Khudik, G. Shvets, T. Ditmire, and M. C. Downer, Nature Commun. **4**, 1988 (2013).
6. H. T. Kim, K. H. Pae, H. J. Cha, I. J. Kim, T. J. Yu, J. H. Sung, S. K. Lee, T. M. Jeong, and J. Lee, Phys. Rev. Lett. **111**, 165002 (2013).
7. T. Kameshima, W. Hong, K. Sugiyama, X. Wen, Y. Wu, C. Tang, Q. Zhu, Y. Gu, B. Zhang, H. Peng, S.-i. Kurokawa, L. Chen, T. Tajima, T. Kumita, and K. Nakajima, Appl. Phys. Exp. **1**, 066001 (2008).
8. S. Karsch, J. Osterhoff, A. Popp, T. P. Rowlands-Rees, Z. Major, M. Fuchs, B. Marx, R. Hörlein, K. Schmid, L. Veisz, S. Becker, U. Schramm, B. Hidding, G. Pretzler, D. Habs, F. Grüner, F. Krausz, and S. M. Hooker, New J. Phys. **9**, 415 (2007).
9. O. Lundh, J. Lim, C. Rechatin, L. Ammoura, A. Ben-Ismail, X. Davoine, G. Gallot, J. P. Goddet, E. Lefebvre, V. Malka, and J. Faure, Nat. Phys. **7**, 219 (2011).
10. N. A. M. Hafz, T. M. Jeong, I. W. Choi, S. K. Lee, K. H. Pae, V. V. Kulagin, J. H. Sung, T. J. Yu, K.-H. Hong, T. Hosokai, J. R. Cary, D.-K. Ko, and J. Lee, Nat. Photon. **2**, 571 (2008).
11. J. S. Liu, C. Q. Xia, W. T. Wang, H. Y. Lu, C. Wang, A. H. Deng, W. T. Li, H. Zhang, X. Y. Liang, Y. X. Leng, X. M. Lu, C. Wang, J. Z. Wang, K. Nakajima, R. X. Li, and Z. Z. Xu, Phys. Rev. Lett. **107**, 035001 (2011).
12. B. B. Pollock, C. E. Clayton, J. E. Ralph, F. Albert, A. Davidson, L. Divol, C. Filip, S. H. Glenzer, K. Herpoldt, W. Lu, K. A. Marsh, J. Meinecke, W. B. Mori, A. Pak, T. C. Rensink, J. S. Ross, J. Shaw, G. R. Tynan, C. Joshi, and D. H. Froula, Phys. Rev. Lett. **107**, 045001 (2011).
13. A. Pak, K. A. Marsh, S. F. Martins, W. Lu, W. B. Mori, and C. Joshi, Phys. Rev. Lett. **104**, 025003 (2010).
14. C. McGuffey, A. G. R. Thomas, W. Schumaker, T. Matsuoka, V. Chvykov, F. J. Dollar, G. Kalintchenko, V. Yanovsky, A. Maksimchuk, K. Krushelnick, V. Y. Bychenkov, I. V. Glazyrin, and A. V. Karpeev, Phys. Rev. Lett. **104**, 025004 (2010).
15. C. Xia, J. Liu, W. Wang, H. Lu, W. Cheng, A. Deng, W. Li, H. Zhang, X. Liang, Y. Leng, X. Lu, C. Wang, J. Wang, K. Nakajima, R. Li, and Z. Xu, Phys. Plasmas **18**, 113101 (2011).
16. S. F. Martins, R. A. Fonseca, W. Lu, W. B. Mori, and L. O. Silva, Nat. Phys. **6**, 311 (2010).
17. J.-L. Vay, C. G. R. Geddes, E. Esarey, C. B. Schroeder, W. P. Leemans, E. Cormier-Miche, and D. P. Grote, Phys. Plasmas **18**, 123103 (2011).
18. M. Tzoufras, C. Huang, J. H. Cooley, F. S. Tsung, J. Vieira, and W. B. Mori, J. Plasma Phys. **78**, 401 (2012).
19. C. B. Schroeder, E. Esarey, C. G. R. Geddes, C. Benedetti, and W. P. Leemans, Phys. Rev. ST Accel. Beams **13**, 101301 (2010).
20. K. Nakajima, A. Deng, X. Zhang, B. Shen, J. Liu, R. Li, Z. Xu, T. Ostermayr, S. Petrovics, C. Klier, K. Iqbal, H. Ruhl, and T. Tajima, Phys. Rev. ST Accel. Beams **14**, 091301 (2011).
21. K. Nakajima, H. Lu, X. Zhao, B. Shen, R. Li, and Z. Xu, Chin. Opt. Lett. **11**, 013501 (2013).
22. N. Blanchot, E. Bar, G. Behar, C. Bellet, D. Bigourd, F. Boubault, C. Chappuis, H. Coïc, C. Damiens-Dupont, O. Flour, O. Hartmann, L. Hilsz, E. Hugonnot, E. Lavastre, J. Luce, E. Mazataud, J. Neauport, S. Noailles, B. Remy, F. Sautarel, M. Sautet, and C. Rouyer, Opt. Express **18**, 10088 (2010).
23. W. Leemans, C. G. R. Geddes, J. Faure, Cs. Tóth, J. van Tilborg, C. B. Schroeder, E. Esarey, and G. Fubiani, Phys. Rev. Lett. **91**, 074802 (2003).
24. S. Kneip, C. McGuffey, J. L. Martins, S. F. Martins, C. Bellei, V. Chvykov, F. Dollar, R. Fonseca, C. Huntington, G. Kalintchenko, A. Maksimchuk, S. P. D. Mangles, T. Matsuoka, S. R. Nagel, C. A. J. Palmer, J. Schreiber, K. Ta Phuoc, A. G. R. Thomas, V. Yanovsky, L. O. Silva, K. Krushelnick, and Z. Najmudin, Nat. Phys. **6**, 980 (2010).
25. K. Nakajima, M. Kando, T. Kawakubo, T. Nakanishi, and A. Ogata, Nucl. Instrum. Methods A **375**, 593 (1996).
26. F. Grüner, S. Becker, U. Schramm, T. Eichner, M. Fuchs, R. Weingartner, D. Habs, J. Meyer-Ter-Vehn, M. Geissler, M. Ferrario, L. Serafini, B. Van der Geer, H. Backe, W. Lauth, and S. Reiche, Appl. Phys. B **86**, 431 (2007).
27. K. Nakajima, Nat. Phys. **4**, 92 (2008).
28. K. Nakajima, A. H. Deng, H. Yoshitama, Nasr A. M. Hafz, H. Y. Lu, B. F. Shen, J. S. Liu, R. X. Li, and Z. Z. Xu, *Free Electron Laser* S. Varró, ed. Vol. 119 (InTech, Rijeka, Croatia, 2012), Ch. 5.

29. K. Nakajima, D. Fisher, D. T. Kawakubo, H. Nakanishi, A. Ogata, Y. Kato, Y. Kitagawa, R. Kodama, K. Mima, H. Shiraga, K. Suzuki, K. Yamakawa, T. Zhang, Y. Sakawa, T. Shoji, Y. Nishida, N. Yugami, M. Downer, and T. Tajima, Phys. Rev. Lett. **74**, 4428 (1995).
30. A. Modena, Z. Najmudin, A. E. Dangor, C. E. Clayton, K. A. Marsh, C. Joshi, V. Malka, C. B. Darrow, C. Danson, D. Neely, and F. N. Walsh, Nature **377**, 606 (1995).
31. N. E. Andreev, L. M. Gorbunov, V. I. Kirsanov, A. A. Pogosova, and R. R. Ramazashvili, Sov. Phys. JETP **55**, 571 (1992).
32. Z. Najmudin, K. Krushelnick, E. L. Clark, S. P. D. Mangles, B. Walton, A. E. Dangor, S. Fritzler, V. Malka, E. Lefebvre, D. Gordon, F. S. Tsung, and C. Joshi, Phys. Plasmas **10**, 2071 (2003).
33. V. Malka, S. Fritzler, E. Lefebvre, M.-M. Aleonard, F. Burgy, J.-P. Chambaret, J.-F. Chemin, K. Krushelnick, G. Malka, S. P. D. Mangles, Z. Najmudin, M. Pittman, J.-P. Rousseau, J.-N. Scheurer, B. Walton, and A. E. Dangor, Science **298**, 1596 (2002).
34. S. P. D. Mangles, C. D. Murphy, Z. Najmudin, A. G. R. Thomas, J. L. Collier, A. E. Dangor, E. J. Divall, P. S. Foster, J. G. Gallacher, C. J. Hooker, D. A. Jaroszynski, A. J. Langley, W. B. Mori, P. A. Norreys, F. S. Tsung, R. Viskup, B. R. Walton, and K. Krushelnick, Nature **431**, 535 (2004).
35. C. G. R. Geddes, C. Toth, J. van Tilborg, E. Esarey, C. B. Schroeder, D. Bruhwiler, C. Nieter, J. Cary, and W. P. Leemans, Nature **431**, 538 (2004).
36. J. Faure, Y. Glinec, A. Pukhov, S. Kiselev, S. Gordienko, E. Lefebvre, J. P. Rousseau, F. Burgy, and V. Malka, Nature **431**, 541 (2004).
37. I. Kostyukov, A. Pukhov, and S. Kiselev, Phys. Plasmas **11**, 5256 (2004).
38. W. Lu, M. Tzoufras, C. Joshi, F. S. Tsung, W. B. Mori, J. Vieira, R. A. Fonseca, and L. O. Silva, Phys. Rev. ST Accel. Beams **10**, 061301 (2007).
39. S. Kalmykov, A. Yi, V. Khudik, and G. Shvets, Phys. Rev. Lett. **103**, 135004 (2009).
40. J. Faure, C. Rechatin, A. Norlin, A. Lifschitz, Y. Glinec, and V. Malka, Nature **444**, 737 (2006).
41. H. Kotaki, I. Daito, M. Kando, Y. Hayashi, K. Kawase, T. Kameshima, Y. Fukuda, T. Homma, J. Ma, L. M. Chen, T. Z. Esirkepov, A. S. Pirozhkov, J. K. Koga, A. Faenov, T. Pikuz, H. Kiriyama, H. Okada, T. Shimomura, Y. Nakai, M. Tanoue, H. Sasao, D. Wakai, H. Matsuura, S. Kondo, S. Kanazawa, A. Sugiyama, H. Daido, and S. V. Bulanov, Phys. Rev. Lett. **103**, 194803 (2009).
42. K. Schmid, A. Buck, C. M. S. Sears, J. M. Mikhailova, R. Tautz, D. Herrmann, M. Geissler, F. Krausz, and L. Veisz, Phys. Rev. ST Accel. Beams **13**, 091301 (2010).
43. A. J. Gonsalves, K. Nakamura, C. Lin, D. Panasenko, S. Shiraishi, T. Sokollik, C. Benedetti, C. B. Schroeder, C. G. R. Geddes, J. van Tilborg, J. Osterhoff, E. Esarey, C. Toth, and W. P. Leemans, Nat. Phys. **7**, 862 (2011).
44. Y. Ehrlich, C. Cohen, A. Zigler, J. Krall, P. Sprangle, and E. Esarey, Phys. Rev. Lett. **77**, 4186 (1996).
45. N. M. Naumova, J. Koga, K. Nakajima, T. Tajima, T. Zh. Esirkepov, S. V. Bulanov, and F. Pegoraro, Phys. Plasma **8**, 4149 (2001).
46. L. M. Chen, H. Kotaki, K. Nakajima, J. Koga, S. V. Bulanov, T. Tajima, Y. Q. Gu, H. S. Peng, X. X. Wang, T. S. Wen, H. J. Liu, C. Y. Jiao, C. G. Zhang, X. J. Huang, Y. Guo, K. N. Zhou, J. F. Hua, W. M. An, C. X. Tang, and Y. Z. Lin, Phys. Plasmas **14**, 040703 (2007).
47. R. F. Hubbard, D. Kaganovich, B. Hafizi, C. I. Moore, P. Sprangle, A. Ting, and A. Zigler, Phys. Rev. E **63**, 036502 (2001).
48. B. Hafizi, A. Ting, P. Sprangle, and R. F. Hubbard, Phys. Rev. E **62**, 4120 (2000).
49. B. Hafizi, A. Ting, R. F. Hubbard, P. Sprangle, and J. R. Penano, Phys. Plasmas **10**, 1483 (2003).
50. C. Chiu, S. Cheshkov, and T. Tajima, Phys. Rev. ST Accel. Beams **3**, 101301 (2000).
51. S. Kneip, S. R. Nagel, S. F. Martins, S. P. D. Mangles, C. Bellei, O. Chekhlov, R. J. Clarke, N. Delerue, E. J. Divall, G. Doucas, K. Ertel, F. Fiuza, R. Fonseca, P. Foster, S. J. Hawkes, C. J. Hooker, K. Krushelnick, W. B. Mori, C. A. J. Palmer, K. Ta Phuoc, P. P. Rajeev, J. Schreiber, M. J. V. Streeter, D. Urner, J. Vieira, L. O. Silva, and Z. Najmudin, Phys. Rev. Lett. **103**, 035002 (2009).
52. D. H. Froula, C. E. Clayton, T. Döppner, K. A. Marsh, C. P. J. Barty, L. Divol, R. A. Fonseca, S. H. Glenzer, C. Joshi, W. Lu, S. F. Martins, P. Michel, W. B. Mori, J. P. Palastro, B. B. Pollock, A. Pak, J. E. Ralph, J. S. Ross, C.W. Siders, L. O. Silva, and T. Wang, Phys. Rev. Lett. **103**, 215006 (2009).
53. T. P. A. Ibbotson, N. Bourgeois, T. P. Rowlands-Rees, L. S. Caballero, S. I. Bajlekov, P. A. Walker, S. Kneip, S. P. D. Mangles, S. R. Nagel, C. A. J. Palmer, N. Delerue, G. Doucas, D. Urner, O. Chekhlov, R. J. Clarke, E. Divall, K. Ertel, P. S. Foster, S. J. Hawkes, C. J. Hooker, B. Parry, P. P. Rajeev, M. J. V. Streeter, and S. M. Hooker, Phys. Rev. ST Accel. Beams **13**, 031301 (2010).

33

Output temporal contrast simulation of a large aperture high power short pulse laser system

Ping Zhu[1,2], Xinglong Xie[1], Xiaoping Ouyang[1], and Jianqiang Zhu[1]

[1] *Joint Laboratory on High Power Laser and Physics, Shanghai Institute of Optics and Fine Mechanics, Chinese Academy of Sciences, Shanghai, China*

[2] *University of Chinese Academy of Sciences, Beijing, China*

Abstract

The work presented in this paper is a study of output temporal contrast degradation by near-field quality deterioration, such as intensity modulation and wavefront deviation, in a large aperture high power short pulse laser system. A two-step focusing algorithm with a coordinate transform based on the Fresnel approximation in the space domain is used for simulating the output focused by an off-axis parabolic mirror. The temporal contrast degradation by intensity modulation and wavefront deviation is analyzed and the influence of the diameter on the temporal contrast degradation is revealed. The simulation and assumption results based on the parameters of the Shen Guang-II laser system are compared with the online experimental temporal contrast data. The near-field quality deterioration might lead to temporal contrast degradation, hindering higher temporal contrast in large aperture high power short pulse laser systems.

Keywords: fill factor of laser beam; temporal contrast; ultrafast optics; wavefront deviation

1. Introduction

With the development of chirped-pulse amplification (CPA) and optical parametric CPA (OPCPA), extremely intense short pulse laser systems are increasingly applied to experimental investigations in 'fast ignition' inertial confinement fusion (ICF), laser–plasma interactions and strong field physics[1]. The European Extreme Light Infrastructure (ELI) has designed a 25 PW, 10^{25} W cm^{-2} single laser beamline (ILE)[2], in which the high temporal contrast required by pre-plasma-free interactions[3] will be one of the major challenges. Various factors resulting in temporal contrast degradation have been analyzed and verified, such as gain saturation[4], asymmetric spectral clipping[5], self-phase modulation [6], amplified spontaneous emission (ASE) in the CPA system[7] and pump distortion-induced noise (PDN)[8], surface-reflection-initiated pre-pulses[9] and parametric fluorescence (PF)[10] in the OPCPA system, etc.

High energy laser systems usually enlarge the beam diameter to reduce the energy density on optical elements to mitigate the limited damage threshold problem, which inevitably leads to worse near-field quality, such as near-field intensity modulation and wavefront deviation. Means have existed for some time to analyze the influence on spatial output characteristics by wavefront deviation[11]. However, the discussed temporal characteristics in the wavefront deviation situation only cover the pulse width[12], leaving the temporal contrast degradation by intensity modulation and wavefront deviation unclear.

In large aperture high power short pulse laser systems, the output is focused by an off-axis parabolic mirror (OAP), and this focusing propagation can be well simulated by a two-step coordinate transformed focusing method, not only taking advantage of using coordinate transformation but also avoiding the contradiction between the space scale and the diffraction limit[13]. Using this method, this paper simulates the output temporal contrast of a large aperture high power short pulse laser system. The temporal contrast degradation by intensity modulation and wavefront deviation is analyzed, the influence of the diameter on the temporal contrast degradation is revealed and the simulation and assumption results based on the parameters of the Shen Guang-II (SG-II) laser system are compared with the online experimental temporal contrast data. The conclusion can be drawn that the near-field quality is one of the possible factors hindering higher temporal contrast in large aperture high power short pulse laser systems.

Correspondence to: P. Zhu, Shanghai Institute of Optics and Fine Mechanics, Chinese Academy of Sciences, No. 390, Qinghe Road, Jiading District, Shanghai 201800, China. Email: zhp1990@siom.ac.cn

2. Theory and simulation

2.1. Model

In practical short pulse applications, the amplified pulse by CPA or OPCPA will be focused onto the target using an OAP after being compressed by a grating compressor. The analogy can be made between the OAP focusing subsystem and an ideal lens, because it linearly reflects propagation in free space without dispersion, chromatic aberration or B-integral nonlinear effects, the spatial intensity distribution of which can be calculated by the Fresnel diffraction integral formula:

$$\begin{aligned} U(x, y, L, \omega) &= \frac{\exp(ikL)}{i\lambda L} \iint_{-\infty}^{\infty} U_0(x_0, y_0, 0) \exp\left(-ik\frac{x^2+y^2}{2f}\right) \\ &\times \exp\left\{\frac{ik}{2L}[(x-x_0)^2 + (y-y_0)^2]\right\} dx_0 dy_0, \end{aligned} \quad (1)$$

where x and y are the coordinates in the observation plane ($z = L$), x_0 and y_0 are the coordinates in the incident plane, $\lambda = 2\pi c/\omega$ is the wavelength of the incident pulse, $k = 2\pi/\lambda$ is the wavevector and f is the focal length of the OAP.

For higher spatial resolution and a more accurate result, on applying the coordinate transforms $x' = x/(1 - L/f)$, $y' = y/(1 - L/f)$ and $z' = z/(1 - L/f)$, Equation (1) becomes

$$\begin{aligned} U'\left(x', y', \frac{fL}{f-L}\right) &= \frac{\exp\left(ik\frac{fL}{f-L}\right)}{i\lambda\frac{fL}{f-L}} \iint_{-\infty}^{\infty} U_0(x_0, y_0, 0) \\ &\times \exp\left\{\frac{ik}{2\frac{fL}{f-L}}[(x'-x_0)^2 + (y'-y_0)^2]\right\} dx_0 dy_0 \\ &= F_{\text{space}}^{-1}\left\{H\left(f_X, f_Y, \frac{fL}{f-L}\right) \cdot F_{\text{space}}[U_0(x_0, y_0, 0)]\right\}, \end{aligned} \quad (2)$$

$$\begin{aligned} H(f_X, f_Y, z) &= \exp(ikz)\exp[-i\pi\lambda z(f_X^2 + f_Y^2)] \quad \text{(Fresnel diffraction)}, \\ &= \exp\left[i2\pi\frac{z}{\lambda}\sqrt{1 - (\lambda f_X)^2 - (\lambda f_Y)^2}\right] \quad \text{(angular spectrum diffraction)}. \end{aligned} \quad (3)$$

Here, $f_X = x/\lambda z$ and $f_Y = y/\lambda z$ are the spatial frequencies in the x and y directions, respectively, and the spatial Fourier transform and its inverse form are introduced to reduce the computing time by the fast Fourier transform (FFT) algorithm:

$$F_{\text{space}}(u) = \iint u(x, y)\exp[-i2\pi(f_X x + f_Y y)]dxdy,$$

$$F_{\text{space}}^{-1}(U) = \iint U(f_X, f_Y)\exp[i2\pi(f_X x + f_Y y)]df_X df_Y. \quad (4)$$

When Equation (2) is applied to focusing propagation ($L = f$), a problem of infinite x', y' and z' is encountered, which can be ingeniously resolved by a two-step simulation, first calculating from $z = 0$ to $z = L$ with coordinate transforms and finally calculating from $z = L$ to $z = f$ without coordinate transforms, in which angular spectrum diffraction theory is applied. The value of L is related to the coordinate magnification, which determines whether there are sufficient effective sampling points in the focal spot area. At the same time, the value of L determines whether the situation meets the requirement for the Fresnel approximation in step one with coordinate transforms. With this method in two spatial dimensions, the calculation of the focusing characteristics of a short wavelength laser beam through a short focus system has been successfully made[13].

Thus, we can use an operator P_{space} to represent the above focusing calculation:

$$U(x, y, z, \omega) = P_{\text{space}}[U(x, y, 0, \omega)]. \quad (5)$$

The output field as a function of time is given by the inverse time Fourier transform:

$$\begin{aligned} U(x, y, z, t) &= F_{\text{time}}^{-1}[U(x, y, z, \omega)] \\ &= \int_{-\infty}^{\infty} P_{\text{space}}[U(x, y, 0, \omega)]\exp(i\omega t)d\omega. \end{aligned} \quad (6)$$

Then, the temporal contrast can be derived by the temporal distribution of the normalized pulse intensity. In this paper, simulation of one dimension in space is used to demonstrate the influence of spatial near-field quality on temporal contrast. The idea of the two-step coordinate transformed space–time propagation simulation is shown in Figure 1.

2.2. Near-field intensity modulation

In large aperture high power laser systems, light beams are always shaped into flattened Gaussian beams (FGBs) for higher energy extraction efficiency and slower nonlinear growth. A FGB of order N and spot size ω_0 has the following normalized field distribution[14]:

$$I_N(r) = \exp\left[-\frac{(N+1)r^2}{\omega_0^2}\right]\sum_{n=0}^{N}\frac{1}{n!}\left(\frac{\sqrt{N+1}r}{\omega_0}\right)^{2n}, \quad (7)$$

where r is the radial coordinate.

In practical applications, the real near-field intensity is always modulated rather than an ideal FGB due to multiple causes, such as diffraction and nonlinear growth. The fill factor (FF) of the laser beam is usually used to describe

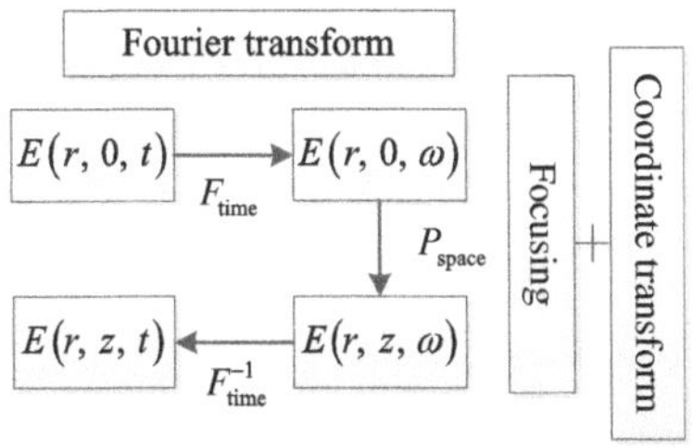

Figure 1. The simulation algorithm of time and space focusing.

the uniformity of the modulated near-field intensity, which is defined as the ratio of the average intensity and the peak intensity in the flattened area:

$$\mathrm{FF} = I_{\mathrm{avg}}/I_{\mathrm{max}}. \tag{8}$$

2.3. Wavefront deviation

The ideal incident wavefront is plane; however, it can be easily deviated by phase shifts from the optical system and optical elements. Phase shifts from the optical system, such as thermal deformation and inhomogeneity of the medium, mainly contribute to dynamic wavefront deviation on a large spatial scale, which can be detected online by a Shack–Hartmann wavefront sensor. According to research carried out at Lawrence Livermore National Laboratory[15, 16], the phase shifts from optical elements, as the main component of static wavefront deviation, can be divided into three spatial frequency regions: figure (0–0.03 mm^{-1}), waviness (0.03–8 mm^{-1}) and roughness (higher than 8 mm^{-1}), which can be expressed by the improved noise model[17]:

$$\phi(x, y) = \sum C_i \cdot R_i(-1,1) * \exp\left\{-\left[\left(\frac{x}{s_{ix}}\right)^2 + \left(\frac{y}{s_{iy}}\right)^2\right]\right\}, \tag{9}$$

where C_i is the noise magnitude, $R_i(-1, 1)$ is a random function distributed between -1 and 1, $*$ represents a convolution operation, s_{ix} and s_{iy} are the spatial scales in the x and y directions, respectively, determining the noise spatial length. The total wavefront deviation is the sum of noise of every spatial length. Common evaluation indices of wavefront deviation are the peak to valley (PV) value, root mean square (RMS), RMS gradient (GRMS) and power spectral density (PSD) curve.

3. Results and discussion

The parameters of the analyzed large aperture high power short pulse model are those of the SG-II laser system[18]. A short pulse of 290 mm near-field diameter (ninth-order FGB) and 5 ps pulse width (Gaussian) at a center wavelength of 1053 nm is focused by an OAP of 800 mm focal length and 320 mm diameter. The two-step coordinate transformed focusing method is used, while the computer samples 2048 points in space from -160 to 160 mm and 2048 points in time from -100 to 100 ps. The value of L is the key parameter of the simulation due to the two reasons mentioned above. Here, we use $L = 0.9997f$–$0.9999f$ to reach the requirements of both high resolution and the Fresnel approximation, then we obtain the convergence results.

3.1. Temporal contrast degradation by intensity modulation

First, the influence of intensity modulation on temporal contrast is analyzed. The model applied here is a ninth-order FGB with a diameter of 290 mm, and its near-field intensity is modulated randomly in three situations for which the FFs are 0.4, 0.6 and 0.8, respectively, as shown in Figure 2(a). The temporal profile of the incident pulse is Gaussian type with a pulse width (FWHM) of 5 ps. No wavefront deviation is introduced during the focusing propagation.

Figure 2(b) shows the temporal contrast degradation by intensity modulation with different FFs. If there is no intensity modulation, the temporal contrast is better than 10^9 (blue dashed line). A pre-pulse can be seen 50 ps before the main pulse when the near-field intensity is modulated, which leads to a poor temporal contrast. Moreover, the front edge of the main pulse becomes moderate 10–20 ps ahead of the peak point, which means that the pulse width of the main pulse will be broadened at the same time. When the intensity is modulated by a FF of 0.6, the temporal contrast will be degraded by more than one order of magnitude to worse than 10^8 in the time window of 50 ps before the main pulse. The tendency can be seen from this figure that the smaller the FF of the modulated intensity is the worse the output temporal contrast is.

3.2. Temporal contrast degradation by wavefront deviation

Second, using the same model as described above, the effect of wavefront deviation on temporal contrast is discussed. At this time the incident intensity has no modulation, but various types of random wavefront deviations are introduced in the focusing propagation, which have different magnitudes and different spatial frequencies. As shown in Figure 3(a), these four wavefront deviations are illustrated in the form of PSD curves.

Figure 3(b) shows the temporal contrast degradation by different types of wavefront deviations. If there is no wavefront deviation, the temporal contrast will be better than 10^9. Comparing the blue line (PV $= 0.5\lambda$) with the green line (PV $= 0.1\lambda$) in the same spatial frequency range (the same minimum spatial period T is 40 mm, which means that the same maximum spatial frequency is 0.025 mm^{-1}), the wavefront deviation with larger PV value will cause

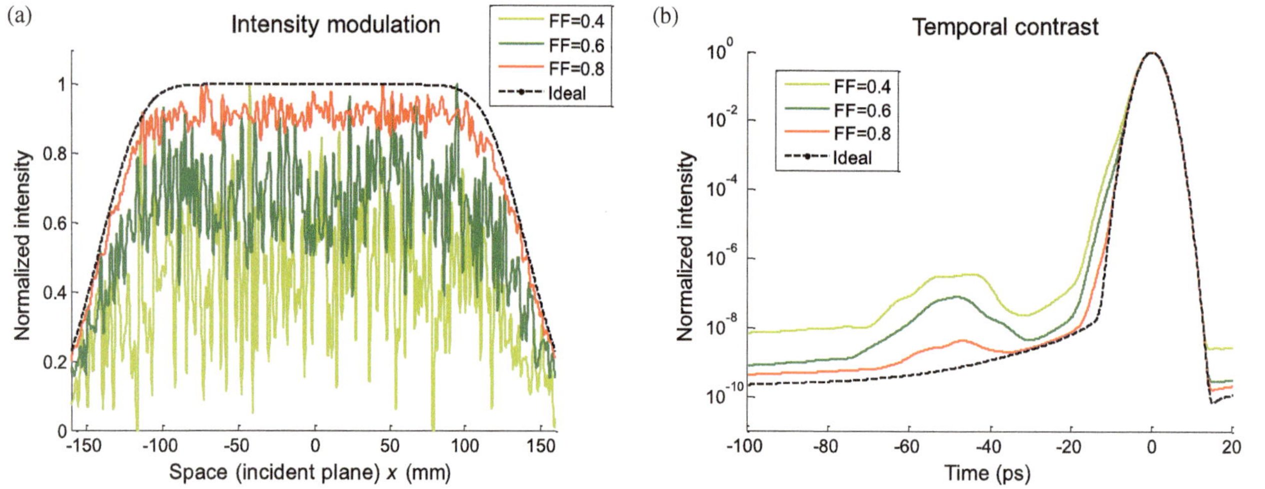

Figure 2. Temporal contrast degradation by intensity modulation. (a) Intensity modulation for different FFs (0.4, 0.6 and 0.8). (b) Temporal contrast for the different intensity modulations. The dashed line is the ideal situation without modulation.

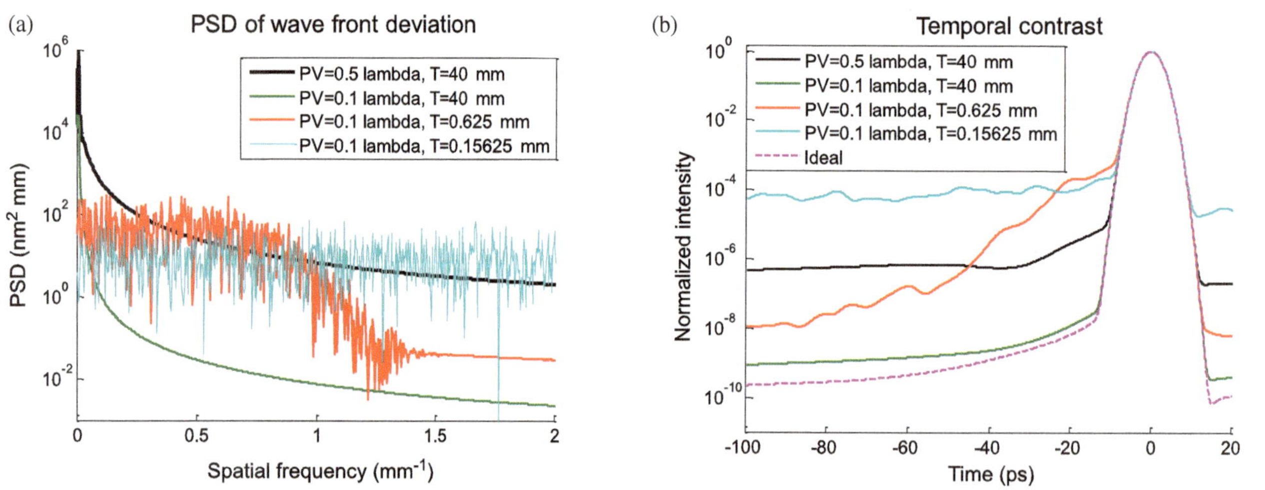

Figure 3. Temporal contrast degradation by wavefront deviation. (a) PSDs of wavefront deviations of different magnitudes and spatial frequencies. (b) Temporal contrasts for different wavefront deviations.

more severe temporal contrast degradation and in this spatial frequency range the temporal contrast will be two orders of magnitude worse than that of the ideal situation.

When the wavefront deviations have the same magnitude ($\mathrm{PV} = 0.1\lambda$) but different spatial frequency ranges, as shown in Figure 3(b), there are large differences between the temporal contrast degradations. The wavefront deviation with the highest maximum spatial frequency of 6.4 mm^{-1} (cyan line; minimum spatial period of 0.15625 mm) causes huge degradation of more than five orders of magnitude from 10^9 to 10^4, while the wavefront deviation with the lowest maximum spatial frequency of 0.025 mm^{-1} (green line; minimum spatial period of 40 mm) induces smaller degradation of one order. Interestingly, temporal contrast degradation by the wavefront deviation with the middle maximum spatial frequency of 1.6 mm^{-1} (red line; minimum spatial period of 0.625 mm) is quite notable within 40 ps before the main pulse; however, it becomes less significant beyond 40 ps before the main pulse. Therefore, higher spatial frequency contributes to temporal contrast degradation in a larger time window and for the same magnitude the wavefront deviation with low maximum spatial frequency has a slight influence on temporal contrast.

3.3. Influence of diameter on temporal contrast degradation

Third, the importance of discussing temporal contrast in a system with large diameter is demonstrated by analyzing the influence of diameter on temporal contrast degradation. In this case, introducing the same magnitude and same spatial frequency wavefront deviation as shown in Figures 4(a) and (b), we consider three models with different diameters: 320 (blue solid line), 160 (green solid line) and 80 mm (red solid line). The focal length is fixed at 800 mm.

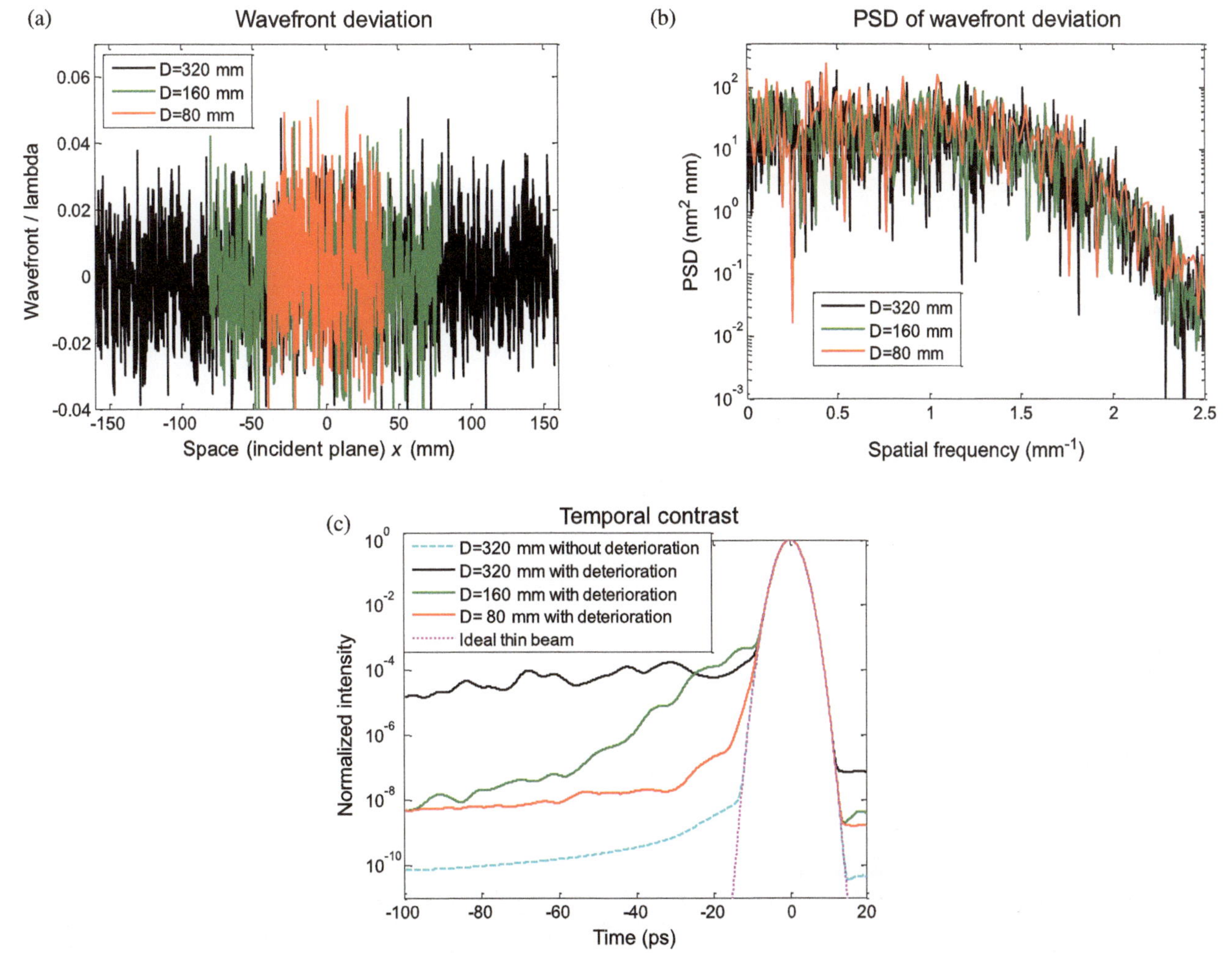

Figure 4. Influence of diameter on temporal contrast degradation. (a) Wavefront deviations for different diameters. (b) PSDs of wavefront deviations for different diameters. (c) Temporal contrasts for different diameters.

The result can be seen in Figure 4(c). The temporal contrast for 80 mm diameter is three orders of magnitude better than that for 320 mm diameter and the model with larger diameter suffers more severe temporal contrast degradation. The temporal contrast is around 10^{10} in the model with 320 mm diameter without any near-field quality deterioration (cyan dashed line). However, for an ideal thin beam that does not need to be focused, the temporal contrast will be promoted to the computer simulation resolution limit, about 10^{30} (purple dotted line). Therefore, temporal contrast degradation deserves more attention in larger aperture high power short pulse laser systems.

3.4. Simulation of temporal contrast degradation compared with experimental measurement results

Finally, the simulation model is combined with the near-field quality measured in the established SG-II large aperture high power short pulse laser system. Figure 5(a) shows the experimental near-field intensity detected in the incident plane and Figure 5(b) shows the experimental intensity modulation (green solid line) and ideal spatial profile (blue dashed line) in one dimension. The calculated FF for the experimental near-field intensity modulation is 0.73. Figure 5(c) shows the experimental dynamic wavefront deviation detected online by a Shack–Hartmann wavefront sensor, which is unable to show waviness or roughness in the intermediate and higher frequency regions. Thus, in this case, no waviness and roughness are introduced into the simulation model. The experimental dynamic wavefront deviation in one dimension and its PSD curve are illustrated as the violet lines in Figures 5(d) and (e). After the propagation through the OAP focusing subsystem, the output temporal contrast can be obtained. In Figure 5(f), the cyan dashed line is the ideal output temporal contrast; the red solid line is the online experimental measurement result and the blue solid line is the calculation result. The online experimental measurement result[19] shows a degraded temporal contrast of 10^4, while the calculation result shows a temporal contrast degradation of less than two orders of magnitude from 10^9. There are two

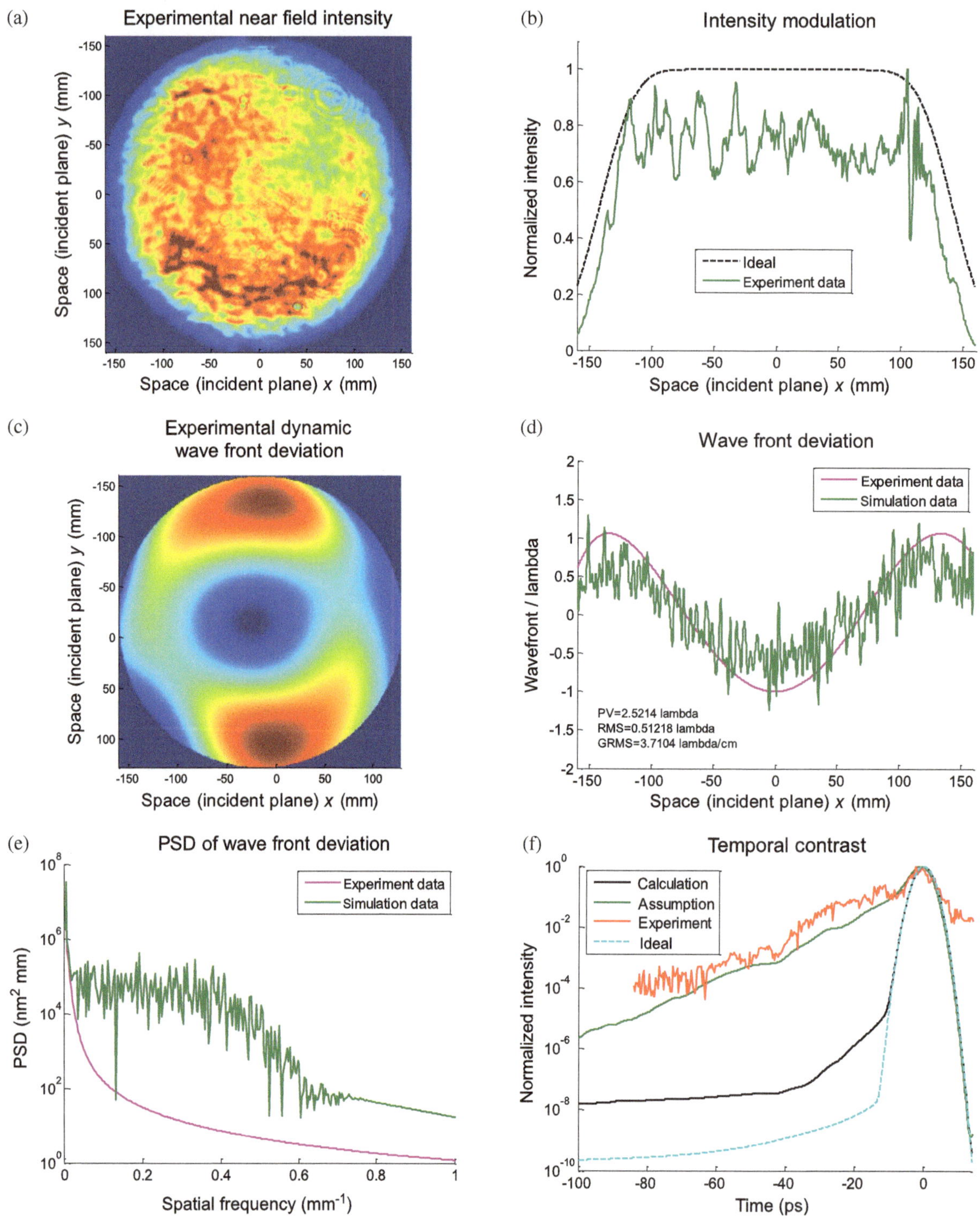

Figure 5. Experimental and simulated intensity modulation, wavefront deviation and temporal contrast degradation. (a) Experimental near-field intensity. (b) Intensity modulation. (c) Experimental dynamic wavefront deviation. (d) Wavefront deviation. (e) PSD of wavefront deviation. (f) Temporal contrast.

reasons why the two results differ. (1) In this model beam quality is considered as the only influencing factor; however, in real experiments there are many other characteristics, such as ASE, PF and spectral modulation. (2) Only near-field intensity and low spatial frequency wavefront deviation are introduced, without considering waviness and roughness.

To introduce the waviness and roughness into the simulation model, an assumption is made based on the SG-II short pulse laser system configuration. In consideration of the cutoff frequency of the spatial filter, a wavefront deviation of waviness and roughness whose maximum spatial frequency is $0.6\ \mathrm{mm}^{-1}$ is added, as shown in Figure 5(d) by the green line. The PV of total wavefront deviation is 2.5λ, the RMS is 0.5λ, the GRMS is 3.7λ cm^{-1} and the PSD curve is shown in Figure 5(e) by the green line. These values are gradually adjusted since there are no online measurement data for the system. After considering the waviness and roughness, there is serious temporal contrast degradation of more than three orders of magnitude from the temporal contrast degradation level due to only intensity modulation and low spatial frequency wavefront deviation. In addition, the assumption result has a similar trend to the online experimental measurement result, in spite of the discrepancy due to some other factors. From this point of view, high spatial frequency wavefront deviations, such as waviness and roughness, influence the temporal contrast more seriously; therefore, waviness and roughness deserve careful attention, and methods of measurement and control of the wavefront deviation within the abovementioned spatial frequency region should be investigated further.

4. Conclusion

In this paper, a two-step coordinate transformed focusing method is used to simulate the output focused by an OAP in a large aperture high power short pulse laser system, and thus the output temporal contrast is calculated. The temporal contrast degradation by intensity modulation and wavefront deviation is analyzed. It is found that intensity modulation with a smaller FF and larger wavefront deviation causes more severe temporal contrast degradation, and this degradation is closely related to the spatial frequency of the total wavefront deviation. Then, the importance of discussing temporal contrast in a system with large diameter is demonstrated by analyzing the influence of diameter on temporal contrast degradation. Finally, with experimental intensity modulation, experimental low spatial frequency wavefront deviation and assumed higher spatial frequency wavefront deviation using the parameters of the SG-II laser system, the output temporal contrast is derived, which shows that waviness and roughness influence the temporal contrast more seriously. Therefore, the conclusion can be drawn that near-field quality deterioration might lead to temporal contrast degradation, hindering higher temporal contrast in large aperture high power short pulse laser systems.

Due to the lack of effective tools to measure intermediate and higher spatial frequency wavefront deviation, the need for online monitoring of the wavefront deviation in these frequency regions in large aperture high power short pulse laser systems is put forward. Improvement of the laser near-field quality might be a new method towards high temporal contrast in large aperture high power short pulse laser systems.

Acknowledgement

The presented work is supported by the National High-Tech Committee of China (Grant no. 2013AA8044010).

References

1. H. Kiriyama, T. Shimomura, M. Mori, Y. Nakai, M. Tanoue, S. Kondo, S. Kanazawa, A. S. Pirozhkov, T. Z. Esirkepov, Y. Hayashi, K. Ogura, H. Kotaki, M. Suzuki, I. Daito, H. Okada, A. Kosuge, Y. Fukuda, M. Nishiuchi, M. Kando, S. V. Bulanov, K. Nagashima, M. Yamagiwa, K. Kondo, A. Sugiyama, P. R. Bolton, S. Matsuoka, and H. Kan, Appl. Sci. **3**, 214 (2013).
2. J.-P. Chambaret, F. Canova, R. L. Martens, G. Cheriaux, G. Mourou, A. Cotel, C. Le Blanc, F. Druon, P. Georges, N. Forget, F. Ple, and M. Pittman, in *Proceedings of CLEO/Quantum Electronics and Laser Science Conference and Photonic Applications Systems Technologies* JWC4 (2007).
3. M. Kaluza, J. Schreiber, M. I. K. Santala, G. D. Tsakiris, K. Eidmann, J. Meyer-ter-Vehn, and K. J. Witte, Phys. Rev. Lett. **93**, 045003 (2004).
4. C. Li, Z. Zhang, and Z. Xu, Acta Opt. Sin. **16**, 299 (1996).
5. Q. Yang, A. Guo, X. Xie, L. Zhang, M. Sun, Q. Gao, M. Li, and Z. Lin, Chin. J. Lasers **35**, 1970 (2008).
6. M. Kalashnikov, A. Andreev, and H. Schönnagel, Proc. SPIE **7501**, 750104 (2009).
7. M. P. Kalashnikov, E. Risse, H. Schönnagel, and W. Sandner, Opt. Lett. **30**, 923 (2005).
8. N. Forget, A. Cotel, E. Brambrink, P. Audebert, C. Le Blanc, A. Jullien, and G. Chériaux, Opt. Lett. **30**, 2921 (2005).
9. J. Wang, P. Yuan, J. Ma, Y. Wang, G. Xie, and L. Qian, Opt. Express **21**, 15580 (2013).
10. C. Manzoni, J. Moses, F. X. Kärtner, and G. Cerullo, Opt. Express **19**, 8357 (2011).
11. A. Guo, H. Zhu, and Z. Yang, Acta Opt. Sin. **33**, 0214001 (2013).
12. Q. Liu, Z. Cen, and X. Li, Acta Opt. Sin. **33**, 0132001 (2013).
13. M. Yang, M. Zhong, and G. Ren, Acta Opt. Sin. **31**, 0507001 (2011).
14. V. Bagini, R. Borghi, F. Gori, A. M. Pacileo, M. Santarsiero, D. Ambrosini, and G. S. Spagnolo, J. Opt. Soc. Am. A **13**, 1385 (1996).
15. C. R. Wolfe and J. K. Lawson, Proc. SPIE **2633**, 361 (1995).
16. J. K. Lawson, J. M. Auerbach, and R. E. English, Jr., Proc. SPIE **3492**, 336 (1999).
17. Y. H. Chen, W. G. Zheng, W. J. Chen, S. B. He, J. Q. Su, Y. B. Chen, and J. Yuan, High Power Laser Part. Beams **17**, 403 (2005).
18. G. Xu, T. Wang, Z. Li, Y. Dai, Z. Lin, Y. Gu, and J. Zhu, Rev. Laser Eng. **36**, 1172 (2008).
19. Y. Wang, X. Ouyang, J. Ma, P. Yuan, G. Xu, and L. Qian, Chin. Phys. Lett. **30**, 024201 (2013).

Permissions

All chapters in this book were first published in HPLSE, by Cambridge Journals; hereby published with permission under the Creative Commons Attribution License or equivalent. Every chapter published in this book has been scrutinized by our experts. Their significance has been extensively debated. The topics covered herein carry significant findings which will fuel the growth of the discipline. They may even be implemented as practical applications or may be referred to as a beginning point for another development.

The contributors of this book come from diverse backgrounds, making this book a truly international effort. This book will bring forth new frontiers with its revolutionizing research information and detailed analysis of the nascent developments around the world.

We would like to thank all the contributing authors for lending their expertise to make the book truly unique. They have played a crucial role in the development of this book. Without their invaluable contributions this book wouldn't have been possible. They have made vital efforts to compile up to date information on the varied aspects of this subject to make this book a valuable addition to the collection of many professionals and students.

This book was conceptualized with the vision of imparting up-to-date information and advanced data in this field. To ensure the same, a matchless editorial board was set up. Every individual on the board went through rigorous rounds of assessment to prove their worth. After which they invested a large part of their time researching and compiling the most relevant data for our readers.

The editorial board has been involved in producing this book since its inception. They have spent rigorous hours researching and exploring the diverse topics which have resulted in the successful publishing of this book. They have passed on their knowledge of decades through this book. To expedite this challenging task, the publisher supported the team at every step. A small team of assistant editors was also appointed to further simplify the editing procedure and attain best results for the readers.

Apart from the editorial board, the designing team has also invested a significant amount of their time in understanding the subject and creating the most relevant covers. They scrutinized every image to scout for the most suitable representation of the subject and create an appropriate cover for the book.

The publishing team has been an ardent support to the editorial, designing and production team. Their endless efforts to recruit the best for this project, has resulted in the accomplishment of this book. They are a veteran in the field of academics and their pool of knowledge is as vast as their experience in printing. Their expertise and guidance has proved useful at every step. Their uncompromising quality standards have made this book an exceptional effort. Their encouragement from time to time has been an inspiration for everyone.

The publisher and the editorial board hope that this book will prove to be a valuable piece of knowledge for researchers, students, practitioners and scholars across the globe.

List of Contributors

Chun Li
Beijing National Laboratory of Condensed Matter Physics, Institute of Physics, CAS, Beijing 100190, PR China

Guo-Qian Liao
Beijing National Laboratory of Condensed Matter Physics, Institute of Physics, CAS, Beijing 100190, PR China

Yu-Tong Li
Beijing National Laboratory of Condensed Matter Physics, Institute of Physics, CAS, Beijing 100190, PR China
IFSA Collaborative Innovation Center, Shanghai Jiao Tong University, Shanghai 200240, PR China

Lei Ren, Dongfeng Zhao
National Laboratory on High Power Laser and Physics, Shanghai Institute of Optics and Fine Mechanics, Chinese Academy of Science, Shanghai 201800, China

Jianqiang Zhu
National Laboratory on High Power Laser and Physics, Shanghai Institute of Optics and Fine Mechanics, Chinese Academy of Science, Shanghai 201800, China

S. Weber
Institute of Physics of the ASCR, ELI-Beamlines, 18221 Prague, Czech Republic

C. Riconda
LULI, Sorbonne Universit´es - Universit´e Pierre et Marie Curie - Ecole Polytechnique - CNRS - CEA, 75005 Paris, France

Guohang Hu
Key Laboratory of Materials for High Power Laser, Shanghai Institute of Optics and Fine Mechanics, Chinese Academy of Sciences, Shanghai 201800, China
University of Chinese Academy of Sciences, Beijing 100049, China

Yueliang Wang
Key Laboratory of Materials for High Power Laser, Shanghai Institute of Optics and Fine Mechanics, Chinese Academy of Sciences, Shanghai 201800, China
University of Chinese Academy of Sciences, Beijing 100049, China

Junxiu Chang
Key Laboratory of Materials for High Power Laser, Shanghai Institute of Optics and Fine Mechanics, Chinese Academy of Sciences, Shanghai 201800, China
University of Chinese Academy of Sciences, Beijing 100049, China

Xiaoyi Xie
Key Laboratory of Materials for High Power Laser, Shanghai Institute of Optics and Fine Mechanics, Chinese Academy of Sciences, Shanghai 201800, China
University of Chinese Academy of Sciences, Beijing 100049, China

Yuanan Zhao
Key Laboratory of Materials for High Power Laser, Shanghai Institute of Optics and Fine Mechanics, Chinese Academy of Sciences, Shanghai 201800, China

Hongji Qi
Key Laboratory of Materials for High Power Laser, Shanghai Institute of Optics and Fine Mechanics, Chinese Academy of Sciences, Shanghai 201800, China

Jianda Shao
Key Laboratory of Materials for High Power Laser, Shanghai Institute of Optics and Fine Mechanics, Chinese Academy of Sciences, Shanghai 201800, China

Baifei Shen
State Key Laboratory of High Field Laser Physics, Shanghai Institute of Optics and Fine Mechanics, Chinese cademy of Sciences, Shanghai 201800, China

Xueyan Zhao
State Key Laboratory of High Field Laser Physics, Shanghai Institute of Optics and Fine Mechanics, Chinese cademy of Sciences, Shanghai 201800, China

Longqing Yi
State Key Laboratory of High Field Laser Physics, Shanghai Institute of Optics and Fine Mechanics, Chinese cademy of Sciences, Shanghai 201800, China

Wei Yu
State Key Laboratory of High Field Laser Physics, Shanghai Institute of Optics and Fine Mechanics, Chinese cademy of Sciences, Shanghai 201800, China

Zhizhan Xu
State Key Laboratory of High Field Laser Physics, Shanghai Institute of Optics and Fine Mechanics, Chinese cademy of Sciences, Shanghai 201800, China

Peng Yuan
Key Laboratory for Laser Plasma (Ministry of Education) and Department of Physics and Astronomy, Shanghai Jiao Tong University, Shanghai 200240, China
Department of Optical Science and Engineering, Fudan University, Shanghai 200433, China

Jingui Ma
Key Laboratory for Laser Plasma (Ministry of Education) and Department of Physics and Astronomy, Shanghai Jiao Tong University, Shanghai 200240, China

Yongzhi Wang
Department of Optical Science and Engineering, Fudan University, Shanghai 200433, China

Jing Wang
Key Laboratory for Laser Plasma (Ministry of Education) and Department of Physics and Astronomy, Shanghai Jiao Tong University, Shanghai 200240, China

Daolong Tang
Key Laboratory for Laser Plasma (Ministry of Education) and Department of Physics and Astronomy, Shanghai Jiao Tong University, Shanghai 200240, China

Liejia Qian
Key Laboratory for Laser Plasma (Ministry of Education) and Department of Physics and Astronomy, Shanghai Jiao Tong University, Shanghai 200240, China

C.B. Edwards
HiPER Project, STFC Rutherford Appleton Laboratory, UK
AWE plc, Aldermaston, UK

C.N. Danson
AWE plc, Aldermaston, UK

Yangshuai Li
National Laboratory on High Power Laser and Physics, Shanghai Institute of Optics and Fine Mechanics, Chinese Academy of Sciences, Shanghai 201800, China
University of Chinese Academy of Sciences, Beijing 100049, China

Jianqiang Zhu
National Laboratory on High Power Laser and Physics, Shanghai Institute of Optics and Fine Mechanics, Chinese Academy of Sciences, Shanghai 201800, China

Xiangyang Pang
National Laboratory on High Power Laser and Physics, Shanghai Institute of Optics and Fine Mechanics, Chinese Academy of Sciences, Shanghai 201800, China

Hua Tao
National Laboratory on High Power Laser and Physics, Shanghai Institute of Optics and Fine Mechanics, Chinese Academy of Sciences, Shanghai 201800, China
University of Chinese Academy of Sciences, Beijing 100049, China

Xiang Jiao
National Laboratory on High Power Laser and Physics, Shanghai Institute of Optics and Fine Mechanics, Chinese Academy of Sciences, Shanghai 201800, China
University of Chinese Academy of Sciences, Beijing 100049, China

Yongzhong Wu
National Laboratory on High Power Laser and Physics, Shanghai Institute of Optics and Fine Mechanics, Chinese Academy of Sciences, Shanghai 201800, China
University of Chinese Academy of Sciences, Beijing 100049, China

Xiong Wang
College of Optoelectronic Science and Engineering, National University of Defense Technology, Changsha 410073, China

Pu Zhou
College of Optoelectronic Science and Engineering, National University of Defense Technology, Changsha 410073, China

Xiaolin Wang
College of Optoelectronic Science and Engineering, National University of Defense Technology, Changsha 410073, China

Hu Xiao
College of Optoelectronic Science and Engineering, National University of Defense Technology, Changsha 410073, China

Lei Si
College of Optoelectronic Science and Engineering, National University of Defense Technology, Changsha 410073, China

Hao Yang
School of Physics and Electronic Engineering, Jiangsu Normal University, Xuzhou 221116, China

Jian Zhang
School of Physics and Electronic Engineering, Jiangsu Normal University, Xuzhou 221116, China

Dewei Luo
School of Electrical and Electronic Engineering, Nanyang Technological University, Singapore 639798, Singapore

Hui Lin
School of Electrical and Electronic Engineering, Nanyang Technological University, Singapore 639798, Singapore

Deyuan Shen
School of Physics and Electronic Engineering, Jiangsu Normal University, Xuzhou 221116, China

Dingyuan Tang
School of Physics and Electronic Engineering, Jiangsu Normal University, Xuzhou 221116, China
School of Electrical and Electronic Engineering, Nanyang Technological University, Singapore 639798, Singapore

Dongfeng Zhao
Shanghai Institute of Optics and Fine Mechanics, Chinese Academy of Sciences, No. 390, Qinghe Road,
Jiading District, Shanghai 201800, China
University of Chinese Academy of Sciences, No. 19A Yuquan Road, Beijing 100049, China

Li Wan
Shanghai Institute of Optics and Fine Mechanics, Chinese Academy of Sciences, No. 390, Qinghe Road,
Jiading District, Shanghai 201800, China

Zunqi Lin
Shanghai Institute of Optics and Fine Mechanics, Chinese Academy of Sciences, No. 390, Qinghe Road,
Jiading District, Shanghai 201800, China

Pin Shao
Shanghai Institute of Optics and Fine Mechanics, Chinese Academy of Sciences, No. 390, Qinghe Road,
Jiading District, Shanghai 201800, China

Jianqiang Zhu
Shanghai Institute of Optics and Fine Mechanics, Chinese Academy of Sciences, No. 390, Qinghe Road,
Jiading District, Shanghai 201800, China

Peiying Zeng
National Laboratory on High Power Laser and Physics, Shanghai Institute of Optics and Fine Mechanics, Chinese Academy of Sciences, Shanghai 201800, China

Baoqiang
National Laboratory on High Power Laser and Physics, Shanghai Institute of Optics and Fine Mechanics, Chinese Academy of Sciences, Shanghai 201800, China

Déan Liu Zhu
National Laboratory on High Power Laser and Physics, Shanghai Institute of Optics and Fine Mechanics, Chinese Academy of Sciences, Shanghai 201800, China

Jianqiang Zhu
National Laboratory on High Power Laser and Physics, Shanghai Institute of Optics and Fine Mechanics, Chinese Academy of Sciences, Shanghai 201800, China

Yanqi Liu
State Key Laboratory of High Field Laser Physics, Shanghai Institute of Optics and Fine Mechanics, Chinese Academy of Sciences, Shanghai 201800, China

Yuxin Leng
State Key Laboratory of High Field Laser Physics, Shanghai Institute of Optics and Fine Mechanics, Chinese Academy of Sciences, Shanghai 201800, China

Xiaoming
State Key Laboratory of High Field Laser Physics, Shanghai Institute of Optics and Fine Mechanics, Chinese Academy of Sciences, Shanghai 201800, China

Yi Xu Lu
State Key Laboratory of High Field Laser Physics, Shanghai Institute of Optics and Fine Mechanics, Chinese Academy of Sciences, Shanghai 201800, China

Cheng Wang
State Key Laboratory of High Field Laser Physics, Shanghai Institute of Optics and Fine Mechanics, Chinese Academy of Sciences, Shanghai 201800, China

X.H. Yuan
Key Laboratory for Laser Plasmas (Ministry of Education) and Department of Physics and Astronomy, Shanghai Jiao Tong University, Shanghai 200240, China
Beijing National Laboratory of Condensed Matter Physics, Institute of Physics, Chinese Academy of Sciences, Beijing 100190, China

Y. Fang
Key Laboratory for Laser Plasmas (Ministry of Education) and Department of Physics and Astronomy, Shanghai Jiao Tong University, Shanghai 200240, China

D.C. Carroll
SUPA Department of Physics, University of Strathclyde, Glasgow G4 0NG, UK
Central Laser Facility, Rutherford Appleton Laboratory, Oxfordshire OX11 0QX, UK

D.A. MacLellan
SUPA Department of Physics, University of Strathclyde, Glasgow G4 0NG, UK

F. Du
Beijing National Laboratory of Condensed Matter Physics, Institute of Physics, Chinese Academy of Sciences, Beijing 100190, China

N. Booth
Central Laser Facility, Rutherford Appleton Laboratory, Oxfordshire OX11 0QX, UK

M. Burza
Department of Physics, Lund University, S-22100 Lund, Sweden

M. Chen
Key Laboratory for Laser Plasmas (Ministry of Education) and Department of Physics and Astronomy, Shanghai Jiao Tong University, Shanghai 200240, China

R.J. Gray
SUPA Department of Physics, University of Strathclyde, Glasgow G4 0NG, UK

Y.F. Jin
Key Laboratory for Laser Plasmas (Ministry of Education) and Department of Physics and Astronomy, Shanghai Jiao Tong University, Shanghai 200240, China

Y.T. Li
Beijing National Laboratory of Condensed Matter Physics, Institute of Physics, Chinese Academy of Sciences, Beijing 100190, China

Y. Liu
Key Laboratory for Laser Plasmas (Ministry of Education) and Department of Physics and Astronomy, Shanghai Jiao Tong University, Shanghai 200240, China

D. Neely
Central Laser Facility, Rutherford Appleton Laboratory, Oxfordshire OX11 0QX, UK

H. Powell
SUPA Department of Physics, University of Strathclyde, Glasgow G4 0NG, UK

G. Scott
SUPA Department of Physics, University of Strathclyde, Glasgow G4 0NG, UK
Central Laser Facility, Rutherford Appleton Laboratory, Oxfordshire OX11 0QX, UK

C.-G. Wahlström
Department of Physics, Lund University, S-22100 Lund, Sweden

J. Zhang
Key Laboratory for Laser Plasmas (Ministry of Education) and Department of Physics and Astronomy, Shanghai Jiao Tong University, Shanghai 200240, China
Beijing National Laboratory of Condensed Matter Physics, Institute of Physics, Chinese Academy of Sciences, Beijing 100190, China

P. McKenna
SUPA Department of Physics, University of Strathclyde, Glasgow G4 0NG, UK

Z.M. Sheng
Key Laboratory for Laser Plasmas (Ministry of Education) and Department of Physics and Astronomy, Shanghai Jiao Tong University, Shanghai 200240, China
SUPA Department of Physics, University of Strathclyde, Glasgow G4 0NG, UK

Tatiana Pikuz
Quantum Beam Science Directorate, Japan Atomic Energy Agency, Kizugawa, Kyoto 619-0215, Japan
Joint Institute for High Temperatures, Russian Academy of Sciences, Moscow 125412, Russia

Anatoly Faenov
Quantum Beam Science Directorate, Japan Atomic Energy Agency, Kizugawa, Kyoto 619-0215, Japan
Joint Institute for High Temperatures, Russian Academy of Sciences, Moscow 125412, Russia
Institute for Academic Initiatives, Osaka University, Suita, Osaka, 565-0871, Japan

Sergey Magnitskiy
International Laser Center of M.V. Lomonosov Moscow State University, Moscow, Russia

Nikolay Nagorskiy
Joint Institute for High Temperatures, Russian Academy of Sciences, Moscow 125412, Russia

Momoko Tanaka
Quantum Beam Science Directorate, Japan Atomic Energy Agency, Kizugawa, Kyoto 619-0215, Japan

Masahiko Ishino
Quantum Beam Science Directorate, Japan Atomic Energy Agency, Kizugawa, Kyoto 619-0215, Japan

Masaharu Nishikino
Quantum Beam Science Directorate, Japan Atomic Energy Agency, Kizugawa, Kyoto 619-0215, Japan

Yuji Fukuda
Quantum Beam Science Directorate, Japan Atomic Energy Agency, Kizugawa, Kyoto 619-0215, Japan

Masaki Kando
Quantum Beam Science Directorate, Japan Atomic Energy Agency, Kizugawa, Kyoto 619-0215, Japan

Yoshiaki Kato
The Graduate School for the Creation of New Photonics Industries, Hamamatsu, Shizuoka 431-1202, Japan

Tetsuya Kawachi
Quantum Beam Science Directorate, Japan Atomic Energy Agency, Kizugawa, Kyoto 619-0215, Japan

H. Tomizawa
RIKEN, SPring-8 Center, Kouto 1-1-1, Sayo, Hyogo 679-5148, Japan

T. Sato
RIKEN, SPring-8 Center, Kouto 1-1-1, Sayo, Hyogo 679-5148, Japan

K. Ogawa
RIKEN, SPring-8 Center, Kouto 1-1-1, Sayo, Hyogo 679-5148, Japan

K. Togawa
RIKEN, SPring-8 Center, Kouto 1-1-1, Sayo, Hyogo 679-5148, Japan

T. Tanaka
RIKEN, SPring-8 Center, Kouto 1-1-1, Sayo, Hyogo 679-5148, Japan

T. Hara
RIKEN, SPring-8 Center, Kouto 1-1-1, Sayo, Hyogo 679-5148, Japan

M. Yabashi
RIKEN, SPring-8 Center, Kouto 1-1-1, Sayo, Hyogo 679-5148, Japan

H. Tanaka
RIKEN, SPring-8 Center, Kouto 1-1-1, Sayo, Hyogo 679-5148, Japan

T. Ishikawa
RIKEN, SPring-8 Center, Kouto 1-1-1, Sayo, Hyogo 679-5148, Japan

T. Togashi
Japan Synchrotron Radiation Research Institute, Kouto 1-1-1, Sayo, Hyogo 679-5198, Japan

S. Matsubara
Japan Synchrotron Radiation Research Institute, Kouto 1-1-1, Sayo, Hyogo 679-5198, Japan

Y. Okayasu
Japan Synchrotron Radiation Research Institute, Kouto 1-1-1, Sayo, Hyogo 679-5198, Japan

T. Watanabe
Japan Synchrotron Radiation Research Institute, Kouto 1-1-1, Sayo, Hyogo 679-5198, Japan

E.J. Takahashi
RIKEN Center for Advanced Photonics, Hirosawa 2-1, Wako, Saitama 351-0198, Japan

K. Midorikawa
RIKEN Center for Advanced Photonics, Hirosawa 2-1, Wako, Saitama 351-0198, Japan

M. Aoyama
Japan Atomic Energy Agency, Umemidai 8-1-7, Kizugawa, Kyoto 619-0215, Japan

K. Yamakawa
Japan Atomic Energy Agency, Umemidai 8-1-7, Kizugawa, Kyoto 619-0215, Japan

S. Owada
The University of Tokyo, Hongo 7-3-1, Bunkyo-ku, Tokyo 113-0033, Japan

A. Iwasaki
The University of Tokyo, Hongo 7-3-1, Bunkyo-ku, Tokyo 113-0033, Japan

K. Yamanouchi
The University of Tokyo, Hongo 7-3-1, Bunkyo-ku, Tokyo 113-0033, Japan

V.V. Semak
Signature Science, LLC, NJ 08234, USA

M.N. Shneider
Department of Mechanical and Aerospace Engineering, Princeton University, NJ 08544, USA

C. Spindloe
Target Fabrication Group, Central Laser Facility, Rutherford Appleton Laboratory, Harwell Science and Innovation Campus, Didcot OX11 0QX, UK

D. Wyatt
Target Fabrication Group, Central Laser Facility, Rutherford Appleton Laboratory, Harwell Science and Innovation Campus, Didcot OX11 0QX, UK

D. Haddock
Target Fabrication Group, Central Laser Facility, Rutherford Appleton Laboratory, Harwell Science and Innovation Campus, Didcot OX11 0QX, UK

I. East
Target Fabrication Group, Central Laser Facility, Rutherford Appleton Laboratory, Harwell Science and Innovation Campus, Didcot OX11 0QX, UK

J.E. Cross
Atomic and Laser Physics, Clarendon Laboratory, University of Oxford, OX1 3PU, UK

C.N. Danson
Atomic and Laser Physics, Clarendon Laboratory, University of Oxford, OX1 3PU, UK
AWE, Aldermaston, Reading RG7 4PR, UK

E. Falize
CEA-DAM-DIF, F-91297 Arpajon, France

J.M. Foster
Atomic and Laser Physics, Clarendon Laboratory, University of Oxford, OX1 3PU, UK AWE, Aldermaston, Reading RG7 4PR, UK

M. Koenig
Laboratoire pour l'Utilisation Des Laser Intense, UMR 7605 CNRS-CEA-Ecole Polytechnique-Universit´e Paris VI, Palaiseau, France

G. Gregori
Atomic and Laser Physics, Clarendon Laboratory, University of Oxford, OX1 3PU, UK

J.P. Zou
Laboratoire pour l'Utilisation des Lasers Intenses, CNRS, Ecole Polytechnique, Palaiseau, France

C. Le Blanc
Laboratoire pour l'Utilisation des Lasers Intenses, CNRS, Ecole Polytechnique, Palaiseau, France

D.N. Papadopoulos
Laboratoire pour l'Utilisation des Lasers Intenses, CNRS, Ecole Polytechnique, Palaiseau, France

G. Chériaux
Laboratoire d'Optique Appliqu´ee, ENSTA ParisTech, CNRS, Palaiseau, France

P. Georges
Laboratoire Charles Fabry, UMR 8501, Institut d'Optique, CNRS, Universit´e Paris Sud, Palaiseau, France

G. Mennerat
CEA, Iramis, SPAM, Saclay, France

F. Druon
Laboratoire Charles Fabry, UMR 8501, Institut d'Optique, CNRS, Universit´e Paris Sud, Palaiseau, France

L. Lecherbourg
Laboratoire pour l'Utilisation des Lasers Intenses, CNRS, Ecole Polytechnique, Palaiseau, France

A. Pellegrina
Laboratoire pour l'Utilisation des Lasers Intenses, CNRS, Ecole Polytechnique, Palaiseau, France
Laboratoire Charles Fabry, UMR 8501, Institut d'Optique, CNRS, Universit´e Paris Sud, Palaiseau, France

P. Ramirez
Laboratoire pour l'Utilisation des Lasers Intenses, CNRS, Ecole Polytechnique, Palaiseau, France
Laboratoire Charles Fabry, UMR 8501, Institut d'Optique, CNRS, Universit´e Paris Sud, Palaiseau, France

F. Giambruno
Laboratoire pour l'Utilisation des Lasers Intenses, CNRS, Ecole Polytechnique, Palaiseau, France
Laboratoire d'Optique Appliqu´ee, ENSTA ParisTech, CNRS, Palaiseau, France

A. Fréneaux
Laboratoire pour l'Utilisation des Lasers Intenses, CNRS, Ecole Polytechnique, Palaiseau, France
Laboratoire d'Optique Appliqu´ee, ENSTA ParisTech, CNRS, Palaiseau, France

F. Leconte
Laboratoire pour l'Utilisation des Lasers Intenses, CNRS, Ecole Polytechnique, Palaiseau, France
Laboratoire d'Optique Appliqu´ee, ENSTA ParisTech, CNRS, Palaiseau, France

D. Badarau
Laboratoire pour l'Utilisation des Lasers Intenses, CNRS, Ecole Polytechnique, Palaiseau, France

J.M. Boudenne
Laboratoire pour l'Utilisation des Lasers Intenses, CNRS, Ecole Polytechnique, Palaiseau, France

D. Fournet
Laboratoire pour l'Utilisation des Lasers Intenses, CNRS, Ecole Polytechnique, Palaiseau, France

T. Valloton
Laboratoire pour l'Utilisation des Lasers Intenses, CNRS, Ecole Polytechnique, Palaiseau, France

J.L. Paillard
Laboratoire pour l'Utilisation des Lasers Intenses, CNRS, Ecole Polytechnique, Palaiseau, France

J.L. Veray
Laboratoire pour l'Utilisation des Lasers Intenses, CNRS, Ecole Polytechnique, Palaiseau, France

M. Pina
Laboratoire pour l'Utilisation des Lasers Intenses, CNRS, Ecole Polytechnique, Palaiseau, France

P. Monot
CEA, Iramis, SPAM, Saclay, France

J.P. Chambaret
CEA, Iramis, SPAM, Saclay, France

P. Martin
CEA, Iramis, SPAM, Saclay, France

F. Mathieu
Laboratoire pour l'Utilisation des Lasers Intenses, CNRS, Ecole Polytechnique, Palaiseau, France

P. Audebert
Laboratoire pour l'Utilisation des Lasers Intenses, CNRS, Ecole Polytechnique, Palaiseau, France

F. Amiranoff
Laboratoire pour l'Utilisation des Lasers Intenses, CNRS, Ecole Polytechnique, Palaiseau, France

Xin-ying Jiang
Research Center of Laser Fusion, CAEP, P.O. Box 919-988, Mianyang 621900, China

Xiong-wei
Research Center of Laser Fusion, CAEP, P.O. Box 919-988, Mianyang 621900, China

Zhen-guo Wang Yan
Research Center of Laser Fusion, CAEP, P.O. Box 919-988, Mianyang 621900, China

Jian-gang Zheng
Research Center of Laser Fusion, CAEP, P.O. Box 919-988, Mianyang 621900, China

Ming-zhong Li
Research Center of Laser Fusion, CAEP, P.O. Box 919-988, Mianyang 621900, China

Jing-qin Su
Research Center of Laser Fusion, CAEP, P.O. Box 919-988, Mianyang 621900, China

Zunqi Lin
National Laboratory on High Power Laser and Physics, Chinese Academy of Sciences, Shanghai 201800, China
Shanghai Institute of Optics and Fine Mechanics, Chinese Academy of Sciences, Shanghai 201800, China

Anle Lei
National Laboratory on High Power Laser and Physics, Chinese Academy of Sciences, Shanghai 201800, China
Shanghai Institute of Optics and Fine Mechanics, Chinese Academy of Sciences, Shanghai 201800, China

Wei Fan
National Laboratory on High Power Laser and Physics, Chinese Academy of Sciences, Shanghai 201800, China
Shanghai Institute of Optics and Fine Mechanics, Chinese Academy of Sciences, Shanghai 201800, China

Shenlei Zhou
National Laboratory on High Power Laser and Physics, Chinese Academy of Sciences, Shanghai 201800, China
Shanghai Institute of Optics and Fine Mechanics, Chinese Academy of Sciences, Shanghai 201800, China

Li Wang
National Laboratory on High Power Laser and Physics, Chinese Academy of Sciences, Shanghai 201800, China
Shanghai Institute of Optics and Fine Mechanics, Chinese Academy of Sciences, Shanghai 201800, China

Yanwen Xia
Research Center of Laser Fusion, CAEP, Mianyang 621900, China

Yue Liang
Research Center of Laser Fusion, CAEP, Mianyang 621900, China

Sen Li
Research Center of Laser Fusion, CAEP, Mianyang 621900, China

Junpu Zhao
Research Center of Laser Fusion, CAEP, Mianyang 621900, China

Zhitao Peng
Research Center of Laser Fusion, CAEP, Mianyang 621900, China

Hongguang Li
Xìan Institute of Optics and Fine Mechanics, Chinese Academy of Sciences, Xi'an 710119, China

Hua Liu
Research Center of Laser Fusion, CAEP, Mianyang 621900, China

Zhihong Sun
Research Center of Laser Fusion, CAEP, Mianyang 621900, China

Kuixing Zheng
Research Center of Laser Fusion, CAEP, Mianyang 621900, China

Xiaofeng Wei
Research Center of Laser Fusion, CAEP, Mianyang 621900, China

Xiang Jiao
Key Laboratory on High Power Laser and Physics, Shanghai Institute of Optics and Fine Mechanics, Chinese Academy of Sciences, Shanghai 201800, China
University of Chinese Academy of Sciences, Beijing 100049, China

Jianqiang Zhu
Key Laboratory on High Power Laser and Physics, Shanghai Institute of Optics and Fine Mechanics, Chinese Academy of Sciences, Shanghai 201800, China

Quantang Fan
Key Laboratory on High Power Laser and Physics, Shanghai Institute of Optics and Fine Mechanics, Chinese Academy of Sciences, Shanghai 201800, China

Yangshuai Li
Key Laboratory on High Power Laser and Physics, Shanghai Institute of Optics and Fine Mechanics, Chinese Academy of Sciences, Shanghai 201800, China
University of Chinese Academy of Sciences, Beijing 100049, China

Hong Jin Kong
Department of Physics, Korea Advanced Institute of Science and Technology (KAIST), 291 Daehak-ro, Yuseong-gu, Daejeon 305-701, Republic of Korea

Sangwoo Park
Department of Physics, Korea Advanced Institute of Science and Technology (KAIST), 291 Daehak-ro, Yuseong-gu, Daejeon 305-701, Republic of Korea

Seongwoo Cha
Department of Physics, Korea Advanced Institute of Science and Technology (KAIST), 291 Daehak-ro, Yuseong-gu, Daejeon 305-701, Republic of Korea

Heekyung Ahn
Department of Physics, Korea Advanced Institute of Science and Technology (KAIST), 291 Daehak-ro, Yuseong-gu, Daejeon 305-701, Republic of Korea

Hwihyeong Lee
Department of Physics, Korea Advanced Institute of Science and Technology (KAIST), 291 Daehak-ro, Yuseong-gu, Daejeon 305-701, Republic of Korea

Jungsuk Oh
Department of Physics, Korea Advanced Institute of Science and Technology (KAIST), 291 Daehak-ro, Yuseong-gu, Daejeon 305-701, Republic of Korea

Bong Ju Lee
Department of Advanced Green Energy and Environment, Handong Global University, Heunghae-eup, Buk-gu, Pohang-si, Gyeongbuk 791-708, Republic of Korea

Soungwoong Choi
Department of Advanced Green Energy and Environment, Handong Global University, Heunghae-eup, Buk-gu, Pohang-si, Gyeongbuk 791-708, Republic of Korea

Jom Sool Kim
Laser Spectronix, 219 Gasan digital 1-ro, Geumcheon-gu, Seoul 153-704, Republic of Korea

Yue Liu
Key Laboratory for Laser Plasmas (MoE) and Department of Physics and Astronomy, Shanghai Jiao Tong University, Shanghai, China

Wei-Min Wang
Beijing National Laboratory of Condensed Matter Physics, Institute of Physics, CAS, Beijing, China

Zheng-Ming Sheng
Key Laboratory for Laser Plasmas (MoE) and Department of Physics and Astronomy, Shanghai Jiao Tong University, Shanghai, China
Department of Physics, SUPA, Strathclyde University, Rottenrow 107, Glasgow, UK

Xueqiong Chen
Fujian Provincial Key Laboratory of Light Propagation and Transformation, College of Information Science & Engineering, Huaqiao University, Xiamen 361021, China

Xiaoyan Li
Fujian Provincial Key Laboratory of Light Propagation and Transformation, College of Information Science & Engineering, Huaqiao University, Xiamen 361021, China

Ziyang Chen
Fujian Provincial Key Laboratory of Light Propagation and Transformation, College of Information Science & Engineering, Huaqiao University, Xiamen 361021, China

Jixiong Pu
Fujian Provincial Key Laboratory of Light Propagation and Transformation, College of Information Science & Engineering, Huaqiao University, Xiamen 361021, China

Guowen Zhang
National Laboratory on High Power Laser and Physics Shanghai Institute of Optics and Fine Mechanics, Chinese Academy of Sciences, Qinghe Road, Jiading District 390, Shanghai 201800, China

Jianqiang Zhu
National Laboratory on High Power Laser and Physics Shanghai Institute of Optics and Fine Mechanics, Chinese Academy of Sciences, Qinghe Road, Jiading District 390, Shanghai 201800, China

Colin Danson

AWE, Aldermaston, Reading RG7 4PR, UK

David Hillier
AWE, Aldermaston, Reading RG7 4PR, UK

Nicholas Hopps
AWE, Aldermaston, Reading RG7 4PR, UK

David Neely
STFC Rutherford Appleton Laboratory, Chilton, Didcot, Oxon OX11 0QX, UK

J. Schreiber
Fakultät für Physik, Ludwig-Maximilians-Universit¨at M¨unchen, Am Coulombwall 1, D-85748 Garching, Germany
Max-Planck-Institut f¨ur Quantenoptik, Hans-Kopfermann-Str. 1, D-85748 Garching, Germany

F. Bell
Fakultät für Physik, Ludwig-Maximilians-Universit¨at M¨unchen, Am Coulombwall 1, D-85748 Garching, Germany

Z. Najmudin
The John Adams Institute, Blackett Laboratory, Imperial College London, London SW7 2AZ, United Kingdom

J.-C. Chanteloup
LULI, Ecole Polytechnique, CNRS, CEA, UPMC, Route de Saclay, 91128 Palaiseau, France

M. Arzakantsyan
LULI, Ecole Polytechnique, CNRS, CEA, UPMC, Route de Saclay, 91128 Palaiseau, France

S. Marrazzo
LULI, Ecole Polytechnique, CNRS, CEA, UPMC, Route de Saclay, 91128 Palaiseau, France

Chiara Liberatore
HiLASE Centre, Institute of Physics ASCR, v.v.i., Za Radnicí 825, 25241 Dolní B˘re˘zany, Czech Republic
Czech Technical University, Th´akurova 2077/7, 160 00 Prague 6, Czech Republic

Klaus Mann
Laser Laboratorium Göttingen (LLG), 37077 Göttingen, Germany

Matthias M¨uller
Laser Laboratorium Göttingen (LLG), 37077 Göttingen, Germany

Ladislav Pina
Czech Technical University, Th´akurova 2077/7, 160 00 Prague 6, Czech Republic

Libor Juha
HiLASE Centre, Institute of Physics ASCR, v.v.i., Za Radnicí 825, 25241 Dolní B˘re˘zany, Czech Republic

Jorge J. Rocca
Department of Electrical and Computer Engineering, Colorado State University, Fort Collins, CO 80523, USA

Akira Endo
HiLASE Centre, Institute of Physics ASCR, v.v.i., Za Radnicí 825, 25241 Dolní B˘re˘zany, Czech Republic

Tomas Mocek
HiLASE Centre, Institute of Physics ASCR, v.v.i., Za Radnicí 825, 25241 Dolní B˘re˘zany, Czech Republic

Andrew Randewich
AWE, Aldermaston, Reading RG7 4PR, UK

Colin Danson
AWE, Aldermaston, Reading RG7 4PR, UK

Kazuhisa Nakajima
Kazuhisa Nakajima, Hyung Taek Kim, Tae Moon Jeong, and Chang Hee Nam
Center for Relativistic Laser Science, Institute for Basic Science (IBS), Gwangju 500-712, Korea

Hyung Taek Kim
Kazuhisa Nakajima, Hyung Taek Kim, Tae Moon Jeong, and Chang Hee Nam
Center for Relativistic Laser Science, Institute for Basic Science (IBS), Gwangju 500-712, Korea

Tae Moon Jeong
Kazuhisa Nakajima, Hyung Taek Kim, Tae Moon Jeong, and Chang Hee Nam
Center for Relativistic Laser Science, Institute for Basic Science (IBS), Gwangju 500-712, Korea

Chang Hee Nam
Kazuhisa Nakajima, Hyung Taek Kim, Tae Moon Jeong, and Chang Hee Nam
Center for Relativistic Laser Science, Institute for Basic Science (IBS), Gwangju 500-712, Korea

Ping Zhu
Joint Laboratory on High Power Laser and Physics, Shanghai Institute of Optics and Fine Mechanics, Chinese Academy of Sciences, Shanghai, China

University of Chinese Academy of Sciences, Beijing, China

Xinglong Xie
Joint Laboratory on High Power Laser and Physics, Shanghai Institute of Optics and Fine Mechanics, Chinese Academy of Sciences, Shanghai, China

Xiaoping Ouyang
Joint Laboratory on High Power Laser and Physics, Shanghai Institute of Optics and Fine Mechanics, Chinese Academy of Sciences, Shanghai, China

Jianqiang Zhu
Joint Laboratory on High Power Laser and Physics, Shanghai Institute of Optics and Fine Mechanics, Chinese Academy of Sciences, Shanghai, China